Topics in Applied Physics
Volume 112

Topics in Applied Physics

Topics in Applied Physics is a well-established series of review books, each of which presents a comprehensive survey of a selected topic within the broad area of applied physics. Edited and written by leading research scientists in the field concerned, each volume contains review contributions covering the various aspects of the topic. Together these provide an overview of the state of the art in the respective field, extending from an introduction to the subject right up to the frontiers of contemporary research.
Topics in Applied Physics is addressed to all scientists at universities and in industry who wish to obtain an overview and to keep abreast of advances in applied physics. The series also provides easy but comprehensive access to the fields for newcomers starting research.
Contributions are specially commissioned. The Managing Editors are open to any suggestions for topics coming from the community of applied physicists no matter what the field and encourage prospective editors to approach them with ideas.

Andreas Schroeder, Christian E. Willert (Eds.)

Particle Image Velocimetry

New Developments and Recent Applications

With 335 Figures and 24 Tables

phys

Andreas Schroeder

Deutsches Zentrum
für Luft- und Raumfahrt (DLR)
Institut für Strömungstechnik
und Aerodynamik
Experimentelle Verfahren
Bunsenstrasse 10
37073 Göttingen
Germany
Andreas.Schroeder@dlr.de

Christian E. Willert

Deutsches Zentrum
für Luft- und Raumfahrt (DLR)
Institut für Antriebstechnik
Triebwerksmesstechnik
Linder Höhe
51147 Köln
Germany
chris.willert@dlr.de

Library of Congress Control Number: 2007939493

Physics and Astronomy Classification Scheme (PACS):
42., 47., 51., 83.

ISSN print edition: 0303-4216
ISSN electronic edition: 1437-0859
ISBN 978-3-540-73527-4 Springer Berlin Heidelberg New York
e-ISBN 978-3-540-73528-1 Springer Berlin Heidelberg New York
DOI 10.1071/978-3-540-73528-1

Springer is a part of Springer Science+Business Media

springer.com

Typesetting: DA-TEX · Gerd Blumenstein · www.da-tex.de
Production: LE-TEX Jelonek, Schmidt & Voeckler GbR, Leipzig
Cover design: eStudio Calamar S. L., F. Steinen-Broo, Girona, Spain

Printed on acid-free paper 57/3180/YL 5 4 3 2 1 0

Preface

The present book provides both a survey of the state-of-the-art of scientific research using particle image velocimetry (PIV) techniques in a wide variety of application fields and also constitutes a synopsis of the main results achieved during the EU funded PIVNET 2 thematic network cooperation. Subtitled with "A European collaboration on development, quality assessment, and standardization of Particle Image Velocimetry for industrial applications" the PIVNET 2 European thematic network has played an important role in the transfer of the PIV technique to industry. Together with PIVNET 1 the network's duration covers a total run time of nearly 10 years during which significant progress on PIV and its applicability was made. The success of this network is due to information exchange, scientific cooperation and synergy effects between the network partners and beyond, stimulated by a multitude of dedicated workshops and actual presentations of PIV in the partner's respective facilities.

The driving force behind these successful networking activities is credited to the initiative and efforts of quite a number of people, who were involved for different periods of time. First of all Michel Stanislas (Laboratoire de Mcanique de Lille) deserves mention as he initiated the cooperation on PIV on the European level by establishing a GARTEUR action group which operated from 1993 to 1995. This activity was later extended to two very successful EC funded research projects: EUROPIV 1 and 2. These projects were devoted to addressing the scientific and technical issues relevant to making the PIV technique operational and feasible in large aeronautical wind tunnels. In the process the need was felt to establish a platform for information exchange on the advantages, prospects and problems of the application of PIV in a much wider range. Thus, the thematic networks PIVNET 1 and later PIVNET 2 were proposed to and subsequently accepted by EC. For more than six years, until the mid-term review of PIVNET 2, both networks were successfully co-ordinated by Jrgen Kompenhans, DLR, before this responsibility was assigned to Andreas Schrder.

The activities of nearly 40 European partners considerably contributed to the fast spreading of the PIV technique in Europe not only to the aeronautical industry including turbomachinery, but also to the car industry, the naval field, the medical and biological fields, as well as household appliances, just to name a few. An important element of conveying the power of the PIV technique to a broader community have been a variety of demonstrations of its use in industrial facilities, where end users could observe and assess 'real' applications. The idea for such demonstrations was conceived during discussions with Dietrich Knrzer, who supported the network for many years as Scientific Officer. In his succession, Rolando Simonini and Andrzej B. Podsadowski both actively continued with the endorsement of the PIVNET activities.

Two further activities should be mentioned as they were initiated within PIVNET and supported through its funding, and which have grown to a much greater extent than initially anticipated. First, the initiative to bring people performing CFD calculations and experimentalists (here the PIV community) together was motivated to foster the understanding of the problems and possibilities on both sides and to encourage the use of 2D-or 3D velocity data as obtained by PIV for the validation of numerical codes. Lately, precisely this type of close cooperation between numerical and experimental people has gained increasing importance and is becoming a standard procedure.

The second significant initiative started within PIVNET is the establishment of the International PIV Challenge by Michel Stanislas (ERCOFTAC SIG 32) together with the Japanese Flow Visualization Society and colleagues from the United States. Up to now, three different challenges provided a common, standardized platform of comparison and benchmarking for PIV evaluation methods as developed by the leading scientific teams and commercial suppliers. Throughout these activities much knowledge was gained and many ideas for further developments were generated. For prospective end-users of the PIV technique the results of these challenges provide a comprehensive source of information for the assessment of PIV evaluation algorithms.

On the whole the activities associated with the PIVNET 1 and 2 thematic networks, namely the way to promote cooperation, the demonstration of the technique's potential, and the establishment of standards, especially at the evaluation algorithms, may serve as example for other areas where methods developed in the laboratory have the potential to be applied in a wide range of industrial applications.

At this point, the PIV technique is widely spread and differentiated into many distinct applications ranging from micro flows to combustion to supersonic flows, for both industrial needs and research. Based on a relatively simple principle PIV has evolved to a highly versatile flow diagnostics tool and has found realization in many sophisticated technical adaptations. The measurement technique along with the associated hard- and software have improved continuously such that PIV has matured to become a reliable and accurate method for "real life" investigations. The partners of the network have made essential contributions in opening new possibilities for PIV in measurements, scientific research and technology. Nevertheless there is still an ongoing process of improvement and extension of the PIV technique towards time resolution, volume-resolved measurements, measurements at micro- and macro-scales, increased accuracy, self-optimizing processing, combinatory measurements with other diagnostics techniques as well as measurements under harsh conditions.

The present book provides a survey of PIV techniques in a variety of application areas corresponding to the workshops organized within PIVNET 2. On this basis outstanding researchers provided a full paper about their specific development, application and/or adaptation of the PIV technique to an experimental investigation within a certain topic of research. In addition, overview articles on the main application topics have been compiled by the workshop organizers or prominent scientists in the respective fields.

Reflecting the network's activities this book is grouped into eleven main topics reviewing the status and potential of the application of PIV to different research fields:

- μPIV and applications to micro systems
- bio-medical flows
- 3D-PIV
- comparison with and validation of CFD
- household appliances
- turbo machinery
- internal combustion
- car industry
- complex aerodynamics
- supersonic flows
- naval applications

The Editors believe that this book serves as a guide through a wide range of research and technology fields in which quantitative flow field data are utilized. It is also an overview of recent improvements and developments in hard- and software and reflects the diversity of applications making use of the powerful PIV technique today. Furthermore this book constitutes the concluding final report of the PIVNET 2 network and shall give valuable information for engineers and physicists facing problems in experimental fluid dynamics using PIV.

The Editors are thankful to all the authors for their efforts in contributing to this book. In the name of the all the involved network partners the Editors acknowledge the financial support of this activity through the European Community. Without doubt this support was a key factor in promoting and disseminating the PIV technique among basic and applied research throughout science and industry.

The compilation and production of this book has been funded by the European Commission within the 5th Framework Programme as part of the PIVNET 2 thematic network (EU GROWTH project G4RT-CT-2002-05081)

Göttingen, Köln
December 2007

Andreas Schroeder
Christian E. Willert

Contents

Measurements and Simulations of the Flow Field in an Electrically Excited Meander Micromixer

Dominik P. J. Barz[1], Hamid Farangis Zadeh[1], and Peter Ehrhard[2]

[1] Research Centre Karlsruhe, Nuclear & Energy Technologies,
P.O. Box 3640, D-76021 Karlsruhe, Germany
dominik.barz@iket.fzk.de

[2] University of Dortmund,
Biochemical & Chemical Engineering, Fluid Mechanics,
Emil-Figge-Str. 68, D-44221 Dortmund, Germany
p.ehrhard@bci.uni-dortmund.de

Abstract. The experimental and numerical verification of the performance of an electrically excited micromixer is the focus of the present work. For the (local) measurement of the flow field within the micromixer we use microparticle image velocimetry (μPIV). Time-dependent and three-dimensional numerical (FEM) simulations, in conjunction with an asymptotic treatment of the electrical double layer, are used as theoretical means. If electroosmotic forces act on the flow, we can, even in straight channel cross sections, resolve complex velocity profiles, which are dominated by electroosmosis close to the walls and by the applied pressure gradient in the channel core. Hence, even flow at walls against the pressure-driven main flow can be observed. Particularly within bends a complex and symmetric flow structure is found, which can be characterized by a number of vortex and saddle points.

1 Introduction

The investigation of mixing and separation processes in microchannels is of great interest with regard to the implementation of such components into lab-on-chip applications. Recent work on mixing, e.g., concentrates on passive mixers, which rely on plane hairpin channels (cf. [1]), on three-dimensional serpentine channels (cf. [2]), or on bas-relief structures on the channel floor (cf. [3]) to achieve centrifugal or chaotic flows, suitable to enhance mixing. Alternatively, active means are employed to induce such secondary flows by, e.g., magnetic forces (cf. [4]) or by electroosmotic forces (cf. [5,6]). Ultimately, all research on efficient mixing or separation in microchannels needs to validate the respective ideas and models. For the experimental validation of such processes, the measurement of velocity and species concentration fields in microchannels appears important.

There are several articles in the literature addressing to some extent the measurement of flow fields in microchannels. The application of laser Doppler anemometry (LDA) in a straight microchannel is discussed, e.g., by [7], featuring a measuring volume of $5\,\mu m \times 10\,\mu m$. A LDA profile sensor is outlined

A. Schroeder, C. E. Willert (Eds.): Particle Image Velocimetry,
Topics Appl. Physics **112**, 1–18 (2008)

by [8], reporting a spatial resolution of 1.6 μm. Such a sensor uses two colors to capture both one velocity component and the position of a tracer within a (long) measuring volume and appears promising for microchannels. *Paul* et al. [9] use scalar image velocimetry (SIV) for pressure-driven (and electrokinetic) flows in capillaries and report reliable measurements at a spatial resolution of typically 20 μm. Molecular tagging velocimetry (MTV) is discussed in [10] and applied to the time-dependent interaction of a vortex ring with a wall, whereas the measuring plane has a few centimeters side length. This technique projects an illumination pattern into the liquid and, by a crosscorrelation technique, determines the offset of this (chemically stored) pattern between two moments in time. This technique, likewise, has great potential for application in microchannel flows. The microparticle image velocimetry (μPIV), finally, is the most widely used method to measure velocity fields in microchannels. Introduced in [11] and [12], it employs an epifluorescent microscope to illuminate the complete volume of a microchannel at two given moments in time. The images of the fluorescent particles are then cross-correlated to obtain offset and velocity fields at a spatial resolution of down to 1 μm in the measuring plane, whereas the "thickness" of the measuring plane can be reduced to 1.5 μm (cf. [11]) or 1.8 μm (cf. [12]). A comprehensive discussion on the spatial resolution of various measuring techniques for the velocity fields in microchannels is conducted in [13]. The μPIV technique has found numerous applications since. Examples are the apparent liquid slip at hydrophobic walls (cf. [14]), the electrokinetic flow excited by dielectric forces (cf. [15]), the mixing of two phases due to hydraulic focusing in microchannels (cf. [16]), and the transition to turbulent flows in straight rectangular (cf. [17, 18]) or straight circular (cf. [19]) microchannels.

We shall concentrate in this chapter on the flow field in an electrically excited micromixer, which has been proposed in [6]. In detail, the authors propose a micromixer comprising a Y-junction and a single meander downstream in the common channel. Mixing is enhanced by applying an alternating electrical field, and hence by superimposing an alternating electroosmotic flow, upon the pressure-driven base flow. The preliminary numerical simulations by [6] demonstrate that, given a reasonable ratio of primary and secondary velocity amplitudes, the time-dependent secondary flow serves to significantly increase mixing quality at the outlet of the device. The present experiments and simulations, therefore, aim to validate the preliminary numerical findings with regard to both velocity and concentration fields. While a concentration-field measuring technique and some concentration measurements are discussed in [20], we shall restrict discussion in this chapter to the velocity field.

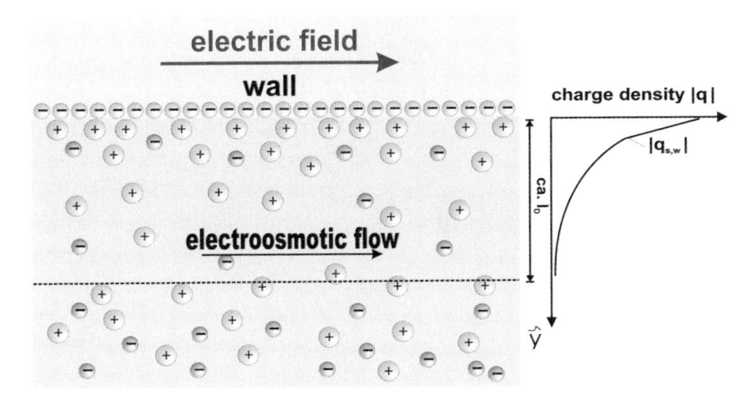

Fig. 1. Physics of electroosmosis

2 Experimental Realization

2.1 Electroosmosis

Before we explain the details of the experimental realization, it appears worthwhile to physically discuss electroosmosis. Given a liquid with mobile charges (e.g., an aqueous solution with ions) adjacent to an electrically insulating solid wall (e.g., glass, plastics), due to chemical/physical interaction we typically find surface charges on the solid. This situation is sketched in Fig. 1. As the surface charges on the solid attract opposite charges out of the liquid, an electrically non-neutral layer of liquid (the so-called electrical double layer, EDL) is the consequence. By applying an electrical field tangentially to the wall, we can introduce electrical forces into this (non-neutral) liquid layer. These forces cause a movement of the liquid – the so-called electroosmotic flow. This effect can be employed in microchannels not only to pump liquids, but likewise to induce secondary flows.

2.2 Experimental Setup

An overview of the experimental setup is presented in Fig. 2. To arrange a flow of liquids through the mixer, we engage a gravity-driven flow with the two inlet reservoirs positioned at a defined height above the outlet reservoir. The liquids are driven through the two inlet channels and leave the micromixer through the common channel. In all cases the flow rates through both inlet channels are identical. As liquid we use deionized water, which has a low electrical conductivity. The low electrical conductivity has two advantages: 1. we minimize electrical currents (due to the applied electrical field) through the liquid, and hence minimize Joule heating, 2. we obtain a thick electrical double layer, and hence pronounced electroosmotic effects. The mass flow rate $\dot{m}$ leaving the common channel is measured by means of a precision balance

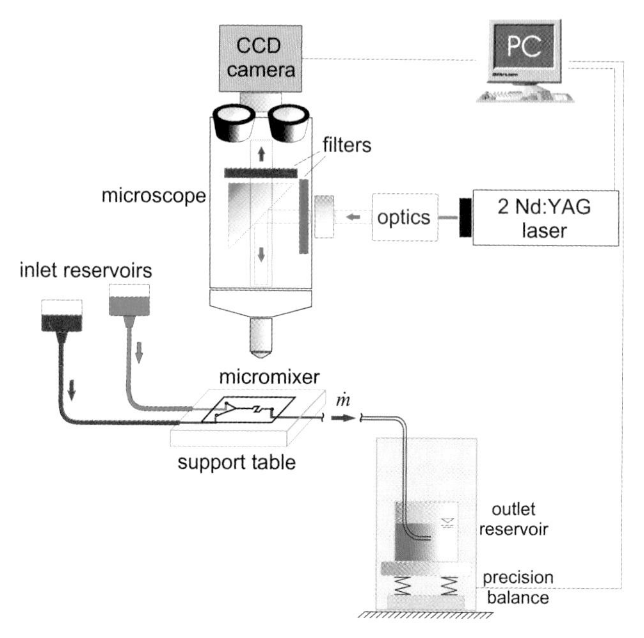

Fig. 2. Sketch of the overall experimental setup

(Sartorius), engaged for time-resolved measurement of the outlet reservoir mass. This balance has an accuracy of ± 0.01 mg and submits the data via a RS232 serial interface to a PC, where LabVIEW (National Instruments) handles the processing of the data. The gravity-driven delivery system is calibrated by mass-flow measurements at several flow rates. First, in the outlet reservoir the evaporation rate of the water is measured. Secondly, after connecting all reservoirs, the levels of the inlet reservoirs are adjusted at different heights above the outlet reservoir, namely in the range $0 \leq \Delta h \leq 100$ mm. The measured mass flow rates $\dot{m}$ are measured and corrected by the evaporation rate (typically $\dot{m}_{\text{ev}} \simeq 20\,\mu\text{g/s}$). This method ensures an accuracy for the mass flow rate of better than $\pm 3\,\mu\text{g/s}$. From the mass flow rates, the volumetric flow rates $\dot{V} = \dot{m}/\rho$ are obtained at known density ρ. This defines the Reynolds number for the rectangular (square) common channel

$$\text{Re} = \frac{\bar{u}d}{\nu}, \tag{1}$$

with the mean velocity $\bar{u} = \dot{V}/d^2$ and the kinematic viscosity of the liquid ν. We use temperature-dependent data for density and viscosity. In summary, the (steady) forced flow through the micromixer will be characterized by the Reynolds number. The Reynolds numbers in (1) is valid for the common channel, whereas both inlet channels are characterized by half the Reynolds number.

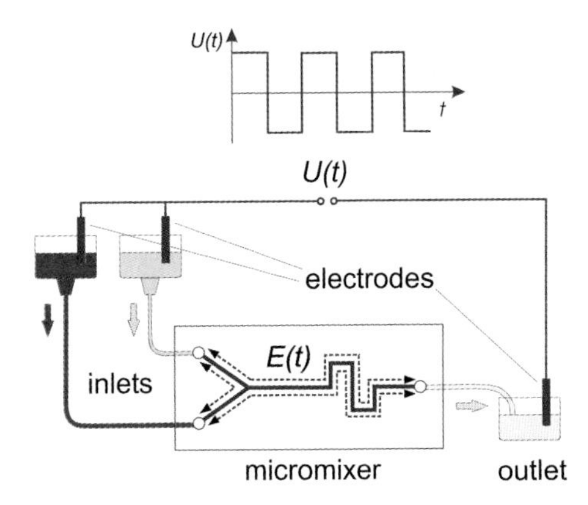

Fig. 3. Arrangement of electrodes for the buildup of the electrical field

2.3 Electrical Excitation

In parallel with the (steady) forced flow, we apply a time-dependent electrical field to set up an electroosmotic flow. The arrangement for the electrical field buildup is sketched in Fig. 3. A DC power supply and an amplifier allow to us apply potential differences of up to 5 kV. Gold electrodes are placed in the inlet and outlet reservoirs. One potential is connected to both inlet reservoirs, the second potential is connected to the outlet reservoir. This arrangement sets up an electrical field, which is roughly directed tangentially to the channel axis. A function generator in conjunction with a relay exchanges the polarity of all electrodes periodically in time at a defined frequency (typically 0.1 Hz). This provides a local electrical field in the channel, alternating in a square-signal fashion. The electrical signal from the function generator is, moreover, used for triggering purposes. In detail, the triggering is necessary to record phase-correctly multiple velocity fields for averaging.

2.4 Micromixer

The micromixer under investigation (cf. Fig. 3) features a Y-form joining of two inlet channels, followed by a straight common channel with a single meander downstream, all realized in FOTURAN glass (cf. [21]). The channels of the Y-mixer and the meander channel, all have square cross sections of $110\,\mu m \times 110\,\mu m$. At the Y-mixer, the merging channels comprise an angle of $40°$, while the meander is located 17 mm downstream of this point. A closeup sketch of the meander is given in Fig. 4.

2.5 Optical Measurement Technique

The micromixer, mounted on the support table, allows for optical access via an inverted microscope (Leica, DMIRM). A number of microscope objectives (Leica, NPLANL, HCPLFL) with different magnification M and numerical

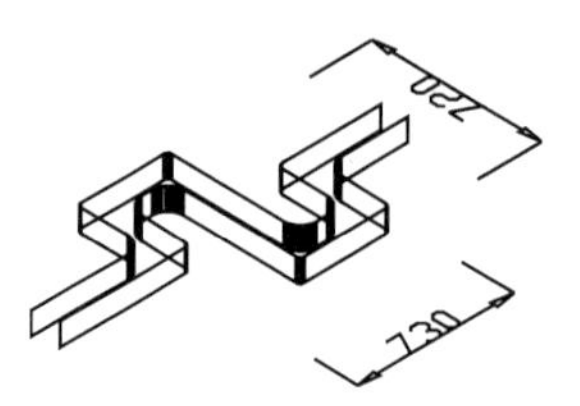

Fig. 4. Closeup sketch of the meander part; measures are given in μm

aperture NA, all accommodated for air, is mounted onto the microscope, depending on the actual field of interest.

We aim to perform flow-field measurements through this microscope by means of the so-called microparticle image velocimetry (μPIV), which is described in detail, e.g., in [11] or [12]. For that reason we engage a fluorescence technique as sketched in Fig. 5. Two Nd:YAG lasers (New Wave, Solo-PIV) provide two pulses of green light ($\lambda_\mathrm{i} = 532\,\mathrm{nm}$), which are expanded and coupled into the coaxial illumination path of the microscope. Hence, we obtain green volume illumination of the microchannel. Within the flow a mixture of fluorescent microspheres of diameter 200 nm and 500 nm (Duke Scientific) is suspended. The density of the microspheres ($\rho = 1.05\,\mathrm{g/cm^3}$) is well adjusted to the water density. The microspheres are customized for emission in the red regime, i.e., at a wavelength around $\lambda_\mathrm{e} \simeq 612\,\mathrm{nm}$. The red light from the microspheres passes the epifluorescent cube within the microscope and is imaged onto the CCD camera. In contrast, green reflected light is blocked by the epifluorescent cube from reaching the CCD camera. We use a high-performance cooled interline CCD camera (PCO, Image Intense) with 1376×1040 pixels and 12-bit readout resolution to record the double frames. The acquired double frames from the camera are transferred to a PC, the software DaVis6.2 (LaVision) is applied for further processing. All timing, synchronization, and control of the camera and the lasers is achieved by a programmable timing unit (PTU) card, installed in the PC, in conjunction with the DaVis software.

We typically subdivide the total area occupied by the fluid into interrogation areas of 16×16 pixels ($4.8\,\mu\mathrm{m} \times 4.8\,\mu\mathrm{m}$), whereas the computation of the displacement vector within each interrogation area of the double frames is based on a crosscorrelation method within the DaVis software. We keep the ratio of microsphere volume and fluid volume in the range 0.05–0.07 % to ensure by this moderate concentration of microspheres a high signal-to-noise ratio. This gives typically around 5 microspheres per interrogation area. We use in all cases ensemble averaging of typically 20–40 single crosscorrelation functions, to improve the accuracy of the measured flow fields (cf. [22]). As we have a periodically alternating electrical field, we can expect likewise a periodically oscillating flow field. Hence, a phase-correct sampling of the flow field during a large number of alternation periods (20–40) is established, which uses the electrical signal from the electrodes for triggering. The fre-

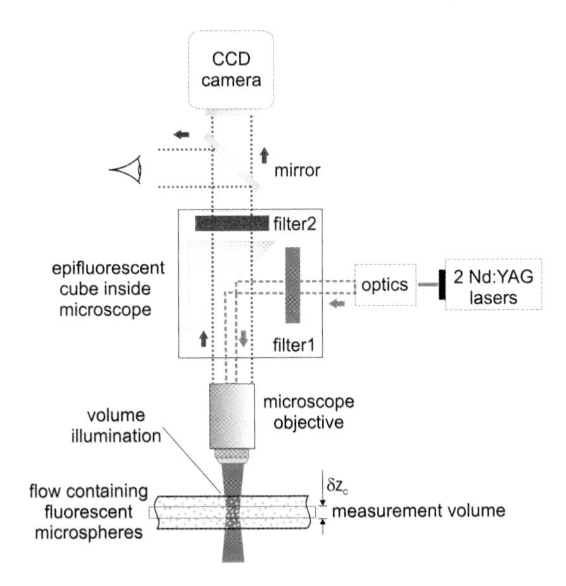

Fig. 5. μPIV setup for microflow investigations

quency of alternation in all cases is 0.1 Hz, i.e., each 5 s the direction of the electrical field is inverted. Accordingly, the double frames are recorded with a small time offset in the range $\Delta t \leq 10\,\mathrm{ms}$ in all cases.

In contrast to conventional PIV, with the thickness of the lightsheet defining the measuring volume, the illumination within the μPIV setup is responsible for an unsharp measuring volume, resulting from the focal plane and the depth of field. Following [22], the thickness of the measuring volume should rather be defined from the so-called depth of correlation than from the pure depth of field. *Meinhart* et al. [22] estimate the depth of correlation by

$$\delta z_{\mathrm{c}} = \frac{3n\lambda_{\mathrm{e}}}{\mathrm{NA}^2} + \frac{2.16 d_{\mathrm{p}}}{\tan\theta} + d_{\mathrm{p}} \,. \tag{2}$$

In (2) n is the refractive index of the medium between the microchannel and the objective, λ_{e} is the emitted light wavelength, NA is the numerical aperture of the objective, d_{p} the particle diameter, and θ is the collection angle of the optical system. For the present measurements the depth of correlation is $\delta z_{\mathrm{c}} \simeq 16\,\mu\mathrm{m}$.

Particles, as used for the μPIV technique, in principle experience electrophoretic forces if subjected to an electrical field. The electrophoretic velocity is checked in a (separate) sealed microchannel of identical cross section and subjected to identical electrical field conditions. For these experiments, particles of diameter $d_{\mathrm{p}} = 0.2\,\mu\mathrm{m}$ and $1.0\,\mu\mathrm{m}$ are used, giving velocity amplitudes of up to $7.7\,\mu\mathrm{m/s}$ and $5.8\,\mu\mathrm{m/s}$, respectively. This choice of particles is made to cover a, preferably wide, range of sizes. In both cases this is smaller than 0.7 % of the electroosmotic velocity amplitude and therefore the electrophoretic contribution can be neglected in the remainder of the chapter.

3 Theoretical Model

The numerical simulations of flow and transport within the micromixer rely on the time-dependent and three-dimensional Navier–Stokes equations and the continuity equation. Hence, for an incompressible Newtonian liquid we have

$$\nabla \cdot \boldsymbol{v} = 0 \,, \tag{3}$$

$$\rho \left[\frac{\partial \boldsymbol{v}}{\partial t} + (\boldsymbol{v} \cdot \nabla) \boldsymbol{v} \right] = -\nabla p + \mu \Delta \boldsymbol{v} - q \nabla \varphi \,. \tag{4}$$

For a poorly conducting liquid, further, the Gaussian law

$$\nabla \cdot (\epsilon_r \nabla \varphi) = -\frac{q}{\epsilon_0} \tag{5}$$

holds. For the electrical charge distribution q within the (thin) electrical double layer (EDL) we can, moreover, invoke the Debye–Hückel approximation

$$q \simeq \frac{q_\zeta}{l_\mathrm{D}} \exp\left(-\frac{z}{l_\mathrm{D}}\right) . \tag{6}$$

Within the above equations we engage a local coordinate system, with the origin on the wall. Hence, x and y are the wall-tangential coordinates and z is the wall-normal coordinate. Further, $\boldsymbol{v}$ is the fluid velocity vector, ρ denotes density, p pressure, μ dynamic viscosity, φ the electrical potential, $\epsilon_0 \epsilon_r$ the respective dielectric properties, q_ζ the charge density at the boundary between shear layer and Gouy–Chapman layer (within the EDL), and l_D the Debye length.

The boundary conditions require no slip and a prescribed electrical potential, according to the Debye–Hückel approximation, at all channel walls. Hence, we have

$$\boldsymbol{v}(x, y, 0) = 0 \,, \tag{7}$$

$$\varphi(x, y, 0) = -\frac{q_\zeta l_\mathrm{D}}{\epsilon_0 \epsilon_r} . \tag{8}$$

We restrict our simulations to the meander segment of the micromixer and exclude the Y-mixer segment. Hence, we need to formulate boundary conditions within the inlet and outlet cross sections. We assume in both cross sections a pressure-driven fully developed and unidirectional flow; for a rectangular channel the developed-flow profile follows a series solution (cf. [23]). Further, the electrical potential at the inlet and outlet cross section is inferred from the time-dependent electrical potential at both electrodes. Hence, we have

$$u_{\text{in}}(y,z) = -\frac{1}{2\mu}\frac{\mathrm{d}p}{\mathrm{d}x}\left[\left(\frac{d^2}{4} - y^2\right) - \frac{8}{d}\sum_{n=0}^{\infty}\frac{(-1)^n}{N_n^3}\frac{\cosh(N_n z)}{\cosh(N_n d/2)}\cos(N_n y)\right], \quad (9)$$

$$v_{in} = 0\,, \qquad w_{in} = 0\,, \quad (10)$$

$$\frac{\partial u_{\text{out}}}{\partial x} = 0\,, \qquad v_{\text{out}} = 0\,, \qquad w_{out} = 0\,, \quad (11)$$

$$\varphi_{\text{in}} = f_1(t)\,, \qquad \varphi_{\text{out}} = f_2(t)\,, \quad (12)$$

whereas $N_n = (2n+1)\pi/d$ and $f_1(t)$, $f_2(t)$ are of square-signal type and phase shifted by π.

We treat the above system (3)–(12) in nondimensional form by introducing the (channel) scales

$$(X,Y,Z) = \frac{(x,y,z)}{d}\,, \quad (U,V,W) = \frac{(u,v,w)}{\bar{u}}\,, \quad (13)$$

$$\tau = \frac{t}{(d/\bar{u})}\,, \quad P = \frac{p}{(\mu\bar{u}/d)}\,, \quad \Phi = \frac{\varphi}{\varphi_0}\,. \quad (14)$$

Thus, lengths are scaled by the width d and velocities by the average (pressure-driven) velocity $\bar{u}$ of the microchannel. Further, a transport time-scale and a viscous pressure scale is used; the electrical potential is scaled by the applied potential difference φ_0. Due to the nondimensionalization, beyond the Reynolds number (cf. (1)), a number of dimensionless groups arise, namely

$$\delta = \frac{l_d}{d} \ll 1\,, \quad \Pi = \frac{q_\zeta \varphi_0}{\mu\bar{u}}\,. \quad (15)$$

The Reynolds number Re characterizes the pressure-driven flow through the micromixer, δ is a ratio of length scales, and Π is the ratio of electrical and viscous forces, characterizing the strength of the electroosmotic flow.

Due to largely different length scales ($\delta \ll 1$), it appears reasonable to seek, by asymptotic means, an inner solution, valid within the EDL, and an outer solution, valid within the channel core; the details of this procedure are given by [6]. The inner solution can be given analytically, the outer solution is obtained by the standard finite-element code FIDAP. It is important to note that, for the outer solution, there remains no need to resolve the EDL numerically. The matching and superposition of both solutions provides an approximation for the solution in the entire domain. The finite-element simulations for the channel core are time dependent and three-dimensional in nature.

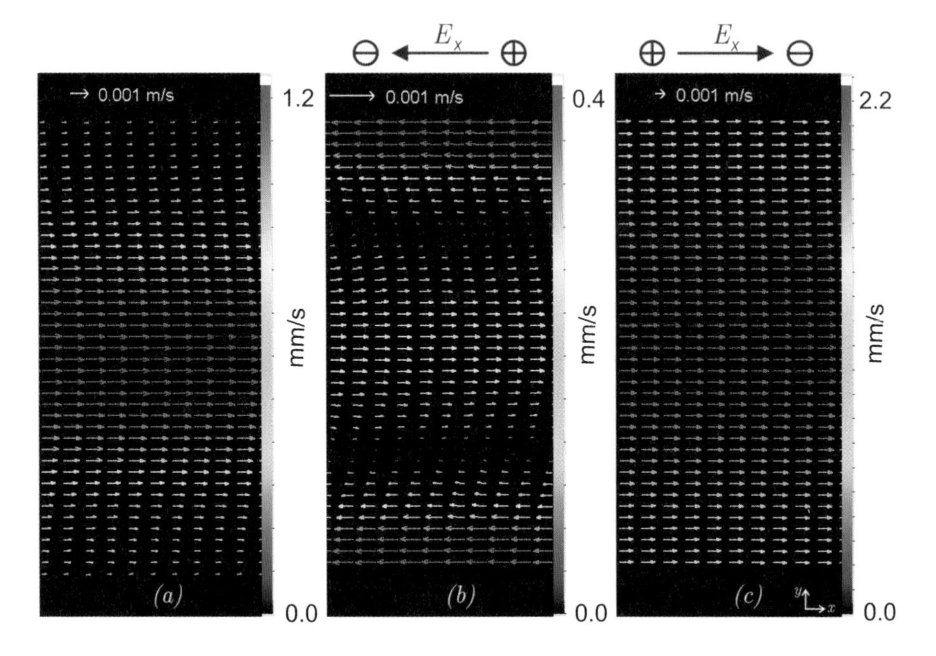

Fig. 6. Velocity fields of the flow at Re $\simeq 0.1$ at the level $z/d = 0.25$ in the straight channel of the micromixer: (**a**) without electrical field, (**b**) subject to an electrical field $E_x \simeq -14.5\,\mathrm{V/mm}$, and (**c**) $E_x \simeq +14.5\,\mathrm{V/mm}$. The velocity amplitudes are color coded, according to the given color tables at each part of the figure

4 Results

The setup, the µPIV measurements, and the numerical simulations are engaged to study the flow in various parts of the micromixer, with and without electrical excitation. The results in this section cover the electroosmotic flow 1. in the straight channel and 2. within the meander bends of the micromixer.

4.1 Electroosmotic Flow in the Straight Channel of the Micromixer

All measurements in this section are taken in the straight channel of the glass micromixer, which is located between the Y-junction and the meander. The precise position is 15 mm downstream of the Y-junction, all cross sections are $110\,\mu\mathrm{m} \times 110\,\mu\mathrm{m}$. We establish a pressure-driven flow by applying a pressure difference of $\Delta p \simeq 157\,\mathrm{Pa}$ between both inlet reservoirs and the outlet reservoir. This results in a Reynolds number of Re $\simeq 0.1$ in the common channel. This weak pressure-driven flow is chosen to have both the pressure-driven flow and the contingent electroosmotic flow at similar velocity amplitudes. Hence, the superposition of both flows makes a real (measurable) difference.

The result for the pure pressure-driven flow at the level $z/d = 0.25$ is given in Fig. 6a. The origin of the coordinate system is on the channel axis, with x pointing downstream along the channel axis and y and z orthogonal to the channel axis, pointing in horizontal and vertical directions. Hence, $z/d = 0.25$ is exactly between the midheight level and the top wall. The velocity vectors in Fig. 6a are based on an ensemble average of the crosscorrelation functions of 40 double frames, with a time interval of $\Delta t = 2\,\mathrm{ms}$ between the two images of each double frame. The microscope objective has a magnification of $20\times$, featuring a depth of correlation (cf. (2)) of $\delta z_{\mathrm{c}} \simeq 16\,\mu\mathrm{m}$ for the given particle size. Interrogation areas of 16×16 pixels ($4.8\,\mu\mathrm{m} \times 4.8\,\mu\mathrm{m}$) are engaged with an overlap of 50 %. We recognize a steady flow profile, parallel to the walls, with a maximum velocity of about 1.2 mm/s in the middle of the channel ($y = 0$), and vanishing velocities at the walls. This profile appears roughly parabolic.

To excite an electroosmotic flow, we apply a potential difference of 1.0 kV between the inlet reservoirs and the outlet reservoir (cf. Sect. 2.3), alternating at a frequency of 0.1 Hz. This causes an electrical field of strength $|E_x| \simeq 14.5\,\mathrm{V/mm}$, directed tangentially to the channel axis and inverting its direction every five seconds. For all measurements we trigger the first image of the double frame at 4.5 s after the switch of polarity, while the time interval between two images of a double frame remains $\Delta t = 2\,\mathrm{ms}$. After each switch of polarity the flow needs less than two seconds to adjust to the new electrical field direction. Hence, between two and five seconds after the switch of polarity a quasisteady flow persist. In summary, we sample phase-correctly multiple measurements of the flow field at the end of the quasisteady period and ensemble average the crosscorrelation functions of 40 double frames. The results are given in Figs. 6b,c.

During the period of a negatively directed electrical field with $E_x \simeq -14.5\,\mathrm{V/mm}$ (cf. Fig. 6b) we find close to both walls of the microchannel electroosmotic flow into the negative x-direction, i.e., against the pressure-driven flow. In the middle of the channel ($y = 0$), due to the pressure field, the flow is in the positive x-direction, again featuring a roughly parabolic profile around the middle region. Hence, negative velocities of up to $u \simeq -0.6\,\mathrm{mm/s}$ are found near the walls, while positive velocities of up to $u \simeq +0.4\,\mathrm{mm/s}$ are found in the middle of the channel. During the period of a positively directed electrical field with $E_x \simeq +14.5\,\mathrm{V/mm}$ we find the flow field given in Fig. 6c. Here, near both walls an electroosmotic movement of the liquid into the positive x-direction is obvious, rapidly rising to velocities of $u \simeq 1.4\,\mathrm{mm/s}$ across a thin wall layer. This leads to larger velocities of up to $u \simeq 2.0\,\mathrm{mm/s}$ in the middle of the channel. The roughly parabolic form of the velocity profile in the channel core appears to be preserved.

It is obvious from Figs. 6b,c that the electrical field causes movement of the liquid near the walls in either direction, depending on the direction of the electrical field. The combination of glass and water is characterized by negative electrical charges at the glass surface (wall). The negatively charged

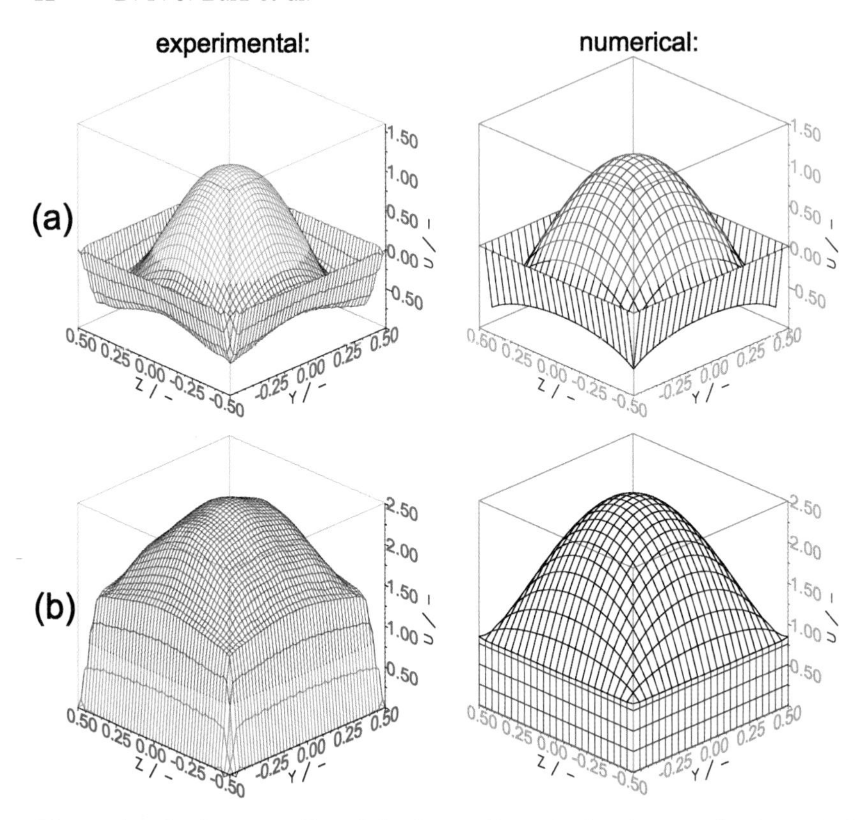

Fig. 7. Axial velocity profiles of the flow at Re $\simeq 0.1$ for the straight channel of the micromixer: (**a**) subject to an electrical field of $E_x \simeq -14.5\,\mathrm{V/mm}$ and (**b**) subject to an electrical field of $E_x \simeq +14.5\,\mathrm{V/mm}$

wall accumulates positive charges (e.g., H_3O^+ ions) next to the wall, leading to a positively charged layer in the liquid – the electrical double layer (EDL). The application of a positively directed electrical field causes forces on the ions in the EDL, which tend to move these ions (and the layers) towards the electrode at the outlet, i.e., in the positive x-direction. The opposite direction of the electrical field inverts the forces onto the wall layer and a complex velocity field, directed at the walls in the negative x-direction and in the middle of the channel in the positive x-direction, is the consequence.

In addition to the measurements in a specific level, we have performed velocity measurements in 15 levels of the microchannel in the range $-0.5 \leq z/d \leq 0.5$. This allows us to infer, e.g., the axial velocity field within the complete cross section of the microchannel. For these measurements all optical and evaluation parameters and procedures remain unchanged, an overlap of the interrogation areas, however, is avoided. In order to verify the measuring technique, the measured flow field of the pure pressure-driven flow has been compared with the analytical series solution for the laminar flow in a rectangular channel, which is given in (9). This comparison reveals good agreement with a deviation of less than ±3 % within the entire channel cross section. It should be noted, though, that velocities exactly at the walls cannot be measured accurately by the μPIV technique due to the finite size of the measuring volume. This is a fundamental problem of volume-averaging measurement techniques, leading to an overestimation of the "wall velocity".

In Fig. 7 we present the measurements and simulations in dimensionless form ($\bar{u} \simeq 0.81$ mm/s) for the cases in which, additionally, an electroosmotic flow is excited by applying an alternating electrical field, as outlined above. For the experimental results in Fig. 7 we manually have to set the velocities at all walls to zero. In the given plane, experimentally we have a spatial resolution of 4.8 μm × 16 μm, while numerically we have 4.6 μm × 4.6 μm. For plotting, the experimental data are linearly interpolated to the visible regular grid. During the period of a negatively directed electrical field of $E_x \simeq -14.5$ V/mm we obtain the flow fields given in Fig. 7a. We recognize a measured profile, which appears, to good approximation, symmetric to both axes, with negative velocities close to the walls and a positive center velocity of about 1 mm/s. It is important to note, though, that the measuring technique does not resolve the EDL, due to the limited spatial resolution. The simulations reveal a velocity profile of similar character, whereas the velocity gradients in the wall layers are steeper, if compared to the experimental profile. This is not surprising, given the difference between averaged (experimental) and local (numerical) data.

During the period of a positively directed electrical field of strength $E_x \simeq +14.5$ V/mm, we find the flow fields given in Fig. 7b. The experimental axial velocity profile remains, to a good approximation, symmetric around both axes and features steep velocity gradients at the walls, caused by electroosmotic forces. The maximum velocity is about 2.4 mm/s in the center of the microchannel. The numerical profile shows even steeper gradients at the walls, whereas the electroosmotic contribution appears smaller in amplitude, if compared to the experimental data. Again, this is due to the averaged data resulting from the measurements. The center amplitude, though, from both the experimental and the numerical profiles are in good agreement.

4.2 Electroosmotic Flow in the Meander Bends of the Micromixer

We shall now concentrate on the flow field within one U-bend of the meander. As in Sect. 4.1, we establish a pressure-driven flow, characterized by the Reynolds number Re $\simeq 0.1$, and superimpose an alternating electrical field of strength $|E_x| \simeq 14.5$ V/mm and frequency 0.1 Hz. The optical and evaluation parameters and procedures are identical to those in Sect. 4.1, namely magnification 20×, $\Delta t = 2$ ms, interrogation area 16 × 16 pixel without overlap, $\delta z_c \simeq 16$ μm, 40 crosscorrelation functions are averaged per level, and 15 levels across the complete depth ($-0.5 \leq z/d \leq 0.5$) of the microchannel are measured. The sampling of multiple double frames is phase-correctly

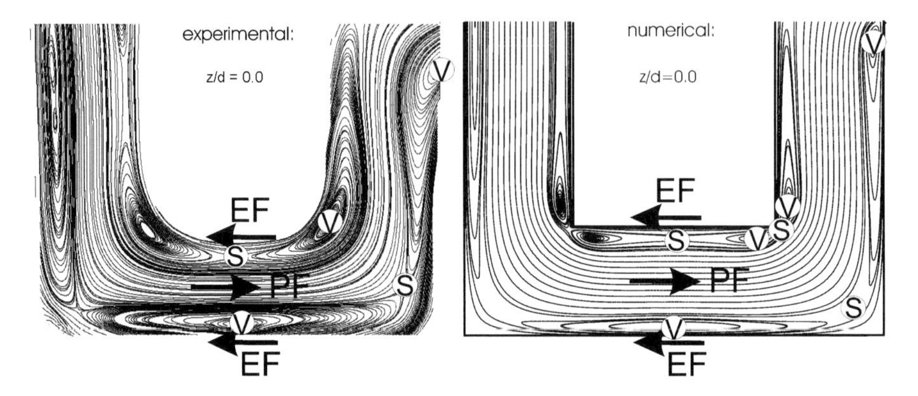

Fig. 8. Streamlines of the flow at Re $\simeq$ 0.1 and for an electrical field of $E_x \simeq -14.5\,\text{V/mm}$ at the level $z/d = 0.0$. The directions of the pressure-driven flow (PF) and the electroosmotic flow (EF) are indicated by *arrows;* saddle (S) and vortex (V) points are indicated in one half of the bend

triggered at the end of the quasisteady period. We focus on the period during which the electrical field is $E_x \simeq -14.5\,\text{V/mm}$. During this period an electroosmotic flow is induced at the walls, which is directed against the pressure-driven flow. This counterflow situation gives rise to highly interesting flow patterns and deserves special attention.

In Fig. 8 we present the flow pattern in the midheight level by means of streamlines. The experimental streamlines are inferred from the (quasisteady) velocity field in this level. For the understanding of the flow pattern, the streamlines present a superior basis over the velocity fields and, moreover, allow us to infer vortex (cf. V) and saddle (cf. S) points within the flow pattern. At the midheight level ($z/d = 0$) we see in the center of the microchannel the pressure-driven flow through the meander from left to right (cf. arrow PF). At the walls, however, the electroosmotic flow is directed against this pressure-driven flow (cf. arrow EF). Hence, a layer of microvortices forms in a narrow zone adjacent to the walls. The whole flow pattern, to good approximation, appears symmetric with respect to the U-bend symmetry line. We can, therefore, characterize the flow by three vortex points and two saddle points (cf. S, V), within the right half of the bend. If we compare with the numerical streamline pattern, the overall picture is similar. In contrast to the experimental situation, however, we do have sharp corners within the numerical simulations. These sharp corners split the single vortex points at the inner (smooth) corners in the experiment into two vortex points, with a saddle point in between. All other numerical singular points (cf. S, V) appear at similar positions within the bend. In summary, the flow topology agrees to a reasonable degree, given the considerable difference of geometry in both experiment and simulations.

If we move towards the bottom or top wall, we expect all liquid to move against the pressure-driven flow from right to left. This is the electroosmotic flow, induced at the bottom and top walls. In summary, in the core of the microchannel the pressure-driven flow is dominant, while near all walls the electroosmotic flow in the opposite direction rules the situation.

5 Summary and Outlook

The present work aims to verify the performance of an electrically excited micromixer. Such a micromixer has been proposed in [6], based on an appropriate model and on preliminary numerical simulations. The basic idea of such a micromixer is to induce electroosmotic forces within the electrical double layer to excite a secondary flow. This requires, firstly, a liquid containing charges (e.g., ions), and secondly, electrically insulating channel walls. Thirdly, an external oscillating electrical field has to be applied. To judge the performance, the important quantities, namely velocity (and concentration) fields, need to be measured and simulated. Although integral quantities such as mass or volumetric flow rate, or pressure drop, are valuable to infer, they are certainly not sufficient for verification purposes.

For the measurement of the velocity field we engage the microparticle image velocimetry (μPIV) technique, for the measurement of the mass flow rate we engage time-resolved precision weighting. The μPIV technique provides two components of the velocity field, for the entire volume of liquid. For that purpose μPIV is applied via shifted focal planes in different levels of the cross section. All measurements are taken time-resolved at time steps of down to 0.25 s. Time-dependent and three-dimensional finite-element simulations for the flow and electrical field in the channel core, in conjunction with an analytical solution for the flow and electrical field within the EDL, are employed on the theoretical side.

For the electrically excited flow within the straight channel part of the micromixer, we infer electroosmotic flow profiles within the complete cross section. For the electroosmotic forces acting in the direction of the forced flow, the flow appears strongly accelerated towards the outlet. For the electroosmotic forces acting against the forced flow, we find backwards flow at the walls and flow towards the outlet in the channel centre. All electroosmotic effects are observed in very thin layers at all walls, whereas velocity profiles within these layers (of several tens of nanometers thickness) cannot be resolved experimentally. Instead, the first measurement point from the wall shows the integral effect of these layers.

Further, we investigate the electrically excited flow through the meander, again by velocity profiles within the complete cross section. For the counterflow situation, within the midheight level both the forced flow towards the outlet and the electroosmotic flow in the opposite direction are observed. Close to all walls, however, the flow is dominated by electroosmotic effects and the forced flow cannot be observed. The flow structure within the meander appears symmetric and can be characterized by means of singular points (vortex and saddle points). Even though the geometries of experiment and simulations do not agree in all details, the flow topology from both approaches agrees reasonably.

Both experiments and simulations – of course – suffer from several imperfections. The experiment differs in two major aspects from the simulations: 1. The µPIV measurements provide velocity data, averaged across volumes of $4.8\,\mu m \times 4.8\,\mu m \times 16\,\mu m$ (at best), while the simulations provide true local data. Here, particularly the averaging along the optical axis appears critical. 2. The meander within the experiments features curved contours of the inner corners of the bends, while the simulations have been performed with sharp corners. This, to some degree, limits the meaning of a direct comparison. Both averaging of the numerical results according to the experimental situation and numerical simulations for the true geometry with rounded corners, are being presently undertaken.

Acknowledgements

The authors would like to thank R. Lindken, R. Matsumoto, and I. Meisel for fruitful discussion.

References

[1] Y. Yamaguchi, F. Takagi, K. Yamashita, H. Nakumura, H. Maeda, K. Sotowa, K. Kusakabe, Y. Yamasaki, S. Morooka: 3-D simulation and visualization of laminar flow in a microchannel with hair-pin curves, AIChE J. **50**, 1530–1535 (2004)
[2] R. Liu, A. Stremler, K. Sharp, M. Olson, J. Santiago, R. Adrian, H. Aref, D. Beebe: Passive mixing in a three-dimensional serpentine microchannel, J. Microelectromech. Syst. **9**, 190–197 (2000)
[3] A. Stroock, S. Dertinger, A. Ajdari, I. Mezić, H. Stone, G. Whitesides: Chaotic mixer for microchannels, Science **295**, 647–651 (2002)
[4] M. Yi, S. Qian, H. Bau: A magnetohydrodynamic chaotic stirrer, J. Fluid Mech. **468**, 153–177 (2002)
[5] S. Qian, H. Bau: A chaotic electroosmotic stirrer, Anal. Chem. **74**, 3616–3625 (2002)
[6] I. Meisel, P. Ehrhard: Electrically-excited (electroosmotic) flows in microchannels for mixing applications, Eur. J. Mech. B/Fluids **25**, 491–504 (2005)
[7] A. Tieu, M. Mackenzie, E. Li: Measurements in microscopic flow with a solid-state LDA, Exp. Fluids **19**, 293–294 (1995)
[8] J. Czarske, L. Büttner, T. Razik, H. Müller: Boundary layer velocity measurements by a laser Doppler profile sensor with micrometre spatial resolution, Meas. Sci. Technol. **13**, 1979–1989 (2002)

[9] P. Paul, M. Garguilo, D. Rakestraw: Imaging of pressure- and electrokinetically-driven flows through open capillaries, Anal. Chem. **70**, 2459–2467 (1998)
[10] C. Gendrich, M. Koochesfahani, D. Nocera: Molecular tagging velocimetry and other novel applications of a new phosphorescent supramolecule, Exp. Fluids **23**, 361–372 (1997)
[11] J. Santiago, S. Wereley, C. Meinhart, R. Beebe, R. Adrian: A particle image velocimetry system for microfluidics, Exp. Fluids **25**, 316–319 (1998)
[12] C. Meinhart, S. Wereley, J. Santiago: PIV measurements of a microchannel flow, Exp. Fluids **27**, 414–419 (1999)
[13] C. Meinhart, S. Wereley, J. Santiago: Micron-resolution velocimetry techniques, in R. Adrian et al. (Ed.): *Developments in Laser Techniques and Applications to Fluid Mechanics* (Springer, Berlin 1999)
[14] D. Tretheway, C. Meinhart: Apparent fluid slip at hydrophobic microchannel walls, Phys. Fluids **14**, L9–L12 (2002)
[15] C. Meinhart, D. Wang, S. Werely: Measurement of AC electrokinetic flows, Biomed. Microdev. **5**, 139–145 (2003)
[16] Z. Wu, N.-T. Nguyen: Rapid mixing using two–phase hydraulic focussing in microchannels, Biomed. Microdev. **7**, 13–20 (2005)
[17] H. Li, R. Ewoldt, M. Olsen: MicroPIV measurements of turbulent flow in square microchannels with hydraulic diameters from 200 μm to 640 μm, Exp. Therm. Fluid Sci. **29**, 435–446 (2005)
[18] W. Wibel, P. Ehrhard: Experiments on the laminar/turbulent transition of liquid flows in rectangular microchannels, in proceedings 5th Int. (ASME) Conference Nanochannels, Microchannels and Minichannels (Puebla, Mexico, June 18–20, 2007 paper no. ICNMMM2007-30037
[19] K. Sharp, R. Adrian: Transition from laminar to turbulent flow in liquid filled microtubes, Exp. Fluids **36**, 741–747 (2004)
[20] R. Matsumoto, H. Farangis Zadeh, P. Ehrhard: Quantitative measurement of depth-averaged concentration fields in microchannels by means of a fluorescence intensity method, Exp. Fluids **39**, 722–729 (2005)
[21] A. Freitag, T. Dietrich, R. Scholz: Glass as a material for microreaction technology, in *Proceedings of IMRET 4, Topical Conference* (Atlanta, GA, May 5-9, 2000) pp. 48–54
[22] C. Meinhart, S. Wereley, M. Gray: Volume illumination for two-dimensional particle image velocimetry, Meas. Sci. Technol. **11**, 809–814 (2000)
[23] A. Ward-Smith: *Internal Fluid Flow – The Fluid Dynamics of Flow in Pipes and Ducts* (Clarendon Press, Oxford 1980)

Index

Characterization of Microfluidic Devices by Measurements with μ-PIV and CLSM

Michael Schlüter, Marko Hoffmann, and Norbert Räbiger

Institut für Umweltverfahrenstechnik, Universität Bremen, 28359 Bremen
michael.schlueter@iuv.uni-bremen.de

Abstract. Microfluidic devices are successfully in use for several applications in chemical engineering and biotechnology. Nevertheless, there is still no breakthrough for microprocess engineering because of a huge lack in understanding of the mechanisms on microscales for momentum transfer, hydrodynamics and mass transfer. Some important questions concern the design of a junction to reach acceptable mixing qualities with minimum pressure drop and narrow residence time distribution even under laminar flow conditions. The micro-particle image velocimetry (μ-PIV) in conjunction with confocal laser scanning microscopy (CLSM) have been used for the characterization of momentum and mass transfer at the Institute of Environmental Process Engineering to evaluate microfluidic devices. The calculation of three-dimensional flow and concentration fields is possible with two-dimensional measurement data for common stationary cases. Streamlines out of velocity gradients and isosurfaces out of fields of the same concentration are providing a helpful impression of the performance of microdevices based on highly reliable measurement data. A quantitative analysis of the velocity and concentration fields allows the calculation of residence-time distribution and mixing quality, which enables the adjustment of microreactor geometries for the demands of chemical, and biochemical reactions.

1 Introduction

Process intensification plays a key role in saving resources, which is one of the main goals in sustainable development. Microfluidic devices are a very interesting tool for process intensification because the contact time between educts is exactly adjustable, backmixing is negligible and the heat transfer for temperature-sensitive reactions is easy to control. Over the past decade, micromixers have been developed for a broad range of applications, such as bioanalytical techniques or the production of organic compounds. The growing demand for flexible multipurpose plants opens a wide field of application for continuous processes in microstructured devices [1]. The design of modern technical applications, e.g., reactions with high selectivity, a precise analysis of local mass transfer and hydrodynamics for different flow regimes, requires microscopic measurements of flow and concentration fields. Although microreaction technology offers diverse possibilities for optimizing processes with highly selective reactions this capability remains largely unused. *Wong* et al. [2] recently gave a general overview of flow phenomena and

A. Schroeder, C. E. Willert (Eds.): Particle Image Velocimetry,
Topics Appl. Physics **112**, 19–33 (2008)

mixing characteristics in T-shaped micromixers and showed the tremendous demand for further experimental data on flow and concentration fields on microscales.

In the last decade particle image velocimetry (PIV), as a nonintrusive technique for measuring flow fields, has been successfully adapted for measuring flow fields within microfluidic devices with micrometer-scale resolution, e.g., [3]. Recently, this nonintrusive diagnostic technique was extended to the infrared waveband [4] with the advantage that measurements can be made directly through silicon without the need for an optical access. For higher temporal resolution, *Shinohara* et al. [5] used a high-speed μ-PIV technique by combining a high-speed camera and a continuous-wave laser in order to investigate transient phenomena in microfluidic devices. Recently, a confocal laser scanning microscope (CLSM) in conjunction with micro-particle image velocimetry (μ-PIV) has been developed by using a dual high-speed spinning disk. While the upper disk is a rotating scanner that consists of 20 000 microlenses, the lower one is a Nipkow disk that consists of 20 000 matching pinholes of 50 μm in diameter. This system shows an optical slicing capability with a true stepwise-resolved μ-PIV vector-field mapping [6]. Another new development is the stereoscopic μ-PIV with an epifluorescence stereo microscope and two synchronized CCD cameras. *Lindken* et al. [7] applied a stereoscopic μ-PIV in the investigation of the three-dimensional flow in a T-shaped mixer.

Although the experimental data of flow velocities in magnitude and direction are essential for calculation and validation of numerical simulations, only a combination of flow and concentration fields enables a full analysis and understanding of phenomena as well as the analysis, development and evaluation of novel microfluidic processes. For direct visualization of flow structures in microchannels, the injection of dye to color a part of the flow field is a common method. As dyes, regular fluorophores are normally used, as their emissions can be stimulated by laser light (e.g., [8]). Contrary to macroscopic flows, it is not practicable to use a thin laser lightsheet because of the tiny dimensions (down to several micrometers) in microdevices. On a microscale, it is essential to illuminate the complete microchannel volume. *Sinton* [9] did a comprehensive study of microscale flow visualization techniques. Recently, *Matsumoto* et al. [10] presented an optical measuring technique for concentration fields in microchannels by averaging the recorded fluorescence intensity along the optical path to provide a height-averaged concentration field. In order to get the three-dimensional information on the concentration field with high spatial resolution the use of confocal microscopy is indispensable. The major advantage of a confocal laser scanning microscope (CLSM) is the possibility of collecting emitted light only from the focus plane. In front of the detector (photomultiplier) a pinhole is arranged on a plane conjugate to the focal plane of the objective. Light emitted from planes above or below the focal plane is out of focus when it strikes the pinhole. Therefore, most of the light cannot pass through the pinhole and does not lead to the distortion of the image [11]. This spatial filtering is the key principle to increase the optical resolution by producing depthwise optical slicing. The illuminating laser can rapidly scan from point to point on a single focal plane, in a synchronized way with the aperture, in order to complete a full-field image on the detector unit. To reconstruct three-dimensional images the scan is repeated for multiple focal planes under the presumption of stationary conditions [6]. *Yamaguchi* et al. [12] used confocal microscopy to visualize fluorescence intensity patterns in cross sections of a meander channel, allowing a qualitative comparison with CFD results. *Stroock* et al. [13] introduced a confocal microscope to determine the mixing quality in a staggered herringbone mixer at small Reynolds numbers. *Ismagilov* et al. [14] used confocal fluorescent microscopy to visualize the fluorescent product formed by reaction between chemical species in a microchannel. Thus, a quantitative description of reaction–diffusion processes near the walls of a channel allows a profound understanding of three-dimensional hydrodynamics and mass transfer.

The given examples show that several groups have performed investigations into two-dimensional or three-dimensional concentration or velocity fields. Most investigations have been done for passive micromixers at low Reynolds numbers (Re < 1), typical for applications like lab-on-a-chip, microarrays, DNA sequencing or micro-total analysis systems (μ-TAS) in biotechnology.

For chemical reactions, much higher Reynolds numbers are usually necessary (above Re > 150) which lead, even in laminar flows, to the generation of vortices with thin layers of different educts and product concentrations. Even for simple T-shaped junctions, hydrodynamic and mass transfer is a three-dimensional phenomenon that is only describable by quantitative data with high resolution in time and space. A deeper understanding of reactor performance is only achievable with the knowledge of both flow and concentration fields. While the flow velocities and directions give the information about residence time and shear stress, the visualized concentration fields allow the evaluation of diffusion lengths and mixing times. This detailed knowledge will allow the prediction of mixing performance as well as the yield and selectivity of different microfluidic devices in the ambitious field called "chemical microprocess engineering".

2 Experimental Setup

The experimental setup used for the characterization of microfluidic devices is shown in Fig. 1. The μ-PIV investigations are performed with a conventional epifluorescence microscope (Olympus BX51WI). Deionized water ($T = 20\,°\mathrm{C}$) enriched with tracer particles is fed out of pressure containers into the microdevice to prevent any pulsation. The mass flow is measured by mass flow meters FI (Bronkhorst High-Tech BV).

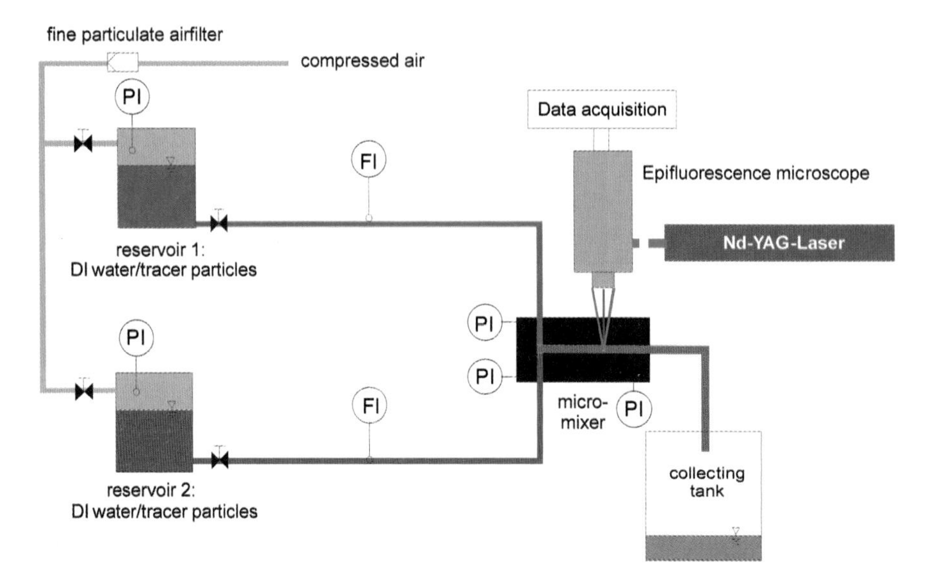

Fig. 1. Experimental setup for flow-field measurements

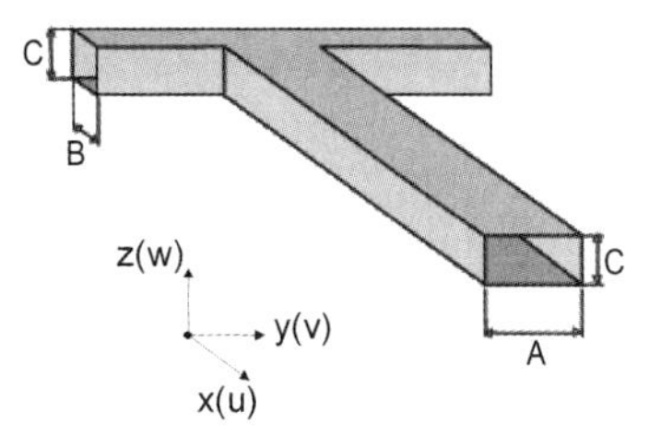

Fig. 2. Geometry of T-shaped micromixer (standard: $A = 200\,\mu\mathrm{m}$, $B = 100\,\mu\mathrm{m}$, $C = 100\,\mu\mathrm{m}$)

The T-mixers with different geometries were provided by the Institute of Microsystem Technology (IMTEK) at the University of Freiburg, Germany [15]. They are made out of silicon sealed by a Pyrex glass lid to allow optical access. The notation for the T-shaped micromixers is $A \times B \times C$ in µm according to Fig. 2.

For µ-PIV nanoparticles with low inertia are used as tracers (polystyrene, particle diameter 500 nm). To detect the invisible particles they are coated with a Rhodamine B layer (Micro Particles GmbH, Berlin) to achieve fluorescence in the laser illumination. Two consecutive laser pulses (wavelength 532 nm; pulse width 5 ns, New Wave Research, Inc.) with a defined pulse distance of 1 µs enable the detection of particle velocities and directions according to the local flow field. A PCO Sensicam QE CCD camera is used

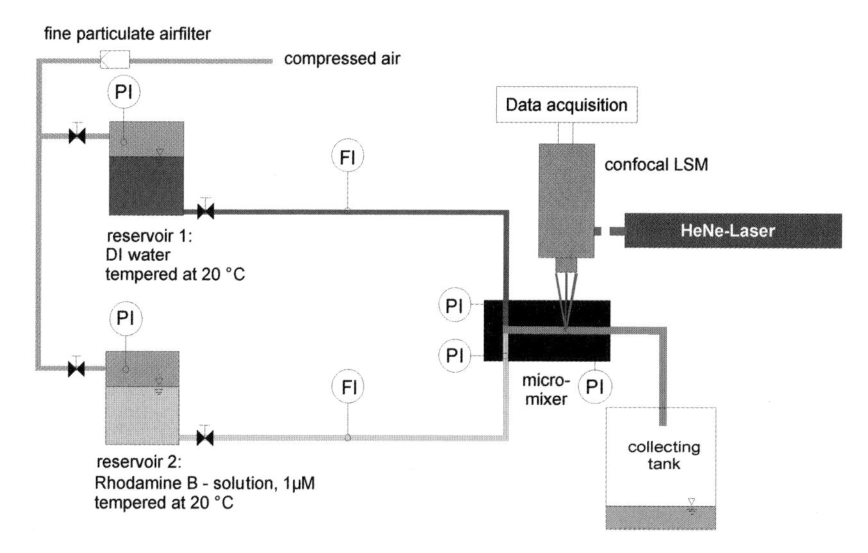

Fig. 3. Experimental setup for concentration-field measurements

for μ-PIV and the data evaluation is performed with a standard crosscorrelation scheme based on a FFT algorithm (μ-PIV system by ILA GmbH, Jülich). The epifluorescence microscope is equipped with a Plan Achromat C Objective (20×/0.4) to achieve a lateral resolution of $7 \times 7\,\mu\text{m}$ with a 32-pixel square interrogation region (half-overlapping) in a measurement depth of approximately 13 μm [16].

For visualization and measurement of local concentration fields a confocal laser scanning microscope (Carl Zeiss LSM 410) is used (Fig. 3). The confocal microscope uses a helium neon laser (543 nm, 1 mW) as excitation source and a 20×/0.5 Plan Neofluar objective. The optical slice thickness is $\approx 8.0\,\mu\text{m}$, the axial resolution $\approx 5.0\,\mu\text{m}$ and the lateral resolution $\approx 0.6\,\mu\text{m}$. A buffer solution (deionized water, pH 8.2 $\mathrm{T} = 20\,^\circ\mathrm{C}$) is fed out of pressure containers into the microdevice to prevent any pulsation, while the mass flow is measured again by mass flow meters FI (Bronkhorst High-Tech BV) with an accuracy of 1 % of measured value.

One inlet stream is enriched with the fluorescent dye Rhodamine B (dissolved in a pH 8.2 buffer solution) in a very low concentration, thus the fluorescence intensity I_f of the fluorochrome can be assumed as proportional to the concentration of the dye c [16]. Thus, the calibration of grayscales with dye concentrations allows a quantitative analysis of the dye distribution in the microdevice as well as its mixing performance.

For a quantitative analysis of mixing performance Danckwerts' quality of mixing

$$\alpha = 1 - \sqrt{\frac{\sigma_M^2}{\sigma_{\max}^2}} \tag{1}$$

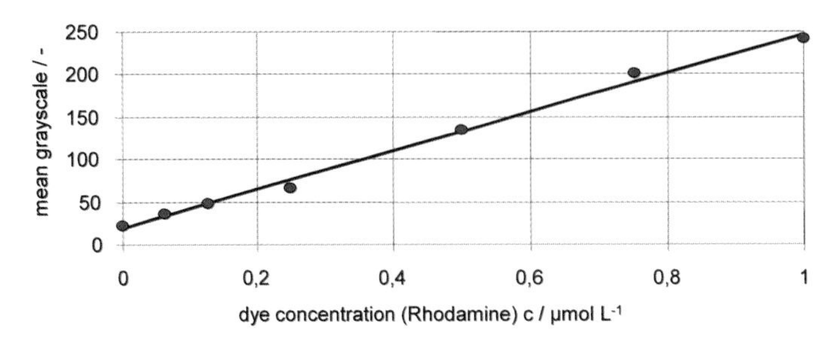

Fig. 4. Grayscale vs. fluorescence dye concentration

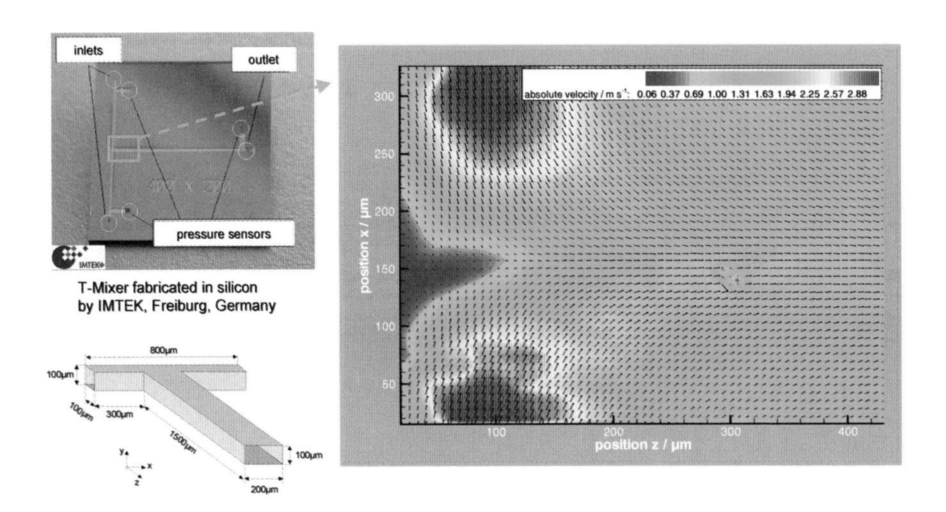

Fig. 5. Plan view of the flow field at the junction of a T-shaped micromixer

can be quantitatively evaluated by means of the standard deviation σ and the maximum standard deviation $\sigma_{\max}$ of concentrations (grayscales) for a cross-sectional area along the channel depth using the Image Processing Toolbox of MATLAB (Mathworks). $\alpha = 0$ corresponds to a totally segregated system, whereas a value of $\alpha = 1$ corresponds to a homogeneous mixture. The image-processing software IMARIS (Bitplane AG, Switzerland) is used to generate a volume-rendered 3D image out of a data set (e.g., 50 slices for a channel 200 µm in depth).

3 Results and Discussion

As stated before, many applications in chemical process engineering as well as biotechnology and analytics require detailed knowledge of the flow field inside microdevices. This becomes obvious in a simple junction that exists in nearly all microfluidic applications. After draft estimation one would expect a rather good mixing between two fluids in a T-shaped junction. Figure 5 shows the flow field measured by µ-PIV in a T-shaped micromixer edged in silicon by the Department of Microsystems Engineering (IMTEK), Freiburg, Germany.

A wide distribution of velocities between 0.06 m/s at the stagnation point and 2.88 m/s in the deflection zones can clearly be seen. The flow field seems to be very regular and symmetrical along the z-axis. A closer view shows the typical hyperbolic profile as expected by Hagen–Poiseuille's law for laminar flow. This would refer to a very poor mixing due to no cross-sectional exchange of fluid elements. As important parameter for the characterization of the flow regime in microdevices is the Reynolds number Re and is given by

$$\mathrm{Re} = \frac{v \cdot d_{\mathrm{h}} \cdot \rho}{\eta}, \tag{2}$$

with the fluid velocity v, the hydraulic diameter d_{h} the density ρ and the dynamic viscosity η of the fluid.

A variation of the Re number leads to minor changes in the velocity profiles as shown in Fig. 6 for Re = 120 and Re = 186, although the flow is still laminar. As indicated with the shadow picture in the background of Fig. 6 the flow field is very different between both flow regimes. It has to be pointed out that the asymmetry of the flow is not influenced by differences in the flow rate between both inlets. Even though the measured value of flow rate varies within 1 % independently for both inletes due to the accuracy of the mass flow meters the flow regime is very stable, as indicated by the sharp picture recorded over a period of 8 s.

Furthermore, a good accuracy and reproducibility of the measurements can be shown by comparsion of measured flow rates and integration over the flow fields of 51 layers along the reactor depth. The difference between mass flow meters and PIV measurements is smaller than 1 % with a standard deviation of less than 2 %. These steady flow conditions show a very high reproducablility and thus enable the combination of µ-PIV and CLSM measurements in a sequential mode. The remaining inaccuracy is mainly caused by the difficulties in measuring the high velocity gradients close to the wall. Therefore, the velocity field in Fig. 6 does not fit the wall conditions reliably.

It has to be pointed out that even if the velocity data is important for quantitative calculation and comparison, an illustrative description of the spatial flow field inside a microdevice is not possible.

An illustrative description and quantitative evaluation of the spatial flow field is more effective when using the laser-induced fluorescence as shown in Fig. 7. Combined with the confocal technique that allows detection of emitted light from a discrete plane it is possible to cut the volume of the microdevice into several slices of 7 µm thickness. By using a commercial image-processing software the reconstruction of a 3D image is possible (Fig. 7). An indispensable condition for using the CLSM is the time invariance of the flow field

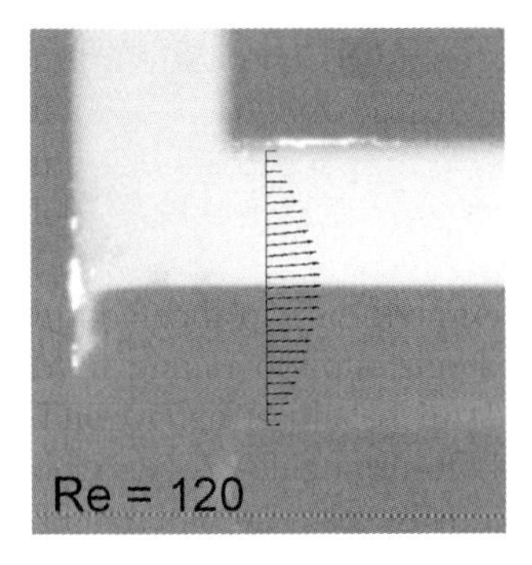

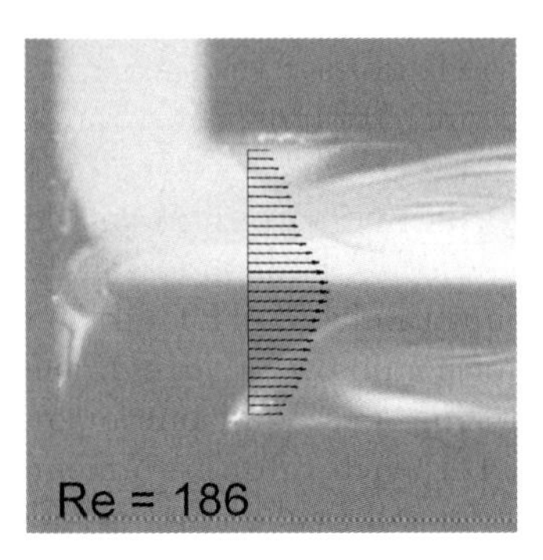

Fig. 6. Velocity profiles at the junction of a T-shaped micromixer for two different flow regimes (measured by μ-PIV)

because the temporal resolution of the scanning technique is in the range of several milliseconds, depending on the spatial resolution. Surprisingly, the flow field in microdevices is stationary and reproducible even though very complex structures occur (Fig. 7, right). The illustrative three-dimensional view into microdevices allows a deep insight into new aspects of mixing. Figure 7 makes it clear that even though only laminar mixing is available in microdevices very fine structures are achievable if the flow regime has been chosen correctly. For example, Fig. 7 shows fine structures with short diffusion lengths down to 3 μm in a T-shaped micromixer. These diffusion lengths are shorter than for almost turbulent flows with large energy input, thus micromixing in microdevices might be more effective than in large facilities. On the other hand, only a reduction of the flow rate to one third causes a tremendous decrease of contact area between two educts and mixing performance (Fig. 7, left).

This can be shown quantitatively with (1) for two T-mixers with different aspect ratio (width mixing channel/height). While the mixing in a device with high aspect ratio is quite poor (Fig. 8, right), an aspect ratio of ≈ 2 enables much higher mixing qualities even for smaller flow velocities (Fig. 8, right). With a higher aspect ratio the development of vortices is supressed and the transition to an engulfment flow occurs at higher velocities. Nevertheless, with one single T-mixer no satisfactory mixing quality is achievable.

Despite the fact that CLSM measurements are helpful for characterizing and designing microdevices and for determining the optimal values for operating parameters, quantitative results for flow velocities, contact times, shear stresses and residence time distributions are only achievable with the help of μ-PIV measurements.

A new method to calculate three-dimensional flow patterns out of two-dimensional PIV data has been recently developed by *Hoffmann* [17]. Hoffmann uses the conservation of mass for incompressible liquids in a closed

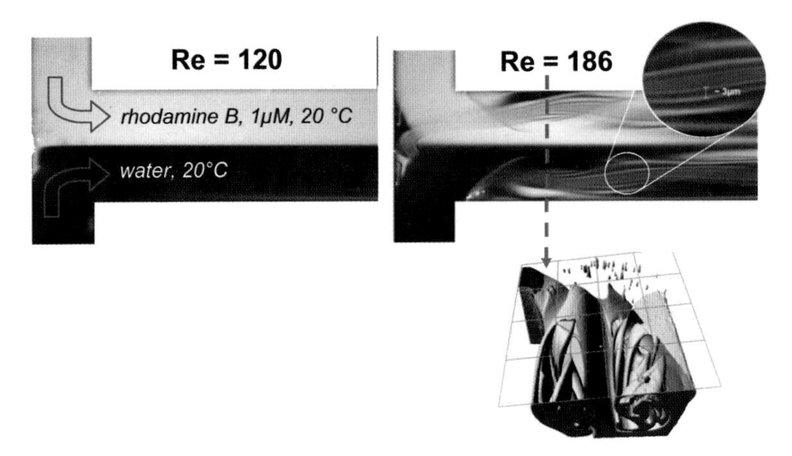

Fig. 7. Plane and spatial distribution of dye in a T-shaped micromixer for two different flow regimes (measured by CLSM)

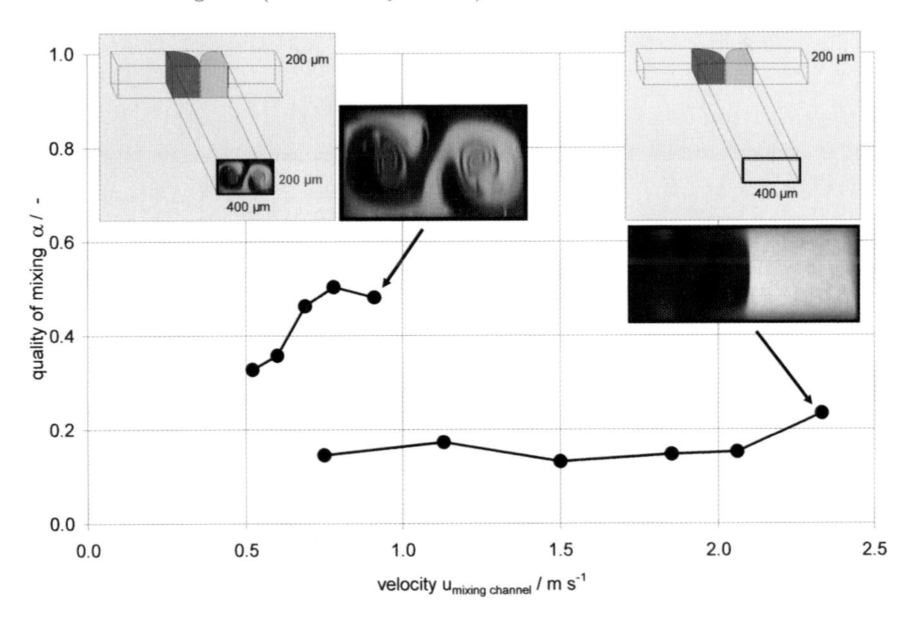

Fig. 8. Quality of mixing α of two mixing devices with different aspect ratio

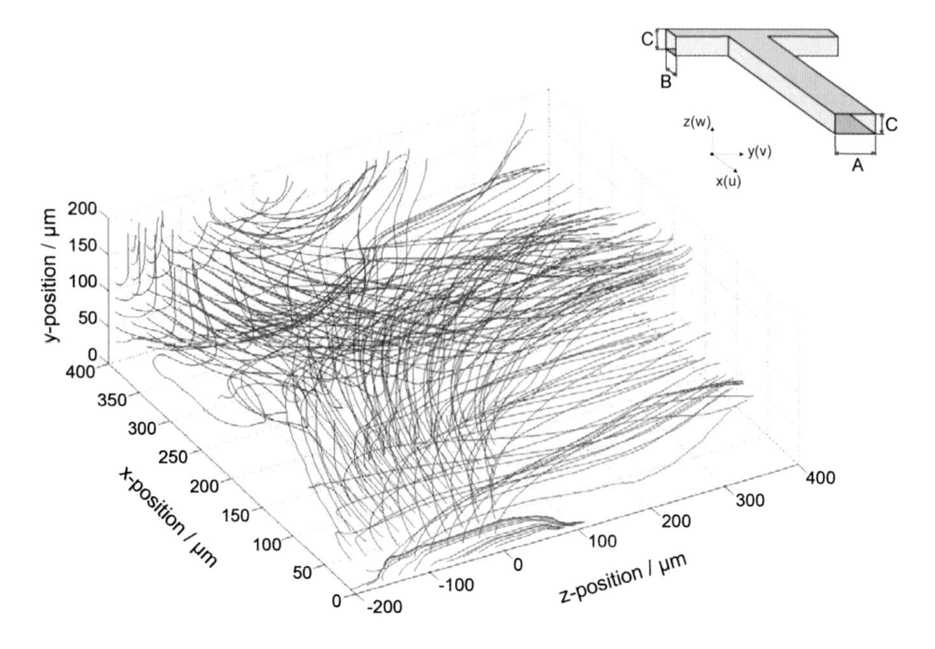

Fig. 9. Velocity streamlines at the junction of a T-shaped micromixer (measured by μ-PIV)

volume with defined inlet and outlet according to the former work of *Feng* et al. [18] for turbulent mixing

$$\frac{\partial u_i}{\partial x_i} = 0 \quad x_i = (x, y, z); \quad u_i = (u, v, w)\,. \tag{3}$$

The missing velocity component w is calculable according to [19] with the continuity equation due to

$$w(x_i, y_j, z_k) = w(x_i, y_j, z_{k-1}) - \int_{z_{k-1}}^{z_k} \left[\frac{\partial u}{\partial x}(x_i, y_j, z) + \frac{\partial v}{\partial y}(x_i, y_j, z) \right] \mathrm{d}z\,. \tag{4}$$

Hoffmann uses a MATLAB algorithm to calculate and evaluate the three-dimensional flow field in micromixers (Fig. 9). The accuracy of the method is evaluated with numerical simulations performed by *Bothe* et al. [20], as well as with three-dimensional measurements carried out with a stereo μ-PIV System by [7].

The knowledge of the three-dimensional flow field allows the calculation of contact times between educts within a microfluidic device as well as residence-time distributions. In combination with the informations about local concentrations of educts and products received from three-

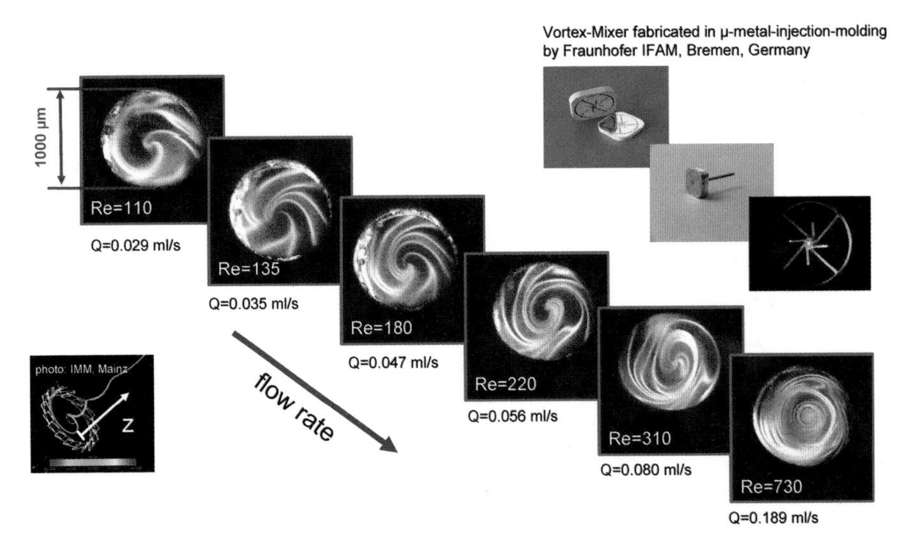

Fig. 10. Plane distribution of dye in a vortex mixer for the synthesis of ionic liquids measured by CLSM [21]

dimensional CLSM measurements, the dependency between the hydrodynamics and chemical/biochemical reactions becomes more apparent and consequently controllable. Therefore, strategies to improve the yield and selectivity will be possible.

The following two examples will show the applicability of the methods for industrial use. For the synthesis of ionic liquids the temperature control of the exothermic reaction is one of the key parameters to ensure a high yield and selectivity. One possibility of achieving a good heat transfer is the use of microreactors with their high surface to volume ratio. A vortex mixer, for example, guarantees a good mixing performance with high heat transfer rates [1]. For the NEMESIS project (BMBF 16SV1970) the Fraunhofer Institute for Manufacturing Technology and Applied Materials Research in Bremen fabricated a vortex mixer out of a metal resistant against most ionic liquids. To check the optimal geometry and flow rate for optimal mixing performance with minimum pressure drop the mixer is fed by water, one inlet stream marked with a fluorescent dye. As shown in Fig. 10 by CLSM, a flow rate of at least 0.08 ml/s should be chosen to achieve a sufficient mixing performance with high contact area between both educts and short diffusion length for micromixing.

Another example concerns the further development of the T geometry by retaining the good mixing conditions in this simple shape. As shown by former investigations [16] it is difficult to achieve a totally mixed flow in a T-shaped micromixer under laminar flow conditions. To ensure a complete mixing in short timescales – which is important for fast chemical reactions

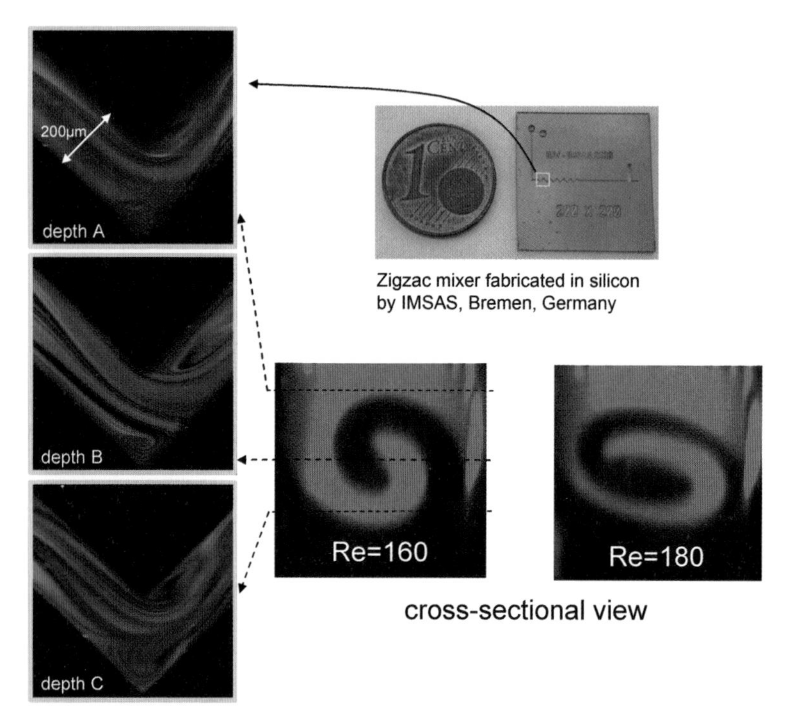

Fig. 11. Plane distribution of dye in a zigzag mixer for parallel consecutive reactions (measured by CLSM)

– a zigzag shape is a reasonable evolution. The zigzag mixer, which is fabricated out of silicon at the Institute for Microsensors, -actuators and -systems, Bremen, Germany (IMSAS) can be characterized by mixing a fluorescent dye into a water stream. Figure 11 shows results of CLSM measurements for three different layers along the depth of the first bend of the mixer as well as the cross section for two different Reynolds numbers. As expected, an increasing formation of vortices becomes apparent with higher Reynolds number, which causes larger contact areas and shorter diffusion lengths with enhanced micromixing.

4 Conclusion

The combined application of µ-PIV and CLSM enables a sophisticated characterization of microdevices. CLSM allows the visualization of flow fields to evaluate flow patterns and mixing performance qualitatively and quantitatively, whereas µ-PIV measurements allow the investigation of velocities, shear stresses, residence-time distributions and contact times. Both techniques together will help to design future generations of microdevices with a much better adaptation of geometries, materials and surface properties to the demands of reactions than with previous methods. Even if reproducible measurements are possible with the given methods, further development has to be done for enhancement of accuracy. In particular, the measurement of high velocity gradients close to walls and high concentrations gradients inside fine structures are still inaccurate with μ-PIV and μ-LIF. Nevertheless, the extended characterization possibilities will bring a clearer insight into the application range of microdevices for a faster and safer selection of microcomponents. This may guide microprocess engineering to an increased acceptance for the accelerated development of sustainable processes.

References

[1] V. Hessel, H. Löwe, A. Müller, G. Kolb: *Chemical Micro Process Engineering, Processing and Plants* (Wiley-VCH, Weinheim 2004) pp. 1–2
[2] S. H. Wong, M. C. L. Ward, C. W. Wharton: Micro T-mixer as a rapid mixing micro-mixer, Sens. Actuators B-Chem. pp. 359–379 (2004)
[3] C. D. Meinhart, S. T. Wereley, M. H. B. Gray: Volume illumination for two-dimensional particle image velocimetry, Meas. Sci. Technol. **11**, 809–814 (2000)
[4] D. Liu, S. V. Garimella, S. T. Wereley: Infrared micro-particle image velocimetry in silicon-based microdevices, Exp. Fluids **38**, 385–392 (2005)
[5] K. Shinohara, Y. Sugii, A. Aota, A. Hibara, M. Tokeshi, T. Kitamori, K. Okamoto: High-speed micro-PIV measurements of transient flow in microfluidic devices, Meas. Sci. Technol. **15**, 1965–1970 (2004)
[6] J. S. Park, C. K. Choi, K. D. Kihm: Optically sliced micro-PIV using confocal laser scanning microscopy (CLSM), Exp. Fluids **37**, 105–119 (2004)
[7] R. Lindken, J. Westerweel, B. Wieneke: Stereoscopic micro particle image velocimetry, Exp. Fluids **41**, 161–171 (2006)
[8] M. H. Oddy, J. G. Santiago, J. C. Mikkelsen: Electrokinetic instability micromixing, Anal. Chem. **73**, 5822–5832 (2001)
[9] D. Sinton: Microscale flow visualization, Microfluid. Nanofluid. **1**, 2–21 (2004)
[10] R. Matsumoto, H. F. Zadeh, P. Ehrhard: Quantitative measurement of depth-averaged concentration fields in microchannels by means of a fluorescence intensity method, Exp. Fluids **39**, 722–729 (2005)
[11] S. Wilhelm, B. Gröbler, M. Gluch, H. Heinz: *Confocal Laser Scanning Microscopy: Principles*, vol. 2–3 (Carl Zeiss 2003) URL: `www.zeiss.de/lsm`
[12] Y. Yamaguchi, F. Takagi, T. Watari, K. Yamashita, H. Nakamuraa, H. Shimizu, H. Maeda: Interface configuration of the two layered laminar flow in a curved microchannel, Chem. Eng. J. **101**, 367–372 (2004)
[13] A. D. Stroock, S. K. W. Dertinger, A. Ajdari, I. Mezic, H. Stone, G. Whitesides: Chaotic mixer for microchannels, Science **295**, 647–651 (2002)
[14] R. F. Ismagilov, A. D. Stroock, P. J. A. Kenis, G. Whitesidesa, H. A. Stone: Experimental and theoretical scaling laws for transverse diffusive broadening in two-phase laminar flows in microchannels, Appl. Phys. Lett. **76**, 2376–2378 (2000)

[15] N. Kockmann (Ed.): *Micro Process Engineering: Fundamentals, Devices, Fabrication, and Applications, Advanced Micro and Nanosystems* (Wiley-VCH, Weinheim 2006)
[16] M. Hoffmann, M. Schlüter, N. Räbiger: Experimental investigation of liquid–liquid mixing in T-shaped micro-mixers using μ-LIF and μ-PIV, Chem. Eng. Sci. **61**, 2968–2976 (2006)
[17] M. Hoffmann, M. Schlüter, N. Räbiger: Untersuchung der Mischvorgänge in Mikroreaktoren durch Anwendung von Micro-LIF und Micro-PIV, Chem. Ing. Tech. **79**, 1067–1075 (2007)
[18] H. Feng, M. G. Olsen, Y. Liu, R. O. Fox, J. C. Hill: Investigation of turbulent mixing in a confined planar-jet reactor, AIChE J. **51**, 2649–2664 (2005)
[19] J. M. M. Sousa: Turbulent flow around a surface-mounted obstacle using 2D-3C DPIV, Exp. Fluids **33**, 854–862 (2002)
[20] D. Bothe, C. Stemich, H.-J. Warnecke: Fluid mixing in a T-shaped micromixer, Chem. Eng. Sci. **61**, 2950–2958 (2006)
[21] M. Schlüter, M. Hoffmann, J. Fokken, N. Räbiger: Experimentelle Methoden zur Charakterisierung von Mikromischern anhand der Vermischungsleistung und Verweilzeitverteilung, in *Proc. Mikrosystemtechnik Kongress 2005* (2005)

Index

Time-Resolved PIV Measurements of Vortical Structures in the Upper Human Airways

Sebastian Große, Wolfgang Schröder, and Michael Klaas

Institute of Aerodynamics, RWTH Aachen University, D-52062 Aachen, Germany
s.grosse@aia.rwth-aachen.de

Abstract. A detailed knowledge of the three-dimensional flow structures in the human lung is an inevitable prerequisite to optimize respiratory-assist devices. To achieve this goal the indepth analysis of the flow field that evolves during normal breathing conditions is indispensable. This study focuses on the experimental investigation of the steady and oscillatory flow in the first lung bifurcation of a three-dimensional realistic transparent silicone lung model. The particle image velocimetry technique was used for the measurements. To match the refractive index of the model, the fluid was a mixture of water and glycerine. The flow structures occurring in the first bifurcation during steady inflow have been studied in detail at different flow rates and Reynolds numbers ranging from $\mathrm{Re_D} = 1250$ to $\mathrm{Re_D} = 1700$ based on the hydraulic diameter D of the trachea. The results evidence a highly three-dimensional and asymmetric character of the velocity field in the upper human airways, in which the influence of the asymmetric geometry of the realistic lung model plays a significant role for the development of the flow field in the respiratory system. The inspiration flow shows large zones with secondary vortical flow structures with reduced streamwise velocity near the outer walls of the bifurcation and regions of high-speed fluid in the vicinity of the inner side walls of the bifurcation. Depending on the local geometry of the lung these zones extend to the next generation of the airway system, resulting in a strong impact on the flow-rate distribution in the different branches of the lung. During expiration small zones of reduced streamwise velocity can be observed mainly in the trachea and the flow profile is characterized by typical jet-like structures and an M-shaped velocity profile. To investigate the temporal evolution of the flow phenomena in the first lung bifurcation time-resolved recordings were performed for Womersley numbers α ranging from 3.3 to 5.8 and Reynolds numbers of $\mathrm{Re_D} = 1050$, 1400, and 2100. The results evidence a region with two embedded counterrotating vortices in the left bronchia. Furthermore, the measurements reveal a strong shear layer in the bronchia that evolves when the flow direction changes from inspiration to expiration. Due to the high intricacy of the natural lung geometry most research has been performed in simplified bifurcation models such that no comparably realistic and detailed experimental study of the flow field within the first bifurcation of the upper human airways has been presented as yet.

A. Schroeder, C. E. Willert (Eds.): Particle Image Velocimetry,
Topics Appl. Physics **112**, 35–53 (2008)

1 Introduction

The lung as the human respiratory organ consists of a repeatedly bifurcating network of tubes with progressively decreasing diameters. The understanding of the highly unsteady and nonlinear flow field within these tubes is of great importance to interpret particle-deposition patterns and thus to develop aerosol drug-delivery systems [1–5]. Furthermore, utilizability and efficiency of artificial lung ventilation highly depends on the velocity and pressure distribution, both of which are associated with a specific ventilation strategy. Despite numerous experimental and numerical investigations having been conducted so far it has to be stated that there still exists a considerable amount of uncertainty concerning the very complex flow field in the human lung. This is mainly due to the high intricacy of the human lung geometry. In other words, results for realistic models of the lung are still rare. The majority of experimental and numerical investigations has been performed using models with a simplified geometry, which might lead to flow fields that deviate considerably from realistic lung flows. For this reason, results could often not explain the fluid-mechanical causes for certain particle-deposition patterns or nonlinear lung-flow phenomena.

One of the earliest publications by *Schroter* and *Sudlow* [6] focused on velocity profiles and flow patterns in symmetric bifurcating geometries. The authors used hot-wire probes and smoke visualization techniques assuming a steady air flow in rigid tubes with smooth surfaces. These assumptions are commonly made with reference to the existing local Reynolds number based on the mean geometry diameters in the range of $\mathrm{Re_D} = 500$ to 2000 and Womersley numbers of about 5 to 25, which are typical values for normal breathing conditions. As far as the influence of the Reynolds number on the characteristics of the flow structures during inspiration is concerned, *Martonen* et al. [7] as well as *Grotberg* [8] report a strong dependence of the flow field on the Reynolds number in the Reynolds number ranges of $\mathrm{Re} = 200$ to 2000 and $\mathrm{Re} = 200$ to 1200, respectively. On the other hand, *Liu* et al. [9] found comparable flow patterns by numerically investigating an asymmetric human lung airway system for Reynolds numbers between 400 and 1600. Furthermore, there is no common consensus on the impact of the entry flow profiles on the flow structures evolving in the upper lung. While some investigators report the bifurcation-flow phenomena to be independent of the inflow conditions, *Yang* et al. [10] report a strong impact of different inflow conditions on the flow in the subsequent bifurcation zones. Since the flow between consecutive bifurcation regions cannot fully develop in a natural lung geometry, the upstream condition within the network of bifurcations influences the flow structures further downstream. Nevertheless, a large number of investigations have been performed for artificial geometries with rather long branches between subsequent bifurcations.

Experimental investigations of *Ramuzat* and *Riethmuller* [11] concentrated on a plane symmetric model of subsequent pipe bifurcations, hence introducing a major simplification of the model geometry. In their experiments, oscillating flow phenomena have been investigated for different Womersley numbers, emphasizing the temporal development of the velocity through successive bifurcations and the quasisteadiness of the flow for low oscillation frequencies.

Recent numerical calculations focused on the analysis of aerosol transport and particle-disposition patterns and have been carried out by numerous groups, e.g., [12] and [13]. *Comer* et al. [1–3] investigated the flow structure in symmetric double bifurcations under steady conditions in a system consisting of three subsequent bifurcations. The Reynolds number based on the pipe diameter varied from $\mathrm{Re_D} = 500$ to 2000. The authors detected strong vortical flow fields especially in the region of the second bifurcation, leading to complex particle distributions and deposition patterns. The flow in a plane and nonsymmetric bifurcation has been calculated by *Liu* et al. [9]. Their simulations comprised the flow field from the fifth to the eleventh branch of the model of *Weibel* [14] resulting in Reynolds numbers in the range of $\mathrm{Re_D} = 200$ to 1600. All the aforementioned models were based on the planar character of the pipe network. Furthermore, most of the studies were performed for highly symmetric models with bifurcations that divide the lower generations' pipe into two subsequent tubular pipes with nearly or exactly likewise diameter.

Recently, experimental and numerical studies have been carried out for nonplanar lung geometries. *Caro* et al. [15] investigated steady inspiratory flow in a symmetric model of the lung with two generations. For planar and nonplanar configurations the results emphasized the different development of the flow fields and volumetric distributions between varying types of geometry. The distribution of the wall-shear stress in the different generations of the bronchia was measured by optically visualizing the reaction between acid vapor and blue litmus coatings in the bronchial system. It was concluded that inspiratory flow in larger human bronchial airways is asymmetric and swirling with implications for all transport processes including those of particles. A similar model has been used in a numerical simulation by *Comer* et al. [1]. The findings show that for the out-of-plane configuration of the different generations of the bronchia the particle-deposition patterns differed from the planar case and possessed higher particle efficiencies for diluted suspensions of inhaled particles where the efficiency is defined as the number of particles deposited in a certain section divided through the total number of particles entering the region.

Numerical results for the most complicated and realistic model of the lung geometry have been presented by van *Ertbruggen* et al. [16]. The results evidenced the highly three-dimensional character of the flow field, emphasizing the importance of nonfully developed flows in the branches due to the relatively short lengths.

Recent particle image velocimetry measurements in a realistic lung model by *Brücker* and *Schröder* [17] revealed large separation zones near the outer wall of the bronchia during inspiration but did not evidence any separation regions during expiration.

The findings in the literature show that physically relevant studies of the bronchial airway anatomy and the corresponding flow field require realistic and thus complex models of the lung geometry. On the one hand, this challenge can be pursued by defining an artificial lung model with high complexity, taking into account the three-dimensional characteristics of the branches, e.g., asymmetry of the bifurcation zones, the "natural" ratio between the cross sections of subsequent generations of branches, the noncircular shape of real bronchia, and the impact of the upstream flow conditions on the mass-flow distributions in subbranches. On the other hand, a realistic lung model can be derived from original, hence exact, lung-geometry data. Modern clinical image-recording techniques allow the exact determination of the lung-geometry at least for the first six to ten bronchial branches. This offers the possibility to manufacture a "negative" kernel of the human airways made from solvable materials, which can be used to generate transparent silicone models ("positive"). This approach finally leads to a realistic model of the human lung providing optical access to the airways. The difficulty in such "natural" models is the lack of symmetry and orientation such that a fully three-dimensional investigation of the flow field is a must to adequately describe the flow structures.

The objective of this study is to perform detailed velocity measurements in the first bifurcation of such a realistic lung model using standard and time-resolved particle image velocimetry (PIV). The first experiments allow us to analyze the flow patterns in the first bifurcation of the human lung consisting of the trachea and the first bronchia generation. Steady inspiration and expiration flows have been studied for a variety of Reynolds numbers based on the trachea diameter D ranging from $\mathrm{Re}_\mathrm{D} = 1250$ to 1700. Since the model extends to the 6th generation the effect of upstream bifurcations on the expiration flow can be investigated. Additionally, oscillating flow with Reynolds numbers ranging from $\mathrm{Re}_\mathrm{D} = 1050$ to 2100 and Womersley numbers α ranging from 3.3 to 5.8, where the Womersley number is defined as $\alpha = 0.5 \cdot D \cdot \sqrt{2 \cdot \pi \cdot f/\nu}$, was investigated. Herein, the quantity D denotes the diameter of the trachea, f is the frequency of the ventilation cycles, and ν represents the kinematic viscosity of the fluid. The time-resolved measurements have been performed in a center plane in the first bifurcation region of the lung geometry, such that the temporal evolution of the flow field in the first lung bifurcation during a complete breathing cycle could be investigated.

In the following sections, first the experimental setup is described in great detail. Then, the results for steady and time-dependent flows are discussed. Finally, some conclusions conclude this chapter.

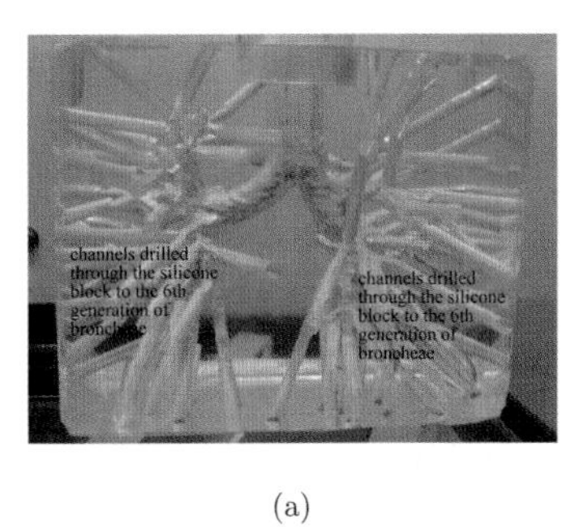

(a)

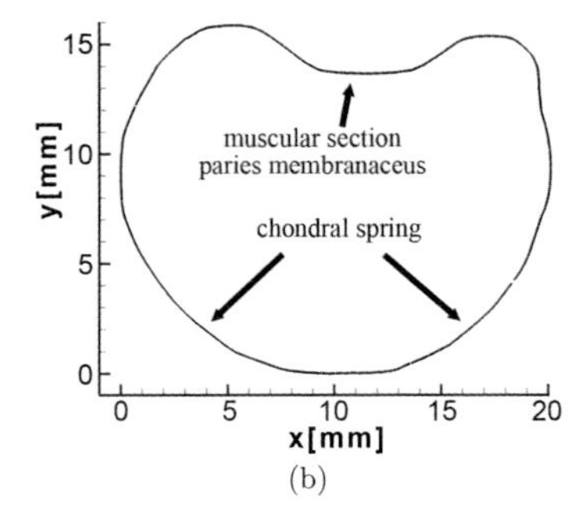

(b)

Fig. 1. (**a**) Picture of the hollow-lung model in silicone. (**b**) Sketch of the cross section of the trachea (plane normal to the average direction of the trachea, oriented 25.0° to the z-direction from Fig. 2)

2 Experimental Setup

2.1 Lung Model

The PIV experiments were conducted in a realistic, three-dimensional lung model that extends to the 6th generation of the bronchial system. It is fabricated from transparent silicone (RTV815) to allow perfect optical access. Figure 1 depicts a picture of the hollow lung model in the silicone block. The complete lung geometry is indicated in Fig. 2 by showing several cross sections.

At inspiration, the fluid enters the lung model through an anatomically shaped trachea as illustrated in Fig. 1. The flow exits the model through channels that have been drilled from the 6th generation's endings through the block sides. The trachea of the model possesses the typical shape of a natural trachea with a dorsal indentation. The real human trachea is stiffened by chondral springs in the ventral part of the cross section. The dorsal, i.e., the back part of the trachea consists of a muscular structure (paries membranaceus) and allows a widening of the trachea.

2.2 In- and Outflow Conditions

The choice of an appropriate inflow condition is crucial for the investigation of the lung flow since the velocity profile in the trachea has a strong impact on the flow field in the first lung bifurcation [10]. To enable a comparison with numerical results and with future computations a defined inflow condition in the trachea was chosen. Most numerical calculations use a classical fully developed laminar profile as the inflow condition. Also, parallel flow profiles have been investigated. To uniquely define the inflow characteristics and to ensure a fully developed velocity profile, an $L = 500\,\mathrm{mm}$ long pipe section with a constant cross section matching that in the upper part of the trachea was modeled. The hydraulic diameter in the model D is approximately

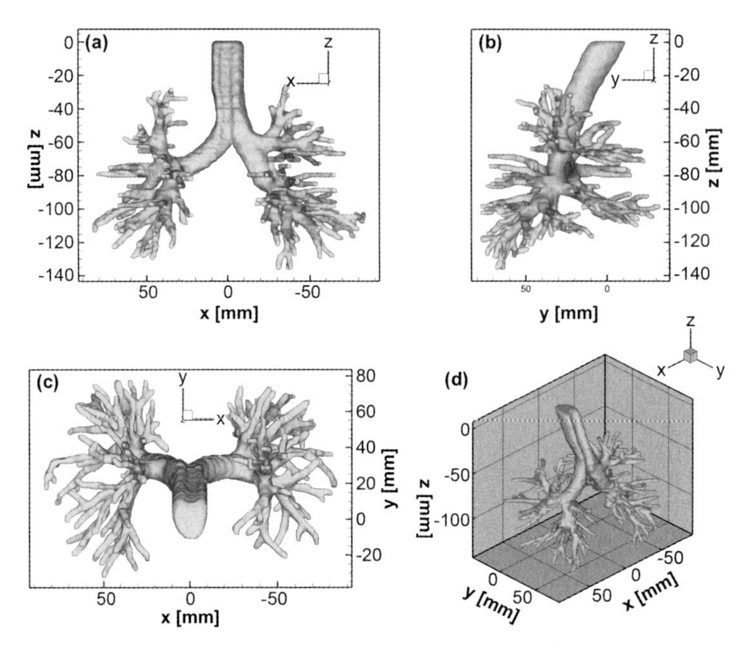

Fig. 2. Illustrations of the realistic lung model. (**a**) dorsal or back view, (**b**) lateral view, and (**c**) caudal or top view; (**d**) isometric view

18.3 mm leading to a ratio of $L/D = 27$. The Reynolds number based on the hydraulic diameter of the trachea ranges from $\mathrm{Re_D} = 1250$ to 1700 for the steady-flow experiments. This results in a fully developed laminar profile at the inlet. Choosing an anatomically shaped trachea instead of a plain pipe flow helps to prevent flow separation in the lung inflow region by avoiding the discontinuous change of the cross section at the junction between the inflow pipe section and the trachea. At any rate, such a geometry would generate unphysical disturbances of the flow field further downstream of the bifurcations.

Note, the human trachea possesses a total length of 10–12 cm, measured from the laryngeal region (cartilage cricoidea) to the first bifurcation (bifurcatio tracheae). Hence, the flow entering the bifurcation will be strongly influenced by the geometry of the upper airways and the perturbations caused by the larynx region. Thus, the real flow in the human body cannot be considered fully developed.

The flow exits into a net of pipes, which have the same diameter as the branches of the 6th generation of the lung model. They were drilled into the silicone block such that they can be considered straight extensions of these branches. Due to the fact that the vertical pressure in the tank complies with the hydrostatic pressure distribution it can be assumed that the outlet pressure in the experiment is nearly constant for all branches of the 6th generation exiting into the tank.

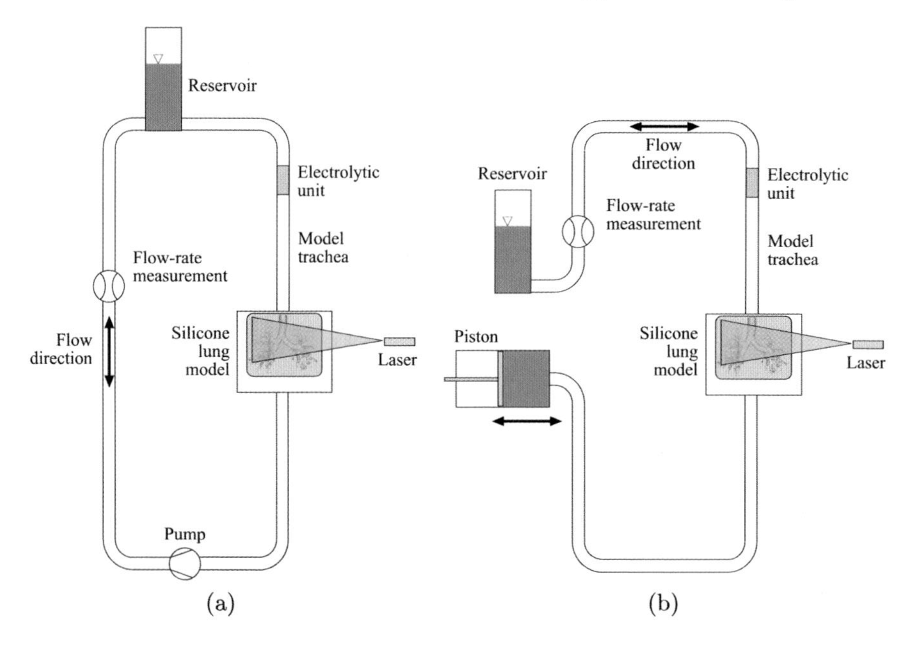

Fig. 3. (**a**) Schematic of the experimental setup for steady flow. (**b**) Schematic of the experimental setup for oscillating flow

Measurements at steady inspiration and expiration have been performed using a closed-circuit flow facility. Note that the experimental setup is identical for both cases except for the flow direction. For either setup the fluid supply for the tank has to be carefully positioned since a homogeneous flow circulation has to be guaranteed for an idealized flow distribution in the lung model. Measurements at oscillating flow, which were designed to simulate a breathing cycle, have been conducted with an open-circuit flow facility. In this case, a sinusoidal fluid movement was realized using a piston driven by an eccentrically supported rod. Figures 3a,b show schematics of the experimental setup for steady and oscillatory flow.

2.3 Flow Parameters

A water/glycerine mixture with a refractive index of $n = 1.44$, being identical to that of the silicone block, was used as flow medium. The dynamic viscosity of the mixture with 60.7 mass per cent glycerine is $\eta = 10.3 \times 10^{-3}\,\mathrm{Pa \cdot s}$ and the density $\rho = 1.153 \times 10^3\,\mathrm{kg/m^3}$ for the temperature during the experiments.

The Reynolds number $\mathrm{Re_D}$ for the steady-flow measurements was varied between 1250 and 1700. The Reynolds number $\mathrm{Re_D}$ for oscillatory flow based on the maximum flow velocity during inspiration was varied between 1050

Table 1. Flow parameters

Parameter	Multi-plane	Steady inspiration	Steady expiration	Oscillating flow
Reynolds number Re_D	1700	1250–1700	1250–1700	1050–2100
Mean trachea bulk velocity U_{mean} (m/s)	0.83	0.60–0.83	0.60–0.83	–
Maximum trachea bulk velocity U_{max} (m/s)	–	–	–	0.50–1.10
Womersley number α	0	0	0	3.27–5.77
Volumetric flow per cycle V (cm^3)	–			90.6–570
Frequency (cycles/min)	–	–	–	10.85–34.5

and 2100. The Womersley number range was 3.3 to 5.8. Hence, typical human breathing cycles under normal conditions can be simulated with this setup. A list of characteristic flow parameters of the different measurements is given in Table 1.

2.4 Tracer Particles

Hydrogen bubbles generated in an electrolytic unit upstream of the trachea were used as tracer particles. Since the amount of bubbles in plain water/glycerine is rather low due to missing electrolytes in the fluid a very small amount of saline additive is mixed into the fluid. The electrolytic unit was especially designed to avoid flow disturbances upstream of the trachea.

The size of the tracer particles depends on the electrolytic voltage, the distance between the metal cathode and anode of the electrolytic unit, and the mean flow velocity through the unit. The distance and voltage were adjusted to achieve an optimized size of the bubbles of 1–5 μm. Due to the laminar flow in the trachea it was necessary to start the electrolytic unit at least 5 min prior to the measurement series to generate a homogeneous distribution of particles. At very low Reynolds numbers it was necessary to switch off the unit during the measurements since bubbles were not torn off the metal surfaces by the low flow velocities but accumulated at the edges of the unit yielding tracer particles that were too large.

The use of hydrogen bubbles was necessary since standard flow particles (PA, etc.) led to heavy particle deposition on the walls of the lung model resulting in strong reflections at the boundaries of the bronchial system and as such in a significantly reduced quality of the images. However, first experiments done with solid particles showed flow structures comparable to those measured with hydrogen bubbles. Nevertheless, the use of solid tracer particles and hence the contamination of the model led to a considerably higher number of invalid vectors in the measured velocity fields. The lifespan of the bubbles is in the range of 5–10 min such that for expiration the electrolytic unit could be kept in the same position in the flow circuit still ensuring a sufficient number of bubbles for the PIV measurements under reversed flow conditions.

2.5 Measurement Equipment and Image Evaluation

In all measurements a standard PIV system was used consisting of a double-cavity Minilite Nd:YAG laser with a nominal power of 25 mJ, a PCO SensiCam QE double-shutter camera with a resolution of 1376 × 1040 px, and a lens with $f\# = 1.8$ to reduce the depth of focus to a minimum and to maximize the available light. To gain some first knowledge of the spatial evolution of the flow field in the first bifurcation during steady inspiration, experiments were performed at a Reynolds number of $\mathrm{Re_D} = 1700$ using a set of parallel lightsheets with a spacing of 1 mm and a field of view of 95.8 × 72.4 mm. Next, 500 images were taken at a frequency of 2 Hz for steady inspiratory and expiratory flow in a streamwise-oriented central plane of the bifurcation. The field of view for these measurements was approximately 74.6 × 56.4 mm. Based upon these results the temporal development of the flow structures in the left bronchia was investigated in detail with time-resolved PIV at a frame rate of 8 Hz and a field of view of approximately 72.4 × 27.7 mm. In this case 200 images were taken for each set of Womersley and Reynolds number. Thus, depending on the specific parameter combination a minimum of 4 to 14 complete inspiration and expiration cycles per experiment were captured.

The data was postprocessed using the commercial VidPIV 4.6 PIV software by ILA GmbH. A multipass correlation was applied starting with a 64 × 64 px interrogation size with 50 % overlap followed by an adaptive cross-correlation with a final 16 × 16 px interrogation window size leading to a vector spacing of 0.7 mm for the time-resolved measurements. Thus, approximately 25 vectors across the diameter of the first generation of bronchia could be achieved. Only weak local reflections from the walls allowed us to exactly determine the bronchial structure.

The dynamic range in the PIV images was up to 4096 grayscales and 4–6 particles per interrogation area led to a high accuracy of the adaptive correlation routines. Due to an achievable subpixel accuracy of less than 0.02 px of the routines under these conditions the accuracy in the velocity determination was better than 0.5 %. To determine the pulse distance at oscillating flow was rather difficult since the setting must be adapted to maximum inspiration or expiration flow velocities as well as to the quasistatic conditions at flow reversal.

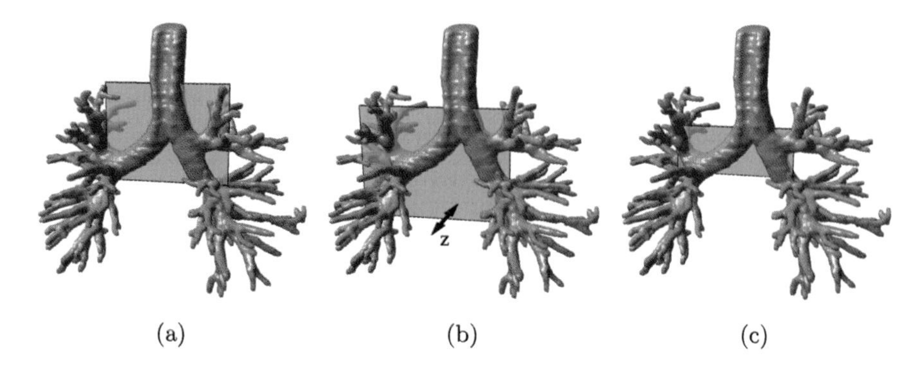

Fig. 4. Measurement planes in dorsal view for (**a**) multiplane measurements (only one plane is illustrated), (**b**) steady, and (**c**) oscillating flows

3 Results

To obtain some knowledge about the global structure of the flow field in the bifurcation of the trachea and bronchia, the first experiments were performed for steady inspiration at a constant Reynolds number of $Re_D = 1700$. Due to the complexity of the velocity field in the bifurcation region the flow was recorded in thirteen parallel, vertical lightsheets with a lateral displacement of 1 mm, hence covering the complete cross section of the bronchia. This allowed the reconstruction of three-dimensional flow phenomena from the two-dimensional velocity distributions. Based on the results of these measurements, experiments were conducted for steady inspiration and expiration at two different Reynolds numbers of $Re_D = 1250$ and $Re_D = 1700$ to determine the influence of the Reynolds number on the flow field. To investigate the temporal and spatial evolution of the flow structures during a breathing cycle measurements were performed for a set of combinations of Reynolds number Re_D and Womersley number α. The analysis of the data showed the impact of these parameters on the characteristics of the flow in the bifurcation region. The oscillating flow was generated by imposing a sinusoidal flow rate with zero mean velocity as described in Sect. 2.

Figure 4 illustrates the positions and sizes of the measurement planes with respect to the lung geometry for all cases investigated. The definitions "left" and "right" used in the following discussion of the findings are defined for the dorsal view on the upper human airways complying with the usual notation in anatomy.

3.1 Multiplane Measurements

Figure 5 illustrates the two-dimensional velocity distributions in the first bifurcation during steady inspiration at a Reynolds number $Re_D = 1700$ in

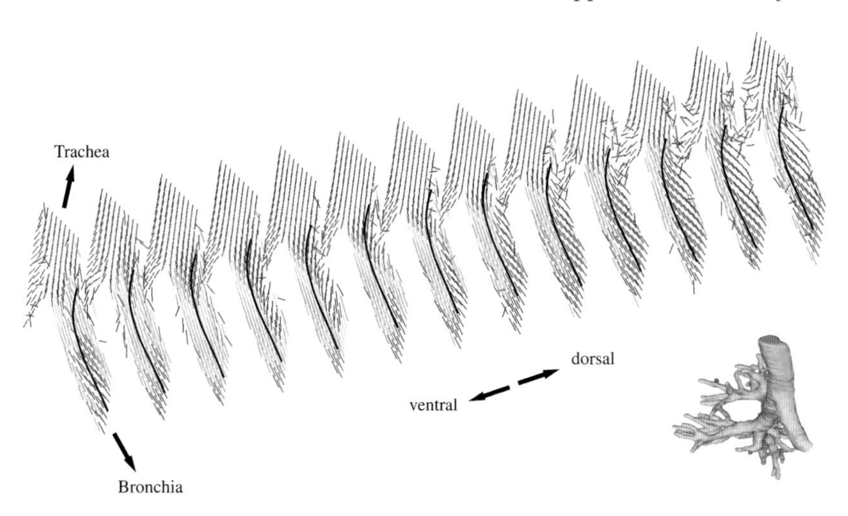

Fig. 5. Two-dimensional velocity distributions in the first bifurcation during steady inspiration at a Reynolds number of $\mathrm{Re_D} = 1700$ in parallel planes. Vector color qualitatively emphasizes u/w. The *black line* represents the extension of the region containing the counterrotating vortices

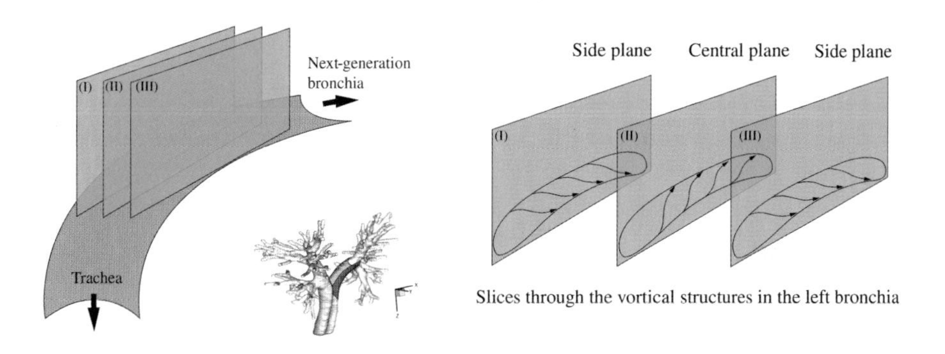

Fig. 6. Schematic of the two-dimensional measurement planes in the first bifurcation at three distinct positions

parallel planes. The vector fields evidence a zone with a strongly reduced streamwise velocity component at the upper wall of the left bronchia, which is slightly skewed in the ventral direction. This asymmetry results most likely from the angle between the central planes of the trachea and bronchia and hence from the torsion of mother and daughter branches of the bifurcation.

Since the terminus "separation" usually denotes a region containing reversed flow, it can not be applied to describe the flow structure identified in the measurements. Although the mean streamwise velocity is strongly reduced within these regions and especially near the side walls, no flow recir-

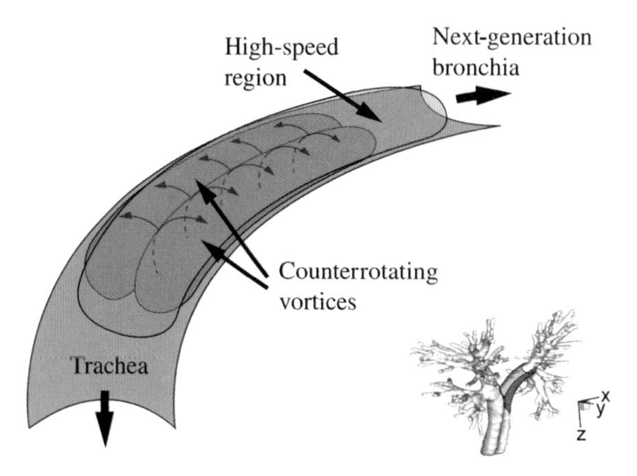

Fig. 7. Schematic of the two counterrotating vortices evolving along the outer bifurcation. The high-speed region is illustrated in *red*

culation could be evidenced. Since the measurement planes cover the whole cross section of the bronchia the three-dimensional shape of the region of secondary flow structures can be determined. The complete flow pattern is depicted in Figs. 6 and 7. The measurements indicate that two counterrotating vortices form inside the region of reduced velocity. The fluid flows along the outer walls of the bronchia to the central plane and then to the centerline towards the high-speed region on the inner curve of the bifurcation. This fluid motion is superposed with fluid transport in the downstream direction generating a spiral-like shape of the vortices. The multiplane measurements also reveal that the measurements only in the central plane allow the reconstruction of the spatial extent of the counterrotating vortices with sufficient accuracy.

3.2 Steady Inspiration and Expiration

Figures 8a,b show a smooth laminar velocity profile from the trachea entering the bifurcation zone at inspiration. The figures illustrate the velocity magnitude and some velocity profiles for selected positions in the bifurcation, which evidence a large zone of strongly reduced velocity in the left bronchia and strongly skewed velocity distributions. This area is located at the inner curve of the bifurcation and extends to the bronchia centerline. As discussed above, two counterrotating vortices exist in this region. These secondary flow phenomena resemble spiral-like structures moving the fluid from the center of the bronchia sidewards to the upper wall of the bifurcation and back to the central region of the bronchia. The fluid captured in these vortical structures is still transported along the bronchia, but with a reduced mean volumetric flow rate. This causes the remaining flow to form a jet-like high-speed region along the lower curve of the bifurcation.

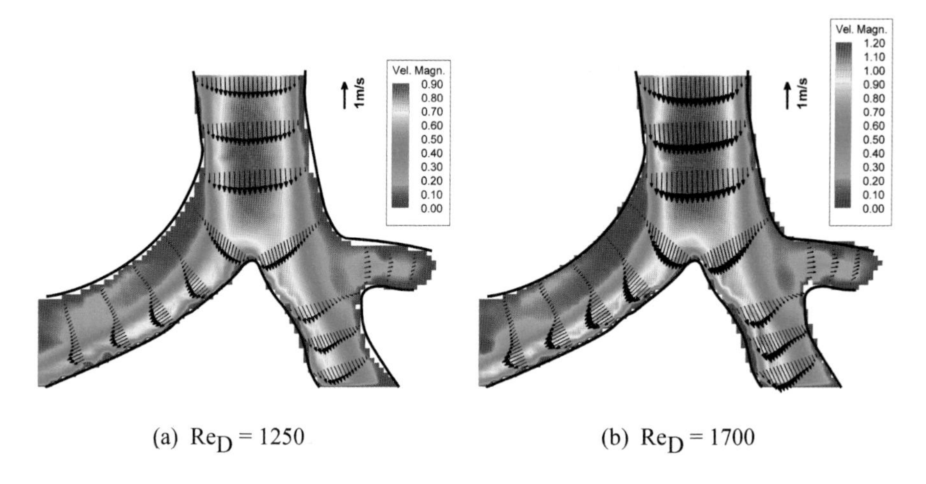

(a) $\mathrm{Re_D} = 1250$ (b) $\mathrm{Re_D} = 1700$

Fig. 8. Velocity profiles at different positions for steady inspiration flow with $\mathrm{Re_D} = 1250$ and $\mathrm{Re_D} = 1700$. Contour levels show the vector magnitude

Figure 9 shows the instantaneous flow pattern evidencing the transport of the fluid in the center plane. A schematic of the overall flow structure in the left bronchia illustrating the counterrotating vortices was already depicted in Fig. 7. As a consequence of the high velocities near the lower wall of the bifurcation a strong shear layer evolves. A similar behavior can be observed in the right bronchia, although the intensity and extension of the corresponding region is less marked compared to that in the left bronchia. This is caused by the smaller angle of deflection of the right bronchia. Regarding the mass-flow distribution at steady inspiration the geometry leads to an unbalanced distribution, i.e., the volume flux in the right lung lobe is larger than in the left lobe.

Figure 10 shows the velocity magnitude and velocity profiles at selected positions for steady expiration. The flow is strongly characterized by the inlet velocity from the subbranches. The velocity profiles in the bronchia possess an M-like shape resulting from the merging fluid jets of the corresponding subbranches. Due to this effect, the flow cannot develop a parabolic velocity profile but rather possesses a two-peak velocity distribution in the radial direction. This velocity profile represents the inflow condition for the next junction during expiration. The flow following the inner curves possesses higher peak values of the velocity magnitude such that a skewed M-form of the velocity profiles develops. Figure 10 also indicates two small regions in the trachea where the fluid cannot follow the extremely strongly curved wall. Here, the fluid is displaced towards the centerline of the trachea where the peak velocity values occur.

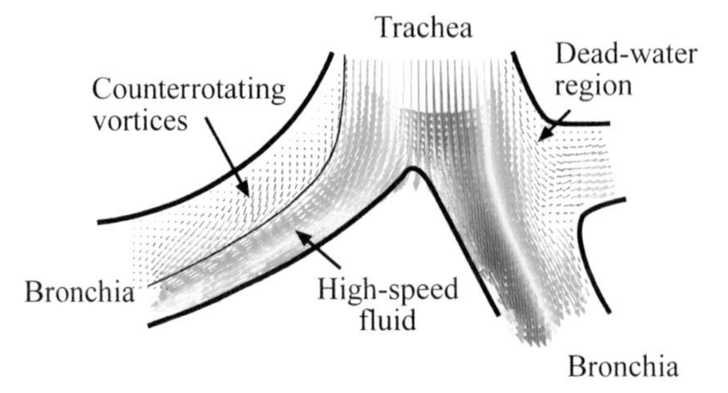

(a) Maximum inspiration

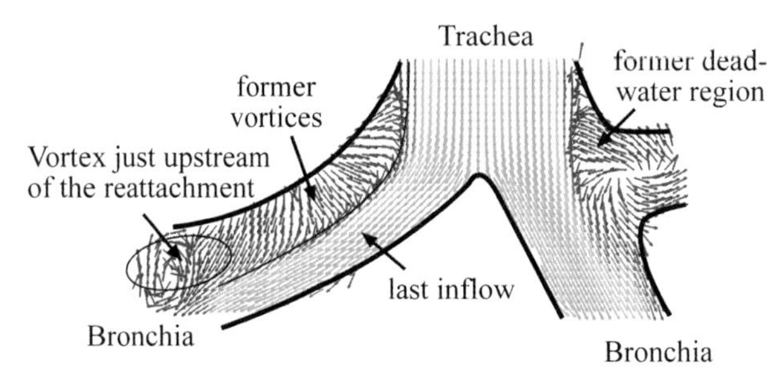

(b) End of inspiration phase

Fig. 9. Flow structure in the center plane of the first bifurcation. (**a**) Vector length and color show the velocity magnitude. (**b**) Vectors possess a uniform length, contours show the velocity magnitude

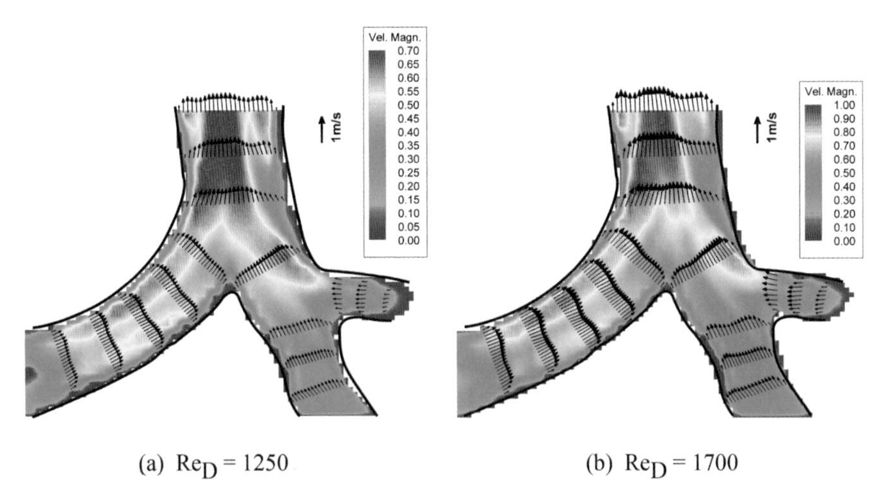

(a) $\mathrm{Re_D} = 1250$ (b) $\mathrm{Re_D} = 1700$

Fig. 10. Velocity profiles at different positions for steady expiration at $\mathrm{Re_D} = 1250$ and $\mathrm{Re_D} = 1700$. Contours show the velocity magnitude

Table 2. Investigated parameter combinations of Reynolds number $\mathrm{Re_D}$ and Womersley number α

Reynolds no. $\mathrm{Re_D}$ (−)	Womersley no. α (−)	Frequency F (cycles/min)	Volumetric flow V ($\mathrm{cm^3}$)
1050	3.27	10.9	282.6
1400	3.27	11.0	376.8
2100	3.27	10.85	565.0
1050	4.70	22.1	136.3
1400	4.70	22.3	181.8
2100	4.70	22.9	272.6
1050	5.77	34.5	90.6
1400	5.77	33.2	120.8
2100	5.77	33.5	181.2

3.3 Oscillating Flow

Oscillating flow measurements have been conducted for a set of parameter combinations of Reynolds number $\mathrm{Re_D}$ and Womersley number α, which are summarized in Table 2. The Reynolds number is based on the maximum bulk velocity at inspiration.

In this phase the Reynolds number increases continuously until it reaches its peak value at 25 % of the breathing cycle. When the Reynolds numbers exceeds a critical level of roughly $\mathrm{Re_D} = 800$ to 1000, the vortical flow structures described in Sect. 3.1 occur in the bronchia. While this region is fed by the near-wall fluid from the trachea, the center fluid from the trachea flows straight through the bifurcation and is split at its inner radius (bifurcatio tracheae) following the lower bronchial contour in small streaks of high-speed fluid. At the lower end of the two counterrotating vortices, i.e., nearly three bronchia diameters downstream of the bifurcation, a strong vortical structure can be identified that increases in size and strength when inspiration comes to its end (Fig. 9). Before expiration starts, the region in which the counterrotating vortices existed begins to roll up from the lower end of the bronchia generating a strong shear layer between the already reversed flow near the outer bifurcation wall and the remaining inflow of the former high-speed zone. Finally, the latter region is completely suppressed and expiration sets in. In the right bronchia, a dead-water region reaches even over the next-generation branch. Here, strong vortical structures can also be identified during the reversal of the flow direction.

During expiration the flow field in the bifurcation and the trachea is strongly influenced by the upstream flow conditions. Neither separation zones nor a dead-water region as found during inspiration can be observed. Only downstream of the trachea does a moderate dead-water region form on the outer walls.

Figure 11 illustrates the time-dependent evolution of the out-of-plane vorticity distribution in the central plane in the region of the first bifurcation for $Re_D = 2100$ and $\alpha = 3.27$. Herein, the vorticity component is defined as $\omega_y = \frac{\partial u}{\partial z} - \frac{\partial w}{\partial x}$, where u is the horizontal and w the vertical velocity component, and x and z are the corresponding cartesian coordinates in the measurement plane.

The illustration of the temporal evolution of the vorticity evidences the spatial extent of the vortical structures during inspiration and expiration by visualizing the shear layer that lies between the vortical flow structures in the bifurcation and the remaining high-speed jet. During inspiration, narrow bands of increased vorticity illustrate the extension of the vortices in the left and right bronchia. As mentioned above, at expiration small dead-water zones can be found in the trachea. Due to an additional mass flow coming from the upper bronchial branch in the right lobe and due to a more pronounced curvature at the wall the secondary structures on the right side of the trachea are stronger than their counterpart on the left side. This can also be clearly observed in Figs. 10a,b. In Fig. 11, i.e., at maximum expiration, a centered region of high vorticity is evidenced in the left bronchia. This shear layer results from the merging jets of the two corresponding subbranches. Furthermore, the results show that, depending on the phase in the flow cycle, the vortical structures along the upper bifurcation wall change strongly in shape and size for low Reynolds numbers. Nevertheless, it can be stated that for Reynolds numbers over approximately $Re_D = 1000$ the counterrotating vortices show the same extension for all investigated cases.

4 Conclusion and Outlook

Velocity measurements of steady inspiration and expiration at two Reynolds numbers of $Re_D = 1250$ and $Re_D = 1700$ were presented. The results evidenced the Reynolds number of the flow to play a minor role for the extension of the flow phenomena. For both Reynolds numbers during steady inspiration, very similar flow structures could be observed. To be more precise, the size of the counterrotating vortical structures did not depend on the Reynolds number as long as this Reynolds numbers stays above a critical level of roughly $Re_D = 800$ to 1000. For the lower Reynolds numbers achieved during the oscillating flow studies the size of the evolving secondary flow structures depends rather strongly on the momentary Reynolds number. A highly intricate flow structure consisting of a spiral-like trajectory was found in a region along the upper wall of the left branch of the first bifurcation. From the presented results it can be clearly concluded that the characteristic length of the bronchial generations is too short to achieve fully developed flow profiles before entering the next generation of bronchia. Therefore, special care has to be taken when flow-rate distributions in this highly complex geometry are to be determined without having determined a complete 3D-3C velocity distri-

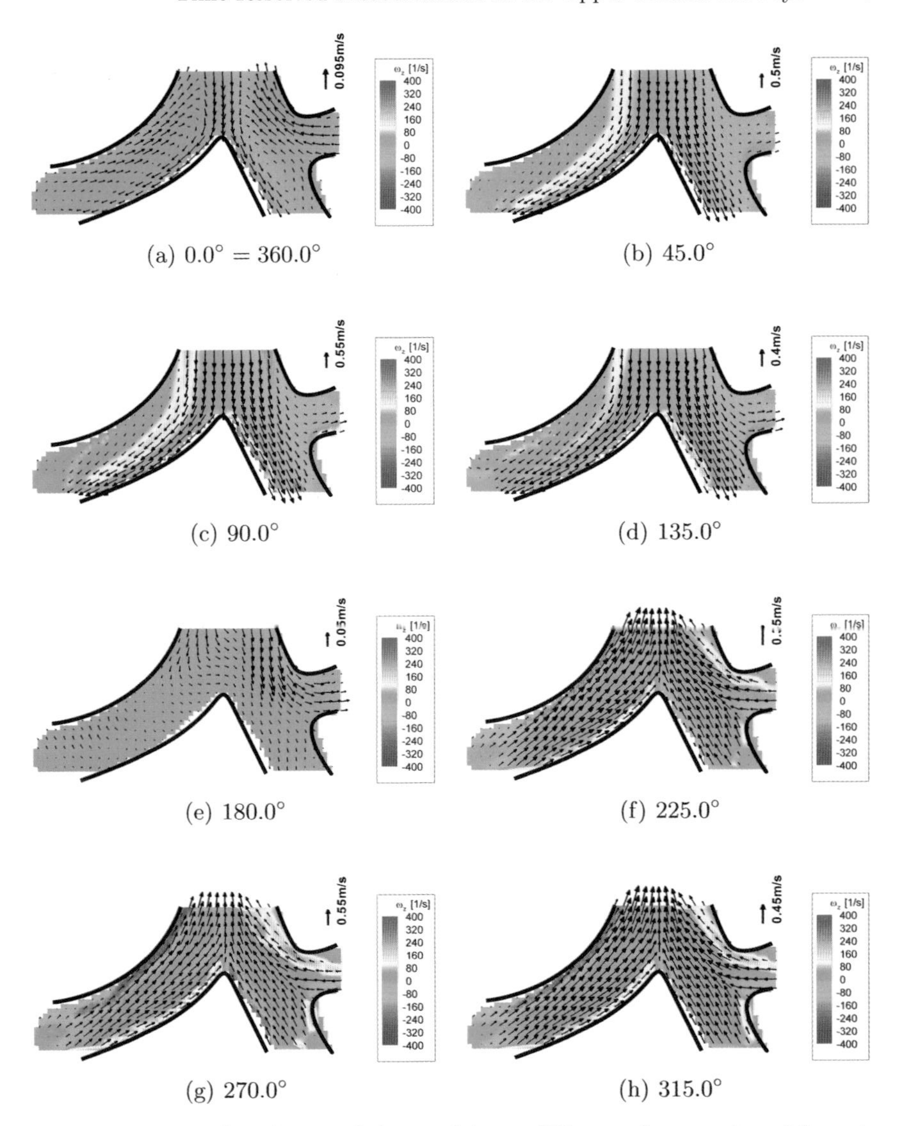

(a) 0.0° = 360.0° (b) 45.0°

(c) 90.0° (d) 135.0°

(e) 180.0° (f) 225.0°

(g) 270.0° (h) 315.0°

Fig. 11. Temporal evolution of the vorticity at different phase angles of the volumetric flow rate. Dorsal view. Vector plot of the local velocities. Only every second vector is displayed. Contours represent the local vorticity. Inspiration: (**b**–**d**), expiration: (**f**–**h**), almost no flow: (**a**, **e**)

bution, e.g., on the basis of stereoscopic PIV measurements. Furthermore, the investigation of the oscillating lung flow indicated the vortical flow regions to strongly change in shape and size due to the local acceleration of the flow for low Reynolds numbers, or in other words, due to the time dependence of the velocity. At expiration, the unsteady and steady flow solutions are more similar than at inspiration since hardly any secondary flow structures are observed.

Further measurements in subsequent generations of the lung geometry will be used to analyze to what extent the upstream flow conditions have an impact on the flow field in corresponding subbranches as a function of the local Reynolds number.

Additionally, a more realistic flow should be examined by incorporating the laryngeal region of the upper human airways in the model such that the flow entering the lung will have experienced likewise upstream conditions as the real human lung flow. Furthermore, it has to be investigated whether or not a defined pressure distribution at the end of the 6th generations' branches and controlled flow volumes for each of these branches possess an influence on the flow field of the upper airways.

Finally, the flow field during high-frequency oscillatory ventilation (HFV) with ventilation frequencies between 5 and 15 Hz will be investigated, since this technique has reached increasing clinical application. It is still to be investigated whether or not a variation of the Womersley number influences the flow structures in the lung for inspiration and expiration.

References

[1] J. K. Comer, C. Kleinstreuer, S. Hyun, C. S. Kim: Aerosol transport and deposition in sequentially bifurcating airways, J. Biomech. Eng.-T ASME **122**, 152–158 (2000)

[2] J. K. Comer, C. Kleinstreuer, C. S. Kim: Flow structures and particle deposition patterns in double-bifurcation airway models. Part 2. Aerosol transport and deposition, J. Fluid Mech. **435**, 55–80 (2001)

[3] J. K. Comer, C. Kleinstreuer, Z. Zhang: Flow structures and particle deposition patterns in double-bifurcation airway models. Part 1. Air flow fields, J. Fluid Mech. **435**, 25–54 (2001)

[4] T. Martonen: Commentary "Effects of asymmetric branch flow rates on aerosol deposition in bifurcating airways", by Z. Zhang, C. Kleinstreuer and C. S. Kim, J. Med. Eng. Technol. **25**, 124–126 (2001)

[5] S. Mochzuki: Convective mass transport during ventilation in a model of branched airways of human lungs, in *Proc. 4th Pacific Symposium on Flow Visualization and Image Processing (PSFVIP4)* (2003)

[6] R. C. Schroter, M. F. Sudlow: Flow patterns in models of the human bronchial airways, Resp. Physiol. **7**, 341–355 (1969)

[7] T. B. Martonen, X. Guan, R. M. Schreck: Fluid dynamics in airway bifurcations: II. secondary currents, Inhal. Toxicil. **13**, 281–289 (2001)

[8] J. B. Grotberg: Respiratory fluid mechanics and transport Processes, Annu. Rev. Biomed. Eng. **3**, 421–457 (2001)

[9] Y. Liu, R. M. C. So, C. H. Zhang: Modeling the bifurcating flow in an asymmetric human lung airway, J. Biomech. **36**, 951–959 (2003)

[10] X. L. Yang, Y. Liu, R. M. C. So, J. M. Yang: The effect of inlet velocity profile on the bifurcation COPD airway flow, Comput. Biol. Med. **36**, 181–194 (2006)

[11] A. Ramuzat, M. L. Riethmuller: PIV investigation of oscillating flows within a 3D lung multiple bifurcations model, in *Proc. 11th Int. Symp. on Applications of Laser Techniques to Fluid Mechanics* (2002) pp. 19.1.1–19.1.10

[12] F. Y. Leong, K. A. Smith, C.-H. Wang: Transport of Sub-micron Aerosols in Bifurcations, Molecular Engineering of Biological and Chemical Systems (MEBCS) (2005)

[13] R. J. Robinson, M. J. Oldham, R. E. Clinkenbeard, P. Rai: Experimental and numerical smoke carcinogen deposition in a multi-generation human replica tracheobronchial model, Ann. Biomed. Eng. **34**, 373–383 (2006)

[14] E. R. Weibel: *Morphometry of the Human Lung* (Springer, Berlin, Heidelberg 1963)

[15] C. G. Caro, R. C. Schroter, N. Watkins, S. J. Sherwin, V. Sauret: Steady inspiratory flow in planar and non-planar models of human bronchial airways, Proc. Roy. Soc. Lond. A **458**, 791–809 (2002)

[16] C. van Ertbruggen, C. Hirsch, M. Paiva: Anatomically based three-dimensional model of airways to simulate flow and particle transport using computational fluid dynamics, J. Appl. Physiol. **98**, 970–980 (2005)

[17] C. Brücker, W. Schröder: Flow visualization in a model of the bronchial tree in the human lung airways via 3-D PIV, in *Proc. of the 4th Pacific Symposium on Flow Visualization and Image Processing (PSFVIP4)* (2003)

PIV Measurements of Flows in Artificial Heart Valves

Radoslav Kaminsky[1,2], Stephan Kallweit[2], Massimiliano Rossi[3], Umberto Morbiducci[3], Lorenzo Scalise[3], Pascal Verdonck[1], and Enrico Primo Tomasini[3]

[1] Hydraulics Laboratory, Institute of Biomedical Technology, Ghent University, Belgium
rado@navier.ugent.be
[2] ILA GmbH, Jülich, Germany
[3] Universita Politecnica delle Marche, Ancona, Italy

Abstract. Through several decades many different models of prosthetic artificial heart valves (PHV) have been designed and optimized in order to enhance hemodynamic properties. These properties are not only material dependent but the major influence results from the mechanical assembly of the particular PHV. For the experimental assessment of the flow through such PHVs particle image velocimetry (PIV) is already an accepted method [1] due to its noninvasive optical approach and accuracy. Here, we present various modifications of PIV in order to explain, compare and realize which method is the most suitable for the quantification of such flows. The choice of the experimental procedure for testing the PHVs is strongly dependent on the optical access of the designed *in-vitro* testing loops simulating the human heart and vascular system. The hardware demand and its configuration for, e.g., stereoscopic PIV is much more complex than standard 2D PIV, therefore the conditions and design of the testing loop have to be realized to allow the desired flow measurement. The flow in heart valves as an unsteady periodically generated flow, can be obtained by averaged phaselocked or measurements with high temporal. The properties, advantages and drawbacks of specific PIV techniques to visualize the flow behind a PHV will be discussed.

1 Introduction

The fluid dynamics through prosthetic heart valves (PHVs) has been shown to play a critical role in the clinical outcome for patients who have undergone cardiac-valve implantation. This is particularly true when mechanical valves are implanted.

Common to all mechanical PHVs is the incidence of:

- thromboembolic complications, primarily due to platelet activation [2, 3], which appears to be intimately related to the nonphysiological flow characteristics of the blood through PHVs. As pointed out by *Bluestein* et al. [4, 5] the vortices shed by a valve's leaflet in the fully open position induce a pathological flow that increases the risk of free emboli formation by activating and aggregating platelets.

A. Schroeder, C. E. Willert (Eds.): Particle Image Velocimetry,
Topics Appl. Physics **112**, 55–72 (2008)

- mechanical trauma to red cells, which could cause the rupture of the membrane of the erythrocytes, leading to hemolysis (a nonexhaustive discussion can be found in [6, 7]).

For the reasons mentioned above, fluid-dynamical characterization must be performed in the risk analysis for those devices.

However, the fluid dynamics of PHVs is particularly complex, as testified by a great number of studies performed by using a laser Doppler anemometry (LDA) technique (see [8], for example), with high spatial gradients and Reynolds shear stresses, especially as far as mechanical heart valves are concerned. The fine-scale characteristics of such flows (e.g., turbulence production, vortex shedding, flow separation, stasis) need experimental techniques with high spatial and temporal resolution for the estimation of the cause of hemodynamically related complications such as hemolysis, platelet activation, and thrombus formation. This is only possible by knowing the precise velocity field and stress distribution obtained by using these techniques.

Laser Doppler anemometry is a widely applied tool for fluid-dynamic investigations that has been used for more than three decades, and it constitutes the gold standard for velocity measurements, by virtue of its peculiar features, (i.e., noninvasivity, high spatial and temporal resolution, excellent accuracy). The application of LDA in the evaluation of PHVs function has been well documented [8]. Estimation of shear-stress-related blood damage in heart-valve prostheses: was assessed in an *in-vitro* comparison of 25 aortic valves [9, 10]. However, the complete analysis of the relevant flow area using the LDA technique is very time consuming, taking longer to investigate both the whole fluid domain, and the whole cardiac working cycle of a PHV. To characterize the fluid dynamics of PHVs in *in-vitro* conditions, alternative to the single point, time-consuming LDA, is particle image velocimetry [11] (a nonexhaustive list of PIV studies on PHVs is summarized in [1]). By virtue of its multipoint measurement characteristics, it furnishes full-field measurement of instantaneous velocity vectors thus allows mapping of the entire velocity flow field, in the fluid domain of interest. Nowadays, PIV has gained sufficient temporal and spatial resolution to estimate velocity fields – depending on the degree of instability in the flow [12] – providing a powerful tool for studying the safety and efficiency of implantable medical devices.

Here we present the application of time-resolved or high-speed and stereoscopic PIV (three-component technique) to study mechanical prosthetic valve fluid dynamics in greater detail. So far these two PIV techniques have been poorly used in the investigation of the challenging issue of PHVs fluid dynamics.

It is expected that time-resolved PIV and 3C stereoscopic PIV characterization, accurately capturing hemodynamically relevant scales of motion, will help the designer to improve PHV performance and endow comprehensive validation with experimental data of fluid dynamics numerical modeling. To achieve the improvement of the PHV design, it is important to clearly understand the phenomena of the flow dynamics according to the different geometry among different PHVs.

2 Materials and Methods

For the experimental study with time-resolved PIV we chose a typical monoleaflet mechanical prosthetic valve with only one occluder, the Björk–Shiley valve with 27 mm tissue annulus diameter, which is no longer implanted for valve replacements but used as part of ventricular-assist devices.

The stereoscopic PIV technique was applied to study the fluid dynamics of the carbomedics valve with 27 mm tissue annulus diameter, a typical bileaflet mechanical prosthetic valve with two flat half-disc leaflets.

2.1 The Testing Loop

Measurements were carried out in pulsatile-flow working conditions, in the testing loop depicted in Fig. 1. A challenge in performing flow measurements downstream of prosthetic heart valves by means of laser-based techniques is to develop a dedicated optical access for carrying out LDA, 2D and 3C PIV measurement. The reason for the design of the testing model in Fig. 1 was to realize an experimental setup particularly designed for studies to be performed with the experimental techniques mentioned above [13–16].

The testing loop mainly consists of a test chamber, an atrial systemic load and a pulsatile-flow volumetric pump. The test bench has an optically clear valve-housing segment to allow unobstructed optical measurements of the aortic flow with different measurement techniques. The test chamber consists of a prismatic box with a trapezoidal section ($73.5 \times 65 \times 50$ mm) in which the 30 mm internal diameter cylindrical glass pipe hosting the prosthesis is mounted. The oblique sides of the box are $45°$ inclined with respect to the central axis of the glass pipe, giving an optical access normal to the view direction of the cameras. This configuration assures the minimum error in the out-of-plane velocity component, when measured with stereoscopic PIV [17].

A Delrin-made coupling system presses the cylindrical glass pipe, maintaining a central alignment of the former. The cylindrical pipe, submerged in a prismatic chamber, is fixed by mechanical compression.

The rest of the testing loop is a three-element windkessel model [18] that simulates the arterial systemic load. A discharge tank mimics the left atrial chamber. The left ventricular function can be simulated at various cardiac outputs and heart rates by means of a volumetric pulse generator, i.e., an electromechanical pushing plate device with a variable stroke volume in the range of 40 to 300 ml. The volume of the left ventricle is approximately 80 ml. This allows the prescription of pulsatile flow through the valve prosthesis. The design of the hydraulic circuit allows the plug-flow velocity pattern entering the valve to be prescribed. Pressure is sampled at opposite sides of the tested

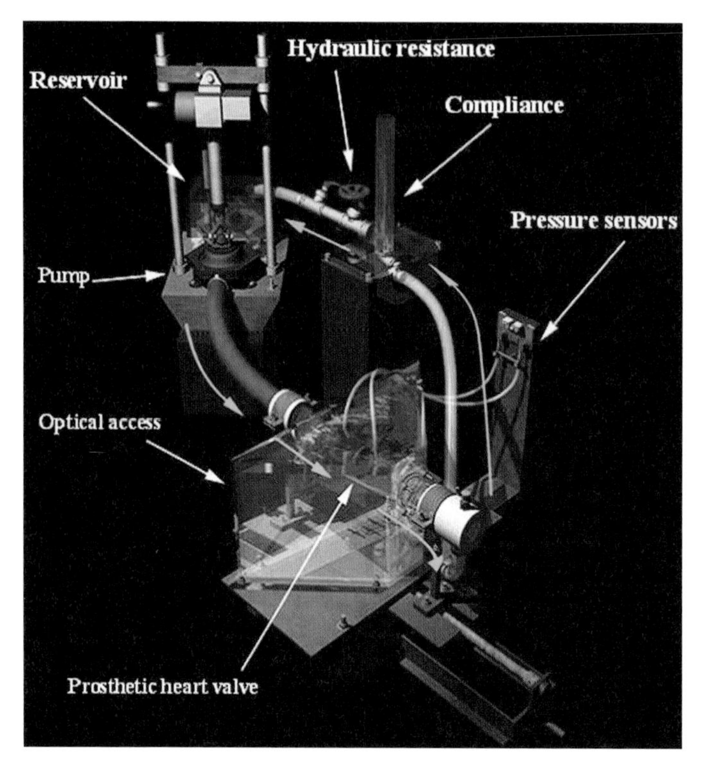

Fig. 1. Schematic illustration of the testing loop used for PIV measurements of flow through prosthetic aortic valves

valve. Pressure taps are placed 40 mm upstream and 100 mm downstream of the valve plane, respectively [13].

Distilled water was used as working fluid ($1.14 \times 10^{-3}\,\mathrm{N \cdot s \cdot m^{-2}}$ dynamic viscosity at 15 °C temperature). As previously stated [13], measurement results could be translated to the physiological domain, taking into account the principle of dynamical similarity theory [19].

We applied the Buckingham theorem, which is the basis of dimensional analysis. The theorem states that if we have an equation (Navier–Stokes) involving a certain number, M, of physical variables, and these variables are expressible in terms of N independent fundamental physical quantities, then the original expression is equivalent to an equation involving a set of P dimensionless variables ($P = M - N$) constructed from the original variables.

A test bench using distilled water as a working fluid is equivalent to the same test conducted using blood, if both Reynolds and Womersley numbers for distilled water and blood tests are equal. To obtain the blood-analog ($4 \times 10^{-3}\,\mathrm{N \cdot s \cdot m^{-2}}$ dynamic viscosity) working conditions – 70 beats/min, $4\,\mathrm{min}^{-1}$ cardiac output, 100 mm Hg mean aortic pressure, we used the same equations stated in [13], corresponding to a mean Reynolds number of 710 and a Womersley number of 20. In this study, we set a 38 % systole/cycle ratio.

The use of a Newtonian working fluid is an approximation widely adopted by researchers involved in *in-vitro* studies on PHVs fluid dynamics. This simplifying assumption is not meaningless, because it is not straightforward to find a working fluid with a shear-thinning, non-Newtonian behavior similar to blood (if available: we recall that the working fluid has to be transparent and with a refractive index the closest to that of the other media crossed by the lightsheet). Furthermore, the eventual use of a blood-analog non-Newtonian working fluid (if available) necessarily involves keeping the temperature constant in the testing circuit.

2.1.1 Time-Resolved PIV: Measurement Technique

The high-speed PIV technique is basically using the same measurement principle as the standard PIV technique, but with an acquisition rate that is much higher.

The combination of a Nd:YLF high repetition rate double-cavity laser with a high frame rate CMOS camera allowed a detailed, time-resolved (up to 10 000 fps depending on the resolution) acquisition of the unsteady flow downstream of the two investigated valve models. The Nd:YLF laser generated the green light ($\lambda = 527$ nm), which was used as a light source for the PIV measurement. In this case we used a double-cavity Nd:YLF laser Pegasus (New Wave Research Inc.) with a maximum repetition rate of 10 kHz and a maximum pulse energy of 10 mJ per pulse.

For the image acquisition a CMOS camera Photron APX RS (Photron Limited, Marlow) with 3000 fps at maximal resolution of 1024×1024 pixels (pixel size = 17 μm^2) was used.

The laser beam was guided from the source to the cylindrical optics via an articulated light arm. The optics converted the laser beam to a lightsheet (LS), a plane with a thickness of about 1 mm, defining the region of interest (ROI) along the plane x–z (Fig. 2). The ROI was placed in the central plane of the glass tube downstream of the valve plane. The LS was orthogonally oriented both to the B-datum of the valve, and to the CMOS camera: a schematic of their mutual position, together with the chosen plane of investigation is shown in Fig. 2. A mirror was placed behind the glass tube to redirect the lightsheet towards the valve to illuminate the region below the leaflets. Red fluorescent ($\lambda_{exc} = 527$ nm, $\lambda_{emis} = 580$–620 nm) microspheres (mean diameter = 50 μm, density = 1.05 g/cm^3) were added to the flow as seeding particles. A red bandpass ($\lambda = 590 \pm 20$ nm) filter placed in front of the CMOS chip eliminated the excitation wavelength and thus the wall reflections of the glass tube and the prosthetic heart valve, allowing only the bandwidth of wavelength to pass through (60 mm f2.8D (Micro Nikkor) camera lens).

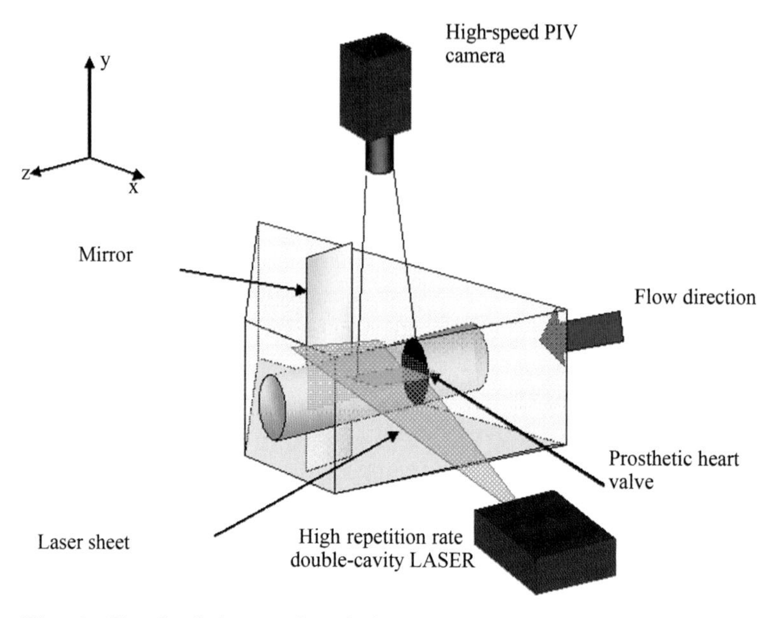

Fig. 2. Sketch of the test bench for time-resolved PIV measurements. The mirror was placed to redirect the light towards the valve in order to illuminate the region behind the leaflets

Commercial VidPIV 46XP (ILA GmbH, Julich) software is used for data evaluation. The calculation of instantaneous velocity vector maps in the region of downstream the PHV was done by means of crosscorrelation of successive images, as follows: a first correlation, including Whitaker peak fit and phase correlation [20] was performed on 32×32 pixel (16 pixel shift) interrogation boxes; then a local median filter was applied, and filtered vectors interpolated; on smoothed data an adaptive correlation function was applied on 16×16 pixel (8 pixel shift) interrogation boxes. This processing step includes subpixel shifting, and window deformation [21, 22] using gray-value reconstruction by B-spline interpolation [23]. The complete algorithm is iteratively performed.

A synchronizer (ILA GmbH, Julich) controlled the acquisition sequence to guarantee a synchronized recording of images to the generated laser pulses. Recording 3000 frames per second, we achieved a measurement's time resolution of 333.33 μs. The frequency resolution of the results is 1500 Hz.

The camera recorded images with a resolution of 1024×512 pixels. As for the spatial resolution, the size of the evaluated area was $355 \times 278\,\text{mm}^2$ with 60×45 vector positions, and a spacing of 0.6 mm corresponding to 9 pixel units. This means that we used a spatial resolution as in "standard" PIV analysis of prosthetic valve fluid dynamics.

The mismatch between the refractive index of the fluid and the surrounding material can be corrected using a nonlinear mapping function. In order

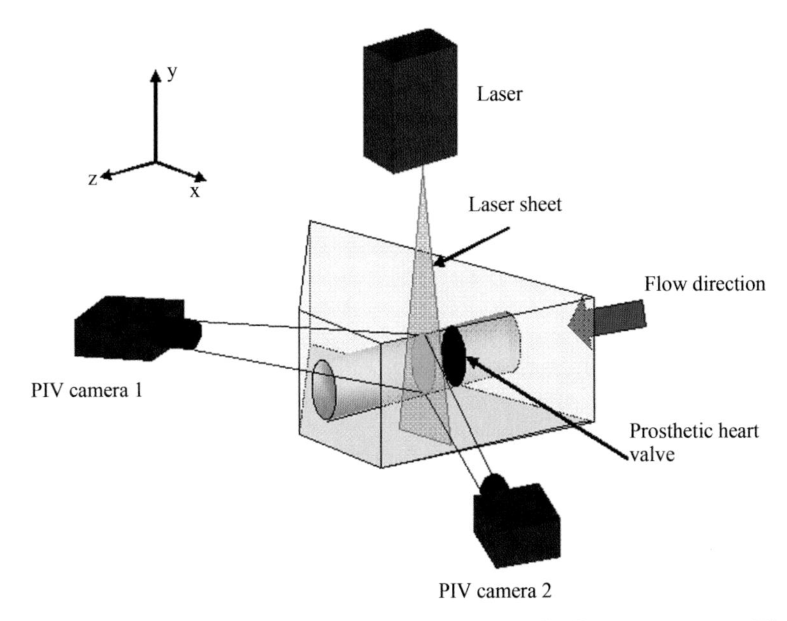

Fig. 3. Sketch of the test bench for stereoscopic PIV measurements. This configuration makes the optical access suitable for optimal 2D and 3D PIV measurements

to minimize optical distortions affecting the data, the camera calibration was done by applying a cubic polynomial mapping grid: datum markers are placed in the middle of crosses that are distorted by the refractive-index mismatch, so the distances (in pixels) among them vary in the distorted image; by knowing the real distance, distortion in the image can be numerically corrected. The limitation of the storage capacity of the CMOS camera (4.5 Gbyte) allowed only the acquisition of 1.3 cycles at 70 bpm.

2.1.2 3D PIV: Measurement Technique

In the 3C stereoscopic-PIV configuration, two CCD cameras were used together with a Nd:YAG double-cavity laser ($\lambda = 532\,\text{nm}$).

The two cameras, capable of recording 1376×1040 pixel double images, were inclined by 45° with respect to the principal direction of the flow (z-axis in Fig. 3). Then, the angle formed by the axes of the two cameras was 90°, assuring an optimal measurement of the crossplane velocity.

The cameras are arranged in the Scheimpflug configuration by means of rotatable-lens and camera-body seats. A mirror was placed to redirect the light towards the valve in order to illuminate the shadowed region behind the leaflets when it opened. The schematic of the test bench for stereoscopic PIV measurements, similar to the one adopted in [13], is displayed in Fig. 3.

The seeding particles used for this measurements were the same as during time-resolved PIV. The images were analyzed with a multigrid technique, starting with a 64×64 pixel correlation window in the first step and a 32×32 pixel correlation window in the second step. The well-known Soloff approach for stereoscopic PIV calibration was followed: a planar target built up with a regularly spaced grid of markers was placed at the position of the lightsheet, and moved by a specified amount in the out-of-plane direction to two or more positions [24]. At each position a calibration function with sufficient degrees of freedom mapped the world xy-plane to the camera planes, while the difference between z-planes provides the z-derivatives of the mapping function necessary for reconstructing the three velocity components. Distortions arising from imperfect lenses or light-path irregularities, e.g., from air/glass/water interfaces are compensated. In this case planes at seven different z-positions, spaced by 0.5 mm (lightsheet thickness equal to 3.5 mm), were taken for the calibration.

Commercial ILA software is used for both data evaluation and stereoscopic PIV calibration. The planes at $z = 1$ and 3 cm were considered in the 3D measurements where $z = 0$ corresponds to the z-coordinate of the valvular plane. For statistical purposes 100 pairs of images have been taken to calculate phase-averaged results.

3 Results

3.1 Time-Resolved PIV: Results

The results presented (velocity and vorticity) were made nondimensional with respect to the reference velocity value (the maximum velocity magnitude at peak systole, V_{ref}), to the diameter of the cylindrical aortic conduit (D), and to the time duration of the cardiac cycle (T).

The analysis of the vortical, spatiotemporally developing flow downstream of the monoleaflet valve model is described in Fig. 4 at nine different time instants, during the cycle. In particular, the PIV-measured velocity vector field evolution around peak systole (late acceleration $t/T < 0.15$, peak systole, early deceleration $t/T > 0.16$) is shown, downstream of the mechanical tilting disk prosthetic valve. The disk valve generates a two-peaked velocity profile, a major orifice and a minor orifice flow. The major orifice flow is directed towards the left wall of the duct, due to the opening angle of the disk, while the minor orifice flow is aligned to the longitudinal axis of the valve conduit. During the acceleration phase, the major flow skews left around peak systole, and tends to realign to the centerline during the deceleration phase. Figure 4 clearly shows both the growth and the spatial displacement of the wake region behind the tilting disk, around peak systole. The maximum orifice flow is confined near the left wall of the duct, due to the boundary action exerted by the minimum orifice expansive jet-like competitive behavior. The wake is located in the region immediately downstream of the valve plane. Moreover, a mixing region is clearly visible in the deceleration phase downstream of the

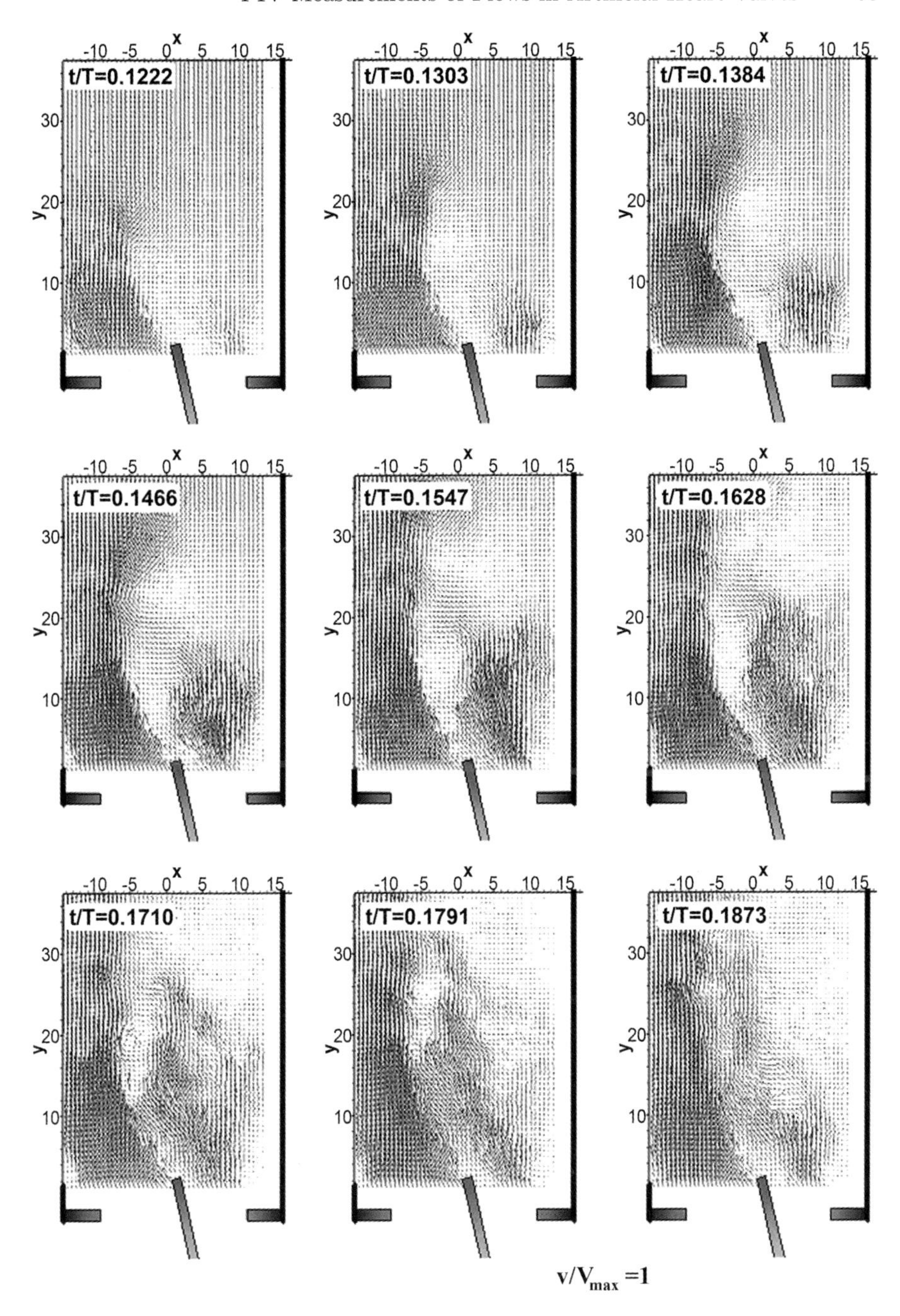

Fig. 4. Sequential presentation of the velocity vector field evolution downstream of the tilting disk valve model around peak systole (late acceleration, peak systole, early deceleration)

valve together with a low-velocity region in the right side of the cylindrical duct.

Figure 5 displays the evolution of the flow field during the closure of the tilting-disk valve. Note the role played by the slow counterrotating vortex that grows in the core region (the great recirculating area in the supravalvular region) of the duct, that pushes the disk to closure.

Notably, improved temporal resolution provides greater detail of the flow compared to “standard” PIV (5 ms interexposure time [1], the interval between two consecutive measured phases, can be considered as representative of conventional PIV analysis for prosthetic valve fluid dynamics), performed with the same spatial resolution and with a dual-head diode-pumped Nd:YLF laser. This allows us to analyze the fluid dynamics of only one cardiac cycle due to the camera-memory limitations. On the contrary, traditional double-pulsed single-cavity laser sources allow us to measure flow fields by acquiring single phases belonging to different phase-averaged cardiac cycles.

3.2 3D PIV: Results

We measured the flow in a cross-sectional plane parallel to the valvular plane, located 1 cm and 3 cm downstream of the valve plane, so that the out-of-plane structures in the velocity field correspond to the main direction of the flow. This choice was suggested by the fact that it has been widely assessed that the mixing process, as the one associated to prosthetic valve fluid dynamics, is intimately connected with the transient of turbulence. The streamwise vortices generated in a jet flow, in addition to ring-type vortices, have been found to mix fluid streams even more efficiently. Good statistics was assured by choosing an adequate sample size. The stereoscopic PIV technique allowed evaluation and comparison of the three-dimensionality of the flow field downstream of two prosthetic valve models, and to catch the specific features of both the jets and the vortical structures in the chosen planes of investigation. The classic three-jets configuration outgoing from the valve around the systolic peak is displayed in Fig. 6, where it is clearly evident that the jets outcoming from the side orifices of the bileaflet PHV reach velocities higher than the jet outcoming from the central orifice.

Figure 7 displays the contour map of the out-of-plane velocity (i.e., the velocity component in the main flow direction) superimposed on the inplane vector field at peak systole. Measured planes are located parallel to the valve plane, 1 and 3 cm downstream of the latter. The three jets outcoming from the valve orifices are clearly evident, together with areas of backflow. It is possible to appreciate also the topology of the inplane flow field in the bulk region, strictly related to the vortices wakes released by the leaflets. It is well observed that the secondary flow plays a significant role.

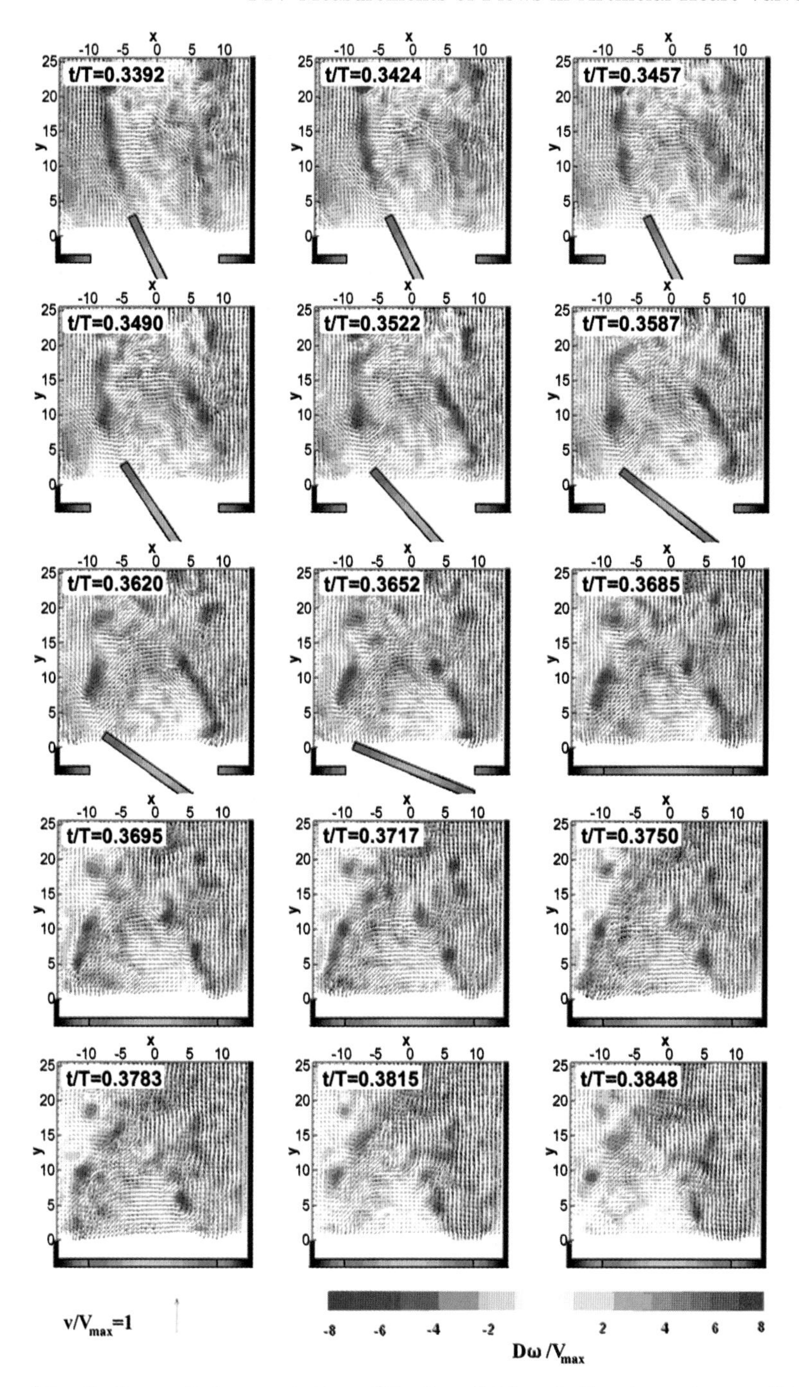

Fig. 5. Sequential presentation of the flow-field evolution downstream of the tilting-disk valve model at closure. A slow counterrotating vortex grows in the core region. Vorticity maps are made nondimensional

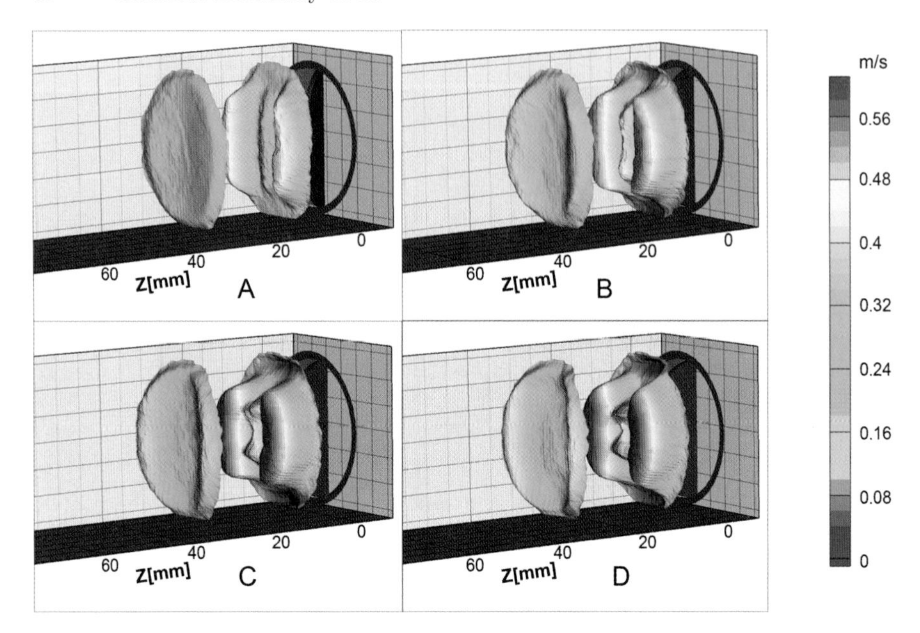

Fig. 6. Stereoscopic PIV: evolution of the flow field downstream of the bileaflet PHV during systole period 1 and 3 cm downstream of the valve plane. The morphology of the velocity component along the main direction of the flow is shown

4 Discussion

4.1 Time-Resolved PIV

Flow through PHVs is a typical example of complex fluid domains dominated by massive unsteady flow separation, and flow patterns rapidly changing in time. The experimental investigation of such complex spatiotemporally developing flows has always been a challenging task.

In-vitro experiments to study the local fluid dynamics inside of artificial devices involve the use of PIV technique [25–27], which has the advantage of performing full-field measurements. However, up to now standard PIV suffered from a lack of temporal resolution, disclosing limitations in the analysis of transient phenomena.

It is worth nothing that in the past PIV investigations of PHVs have been carried out with interexposure times higher than 1 ms (from 10 to 1 ms, typically 5 ms [1, 24, 28, 29]. This is due to the fact that in the past, acquisition systems used single-cavity laser light sources for studies of PHVs flow dynamics: these systems can accelerate, if necessary, their repetition rates by emitting double pulses, with part of the cavity's energy released in the first pulse and the remainder in the second. This approach, however, introduces

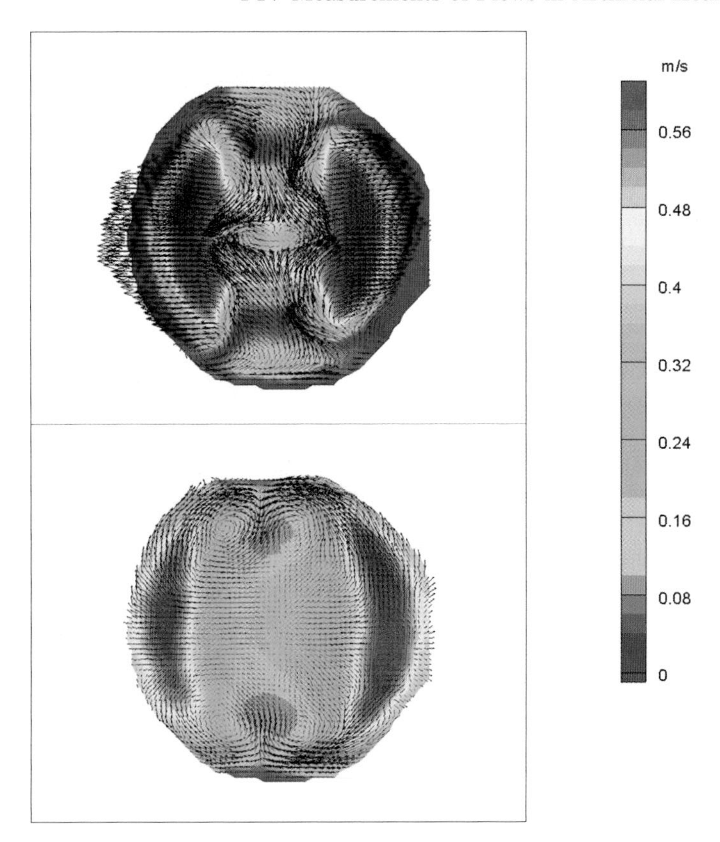

Fig. 7. Stereoscopic PIV: contour map of the out-of-plane (axial) velocity superimposed on the inplane vector field behind the bileaflet valve peak systole. The measurement plane is 1 and 3 cm downstream of the valve plane. 2D view presents color-scaled absolute velocity and vectors indicate in plane velocity as a result of u and v components

restrictions in the allowable lower time pulse separations, pulse energy and beam quality.

The time-resolved PIV technique allows us to overcome limitations: the use we made of a dual-head diode-pumped Nd:YLF source configuration allowed both to generate pulses with better energy distribution, and to accommodate a broader range of experimental conditions than is possible with double-pulsed single-cavity arrangements.

The use of the time-resolved PIV technique, by assuring high sampling rate, made it possible to fully resolve the temporal evolution of the flow downstream of prosthetic valves and to map their complicated dynamics, characterized by very high nonhomogeneity and unsteadiness, which at conventional sampling rates would not be evident.

Furthermore, time-resolved PIV, when applied to the study of PHVs fluid dynamics, allows the analysis of complete, single working cycles: this cannot be done with conventional PIV, where an "equivalent cycle" is measured by acquiring single phases belonging to different cardiac cycles. It can be realistically expected that time-resolved PIV characterization, accurately capturing all hemodynamically relevant scales of motion, will be helpful in the design of reliable mechanical heart valves, thus improving their performance.

At the moment, a possible limitation in the application of time-resolved PIV to the study of prosthetic valve fluid dynamics might be that the measured flow fields do not represent an average (i.e., the dimension of our sample size is one), in consequence of the high costs in terms of storage. Thus, in order to resolve the turbulence in detail associated with PHVs (e.g., Reynolds stresses calculation) great efforts have to be made, in terms of storage: besides supplying instantaneous flow-field information, PIV measurements are also used to obtain true-mean flow-field and turbulence statistics by collecting a large number of vector maps and ensemble averaging them for that purpose.

In the future, measurement of the ensemble-averaged time-resolved velocity vector components with the reachable minimal experimental uncertainty could yield an estimation of turbulence descriptors characterized by greater accuracy. This could allow us to gain knowledge on the turbulent timescale of interest in PHVs fluid dynamics. In fact, as recently stated [30] while for the spatial turbulent scale the ideal resolution is set by the dimensions of red cells and platelets (i.e., a blood corpuscle-smallest eddies interaction takes place, whenever the size of red cells/platelets and smallest eddies are comparable), how to define the ideal temporal resolution is still open to conjecture. There are two approaches that could be followed:

- the one followed in LDA measurements, where cycle-resolved analysis was applied to obtain turbulence data. Several researchers divided the cardiac cycles into segments, considering each segmental duration (30 ms in [31]) long enough to provide statistical validity of the results, and short enough to be considered local stationary during calculation: only in this way is it possible to determine, indirectly (due to the lack of procedures for evaluating these scales in unsteady flows), the Kolmogorov length scale of turbulence, the integral timescale (i.e., the biggest turbulence timescale), and the Taylor microtimescale.
- to regard blood damage an all-or-none phenomenon, setting the ideal temporal resolution as the lowest time of exposure to the higher stress value.

Both approaches are not fully persuasive: the former, because the choice of the segmental duration is based on an assumption not completely proven (turbulence is varying during the cardiac cycle, and the duration of the local stationarity for the phenomenon in consequence); the latter, because it does not account for sublethal damage to blood cells.

One area for development is the extension of time-resolved PIV to measure velocities through a volume: higher camera and laser repetition rates may lead not only to further temporal information, but also to increase volume-velocity information.

4.2 Stereoscopic PIV

The flow field downstream of a prosthetic heart valve is fully three-dimensional, so stereoscopic PIV seems to be eligible for the evaluation of the flow pattern behind implantable devices [17].

Recently, a stereoscopic PIV technique has been proposed as a measurement solution for a deep and complete analysis of PHVs' fluid dynamics in commercial circulatory loop models [32, 33].

The accuracy in the determination of the out-of-plane displacement (i.e., velocity) is related to the angle between the two cameras [28]. The larger this angle (up to 90°), the more accurate is the determination of the out-of-plane displacement. With a restricted optical access, a smaller angle must be used, and a cost of accuracy in determining the third velocity component must be paid.

As recently assessed by *Grigioni* and colleagues [34], several problems arise in performing 3D PIV on heart-valve substitutes, due to the fact that with a commercial test facility the possibility to configure the setup in order to attain an angle of 90° between the probes' axes, without image distortions is not guaranteed. This limitation was overcome in the present study, due to the great modularity and maneuverability of the circulatory loop model used, which can be considered particularly useful from a validation point of view.

However, in the application of the PIV technique as a reliable measurement solution for the complete analysis of prosthetic heart-valves-related fluid dynamics, several methodological and technological problems are still open. Among them, the high computational and storage costs for stereoscopic PIV, to fully characterize the whole 3D flow field through PHVs, as recently underlined by *Grigioni* et al. [34]. The evaluation of second-order moments of turbulence, i.e., variance, Reynolds stresses, etc., is strongly dependent on the accuracy [35] and could require the acquisition of a great number of images per measurement plane.

The visualization of the three-dimensional flow field downstream of a mechanical heart valve by stereoscopic PIV provides the best opportunity to examine the nonstationary flow behavior during a heart-valve cycle. Flow stagnation may be considered for thrombosis prediction, and the three-dimensionality of the vortices shed by the leaflets may be investigated in more detail, being involved in platelet activation and emboli formation.

References

[1] M. Grigioni, C. Daniele, G. D'Avenio, U. Morbiducci, C. Del Gaudio, M. Abbate, D. D. Meo: Innovative technologies for the assessment of cardiovascular medical devices: state of the art techniques for artificial heart valves testing, Expert Rev. Med. Dev. **1**, 89–101 (2004)

[2] L. H. Edmunds, Jr.: Is prosthetic valve thrombogenicity related to design or material?, Texas Heart Inst. J. **23**, 7 (1996)

[3] A. Renzulli, G. Ismeno, R. Bellitti, D. Casale, M. Festa, G. A. Nappi, M. Cotrufo: Long-term results of heart valve replacement with bileaflet prostheses, J. Cardiov. Sur. **38**, 241–247 (1997)

[4] D. Bluestein, W. Yin, K. Affeld, J. Jesty: Flow-induced platelet activation in mechanical heart valves, J. Heart Valve Dis. **13**, 8 (2004)

[5] D. Bluestein, Y. M. Li, I. B. Krukenkamp: Free emboli formation in the wake of bi-leaflet mechanical heart valves and the effects of implantation techniques, J. Biomech. **35**, 1533–1540 (2002)

[6] M. Grigioni, U. Morbiducci, G. D'Avenio, G. D. Benedetto, C. Del Gaudio: Proposal for a new formulation of the power law mathematical model for blood trauma prediction, Biomech. Model. Mechanobiol. **4**, 249–260 (2005)

[7] M. Grigioni, C. Daniele, U. Morbiducci, G. D'Avenio, G. D. Benedetto, V. Barbaro: The power law mathematical model for blood damage prediction: Analytical developments and physical inconsistencies, J. Artif. Organs **28**, 467–475 (2004)

[8] A. A. Fontaine, J. T. Ellis, T. M. Healy, J. Hopmeyer, A. P. Yoganathan: Identification of peak stresses in cardiac prostheses. a comparison of two-dimensional versus three-dimensional principal stress analysis., ASAIO J. **42**, 3 (1996)

[9] V. Barbaro, M. Grigioni, C. Daniele, G. D'Avenio, G. Boccanera: 19 mm sized bileaflet valve prostheses' flow field investigated by bidimensional laser Doppler anemometry (part I: Velocity profiles), Int. J. Artif. Organs **20**, 622–8 (1997)

[10] J. T. Ellis, T. M. Healy, A. A. Fontaine, M. W. Weston, C. A. Jarret, R. Saxena, A. P. Yoganathan: An in vitro investigation of the retrograde flow fields of two bileaflet mechanical heart valves, J. Heart Valve Dis. **5**, 600–606 (1996)

[11] R. J. Adrian: Particle-imaging techniques for experimental fluid mechanics, Ann. Rev. Fluid Mech. **23**, 261–304 (1991)

[12] A. K. Prasad, R. J. Adrian, C. C. Landreth, P. W. Offutt: Effect of resolution on the speed and accuracy of particle image velocimetry interrogation, Exp. Fluids **13**, 105–116 (1992)

[13] M. Marassi, P. Castellini, M. Pinotti, L. Scalise: Cardiac valve prosthesis flow performances measured by 2D and 3D-stereo particle image velocimetry, Exp. Fluids **36**, 176–186 (2004)

[14] M. Rossi, U. Morbiducci, L. Scalise: Laser based measurement techniques applied to the study of prosthetic mechanical heart valves fluid dynamics, Int. J. Artif. Organs **28** (2005)

[15] R. Kaminsky, M. Rossi, U. Morbiducci, L. Scalise, P. Castellini, S. Kallweit, P. Verdonck, M. Grigioni: 3D PIV measurement of prosthetic heart valves fluid dynamics, Int. J. Artif. Organs **28** (2005)

[16] R. Kaminsky, M. Rossi, U. Morbiducci, L. Scalise, P. Castellini, S. Kallweit, P. Verdonck, M. Grigioni: Time resolved PIV technique for high temporal resolution measurements of prosthetic heart valves fluid dynamics, in *ESAO, XXXIIth Annual Congress* (2005)

[17] A. K. Prasad, R. J. Adrian: Stereoscopic particle image velocimetry applied to liquid flows, Exp. Fluids **15**, 49–60 (1993)

[18] N. Westerhof, G. Elzinga: Pressure and flow generated by the left ventricule against different impedances, Circ. Res. **32**, 178–186 (1973)

[19] L. J. Temple, R. Serafin, N. G. Calvert, J. M. Drabble: Principle of fluid mechanics applied to some situations in the human circulation and particularly to testing of valves in a pulse duplicator, Thorax **19**, 261 (1964)

[20] M. Wernet: Symmetric phase only filtering: A new paradigm for DPIV data processing, Meas. Sci. Technol. **16**, 601–618 (2005)

[21] F. Scarano: Iterative image deformation methods in PIV, Meas. Sci. Technol. **13**, R1–R19 (2002)

[22] F. Scarano: On the stability of iterative PIV image interrogation methods, in *12th International Symposium on Applications of Laser Techniques to Fluid Mechanics* (2004)

[23] P. Thévenaz, T. Blu, M. Unser: Image interpolation and resampling, in I. N. Bankman (Ed.): *Handbook of Medical Imaging, Processing & Analysis* (Academic Press, San Diego 2000) pp. 393–420

[24] S. M. Soloff, R. J. Adrian, Z. C. Liu: Distortion compensation for generalized stereoscopic particle image velocimetry, Meas. Sci. Technol. **8**, 1441–145 (1997)

[25] W. L. Lim, Y. T. Chew, T. C. Chew, H. T. Low: Pulsatile flow studies of a porcine bioprosthetic aortic valve in vitro: PIV measurements and shear-induced blood damage, J. Biomech. **34**, 1417–27 (2001)

[26] P. Browne, A. Ramuzat, R. Saxena, A. P. Yoganathan: Experimental investigation of the steady flow downstream of the St. Jude bileaflet heart valve: a comparison between laser Doppler velocimetry and particle image velocimetry techniques, Ann. Biomed. Eng. **28**, 39–47 (2000)

[27] D. Bluestein, E. Rambod, M. Gharib: Vortex shedding as a mechanism for free emboli formation in mechanical heart valves, J. Biomech. Eng. **122**, 125–34 (2000)

[28] C. Brucker, U. Steinseifer, W. Schroder, H. Reul: Unsteady flow through a new mechanical heart valve prosthesis analysed by digital particle image velocimetry, Meas. Sci. Technol. **13**, 1043–1049 (2002)

[29] K. B. Manning, V. Kini, A. A. Fontaine, S. Deutsch, J. M. Tarbell: Regurgitant flow field characteristics of the St. Jude bileaflet mechanical heart valve under physiologic pulsatile flow using particle image velocimetry, Int. J. Artif. Organs **27**, 6 (2003)

[30] R. Kaminsky, U. Morbiducci, M. Rossi, L. Scalise, P. Verdonck, M. Grigioni: Time resolved PIV technique allows a high temporal resolution measurement of mechanical prosthetic aortic valves fluid dynamics, Int. J. Artif. Organs **30** (2007)

[31] J. S. Liu, P. C. Lu, S. H. Chu: Turbulence characteristics downstream of bileaflet aortic valve prostheses, J. Biomech. Eng. **122**, 118–24 (2000)

[32] H. L. Leo, L. P. Dasi, J. Carberry, H. A. Simon, A. P. Yoganathan: Fluid dynamic assessment of three polymeric heart valves using particle image velocimetry, Ann. Biomed. Eng. **34**, 936–952 (2006)

[33] U. Morbiducci, G. D'Avenio, C. Del Gaudio, M. Grigioni: Testing requirements for steroscopic particle image velocimetry measurements of mechanical heart valves fluid dynamics, in *Proc. III Workshop BioFluMen* (2004) pp. 21–27
[34] M. Grigioni, U. Morbiducci, G. D'Avenio, D. D. Meo, C. Del Gaudio: Laser techniques to study prosthetic heart valves fluid dynamics, in *Recent Research Developments in Biomechanics* (Transworld Research Network, Trivandrum 2005) pp. 79–106
[35] O. Uzol, C. Camci: The effect of sample size, turbulence intensity and the velocity field on the experimental accuracy of ensemble averaged PIV measurements, in *4th International Symposium on Particle Image Velocimetry* (2001)

Index

Particle Image Velocimetry in Lung Bifurcation Models

Raf Theunissen and Michel L. Riethmuller

von Karman Institute for Fluid Dynamics,
Chaussée de Waterloo 72, 1640 Rhode St.-Genèse, Belgium
{raf.theunissen,riethmuller}@vki.ac.be

Abstract. To better understand the human pulmonary system and governing aerosol deposition mechanisms within the lung, an accurate description of the airflow in the conductive and respiratory pulmonary airways is considered to be essential. *In-vivo* measurements are deemed impossible due to the small scales in the lung structure. Though numerical simulations can improve the insight, validation of the results is mostly lacking, especially in the extraction of particle trajectories. Numerical and experimental studies have been performed at the von Karman Institute for Fluid Dynamics on single and multiple bifurcation models representing simplifications of the lung system. Both steady and oscillating flows have been studied for the upper lung airways, while steady flow conditions were imposed in the modeling of the alveolar zones. Further experiments consisted in the extraction of particle trajectories in the models of the respiratory airways. This chapter presents an overview of the conducted measurement campaigns complemented with the main observations and results.

Nomenclature

α	Womersley number
μ	Dynamic viscosity expressed in $\mathrm{Pa \cdot s}$
ρ	Density expressed in $\mathrm{kg/m^3}$
ω	Frequency of oscillation expressed in Hz
D	Airway diameter
z	Generation number
CCD	Charge-coupled device
CFD	Computational fluid dynamics
FOV	Field of view
LDV	Laser Doppler velocimetry
PIV	Particle image velocimetry
PTV	Particle tracking velocimetry
Re	Reynolds number
VKI	von Karman Institute for Fluid Dynamics
YFT	Young's fringe technique

A. Schroeder, C. E. Willert (Eds.): Particle Image Velocimetry,
Topics Appl. Physics **112**, 73–101 (2008)

1 Introduction

Aerosol particles, particularly small particles that can reach the gas-exchange surfaces of the alveolar region of the lung, are increasingly recognized either as possible health risks in the environment or as diagnostic and therapeutic tools in medical research. To identify the factors that may be contributing to breathing disturbances and to comprehend pollution effects on the alterations in breathing patterns, a better understanding of the human pulmonary system is needed. Besides serving as a systematic treatment, the use of aerosols has become a mainstay in the diagnosis, prevention or control of lung diseases. Pulmonary drug delivery has the advantage that it can deliver drugs directly to the region of therapeutic treatment, hence requiring lower medicine doses. Due to the poor spatial targeting, therapies aiming at delivering aerosolized drugs directly to the regions deeper inside the lung, the so-called alveolar regions, suffer, however, from a lower efficiency and possible side effects. Thus, the concept of spatial targeting requires the knowledge of the nature of the aerosol being delivered and its behavior in the lung. As a result, fields to be investigated are mostly flows in upper lung airways under unsteady conditions where the initial particle deposition takes place due to inertial impact and flows in alveolar regions where gas exchange takes place.

Morphologically, the lung is a complex network of successive dichotomic bifurcations where the airways from the trachea to the alveolar zone become shorter and narrower as they penetrate deeper into the lung. As a result, *in-vivo* investigations of pulmonary flows are not possible, and *in-vitro* experiments in models have to be performed or one relies on numerical calculations. In particular, studies related to aerosol deposition in the alveolar pulmonary airways have so far been mostly restricted to numerical studies, which lack further experimental validation. Based on the reigning Reynolds number in each generation, three flow regimes can be distinguished in the global bifurcation network; a turbulent regime, a laminar one and a regime in which diffusion in the airways is the main transport mechanism. Combined with the fact that the limit between the different types of flow cannot be defined precisely, it becomes impossible to model the whole lung system, either numerically or experimentally. Simplifications of the pulmonary system are therefore needed.

The von Karman Institute (VKI) adopted the methodology of investigating single and multiple bifurcations keeping in mind the local *in-vivo* flow conditions. A complete description of steady and oscillating flow conditions in bifurcations has been performed at the VKI by experimental and numerical modeling, simulating the different pulmonary areas. Recent activities involved the experimental tracking of particles under flow conditions similar to those found in the alveolar lung regions. This chapter will start with a short description of the pulmonary physiology. Next, an overview will be presented on the experimental and numerical campaigns that took place focusing on the constructed facilities, the applied metrology and obtained velocity or particle tracking results.

At this point the authors would like to stress once more the fact that the VKI models were strong simplifications of the true long bifurcation. As such, the obtained findings did not exactly reflect did not exactly reflect the true lung flows but rather those particular to the studied geometries.

Nevertheless, the experimentally obtained velocity data allow a first general understanding of the pulmonary flow dynamics and may serve as an important tool in the validation of numerical codes. Properly validated, these computational models can then be applied to more complex lung structures with sufficient confidence.

2 Pulmonary Physiology

A closer look at the structure of the respiratory system reveals the lungs to be a network of small bifurcations (Fig. 1, left). This complex structure makes the study of gas flow and transport phenomena impossible without any reasonable simplifications. The earliest and most widespread anatomical model is that from [1], consisting of a symmetric, dichotomously branching network of cylindrical tubes with 23 generations (Fig. 1, right). All branches are classified by generations whereby a further distinction is made in the upper (conducting) airways, represented by generations 0 to 16, while the lower (transitional and respiratory) zones extend over generations 17–23. The lower airways are again subdivided into respiratory bronchioles, the alveolar ducts and alveolar sacs.

Based on the diameter in each generation and standard air properties at body temperature (37 °C), the Reynolds number can be calculated for each generation at normal respiration (breathing frequency = 0.25 Hz, flow rate at mouth = 0.5 l/s) with

$$\mathrm{Re} = \frac{\rho\, U D}{\mu}\,. \tag{1}$$

In straight tubes the critical Reynolds number to classify a flow turbulent or laminar is about 2300. For bifurcational flows, however, the critical Reynolds number depends on the morphology of the bifurcations and is not well established yet. Lung flow is also of oscillatory nature. This unsteady flow behavior is characterized by the nondimensional Womersley number (α), which is based on the frequency of oscillation (ω) and the airway diameter (2) with

$$\alpha = \frac{D}{2}\sqrt{\frac{\omega\rho}{\mu}}\,. \tag{2}$$

The Womersley number is a quantification of the quasisteadiness that represents the ratio of the fully developed viscous layer thickness (equal to $D/2$

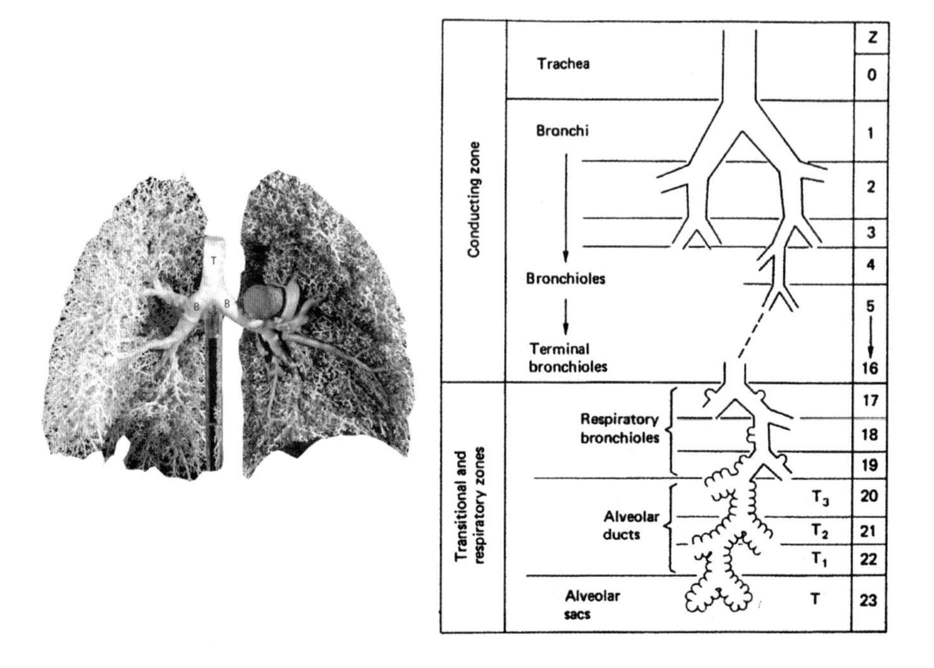

Fig. 1. *Left:* Resin cast of an adult lung. T: trachea, B: bronchial tree. *Right:* Weibel's classification of the lung into generations

for Poiseuille flow) relative to the oscillating boundary layer thickness. Womersley values larger than unity indicate therefore that quasisteadiness cannot be assumed (Table 1).

3 PIV Measurements in Bifurcation Models

3.1 LDV and PIV in a Single 3D Bifurcation Model

To model the 16th airway generation, a single bifurcation was built in 1989. The Pyrex glass model consisted of circular tubes with asymmetrical branching angles corresponding to [2] as 37° and 63° (Fig. 2). The diameters were chosen in accordance with the parent–daughter diameters ratio based on Weibel's measurements and equaled 2.57 cm for the parent branches and 2.38 cm for the two daughter tubes. These dimensions gave a ratio of diameters equal to 0.926. Lengths of the branches were scaled with the diameter by a factor of eight in the parent tube and ten for the daughter tubes. As such, fully established flow could be assumed as boundary conditions both for the inlet and outlet [3] during the numerical computations. A glass-blowing technique enabled the connection of the three tubes in the construction of the complex geometry. Irregularities in the connection were considered to be

Table 1. Reynolds and Womersley numbers in the airways for normal respiration ($f = 0.25$ Hz, 0.5 l/s flow rate at mouth) according to Weibel's morphology

Generation z	D mm	U mm/s	Re	α
...	...	...	...	...
4	4.50	1 965	531.4	0.69
5	3.50	1 624	341.6	0.54
6	2.80	1 269	213.5	0.43
7	2.30	940	130.0	0.35
...	...	...	...	...
17	0.45	17	0.54	0.08
18	0.5	10	0.29	0.08
19	0.47	5	0.15	0.07
20	0.45	3	0.08	0.07
21	0.43	2	0.04	0.07
...	...	...	...	...

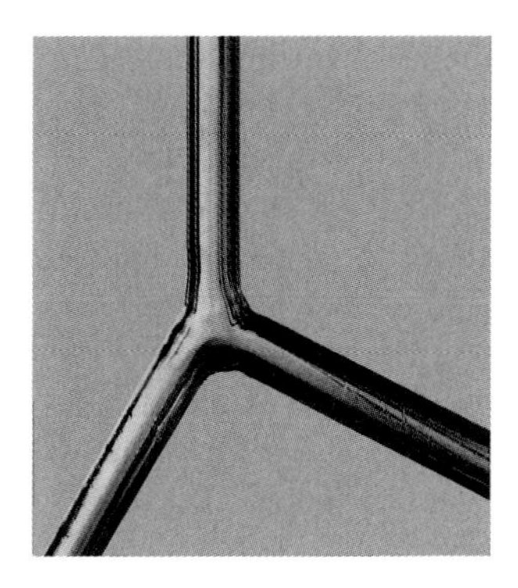

Fig. 2. Pyrex glass model of the single three-dimensional bifurcation

representative of real lung irregularities. By using pure glycerol with a viscosity of $1120 \times 10^{-6}\,\text{m}^2/\text{s}$ as carrier fluid, a diameter-based Reynolds number of 1 in the parent tube could be achieved at an average velocity of 4.3 cm/s. An extra advantage of using glycerin was the small discrepancy in refractive index with glass (i.e., $n \approx 1.4$), minimizing refractions and reflections of the laser light near the tubular walls.

Velocity data of the inspiratory flow was extracted by means of laser Doppler velocimetry (LDV) and particle image velocimetry (PIV) using Young's fringe technique (YFT) for image processing. The emitting light source in the case of LDV was an argon-ion Spectra Physics laser with a nominal power of 1.5 W and a wavelength of 514.5×10^{-6} m. A polarizing filter was placed at the output of the laser, followed by a beam splitter separating the two beams by 50 mm apart. A converging lens with a focal length of 200 mm made the beams cross. Metallic-coated TSI particles with a mean diameter

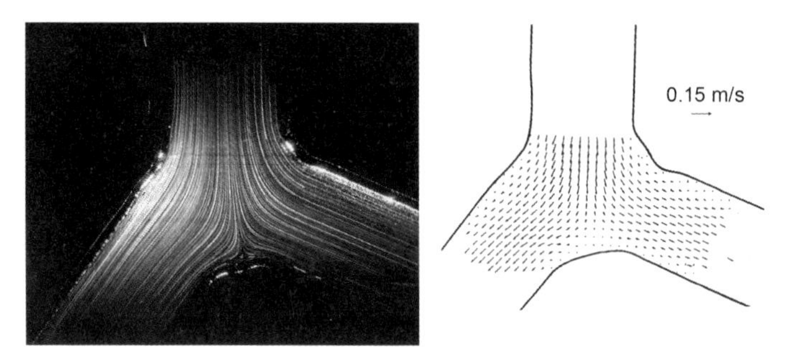

Fig. 3. (*Left*) Flow visualization in the symmetry plane of the bifurcation at Re = 1. (*Right*) PIV measurements in the coronal plane of the asymmetrical bifurcation

of 4.0 μm served as seeding (concentration of about 2×10^{-11} particles/m^3). Having passed an objective of 50 mm, the collected light was received by a photodetector, feeding the processing system to analyze the Doppler frequency [4]. The optical arrangement for particle image velocimetry using Young's fringe technique, the predecessor of the current particle image velocimetry systems, consisted of a CW argon laser combined with a converging and spherical lens to generate the laser sheet. The pulsed light source comprised a continuous CW laser and mechanical rotating slotted disk shutter. A photographic camera recorded the scattered light intensities of the small seeding particles with two successive images superimposed on a single frame. Processing was performed using Young's fringe technique and digital analysis of the fringes. This was the standard technique of PIV processing at the time. Since separation zones were expected, a prescribed shift of the second image was imposed using a rotating mirror.

Flow visualization evinced the laminar character of the flow and the stagnation zone situated at the apex of the bifurcation (Fig. 3, left). As a result of the low Reynolds number and smooth walls, neither recirculation nor separation appeared. Nondimensional axial velocity profiles measured in the coronal plane are presented in Fig. 4 both for LDV and PIV. A good agreement between the two techniques was observed (average difference of 4%) with PIV having the additional advantage of being a whole field measurement technique (Fig. 3, right). Velocity profiles were fully established close to the bifurcation, demonstrating the domination of viscous effects for low Reynolds numbers [3].

Computation of the re-establishment length at higher Reynolds numbers using numerical codes attested the length of the tubes to be too short for a fully developed flow when entering the following bifurcation. As such, all bifurcations had a mutual influence and the use of a network of bifurcations would greatly improve the validity of both experimental and numerical simulations.

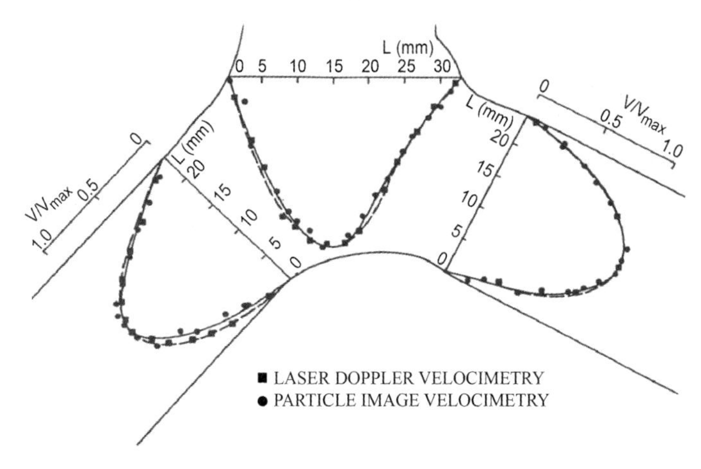

Fig. 4. Comparison of LDV and YFT measurements

3.2 LDV and Digital PIV in a Multiple 2D Bifurcations Model

Having proven the reciprocal influence of the bifurcations, the previous investigations on the 3D single bifurcation were extended to a system of three 2D generations. The choice of the pulmonary zone to study was the laminar flow zone extending between generations 4 to 7, relating to Reynolds numbers ranging from 200 to 1000 [5].

One of the purposes of the model was, furthermore, to verify the results of an earlier CFD code (1993), developed with the intention to simulate the flow in a multiple bifurcation model. Following the two-dimensional multiblock method [6] applied in the numerical algorithm, the experimental model was built in plexiglas to make it optically accessible, whilst using rectangular tubes to have the most practicable approach of the two-dimensional geometry (Fig. 5, left). The diameter related to the width of the tubes and followed Weibel's classification [1] for the first tube (3 cm). The choice of bifurcation angle (35°) and diameter ratio between two successive generations (0.8) culled from the model defined by [7]. Zero gradients of the flow properties were obtained by extending the model three times the diameter in depth. This extension minimized the border effects in the central section of the tubes and admitted a bidimensional established flow in the plane parallel to the length of the model. Given the target Reynolds numbers, water could be used as carrier fluid, whereby equal flow rates were imposed at the four outlets.

Steady Inspiratory Flow

The first measurements were performed with a TSI laser Doppler velocimeter in 1996. With a maximum output power of 5 W, an argon laser provided the laser beam. A beam splitter produced two pairs of shifted and unshifted beams with different wavelengths, allowing the pointwise retrieval of the two

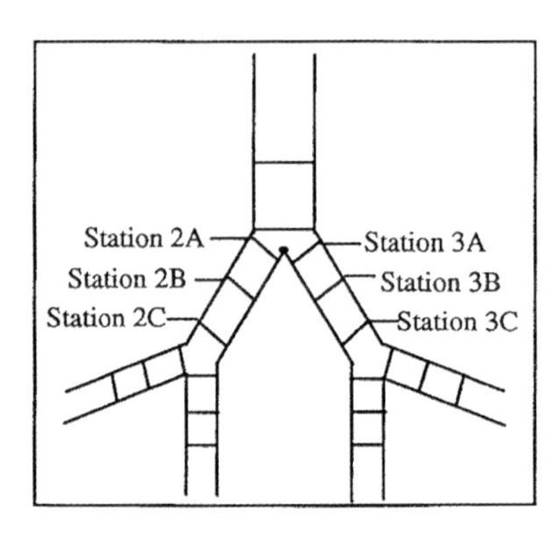

Fig. 5. *Left:* Plexiglas 2D model of multiple symmetric bifurcations. *Right:* Locations of extracted LDV velocity profiles

displacement components. The lens of the probe had a focal length of 260 mm, returning a probe volume of approximately $0.45 \times 10^{-3}\,\text{mm}^3$. A signal processor extracted and digitized the data from the multichannel receiver, after which the data was analyzed and displayed through a specialized software package.

Figures 6 and 7 present the nondimensionalized velocity profiles at different locations in the second generation (Fig. 5, right) for a Reynolds number of 410 based on the diameter of the parent tube. With an average velocity of 1.37×10^{-2} m/s in the first generation, velocity profiles in the subsequent branches were skewed towards the inner wall. With downstream distance the maximum in the skewed profiles moved towards the center of the branch. Clearly the velocity profiles had not re-established due to the short length of the tubes, implying that the bifurcations could not be thought of as independent of one another. Measurements in the second generation further revealed the flow to have the tendency to re-establish quickly where the velocity was smallest and showed the presence of a recirculation zone near the outer wall when passing a bifurcation (i.e., negative velocities in Fig. 7, right). At higher Reynolds numbers the experimental velocity profiles became very disturbed, especially in the recirculation zones. Comparison with numerical results showed an acceptable agreement even though the recirculation zones were larger than in the experimental investigation (Fig. 7, right).

Particle image velocimetry (PIV) was subsequently applied to further study the fluctuations in the recirculation zones [8], allowing both spatial and temporal information of the entire flow field. Since the experiments were performed with water, the seeding particles had to be neutrally buoyant with respect to the experimental fluid. The seeding used in the setup consisted of bee pollen with spherical diameters of approximately 40 μm, compromising between being large enough for sufficient illumination and not too large for

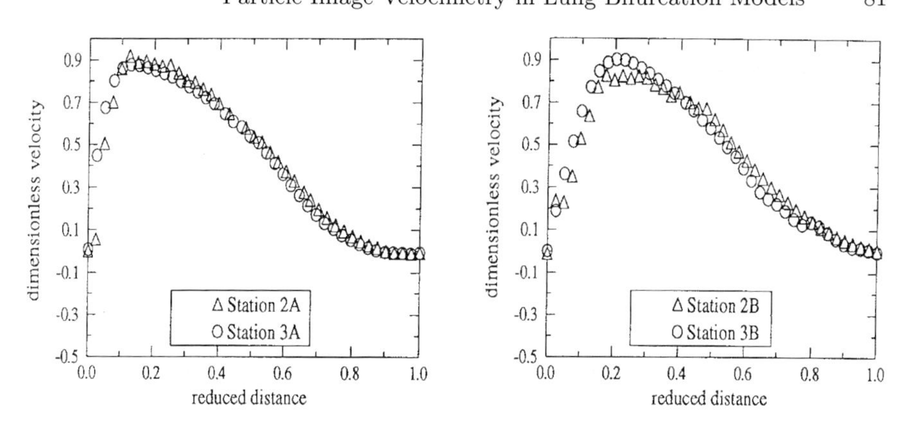

Fig. 6. Extracted velocity profiles with LDV at Re = 410 with equal flow rates (*left*) at the entry (*right*) at the midlocation of the first daughter tubes. See Fig. 5 for locations

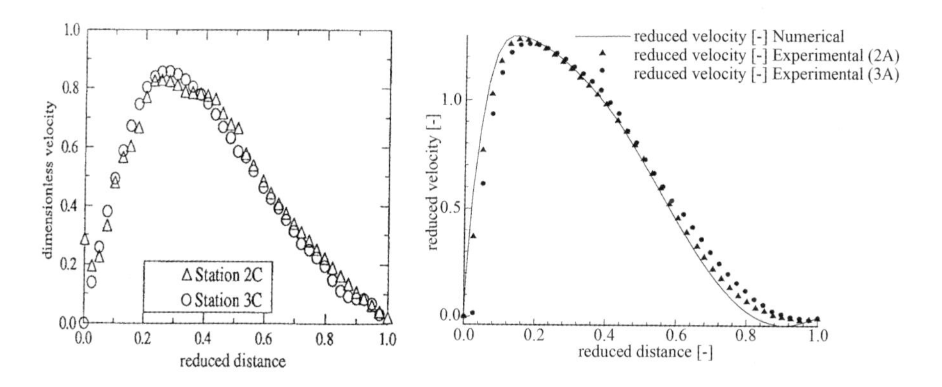

Fig. 7. Extracted velocity profiles with LDV in the 2D multiple bifurcation model at Re = 410 with equal flow rates at all outlets (*left*) upstream of the second bifurcation (*right*) comparison between numerical and experimental velocity profiles downstream of the first bifurcation. See Fig. 5 for locations

a good system resolution. Whereas for low speeds a CW argon laser was previously used to illuminate the tracers with a mechanical shutter generating the light pulses, the introduction of video cameras to record the scattered intensities required acousto-optical systems that could easily be synchronized with the video cameras. Reynolds numbers ranged between 200 and 1000, leading to velocities larger than 2.5 cm/s. At such displacements it was not possible to use a direct video recording of the tracer motion. Therefore, a new technique of PIV was developed using a continuous laser and a Bragg cell as an optical shutter. The acousto-optic Bragg cell allowed us to shutter the laser

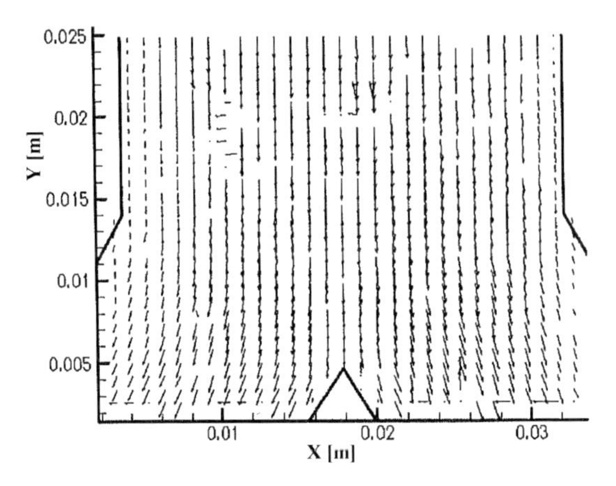

Fig. 8. Velocity field obtained with PIV for Re = 850 at the level of the first bifurcation in the 2D multiple bifurcations model

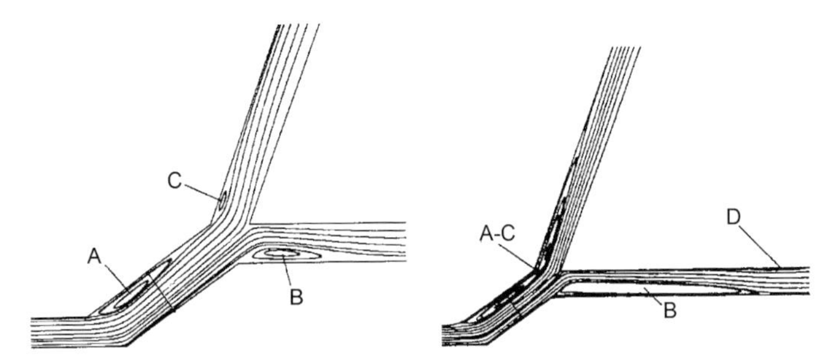

Fig. 9. Numerical streamlines indicating recirculation zones in the 2D multiple bifurcations model for 300 < Re < 400, (*left*). 1000 < Re < 1200 (*right*)

beam, creating single and periodic laser pulses, and allowed the modulation of the laser beam in intensity and position. Analysis of subareas of the acquired recordings was performed with a basic crosscorrelation algorithm [9].

An example of the velocity field measured at the level of the first bifurcation for a Reynolds number of 850 is shown in Fig. 8. In this figure, the velocity profile is Poiseuille-like. The recirculation zone appeared on the external wall of the bifurcation, even though at the internal wall, close to the angle of the bifurcation the velocity increased.

Numerical Simulation of Inspiratory Flow

Some important conclusions could be drawn from numerical simulations [10] on the same two-dimensional model when varying the Reynolds number (Fig. 9).

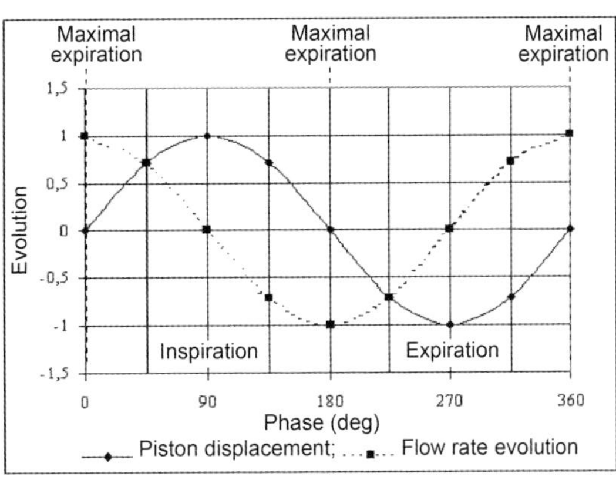

Fig. 10. *Left:* Pistons system used to generate oscillating flow. *Right:* Sinusoidal flow rate: piston and velocity evolution

The simulations made clear the existence of multiple recirculation zones where the number, position and size were dependent on the Reynolds number. In particular, four separation zones were detected (Fig. 9); zone A near the external wall in the first daughter branch, zone B near the external wall in the median distal branch, zone C near the external wall in the lateral distal branch and zone D near the internal wall in the median distal branch. Both zones A and B were present for Reynolds numbers around 200. Starting from a Reynolds number of 300 zone C appeared. The size of the recirculation zones increased with Reynolds number. At Re = 600, zone A joined zone C to form a large recirculation bubble, while zone B continued to enlarge. Finally, for Re = 1200, far downstream of the second bifurcation zone D appeared.

Sinusoidal Oscillating Flow

So far, only the inspiration phase had been simulated. To determine the effects of one bifurcation on the others under realistic conditions, unsteady two-dimensional flow was created. Pistons were placed at the four exits of the two-dimensional bifurcation model. A stepper motor (Fig. 10, left) controlled their positions such that they would follow step functions, thus imposing sudden reversal flow, or a sinusoidal flow rate.

Initial experiments aimed at measuring the time response of an instantaneously reversed flow at several positions within the bifurcation. Based on comparisons of the time history with measurements of the steady profiles an estimate of the development time could be made for each flow condition to determine the quasisteadiness of the flow [11]. It was shown that the flow could be considered quasisteady for Womersley numbers smaller than five. The establishment time of the flow significantly increased by the fact that the flow could not reach an equilibrium between two bifurcations. Subsequently, unsteady PIV measurements were performed [12] with a piston movement ensuring a sinusoidal flow rate around a nil mean velocity value (Fig. 10, right). The Reynolds number at maximum inspiration equaled 300 with an accompanying Womersley number of 11.4 in the parent branch.

Performing the experiments in water necessitated appropriate seeding. Here, seeding consisted of polyamide particles called Vestosint from Hüls GmbH, with a spherical diameter of approximately 25 μm and a density of 1.01 g/cm^3. The PIV system used a 5-W continuous argon laser feeding a fiber optic cable. A lasersheet was formed by passing the laser beam through a cylindrical lens at the end of the cable. The CCD camera's electronic shutter controlled short exposures of the video images, ensuring truly round particle images instead of streaks. The introduction of a grabbing card enabled the acquisition of up to 1000 successive images allowing us to obtain both temporal and spatial evolution of the entire flow field. A crosscorrelation algorithm processed two successive images, incorporating window refinement and distortion within an iterative structure [13].

Within each branch, periodicity in the velocity profiles could be evinced; the outlet conditions of one period served as inlet conditions of the next one. However, asymmetry in the flow was observed, indicating the strong dependency of the flow to any disturbance.

Steady Streaming Displacement

As a result of the velocity profile during inspiration being completely different from the profile at expiration, a fluid element near the bifurcation did not remain in the same segment but moved by a finite amount after each period (Fig. 11, left). This phenomenon, dependent on the element's special location and both the Reynolds and Womersley number, was introduced as *steady streaming displacement* and put into evidence by tracking particles to determine their displacement. The knowledge that the spatial locations of inhaled particles is related to the manner of respiration and number of breathing cycles is of special interest in studies involving inhaled aerosol motion such as, e.g., pulmonary drug delivery [14]. An example of particle displacement for Reynolds and Womersley numbers of, respectively, 280 and 11.4 is presented in Fig. 11, right.

3.3 PIV in a Multiple 3D Bifurcations Model

Though meaningful information had been extracted from the two-dimensional configuration, it did not represent the complex three-dimensional flow structures that exist in bifurcating airways [10]. A three-dimensional model was

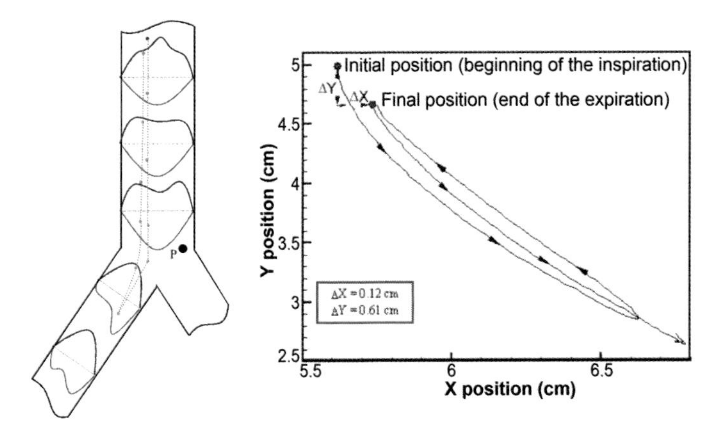

Fig. 11. *Left:* Example of particle trajectory after inspiration (*red profiles*) and expiration (*blue profiles*). *Right:* Measured particle positions of point P during breathing cycle at a frequency of 0.1 Hz and Re = 280 in the 2D bifurcation model

therefore fabricated in 2000 (Fig. 12, bottom left), scaled up from physiological data preserving the geometric characteristics of each bifurcation. The model had been designed following the description in [1] concerning the lung geometry, focusing on generations 5 to 7. The characteristic dimensions such as the length diameter branch ratio equal to 3.2, the branching diameter ratio between two successive generations of bifurcations equal to 0.8 and the branching angle of 70° had all been chosen according to the description made in [7] and reported in [15].

Great care was taken in the design of the transition zone between the parent and daughter branches, which was subdivided into an elliptical and a carinal region (Fig. 12, top left). In the elliptical region of the transition zone, the cross section evolved in shape from circular (section a) to elliptical (section b), keeping the cross-sectional area constant. The carinal region (section c) had a complex three-dimensional shape exhibiting progressive carinal indentation (area increase by 13 %) and ended at the circular onset of the daughter branch (section d). Performing investigation of the flows by means of laser measurement techniques implied the model to have optical access. Moreover, the study of the axial and secondary flows required the model to be completely transparent and have optical access from any direction under study. For this reason the model was built with a novel casting technique (Fig. 13). The casting procedure used an alloy with a low melting point and a mixture of silicone and curing agent. The particularity of the silicone rubber was its transparency, which allowed good optical access.

The experimental setup was designed to simulate both steady and oscillating flows, the latter being generated with the piston system previously presented in Fig. 10, left. During the presented experiments a Poiseuille-like inlet profile was assumed, even though the velocity profile at the inlet of the

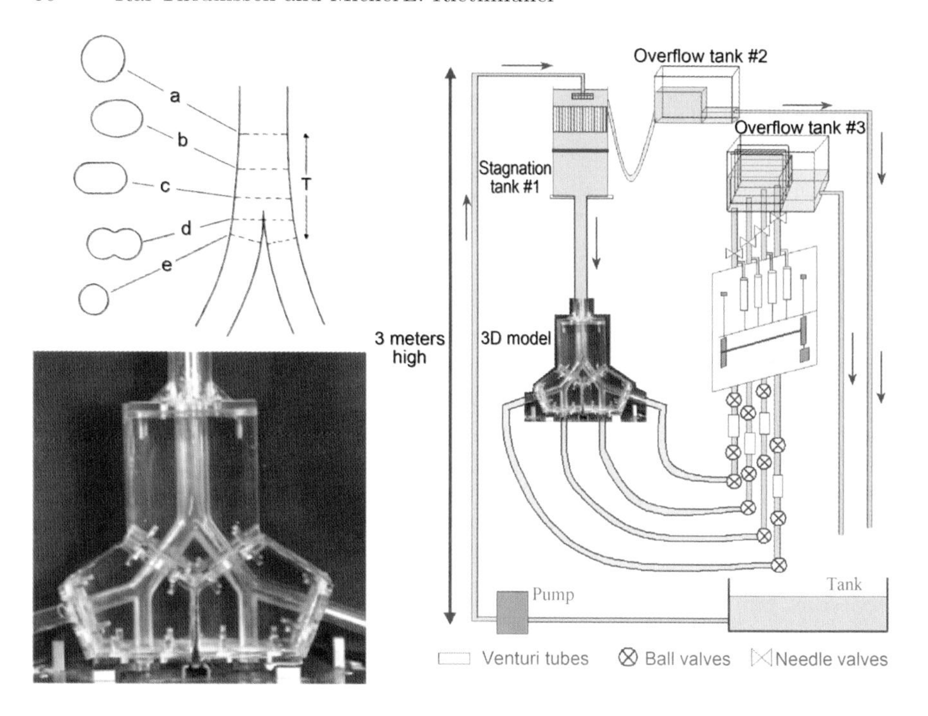

Fig. 12. (*Left top*) Details of the transition region. *Left bottom:* Three-dimensional model of multiple bifurcations. *Right:* Experimental setup

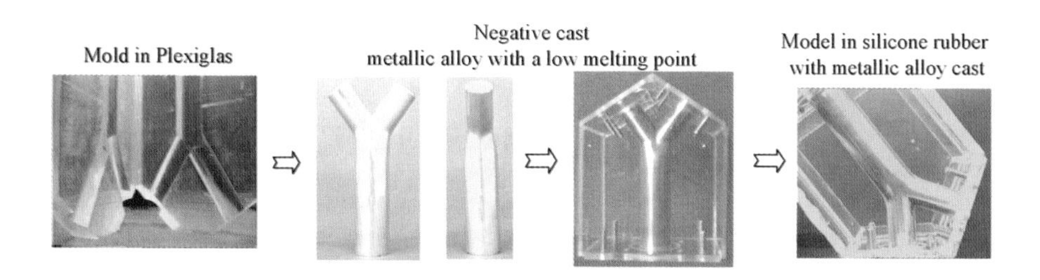

Fig. 13. Casting technique to produce the 3D bifurcations model

5th generation was demonstrated numerically and experimentally to be of influence on the entire flow [16]. A stagnation tank placed at the top of the test section created a smooth entrance flow, leading to a better developed flow. To obtain the necessary gravity-induced flow rates, the stagnation tank was placed at a height of three meters, whereby valves placed downstream of the test section regulated the flow rates. Initial experiments were conducted with water. To avoid possible problematic reflections at the wall and particularly at the carina, fluorescent particles (10 µm) seeded the flow. A mixture

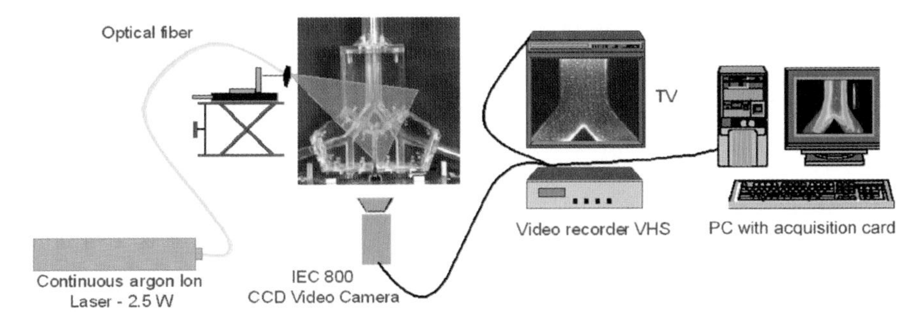

Fig. 14. Time-resolved PIV setup

of glycerin and water was used as carrier fluid in later experiments, with the advantage of having the same refractive index as the silicone model, while having a high enough viscosity to simulate the targeted Reynolds numbers. Dealing with maximum velocities ranging from 1.5 to 1.6 cm/s in the parent tube admitted the use of a continuous laser sheet and a CCD camera directly linked to a video recorder to continuously acquire images at 25 Hz (Fig. 14).

The processing of two images was performed with a crosscorrelation algorithm [13]. The method iteratively incorporated a progressive interrogation window refinement as well as validation. It, furthermore, applied a relative deformation between corresponding windows according to a predicted distribution of velocity and of the local velocity gradient.

Steady Inspiratory and Expiratory Flow

Velocity fields were extracted for steady inspiratory and expiratory flows at Reynolds numbers of 400 and 1000 (Fig. 15) keeping an equal distribution of flow rates in the four branches. The unsteadiness at high Reynolds numbers already present in the 2D model reappeared, indicating it to be linked to the flow regime rather than the geometry of the bifurcation.

Sinusoidal Oscillating Flow

In the case of oscillating breathing patterns, the flow topology in the multiple bifurcation model depended on both the Reynolds number and Womersley values. For this reason, different combinations of both parameters were studied [17]. In Fig. 16 the temporal evolution of the flow in the parent branch (stations 4 and 3 in, respectively, Fig. 15, left and right) is presented as a function of the phase of the flow for a constant Reynolds number of 1000, whilst varying the breathing frequency.

All profiles displayed a two-peak shape in the parent branch at maximal expiration. Going through the inspiration phase the flow rate decreased with

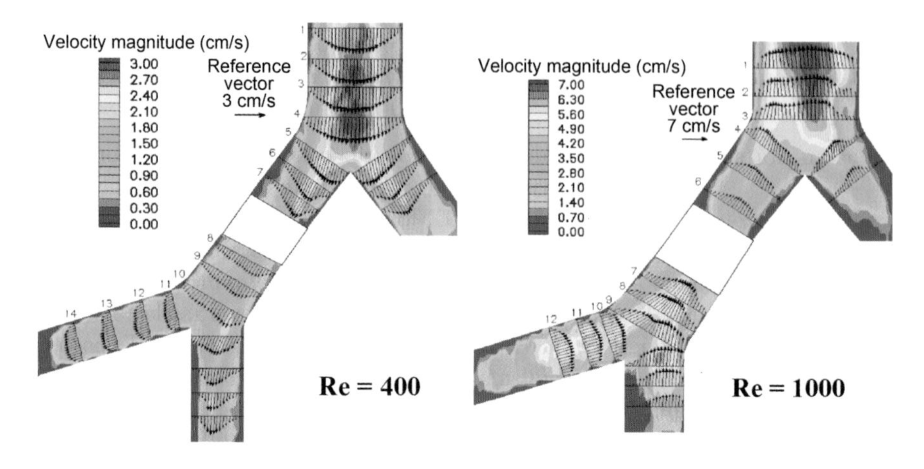

Fig. 15. Velocity field in the 3D multiple bifurcations model obtained with PIV. *Left:* Steady inspiration at Re = 400. *Right:* Steady expiration at Re = 1000

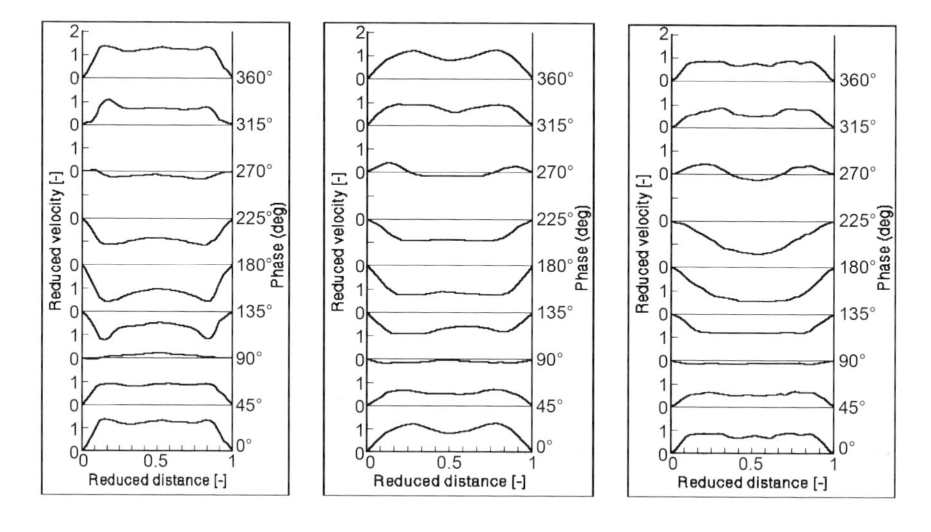

Fig. 16. Temporal evolution of the flow at station 4 (upstream of the first bifurcation) for different Womersley numbers at constant Reynolds number Re = 1000, $\alpha = 16.1$ (*left*), Re = 1000, $\alpha = 8.1$ (*middle*), Re = 1000, $\alpha = 3.6$ (*right*)

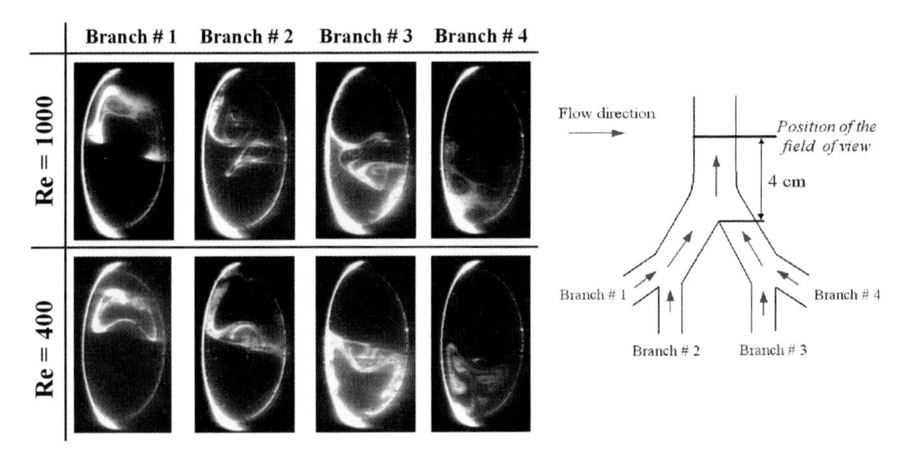

Fig. 17. Flow visualization of the expiratory flow at Re = 400 and Re = 1000, 4 cm above the first bifurcation. Color dye was injected in each branch separately. As such, the influence of each branch could be investigated. The camera was placed at an angle of 45° with the vertical axis, as such cross sections are elliptical

a convex velocity profile reaching positive values for the largest Womersley value, while the other profiles kept their two-peak shape. At maximum inspiration the profile could either have two peaks, be flat or be parabolic, depending on the breathing frequency. In the expiration phase the boundary layer reacted first (phase = 270°) in the parent branch and all velocity profiles yielded two peaks. For high Womersley numbers the fluid acceleration was high and inertial forces dominated the central tubular zones. At smaller oscillation frequencies the flow had more time to react and to behave in a quasisteady manner. By means of numerical investigations on the 3D model using the commercial software package FLUENT© the spatial cyclic behavior was evidenced [16]. Similar experimental observations were made revealing the similarity between the velocity profiles at different stations within the model. This characteristic of the flow did not exist for the smaller Womersley numbers, because then the flow tended to behave more like a quasisteady flow with skewed velocity profiles in the daughter tubes. The numerical simulations, furthermore, indicated the establishment of two symmetric vortices with respect to the bifurcation plane. By putting the laser sheet perpendicular to the flow direction, these strong secondary flows could be visualized during expiration (Fig. 17).

Steady Streaming Displacement

Because of the temporal series used for the time-resolved PIV measurements, chronological velocity vector fields were obtained from which a virtual particle could be traced during a complete breathing cycle, thus visualizing the

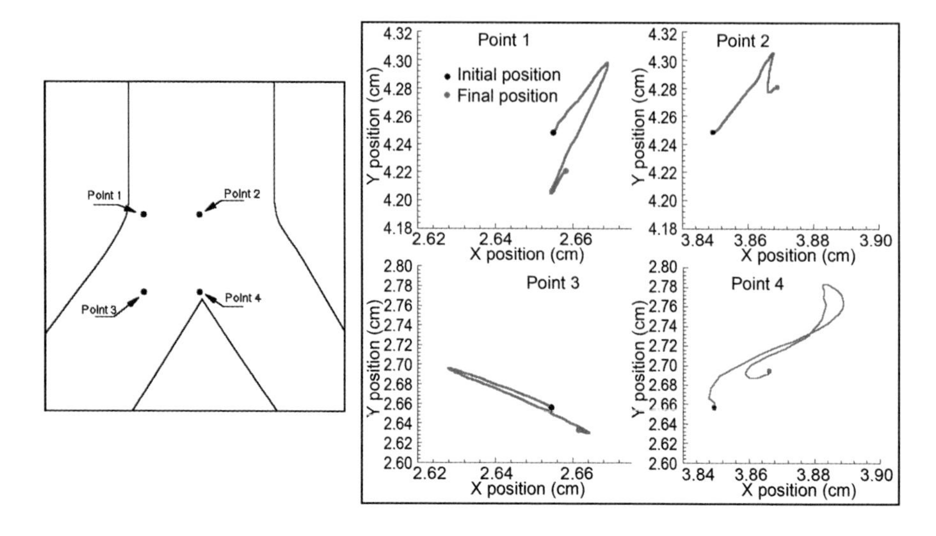

Fig. 18. Illustration of steady streaming displacement at 4 different points for Re = 1000 and $\alpha = 16.1$. *Left:* Position of the points. *Right:* Spatial displacement after one cycle

phenomenon of steady streaming displacement. Particles close to the wall in the parent branch had the tendency to fall down (Fig. 18), whereas the particles at the same height but located in the middle of the parent tube were going up. Particles at point 3 exhibited a motion towards the bifurcation wall, whereas particles at the carina (point 4) showed a more perturbed trajectory. The latter indicated particle deposition due to inertial impact to occur most commonly in the carinal region.

Stokes Flow at Steady Inspiration

Though the 3D bifurcation was originally designed to model the upper airways, the test section was also used to simulate the lower alveolar airways. Acting as a feasibility study for future experiments on more complex models, the intention of the performed study was not to reflect the exact behavior of the human lung, but to take into account as many essential characteristics as possible. Targeting the 14–16th airway generations, a Reynolds number of 2.24 was established in the main tube by using pure glycerin as carrier fluid. Again, the negligible difference in refractive index with the silicone model minimized reflections near interfaces. Conventional flow meters did not function properly when working with viscous fluids under Stokes conditions. For this reason, flow meters were developed for the given range of flow rates (order of 1 to 2 cl/s) based on the principle of a paddle wheel [18], relating the number of rotations of the wheel to the flow rate. Flow rates were kept constant for all four outlets.

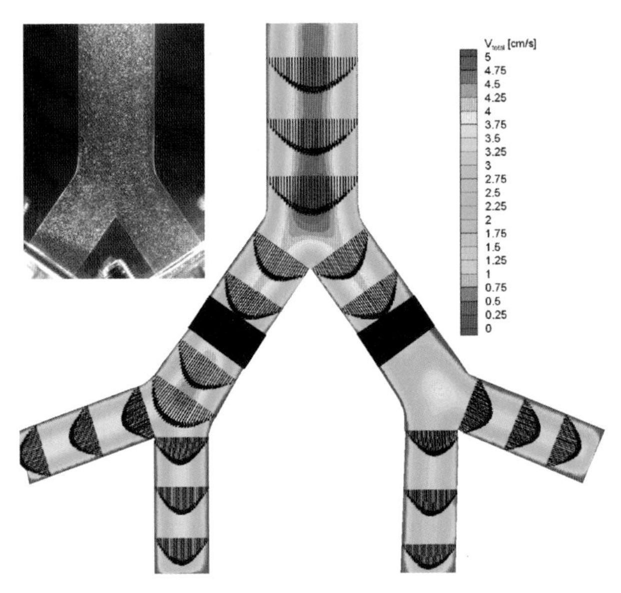

Fig. 19. Velocity field obtained with PIV in the 3D multiple bifurcations model at Re = 2.24. *Top:* PIV image of the lung bifurcation wherein tracers consist of air bubbles

Air bubbles seeded the flow during PIV experiments. These bubbles had relaxation times small enough to be considered as ideal, while at the same time sufficiently scattering the laser light (Fig. 19, top). Given the low velocities encountered, a continuous laser could be used as the light source, while imposing a shutter time of the order of one millisecond to ensure the capture of frozen images of the tracers. An external trigger enforced the time separation between images to minimize uncertainties in the velocity calculation.

Closing in to the dichotomy the flow decelerated and a stagnation point was formed (Fig. 19). The pocket of relatively higher fluid velocity upstream was due to the shape of the cross-sectional area when passing from the parent tube to the daughter tubes (Fig. 12, top left). Given the low flow rates and high viscosity of the fluid, the flow quickly re-established itself after passing a bifurcation. Compared to high Reynolds flows, velocity profiles were no longer skewed towards the inner wall but were almost identical to a Poiseuille-shape at every location.

3.4 PIV in a Single Alveolated Bend

The three-dimensional model of the alveolated bend (2005) represented a "half" bifurcation of the 21st generation (Fig. 20, left). The model consisted of two alveolated ducts joined by a nonalveolated bend. Each duct consisted of a cylindrical lumen 1.5 cm in diameter surrounded by an annulus divided into three alveoli 3 cm in diameter). In total, the model featured six toroidal alveoli

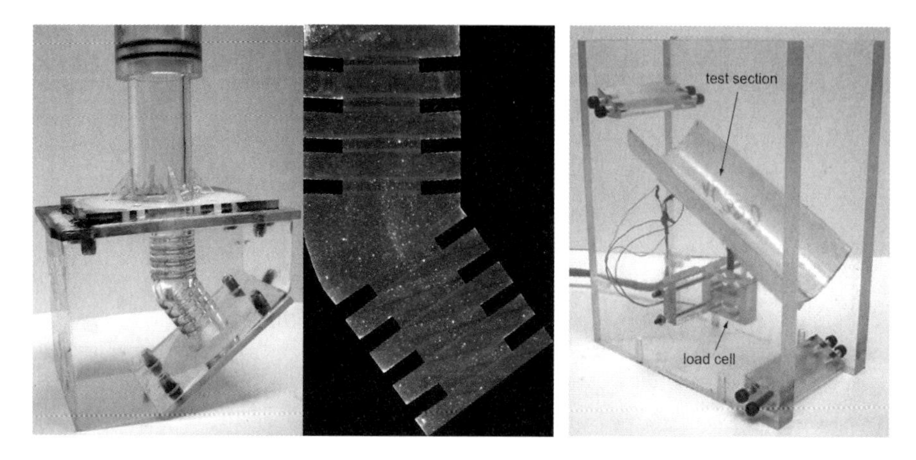

Fig. 20. *Left:* Transparent silicone model of the alveolated tubular bend. *Right:* VKI flow balance

and a transitional zone. The two ducts were joined under an angle of 35°, which is half the branching angle between two bifurcating airways [1]. The dimensions of the 21st airway generation were roughly scaled up by a factor of 50 to allow *in-vitro* measurements. As velocity measurements using PIV had to be performed as well as extraction of paths of injected aerosols, the model needed to be optically accessible. The same casting procedure as described for the three-dimensional multiple bifurcation model was applied. The model did not incorporate oscillating wall motion but could be thought of as a simplified model of alveolar ducts made of central lumens surrounded by alveolar cavities. Such a model was acceptable in the context of this study as the main objective was to retrieve information about the influence aerosol characteristics have on their deposition pattern in low Reynolds number flow conditions.

To match the Reynolds number of 0.07 based on the inner diameter, silicone oil served as carrier fluid, allowing, because of its high viscosity, flow rates of 0.84 ml/s. The experimental setup was built on the basis of the existing physiological flow facility used in previous campaigns (Fig. 12, right). A new flow meter developed at the VKI determined the flow rate based on the measurement of the liquid's weight on a slide (Fig. 20, right). The transducer amplified and translated the strain gauges' signal into voltage, which was related to the flow rate by a preliminary calibration.

Twenty-micrometer iron particles mixed with the fluid acted as seeding. The relatively large necessary separation times and small flow velocities enabled the use of a continuous illumination source [19]. An existing crosscorrelation algorithm [13] was elaborated to include ensemble correlation [20]. The former incorporated window distortion and reduction of the interrogation windows within an iterative structure. The adapted algorithm allowed

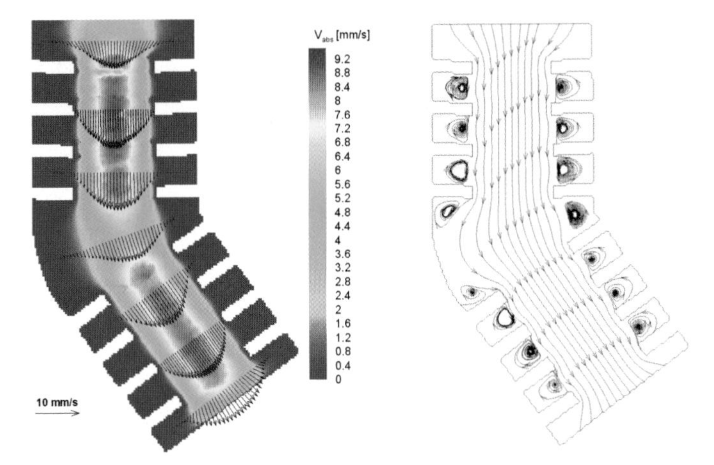

Fig. 21. *Left:* Velocity field obtained with PIV in the alveolated bend model at Re = 0.069. *Right:* corresponding streamline pattern

the extraction of valuable velocity information with relatively high spatial resolution even in regions of poor image quality and low seeding.

A curvilinear separation streamline at the alveolar openings characterized the flow field and indicated little convective change with the lumen flow [21]. Two regions of high velocity in the lumen upstream and downstream of the bend were identified (Fig. 21, left). In the center of the tube, measured velocity profiles showed good agreement with the theoretical Poiseuille flow. Tubular vortices were found in the corners of the outer radius of the bend that then merged into a larger vortex when reaching the inner radius (Fig. 21, right). The velocity inside the alveoli was about two orders of magnitude smaller compared to the mean lumen velocity (9.2 mm/s). Through the advanced interrogation methodology even the slow-rotating fluid elements, located at the center of each cavity could be properly identified.

The obtained results were subsequently used as a validation tool for numerical simulations performed in conjunction with the University of California, San Diego [22] in a computational model that was an exact replica of the experimental one. Experimental data was on average 5 % lower than numerical results, which was thought of to be an effect of the flexibility of the model during the experimental runs (Fig. 22). The agreement between the two approaches not only indicated the measurements to be of good quality, but also suggested the computational model to correctly predict the flow behavior in the complex structure. The latter observation evidenced its usefulness for further predictions in alveolar models.

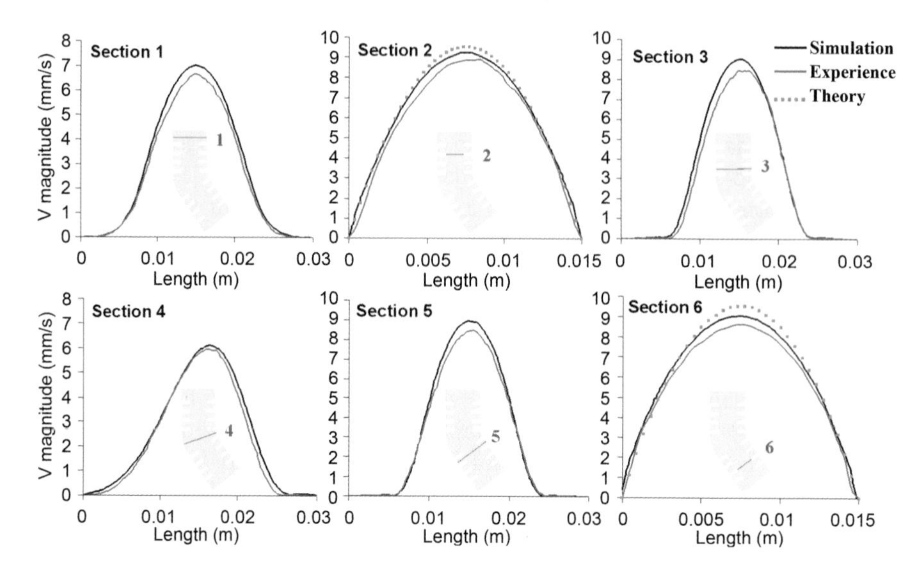

Fig. 22. Extracted velocity profiles at different locations in the alveolated bend. Comparison between experimental, numerical and theoretical Poiseuille profiles. The orientation of the axes is such that zero length corresponds to the inner side of the bend

4 Particle-Deposition Measurements in Respiratory Airway Models

Having identified the hazardous and beneficial influences of aerosols, numerous numerical studies on the aerosol behavior within the alveolar regions have been performed including particle deposition (Fig. 23), but all lack an experimental validation [23]. One of the main objectives was therefore the initialization of an experimental campaign focusing on the extraction of experimental aerosol trajectories, later to be used in the validation of numerical calculations.

4.1 PTV in a Multiple 3D Bifurcations Model

Experiments were conducted in the 3D multiple bifurcation model under reigning Stokes flow [18]. As the starting point, the behavior of an *in-vivo* aerosol particle of 1 μm diameter, 1000 kg/m^3 in density and having a terminal velocity in still air of 3.3×10^{-5} m/s was chosen to be modeled. To exactly replicate the aerosol behavior two criteria had to be satisfied; Reynolds similarity and a constant ratio of the particle settling velocity and fluid velocity. Satisfying both conditions and keeping in mind the characteristic data of the real aerosol resulted in having to use particles with densities up to

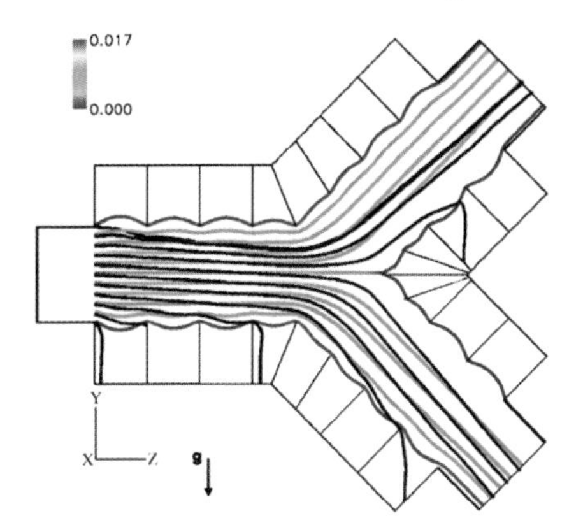

Fig. 23. Numerical particle tracks of 5 μm particles (*black lines*) superimposed on streamlines (*colored lines*) for generations 20–21 (taken from [24])

134×10^3 kg/m^3 in the experiments. As both the Reynolds similarity and velocity ratios for the particle clearly could not be satisfied simultaneously, lead particles of 1 mm and 2 mm diameter were employed instead. These particles had equivalent aerosol diameters of 20 and 39 μm respectively. Although neither of the two similarity conditions were now fulfilled, the main objective could still be reached, being the extraction of experimental results serving for numerical validation purposes.

Particle trajectories were extracted from the time-resolved recordings with particle tracking velocimetry. Incorporated in the algorithm was the four-frame tracking method, followed by a linear extrapolation of previous determined particle locations to serve as a predictor for the next position [18]. None of the particles were capable of following the geometry of the first bifurcation due to gravitational and inertial influences (Fig. 24, left), but would deviate from streamlines or cross them. The larger the particle the larger the angle of the trajectory (Fig. 24, right), indicating that larger particles would impact earlier on the wall, as expected. Based on the known particle positions, velocity lag could be calculated and hence the particle Reynolds number. The latter turned out to be scattered around 0.15 for the 2 mm particles. Close to the outer wall, particles underwent an acceleration when nearing the carinal region, after which they decelerated to zero and impacted the inner bifurcation wall.

4.2 PTV in a Single Alveolated Bend

The aim was to perform an initial study on the aerosol behavior in the 21st airway generation using the model of the alveolated bend. As was the case for the 3D multiple bifurcations model, neither Reynolds similarity nor gravitational similarity (i.e., constant particle-fluid-velocity ratio) could be satisfied simultaneously for the given *in-vivo* aerosol characteristics. It was therefore

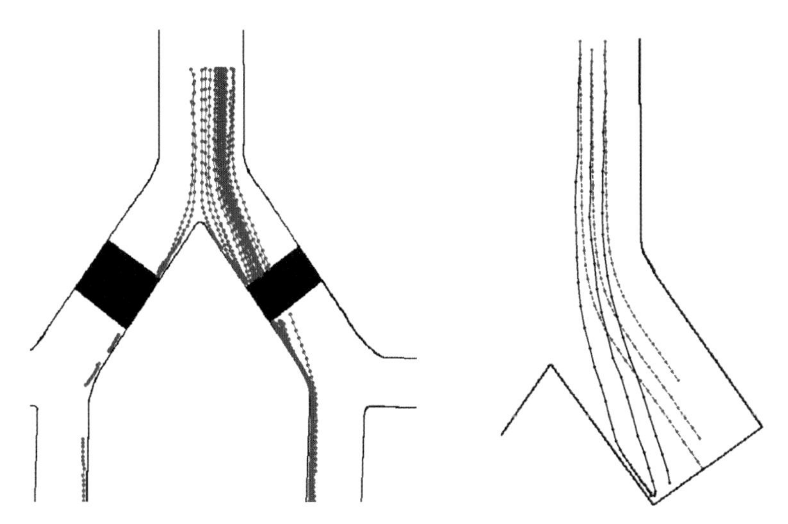

Fig. 24. Resulting trajectories for lead particles in the multiple 3D bifurcation model for Re = 2.24 (*left*) 1-mm particles (*right*) trajectories of 2-mm particles (*black*) compared with streamlines (*blue*) around the first bifurcation

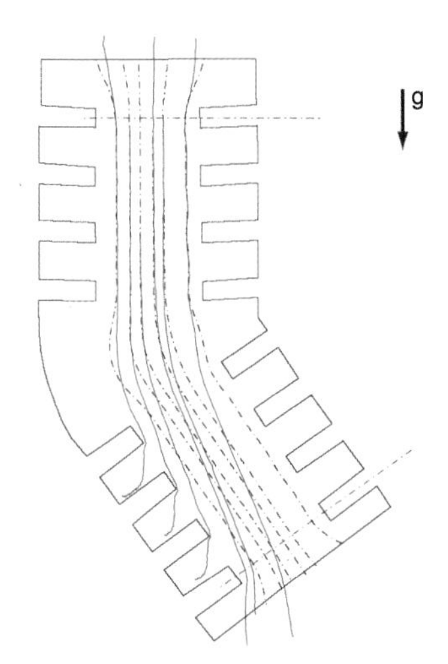

Fig. 25. Particle trajectories for 1.2 mm iron particles (*blue full lines*) superimposed on streamlines (*dashed black lines*) in the alveolated bend for Re = 0.07

opted to concentrate only on the imposed velocity ratio between terminal velocity and flow velocity. Spherical iron particles of 0.5 mm and 1.2 mm diameter were chosen with corresponding aerosol diameters of 5.3 μm and 12.8 μm, respectively, because of their good scattering ability and magnetic attraction. The latter enabled the introduction of the particle to the flow and their collection after passing the flow cycle.

Two-Dimensional Trajectories

Results of two-dimensional PTV confirmed none of the 1.2-mm particles to be capable of following the geometry of the lumen tube [21]. Instead, after passing the entrance of the bend, trajectories deviated from the flow streamlines under the influence of gravity and inertia (Fig. 25). Once the particle entered the lower lumen they moved along a linear path and were almost mutually parallel. Here, they continued to deviate from streamlines and could either enter the lower alveoli or impact the outer lumen wall. All trajectories showed a curvilinear behavior, indicating gravity to be the predominant deposition mechanism. The almost constant particle Reynolds numbers implied the velocity lag to be constant and hence also the drag force. Furthermore, the Stokes number was of the order of 10^{-4}, indicating that viscous forces dominated the inertial forces.

Three-Dimensional Trajectories

In order to obtain fully three-dimensional particle trajectories a two-camera approach was applied, whereby the cameras' optical axes were 90° apart [25]. A focal length of 50 mm was chosen for both cameras in order to minimize variation in magnification with distance to the camera. From the two fields of view (FOV), the three-dimensional particle positions could be constructed. For the given dimensions of the experimental model, it was found that the change in depth affected the magnification only by as much as 0.5 %. Therefore, corrections to the particle tracking methodology, adapted to three dimensions, were deemed unnecessary. Initially, a laser with a spherical lens was unable to illuminate the entire volume. For this reason the setup was equipped with a 1-kW photography lamp. To obtain a homogeneous light intensity distribution in the FOV a semitransparent white sheet of paper was placed in front of the light source. No longer having Gaussian-shaped intensity distributions, an accurate positioning of the particle center was no longer possible leading to inaccuracies of the order of 1 pixel in the localization depending on the particle diameter used.

Based on the initial measured trajectories (Fig. 26), the same conclusions for the particle behavior yielded as for the two-dimensional experiments. However, the front view now reveals that trajectories were fully linear even when passing the bend (Fig. 26, right) indicating that all trajectories were two-dimensional. Trajectories in the front view were slightly inclined with respect to the geometrical boundaries, caused by a slight misalignment in the determination of the edges of the model. Further experiments will result in a more detailed description of the full 3D motion of particles in the alveolated bend.

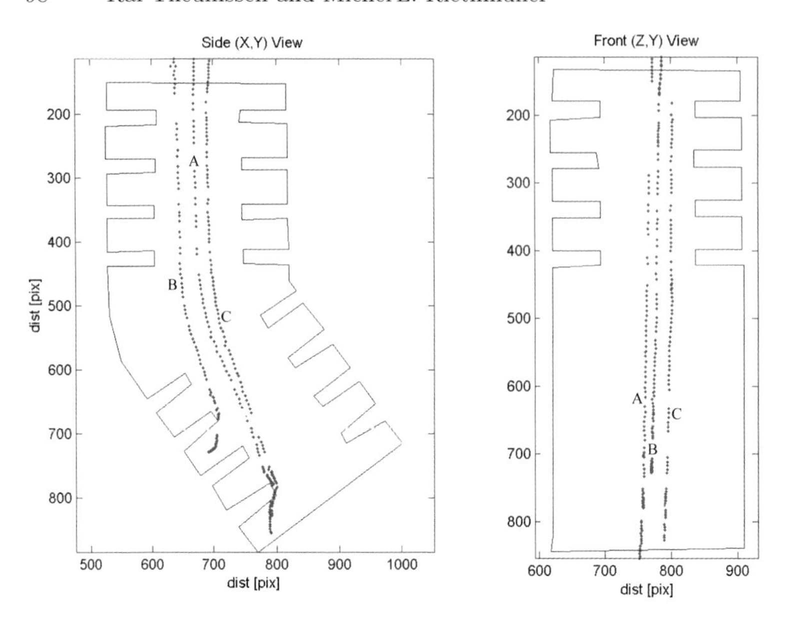

Fig. 26. Particle trajectories for 1.2 mm iron particles in the alveolated bend for Re = 0.07. *Left:* side view, *right:* front view

5 Conclusions

To enhance the understanding of pulmonary diseases and therapeutic pulmonary drug-delivery, investigations of the fluid flow in lung bifurcations are needed. Results will primarily serve to obtain a more profound understanding of the pulmonary fluid dynamics, allowing the enhancement of numerical and experimental models. Extracted velocity data will also be useful in the validation of numerical codes.

A complete description of steady flow in a single 3D bifurcation had been previously performed at the von Karman Institute by experimental and numerical modeling. As a result of this study, it had been shown that the first bifurcation influenced the flow in the second and in the third bifurcation when the length of the second one was not sufficiently long.

The findings on the single bifurcation led to the extension of the investigations to a network of bifurcations incorporating three generations. This first approach to the investigation of the steady and oscillating flow within 2D multiple bifurcations allowed us to validate a numerical code developed at the VKI.

To experimentally investigate flows within three-dimensional multiple bifurcations, a 3D model had been built with an innovative casting technique. The particularity of the silicone model was its transparency, allowing high-quality optical access. The design of the model employed physiological data, preserving the geometric characteristics of each bifurcation. The scaled experimental model respected the dynamic similarity with the two main nondimensional parameters; the Reynolds and Womersley numbers. Using particle image velocimetry, velocity data in the entire flow field could be acquired both for steady and oscillating flow conditions. Modifications made the technique particularly suitable for low-speed flows. Under steady conditions, inspiratory and expiratory flows as well as oscillating flows were investigated for different Reynolds numbers based on the diameter of the parent branch and on the mean velocity in the parent branch. Sinusoidal movement of the piston system placed downstream of the bifurcation controlled the oscillating flow rates. For high Reynolds number ($\text{Re} \approx 1000$), the velocity profile at the maximum inspiration was not completely established in comparison with the velocity profile for lower Reynolds numbers ($\text{Re} \approx 400$). These results led to the conclusion that the quasistationarity of the flow was highly dependent on the Reynolds and Womersley numbers. The experimental data, and enforced by numerical results, allowed the demonstration of the spatial cyclic behavior of the flow. Furthermore, thanks to the time-resolved character of the PIV measurements, steady streaming displacement was brought into evidence showing the lack of periodicity in particle trajectories.

The same 3D model was used in a later stage to investigate aerosol deposition under the Stokes conditions encountered in the alveolar lung regions ($\text{Re} \approx 2.24$). PIV results under steady flow proved the flow to be fully developed at every location. A preliminary theoretical analysis evinced the impossibility to exactly simulate *in-vivo* aerosol behavior. As the experimental results would serve in the validation process, the use of larger particles was opted for. By means of particle tracking velocimetry, particle trajectories were extracted.

As a next step in the investigation of the alveolar lung regions, PIV and PTV measurements were performed in a cylindrical bend surrounded by toroidal alveoli ($\text{Re} \approx 0.07$). PIV results were in good agreement with numerical simulations, showing Poiseuille-like profiles in the lumen flow with recirculating structures inside the alveoli and tubular vortices in the bend. So far, all particle-tracking experiments showed gravity to be the dominant deposition mechanism.

Current work involves further experimental investigation of the three-dimensional flow character and particle trajectories. Experiments are planned to take place on a three-dimensional alveolated bifurcation, augmenting the complexity of the flow and approaching a more realistic representation of the *in-vivo* structures encountered in the alveolated pulmonary areas.

References

[1] E. R. Weibel: *Morphometry of the Human Lung* (Springer, Berlin, Heidelberg 1963)

[2] R. F. Phalen, H. C. Yeh, G. M. Schum, O. G. Raabe: Application of an idealised model to morphometry of the mammalian tracheobranchial tree, Anat. Rec. **190**, 167–176 (1978)

[3] P. Corieri: *Experimental and numerical investigation of flows in bifurcations within lung airways*, Ph.D. thesis, von Karman Institute for Fluid Dynamics, Université Libre de Bruxelles and Rheinisch-Westfälische Technische Hochschule Aachen (1994)

[4] H.-J. Pfeifer: Fundamentals of signal processing, in *Laser Velocimetry*, von Karman Institute Lecture Series (von Karman Institute 1991)

[5] A. Ramuzat: *Experimental investigation of flows within 2D bifurcations*, VKI project report, von Karman Institute for Fluid Dynamics (1996)

[6] F. Wilquem: *A 2D multiblock approach to solve viscous flows in lung bifurcations*, VKI project report, von Karman Institute for Fluid Dynamics (1993)

[7] J. Hammersley, D. Olson: Physical models for the smaller pulmonary airways, J. Appl. Physiol. **72**, 2402–2414 (1992)

[8] A. Ramuzat, M. L. Riethmuller, L. Angelucci: Particle image velocimetry using Bragg cell as optical shutter, in *Proc. 17th Int. Congress on Instrumentation in Aerospace Simulation Facilities (ICIASF)* (1997)

[9] J. B. Moens: *Development of a new D-PIV technique to investigate the flow behind a bluff body*, VKI project report, von Karman Institute for Fluid Dynamics (1995)

[10] F. Wilquem, G. Degrez: Numerical modelling of steady inspiratory airflow through a three-generation model of the human central airways, J. Biomech. Eng. **119**, 52–65 (1997)

[11] A. Ramuzat, S. Day, M. L. Riethmuller: Steady and unsteady LDV and PIV investigations of flow within 2D lung bifurcations model, in *Proc. 9th Int. Symposium Applications of Laser Techniques to Fluid Mechanics* (1998)

[12] A. Ramuzat, H. Richard, M. L. Riethmuller: PIV investigation of unsteady flows within lung bifurcations, in *Proc. 8th Int. Conf. Laser Anemometry, Advances and Applications* (1999)

[13] F. Scarano, M. L. Riethmuller: Iterative multigrid application in PIV image processing, Exp. Fluids **26**, 513–523 (1999)

[14] F. R. Haselton, P. W. Scherer: Flow visualization of steady streaming in oscillatory flow through a bifurcating tube, J. Fluid Mech. **123**, 315–333 (1982)

[15] T. J. Pedley: Pulmonary fluid dynamics, Ann. Rev. Fluid Mech. **9**, 229–274 (1977)

[16] C. Raick, A. Ramuzat, P. Corieri, M. L. Riethmuller: Numerical and experimental study of spatial periodicity of steady air flows in pulmonary bifurcations, in *Proc. 7th Triennal Int. Symposium Fluid Control Measurement and Visualization* (2003)

[17] A. Ramuzat, M. L. Riethmuller: PIV investigation of oscillating flows within a 3D lung multiple bifurcations model, in *Proc. 11th Int. Symposium on Application of Laser Techniques to Fluid Mechanics* (2002)

[18] R. Theunissen: *Experimental investigation of aerosol deposition in lung airways*, VKI project report, von Karman Institute for Fluid Dynamics (2004)

[19] R. Theunissen, N. Buchmann, P. Corieri, M. L. Riethmuller, C. Darquenne: Experimental investigation of aerosol deposition in alveolar lung airways, in *Proc. 13th Int. Symposium on Applications of Laser Techniques to Fluid Mechanics* (2006)

[20] S. T. Wereley, L. Gui, C. D. Meinhart: Advanced algorithms for microscale particle image velocimetry, AIAA J. **40**, 1047–1055 (2002)

[21] N. Buchmann: *Experimental modeling of aerosol particles within lung bifurcations*, VKI stagaire report, von Karman Institute for Fluid Dynamics (2005)

[22] C. van Ertbruggen, N. Buchmann, R. Theunissen, P. Corieri, M. L. Riethmuller, C. Darquenne: Flow in a 3D alveolated bend: Validation of CFD predictions with experimental results, in *5th World Congress of Biomechanics* (2005)

[23] C. Darquenne: *Numerical and experimental investigation of aerosol transport and deposition in the human lung*, Ph.D. thesis, Université Libre de Bruxelles and von Karman Institute for Fluid Dynamics (1995)

[24] L. Harrington, K. G. Prisk, C. Darquenne: Importance of the bifurcation zone and branch orientation in simulated aerosol deposition in the alveolar zone of the human lung, J. Aerosol Sci. **37**, 37–62 (2006)

[25] M. Bilka: *Experimental investigation of 3D Aerosol Motion within an alveolated duct*, VKI project report, von Karman Institute for Fluid Dynamics (2006)

Tomographic 3D-PIV and Applications

Gerrit E. Elsinga[1], Bernhard Wieneke[2], Fulvio Scarano[1], and Andreas Schröder[3]

[1] Department of Aerospace Engineering, TU-Delft, The Netherlands, {g.e.elsinga,f.scarano}@tudelft.nl
[2] LaVision GmbH, Göttingen, Germany, bwieneke@lavision.nl
[3] Institut f. Aerodynamik und Strömungstechnik, DLR Göttingen, Germany, andreas.schroeder@dlr.de

Abstract. Tomographic particle image velocimetry is a 3D PIV technique based on the illumination, recording, reconstruction and analysis of tracer-particle motion within a three-dimensional measurement volume. The recently developed technique makes use of several simultaneous views of the illuminated particles, typically 4, and their three-dimensional reconstruction as a light-intensity distribution by means of optical tomography. The reconstruction is performed with the MART algorithm (multiplicative algebraic reconstruction technique), yielding a 3D distribution of light intensity discretized over an array of voxels. The reconstructed tomogram pair is then analyzed by means of 3D crosscorrelation with an iterative multigrid volume-deformation technique, returning the three-component velocity vector distribution over the measurement volume. The implementation of the tomographic technique in time-resolved mode by means of high repetition rate PIV hardware has the capability to yield 4D velocity information. The first part of the chapter describes the operation principles and gives a detailed assessment of the tomographic reconstruction algorithm performance based upon a computer-simulated experiment. The second part of the chapter proposes four applications on two flow cases: 1. the transitional wake behind a circular cylinder; 2. the turbulent boundary layer developing over a flat plate. For the first case, experiments in air at $\mathrm{Re_D} = 2700$ are described together with the experimental assessment of the tomographic reconstruction accuracy. In this experiment a direct comparison is made between the results obtained by tomographic PIV and stereo-PIV. Experiments conducted in a water facility on the cylinder wake shows the extension of the technique to time-resolved measurements in water at $\mathrm{Re_D} = 540$ by means of a low repetition rate PIV system. A high data yield is obtained using high-resolution cameras (2k $\times$ 2k pixels) returning 650k vectors per volume. Measurements of the turbulent boundary layer in air at $\mathrm{Re}_\theta = 1900$ provide a clear visualization of streamwise-aligned low-speed regions as well as hairpin vortices grouped into packets. Finally, in similar flow conditions the boundary layer is measured using a high repetition rate PIV system at 5 kHz, where the spatiotemporal evolution of the flow structures is visualized revealing a mechanism for the rapid growth of a Q2 event, possibly associated to the generation of hairpin-like structures.

A. Schroeder, C. E. Willert (Eds.): Particle Image Velocimetry,
Topics Appl. Physics **112**, 103–125 (2008)

1 Introduction

The measurement of the instantaneous three-dimensional velocity field is of great importance in fluid mechanics and particularly in turbulence research since it reveals the complete topology of turbulent coherent structures. Moreover, three-dimensional measurements are advantageous for those situations where the flow does not exhibit specific symmetry planes or axes. The advent of PIV and its developments (stereo-PIV, [1], dual-plane stereo-PIV, [2]) showed the capability of the PIV technique to quantitatively visualize complex flows.

Several methods have been proposed to extend the PIV technique to a three-dimensional measurement technique. Scanning lightsheet [3], holography [4], 3D PTV [5] and very recently tomographic particle image velocimetry (tomographic-PIV, [6]) are the most popular approaches (see *Arroyo* and *Hinsch* in the present book for a complete review). In particular, the development of the tomographic-PIV technique was motivated by the need for a simple optical arrangement such as those used in the photogrammetric approach combined with a robust particle-volume reconstruction procedure not relying upon particle identification as opposed to the particle-tracking techniques. The latter is achieved by means of optical tomography. The resulting procedure offers the capability to handle relatively high seeding densities without compromising the robustness of the measurement.

The present work describes the working principle of tomographic PIV including a numerical assessment of its performances. The application to real experiments has two objectives: first, the method's accuracy can be assessed under actual experimental conditions by a comparison with stereo-PIV; and secondly, the applications to the turbulent cylinder wake and the turbulent boundary layer directly show the type of flow information that can be obtained. Finally, the extension of the technique to time-resolved measurements (4D) in air and water flows is discussed and two applications are presented showing the time evolution of the 3D Karman wake behind a circular cylinder in water and the development of hairpin-like structures in a turbulent boundary layer.

2 Principles of Tomographic PIV

The working principle of tomographic PIV is schematically represented in Fig. 1. Tracer particles immersed in the flow are illuminated by a pulsed light source within a three-dimensional region of space. The scattered light pattern is recorded simultaneously from several viewing directions using CCD cameras. The Scheimpflug condition between the image plane, lens plane and the midobject plane is fulfilled by means of camera-lens adapters with two degrees of freedom. The particles within the entire volume are imaged in focus, which requires a relatively small lens aperture. The 3D particle distribution (the object) is reconstructed as a 3D light-intensity distribution from its projections on the CCD arrays. This reconstruction problem is of inverse nature and in general underdetermined, meaning that a single set of projections (viz. images) can result from many different 3D objects. Determining the most likely 3D distribution is the topic of tomography [7], which is addressed in the following section. Then, the particle displacement (hence velocity) within a chosen interrogation volume is obtained by the 3D cross-correlation of the reconstructed particle distribution at the two exposures. The crosscorrelation algorithm is based on the crosscorrelation analysis with the iterative multigrid window (volume) deformation technique (WIDIM, [8]) extended to three dimensions.

The relation between image (projection) coordinates and the physical space (the reconstruction volume) is established by a calibration procedure similar to what is used in stereoscopic PIV or 3D-PTV. Each camera records images of a calibration target at several positions in depth throughout the volume, from which the calibration procedure returns the viewing directions and field of view. The tomographic reconstruction relies on accurate triangulation of the views from the different cameras. In particular, to successfully reconstruct particle images the accuracy requirement for the calibration is a fraction of the particle image size (typically below 0.4 pixels according to [6]). An adequate mapping function from physical space to the image coordinate is provided by either a pinhole model [9] or third-or-higher-order polynomials in x and y and linear along z [10]. Procedures developed for the *a-posteriori* correction of calibration errors (self-calibration, [11]) have been shown to further increase the robustness of the reconstruction with respect to experimental uncertainties of the optical system (i.e., relative camera position).

2.1 Tomographic Reconstruction Algorithm

The novel aspect introduced with the tomographic-PIV technique is the reconstruction of the 3D particle distribution by tomography, which therefore deserves discussion in some more detail. Before the 3D intensity distribution can be reconstructed it is necessary to produce a linear mathematical model of the imaging system. In the present model (Fig. 2) the measurement volume is discretized as a 3D array of cubic voxel elements with physical coordinates (X, Y, Z). The voxel intensity distribution is then defined as $E(X, Y, Z)$. Then, the projection of this volume intensity $E(X, Y, Z)$ onto the ith image pixel with image coordinates (x_i, y_i) returns the pixel intensity $I(x_i, y_i)$, which is known from the recorded images. The relation between the 3D light intensity and the image intensity is written in the form of a linear equation as:

$$\sum_{j \in N_i} w_{i,j} E(X_j, Y_j, Z_j) = I(x_i, y_i) , \tag{1}$$

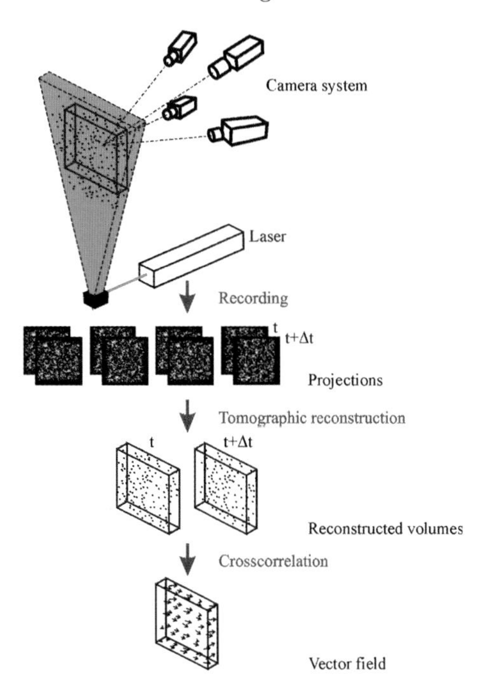

Fig. 1. Principle of tomographic PIV

where N_i is the voxels neighborhood (typically a cylinder of 3×3 voxels cross section) around the line of sight of the ith pixel (x_i, y_i) through the volume. The weighting coefficient $w_{i,j}$ describes the contribution of the jth voxel with intensity $E(X_j, Y_j, Z_j)$ to the pixel intensity $I(x_i, y_i)$ and is calculated as the intersecting volume between the voxel and the line of sight (having the cross-sectional area of the pixel) normalized with the voxel volume. A unit ratio between the size of the pixel projection and the voxel element is assumed for simplicity, however, different values may be allowed. In particular, values lower than one have the advantage of reducing the size of the discretized volume for reconstruction and interrogation.

The iterative tomographic reconstruction algorithm MART (multiplicative algebraic reconstruction technique, [7]) is used to solve the system of (1). Starting from an initial guess ($E(X, Y, Z)^0$ uniform and nonzero) the object $E(X, Y, Z)$ is updated at each full iteration in a loop over each pixel i from all cameras and each voxel j as:

$$E(X_j, Y_j, Z_j)^{k+1} = E(X_j, Y_j, Z_j)^k \left(I(x_i, y_i) \Big/ \sum_{j \in N_i} w_{i,j} E(X_j, Y_j, Z_j)^k \right)^{\mu w_{i,j}}, \quad (2)$$

where $\mu \leq 1$ is a scalar relaxation parameter (usually set to 1 for fastest convergence). The magnitude of the update is determined by the ratio of the measured pixel intensity I with the projection of the current object

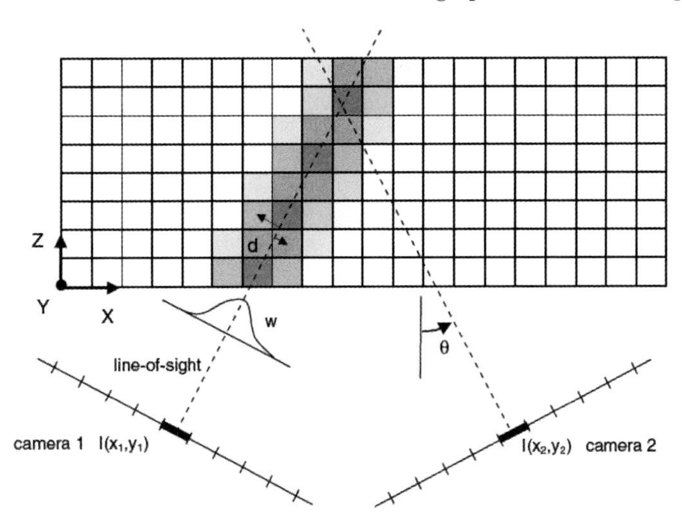

Fig. 2. Representation of the imaging model used for tomographic reconstruction. In this *top view* the image plane is shown as a line of pixel elements and the measurement volume is a 2D array of voxels. The *gray level* indicates the value of the weighting coefficient ($w_{i,j}$) in each of the voxels with respect to the pixel $I(x_1, y_1)$

$\sum_{j \in N_i} w_{i,j} E(X_j, Y_j, Z_j)^k$. The exponent again ensures that only the elements in $E(X, Y, Z)$ affecting the ith pixel are updated. Furthermore, the multiplicative MART scheme requires that E and I are positive-definite.

The iterative process has a relatively rapid convergence and can be stopped after 4 to 5 iterations without significant loss of accuracy. This is particularly interesting since the reconstruction procedure is computationally intensive. Moreover, experimental results show that additional iterations do not yield any significant change in the measured velocity vector field.

2.2 Numerical Assessment of Performances

The performance of the tomographic method was assessed by numerical simulations. Details are given in [6]. A domain with reduced dimensionality was adopted to simplify the computation and to ease the evaluation of results. Therefore, the 3D volume was reduced into a 2D slice and the images from the 2D CCDs were replaced by intensity profiles as issued by linear CCD arrays. This situation corresponds to that displayed in Fig. 2. A 2D particle field in the coordinate system (X, Z) of $50 \times 10\,\text{mm}^2$ is recorded by linear array cameras with 1000 pixels each. However, the qualitative information obtained by this procedure can be directly applied to the case of a 3D domain imaged by 2D images.

Figure 3 presents a detail of a reconstructed particle field by the MART algorithm. Distinct particles are returned at the correct position with an

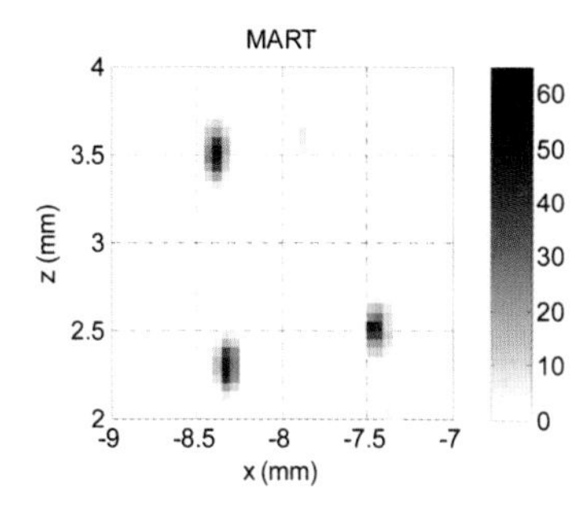

Fig. 3. Detail of the MART reconstruction showing individual particles. *Circles* indicate actual particle positions

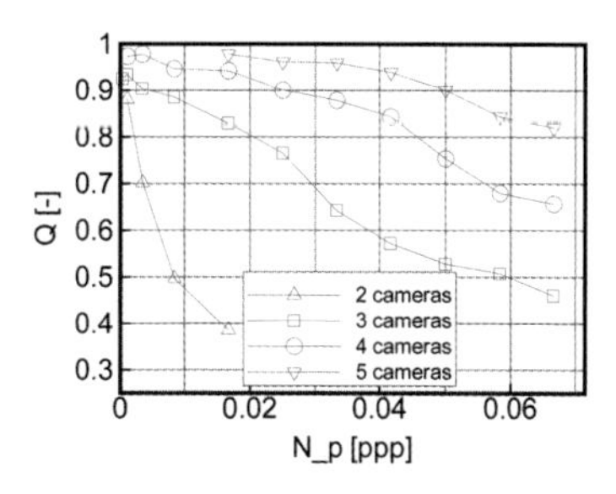

Fig. 4. Reconstruction quality Q as a function of number of cameras and particle image density

intensity distribution slightly elongated in depth depending on the viewing directions of the cameras. The reconstruction also contains a number of false intensity peaks ("ghost particles"), which do not correspond to actual tracers (e.g., $x = -7.8$, $y = 3.6$ in Fig. 3). They occur when the particles cannot be located unambiguously inside the volume [5, 12], and their number increases with the number of actual particles, the particle image diameter and the depth of the measurement volume. However, as is also clear from the figure, the intensity of the ghost particles is low when compared to that of the actual particles and therefore their effect on the crosscorrelation peak and the velocity measurement is limited.

To quantify the relation between the accuracy of the tomographic reconstruction and experimental parameters, the reconstructed object $E_1(X, Z)$ is compared with the exact distribution of light intensity $E_0(X, Z)$ where the particles have been modeled by a Gaussian intensity distribution. The reconstruction quality Q is defined as the normalized correlation coefficient of the exact and reconstructed intensity distribution according to:

$$Q = \frac{\sum_{X,Z} E_1(X,Z) \cdot E_0(X,Z)}{\sqrt{\sum_{X,Z} E_1^2(X,Z) \cdot \sum_{X,Z} E_0^2(X,Z)}} . \tag{3}$$

A direct estimate of the correlation coefficient to be expected in the cross-correlation analysis of the reconstructed objects is given by Q^2.

The parametric study shows that the accuracy of the tomographic reconstruction is largely determined by two experimental parameters: the particle image density (in particles per pixel, ppp, in the recorded images) and the number of cameras. The former is a measure of the amount of signal contained in the measurement volume and the latter expresses how many 2D views are available of the given 3D particle distribution. Figure 4 shows the reconstruction quality Q as a function of the number of cameras and the particle image density. The results of the simulation performed in a 2D domain have been translated to the case of a 3D measurement taking into account the additional dimension by dividing the particle image density by the particle image diameter (3 pixels). For a seeding density of 0.05 ppp the 2-camera system proves to be largely insufficient, whereas Q rapidly increases going to 3, 4 and 5 cameras. The additional camera provides additional information on the object, which increases reconstruction accuracy. Conversely, for a fixed number of cameras increasing the particle density produces a larger percentage of ghost particles, consequently decreasing the reconstruction quality. On the other hand, a larger number of particles allows a higher spatial sampling rate of the flow, returning a higher spatial resolution, which is usually desired. Based on the simulation results (Fig. 4) the maximum imaged particle density yielding an acceptable reconstruction quality ($Q > 0.75$) is 0.05 particles per pixel (50 000 particles per megapixel) for a 4-camera system, which is rather advantageous with respect to particle-detection and -tracking procedures typically yielding 5000 to 10 000 particles per megapixel [5, 13].

Furthermore, background light and image noise have a strong impact on the reconstruction, because the background light will be reconstructed inside the measurement volume affecting the accuracy of the particle pattern. As a result, the reconstruction quality Q strongly deteriorates with increasing random image noise. However, image preprocessing operations, such as background subtraction, can be applied to reduce the mentioned effects. Furthermore, intensity normalization and the convolution with a 3×3 kernel with Gaussian distribution are commonly applied in real experiments to account for differences in sensitivity between cameras and to further reduce uncorrelated pixel noise.

A 3D numerical simulation of a vortex ring flow has been performed to validate the procedure and to assess the velocity measurement accuracy. The $35 \times 35 \times 7\,\text{mm}^3$ measurement volume contains 24 500 tracer particles, which are imaged onto four 700×700 pixel cameras resulting in a particle image density of 0.05 ppp. The maximum particle displacement between the two exposures is 2.9 voxels. The volume is reconstructed at $700 \times 700 \times 140$ voxels resolution returning a reconstruction quality factor Q of 0.75 as predicted in Fig. 4. The measured displacement field contains $66 \times 66 \times 10$ vectors, which are obtained from a crosscorrelation analysis of the reconstructed volumes with 41^3 interrogation volumes at 75 % overlap. The vector field is shown in Fig. 5, left, where the overall motion is well captured. Furthermore, the isovorticity surface shows the expected torus shape.

The crosscorrelation coefficient (> 0.56) and the average signal-to-noise ratio (first-to-second highest peak) approaching 5 indicate a high confidence

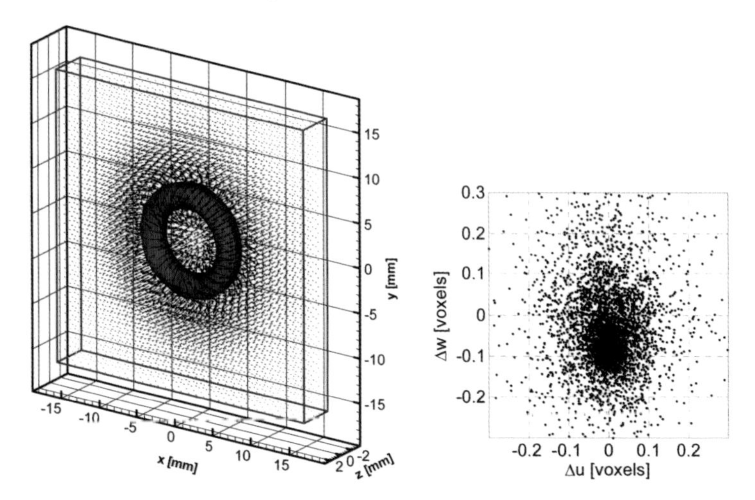

Fig. 5. Velocity field and isovorticity surface (0.13 voxels/voxel) from the simulated vortex ring (*left*) and scatter plot of the measurement error for the u-component and depth component (w) of the particle displacement (*right*)

level for the measurement. The final measurement accuracy is presented in a scatter plot of the error in the u- and w-component of the particle displacement (Fig. 5, right), where 90 % of the vectors have an absolute error smaller than 0.10 voxels for the u- and v-component and less than 0.16 voxels for the w-component. The larger uncertainty in depth is ascribed to the viewing configuration with angles of 30° wrt the z-axis, which agrees with former studies on stereoscopic PIV. Furthermore, the asymmetry on the w-component should not be interpreted as a bias error resulting from the tomographic reconstruction, because it also appears when the exact light distribution $E_0(X, Y, Z)$ is crosscorrelated and results from the combination of the limited spatial resolution (modulation error) and the asymmetric distribution of the w-component within the measured volume.

3 Applications to Circular Cylinder Wakes

3.1 Experimental Procedure

The first application of the tomographic-PIV technique has been performed measuring the cylinder wake flow in a low-speed open-jet wind tunnel of Delft University of Technology [6, 12, 14]. The wind tunnel has a $40 \times 40\,\text{cm}^2$ square crosssection in which a circular cylinder of 8 mm diameter is placed horizontally or vertically (Fig. 6). The free-stream velocity is 5 m/s with a corresponding Reynolds number of 2700 based on the cylinder diameter.

The flow is seeded with 1-μm droplets at a concentration resulting in a particle image density in the range of 0.02–0.08 ppp. The illumination source is a Quanta Ray double-cavity Nd:YAG laser from Spectra-Physics with a pulse energy of 400 mJ. A slit is added in the path of the laser lightsheet cutting the low-intensity side lobes from the light profile to obtain a sharply defined illuminated volume of 8 mm thickness. The time separation between exposures is 35 µs, yielding a free-stream particle displacement of 0.18 mm corresponding to 3.2 voxels. The particle images are recorded at 18.3 pixels/mm resolution using 4 CCD cameras (PCO Sensicam QE and LaVision Imager Intense) with 1376×1040 pixels and 12-bit quantization. The views are arranged from the vertices of a rectangle as shown in Fig. 6. The viewing directions with respect to the cylinder axis (z-direction) are $\pm 22°$ in the vertical direction and $-10°$ and $+20°$ in the horizontal directions. The cameras are equipped with Nikon lenses using $f/8$ and Scheimpflug adapters to have the entire volume within the depth-of-focus. The imaging system is calibrated by recording images of a plate with 15×12 marks (crosses) at three depth positions separated by 4 mm. At each calibration Z-location the relation between the physical coordinates (X, Y) and image coordinates is described by a 3rd-order polynomial fit with an accuracy of approximately 0.2 pixels. Linear interpolation is used at intermediate Z-locations.

The reconstruction process is improved by means of image preprocessing with background-intensity removal, particle-intensity equalization and a Gaussian smooth (3×3 kernel size), as described in the previous section.

The intensity distribution in the measurement volume is reconstructed at 18.2 voxels/mm resolution using the MART algorithm with 5 iterations. The dimensions of the reconstructed volume are 36.5(length) $\times$ 35.8(height) $\times$ 11(depth) mm corresponding to $667 \times 654 \times 203$ voxels. Note that the reconstructed volume depth is slightly larger than the lightsheet thickness to ensure that the complete lightsheet is contained, which is important for accurate reconstruction. Figure 7 shows an example of a reconstructed volume. The lightsheet position can be clearly identified within the reconstructed volume due to the application of a slit in the light path. Further analysis of the reconstructed volume yields an estimated ratio of actual particles over ghost particles of 2 for a particle image density of 0.05 ppp [12].

The particle displacement is obtained using a 3D FFT-based crosscorrelation algorithm with iterative multigrid and window deformation [8]. The analysis returned $77 \times 79 \times 15$ velocity vectors using an interrogation volume size of 31^3 voxels with 75 % overlap. Data validation based on a signal-to-noise ratio threshold of 1.2 and on the normalized median test with a maximum threshold of 2 [15] returns 5 % spurious vectors. The average signal-to-noise ratio and normalized correlation coefficient are 3.3 and 0.6, respectively, for a particle image density of 0.05 ppp.

3.2 Results

An instantaneous velocity and vorticity field is shown in Fig. 8, left. The separated shear layer is visible in the form of a vorticity sheet emanating from

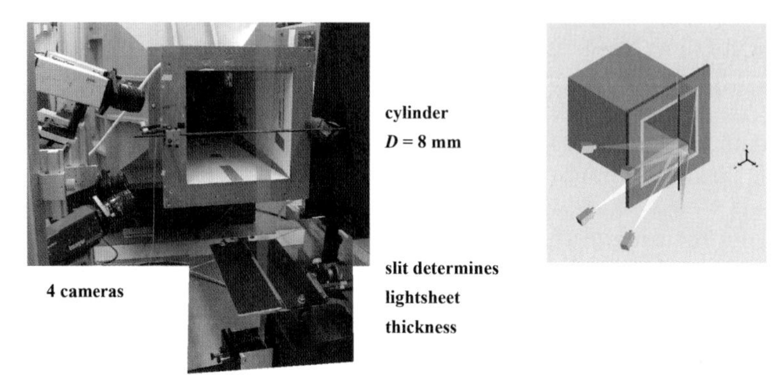

Fig. 6. The optical arrangement for the tomographic-PIV experiments in the low-speed wind tunnel. *Left:* horizontal-cylinder experiments. *Right:* schematic of vertical-cylinder experiments

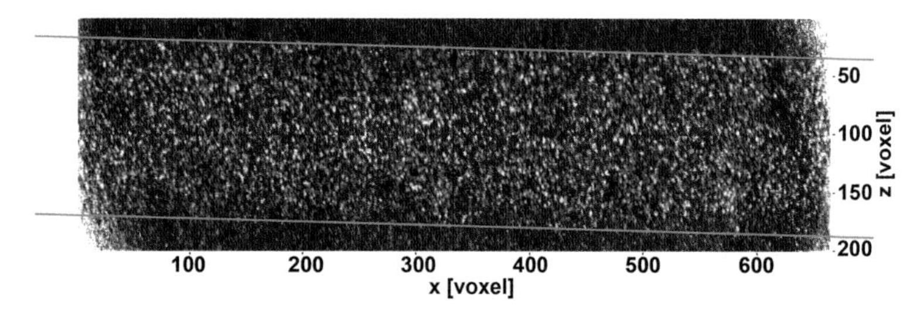

Fig. 7. Top view of the reconstructed volume showing the light intensity integrated in the y-direction. The *green lines* indicate the position of the lightsheet

the upper and lower sides of the cylinder. On the bottom side of the wake a counterclockwise roller is being formed as also indicated by the swirling pattern of the instantaneous velocity vectors. The previously shed primary roller can be identified with the roll-up of the vorticity sheet on the upper side just downstream of the first primary vortex. Finally, a third Kármán vortex with the same rotation sign as the first one is visible downstream at the bottom of the measurement volume. Even in this relatively thin view of the cylinder wake a significant three-dimensional behavior can be observed with a secondary roller interconnecting the second and the third vortex oriented approximately at 45° and exhibiting a vorticity level comparable with the primary rollers. At the present Reynolds number the shear layers separating from the cylinder are transitional and three-dimensionality on the scale of the Kármán vortices is expected [16]. The isosurfaces of vorticity stretching vector magnitude ($|\vec{\omega} \cdot \overline{\overline{\nabla V}}|$, Fig. 8, right) show that the stretching activity is concentrated in the core of the secondary roller and indicates that these structures are responsible for the increase and reorientation of the vorticity between the main rollers.

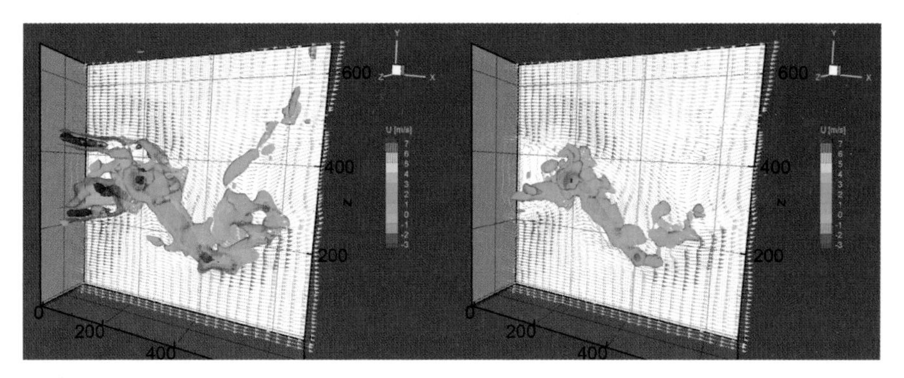

Fig. 8. Instantaneous flow-field snapshot. *Left:* vorticity vector magnitude isosurfaces ($|\omega| = 2 \times 10^3\,\mathrm{s}^{-1}$, *green*; $|\omega| = 4 \times 10^3\,\mathrm{s}^{-1}$, *red*) and velocity vectors in the midsection of the measurement volume. Vectors color code the streamwise velocity component. *Right:* vorticity stretching vector magnitude isosurfaces ($|\vec{\omega} \cdot \overline{\overline{\nabla V}}| = 5 \times 10^6\,\mathrm{s}^{-2}$, *green*; $|\vec{\omega} \cdot \overline{\overline{\nabla V}}| = 15 \times 10^6\,\mathrm{s}^{-2}$, *red*)

To improve the visualization of the structural organization of the flow the spanwise and the combination of streamwise and y-component of vorticity are color-coded as shown in Fig. 9. The four uncorrelated snapshots show consecutive phases of the vortex-shedding cycle. The transition from snapshot (a) to (b) marks the separation of a Kármán vortex from the upper shear layer with stretching of the secondary vortex between this Kármán vortex and the new vortex forming from the lower shear layer. This observation is consistent with the large value for the vorticity stretching magnitude at this location in Fig. 8, right, which is computed from snapshot (a). The snapshots (c) and (d) show further stretching and curling of the secondary vortices around the Kármán vortex as it is convected to $x/D = 4.2$ and 4.4, respectively. Moreover, the normalized vorticity level in the Kármán vortices $\omega_z D/u_\infty = 2.2$ agrees closely with planar PIV measurements in the Reynolds number range 2000 to 10 000 from [17], who report an average normalized peak vorticity of 2.1 at $x/D = 3$.

The spanwise organization of the flow structure becomes more visible from the vertical-cylinder experiments (Fig. 10). The secondary vortex structures (blue and red depending on the orientation of vorticity in the streamwise direction) appear to be organized in counterrotating pairs yielding a quasiperiodic behavior. From a visual inspection of 100 snapshots the normalized spatial wavelength λ_z/D is estimated at 1.2 [14], which is in good agreement with [17] and slightly in excess of the reported values at lower Reynolds number [16]. The effect of the secondary structures is to first distort the primary Kármán vortices, as clearly seen in Fig. 10, right at $x/D = 4.5$, and finally cause breakup of the primary vortices.

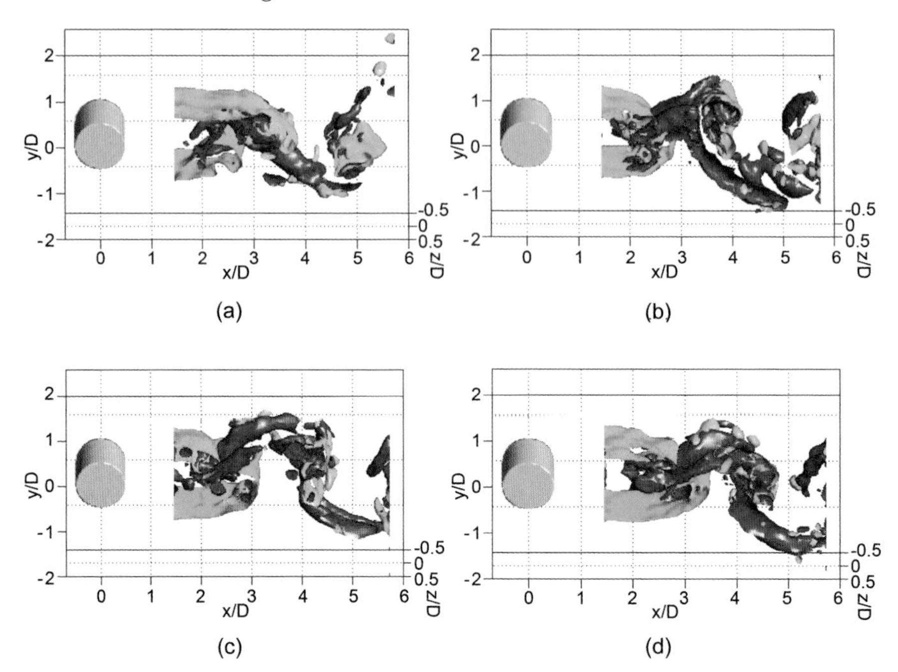

Fig. 9. Instantaneous vorticity isosurfaces showing four consecutive phases of the vortex shedding cycle ($\omega = 1.4 \times 10^3\,\mathrm{s}^{-1}$. Color coding: *cyan* $\omega_z < 0$; *green* $\omega_z > 0$; *blue* $\sqrt{\omega_x^2 + \omega_y^2} \cdot \mathrm{sign}(\omega_x) < 0$; *red* $\sqrt{\omega_x^2 + \omega_y^2} \cdot \mathrm{sign}(\omega_x) > 0$

The returned spatial organization of the flow and vorticity level compare well with what is reported in the literature, as shown above, so that the current results demonstrate that the instantaneous structures in the flow can be captured in a 3D volume with tomographic PIV.

3.3 An Experimental Assessment: Comparison with Stereo-PIV

In the same experimental configuration the accuracy of the tomographic-PIV technique is investigated by comparing the flow statistics with measurements made with stereoscopic PIV. The ensemble for both techniques consists of 100 snapshots obtained in the horizontal-cylinder configuration (Fig. 6, left). The limited ensemble size is due to the fact that tomographic reconstruction and interrogation operations are time consuming. Stereo-PIV results are obtained using two cameras from the above imaging system and a 2-mm thick lightsheet. Self-calibration on the particle images [11] is applied to eliminate lightsheet misalignment errors in stereo-PIV. To compare the results, vertical profiles (at $x/D = 2.6$) of the mean and RMS fluctuations of the streamwise velocity component are extracted from the data (Fig. 11). The overall agreement between stereo- and tomographic PIV is good, with the largest

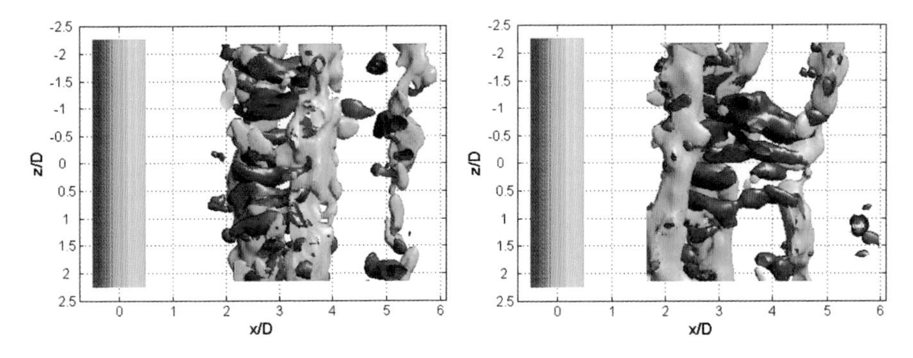

Fig. 10. Instantaneous vorticity isosurfaces from vertical-cylinder experiments. Color coding as in Fig. 9

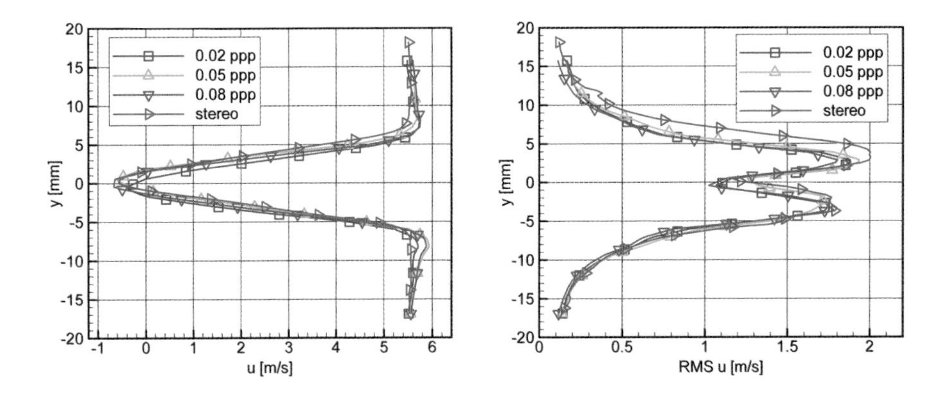

Fig. 11. Profiles of the u-component of velocity at $x/D = 2.6$, mean (*left*) and RMS (*right*). Comparing tomographic-PIV at different seeding densities with stereo-PIV

difference in the mean u of 0.50 m/s (corresponding to 0.30 voxel particle displacement). The RMS profiles for the u-component agree within 0.30 m/s. *Wieneke* and *Taylor* [18] confirmed this level of agreement based on measurements of a vortex ring. It is interesting to observe that the tomographic results show very little variations with respect to the particle image density in the range of 0.02 to 0.08 ppp for the present experiment. Further details of this experimental assessment are given in [12].

3.4 Time-Resolved Measurements in Water Flow

The dynamics of the primary and secondary vortices in the cylinder wake are investigated further by means of time-resolved tomographic-PIV experiments, which were carried out in a water flow at several $\mathrm{Re_D}$ ranging between 180 and 2520 with $D = 12\,\mathrm{mm}$. A more detailed description of the experiments can be found in [19]. The cylinder axis was aligned with the 16-mm thick lightsheet as in the previous section (Fig. 6, right). Due to the low velocity in water (of the order of a few cm/s) a recording rate of 10 image pairs/s proved to be sufficient to follow the time evolution of the large scales in the flow. The 30-μm latex particles were illuminated with a Quantel CFR-200 laser providing pulses of 200 mJ at a rate up to 30 Hz. Four high-resolution CCD cameras (Imager Pro X, 2048 × 2048 pixels, 14-bit) were used to record the particle images. The crosscorrelation analysis with 48^3 voxel interrogation volumes at 75 % overlap returned $174 \times 117 \times 32$ vectors in the $88(\mathrm{L}) \times 59(\mathrm{H}) \times 16(\mathrm{D})\,\mathrm{mm}^3$ measurement volume.

A sample of the time evolution of the vortical flow structures is shown in Fig. 12 for the case of $\mathrm{Re}_\mathrm{D} = 540$. Six snapshots separated by 200 ms (15 % of the shedding period) each are presented. The vorticity magnitude in the secondary flow structures is comparable with or higher than that of the Kármán vortices, which are no longer visible downstream of $x/D = 4$ at the present level of display. The secondary vortices, however, show a significant persistence even up to the edge of the observable domain, as shown by the evolution of three of those structures in Fig. 12 (indicated by the arrows in the first and last snapshot). The streamwise coherence length of the secondary vortices locally exceeds the distance between the main rollers, indicating a 3D organization that can be assimilated to Mode B [16]. However, some uncontrolled disturbances in the free stream do not allow us to draw conclusions about the flow statistical properties.

4 Application to Turbulent Boundary Layers

4.1 Coherent Motion

A fully developed turbulent boundary layer at $\mathrm{Re}_\theta = 1900$ ($\delta_{99} = 24\,\mathrm{mm}$ and 9.7 m/s free-stream velocity) was investigated with the objective of inferring the 3D spatial organization of the flow along its entire height. A complete description of the results is given by [20]. The boundary layer develops along the bottom wall of the low-speed tunnel, where the point of transition is fixed by a tripping wire placed 1 m upstream of the measurement area.

The measurement of coherent motion in a turbulent boundary layer is a challenging task, because it requires a high dynamic range and good spatial resolution in order to resolve the velocity fluctuations. Moreover, the measurement in the proximity of solid interfaces poses several technical problems associated with stray light reflection. While the RMS fluctuations in the cylinder wake are of the order of the free-stream velocity, in the boundary layer they reduce to approximately 5 % of the mean velocity. Hence, the time separation between the laser pulses was set at 100 μs, yielding a 26-voxel particle displacement in the free stream. The resolution in the measurement

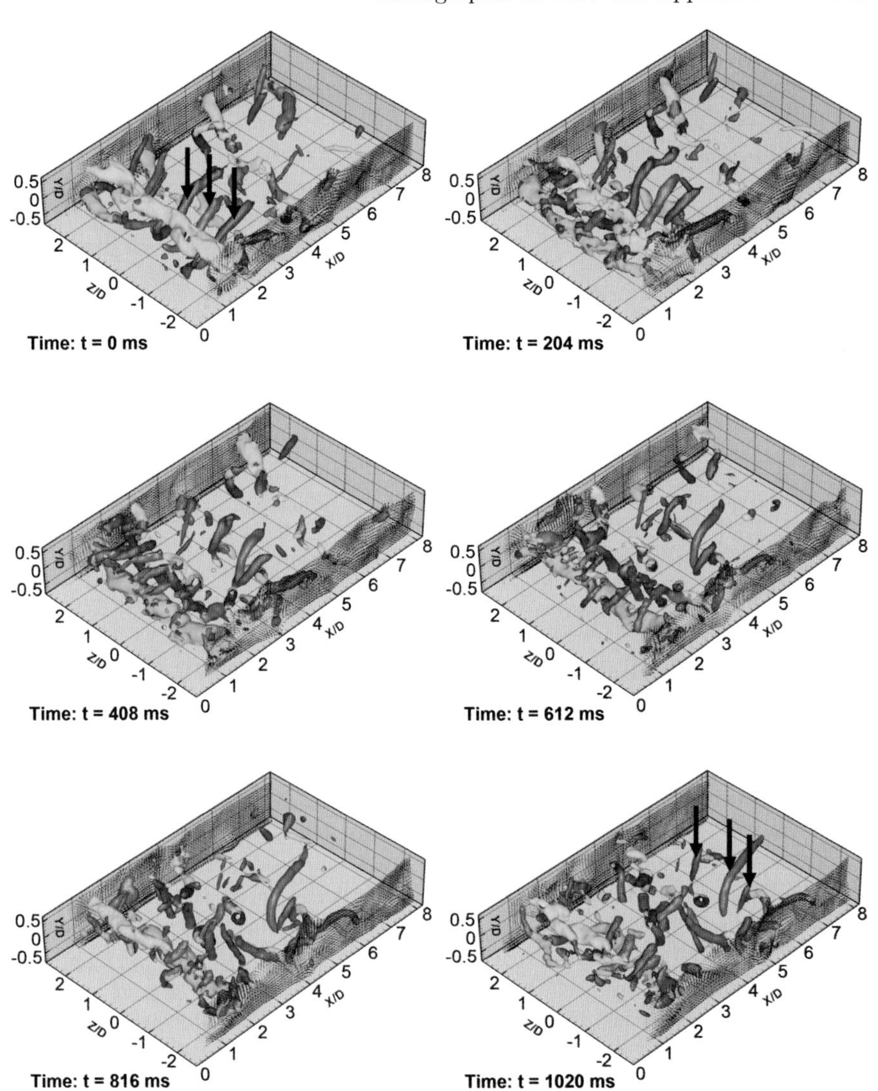

Fig. 12. Time series at $\mathrm{Re_D} = 540$ showing vortical structures detected using the λ_2-criterion. Color coding depending on the dominant component: when ω_z is dominant the color is cyan ($\omega_z > 0$) or green ($\omega_z < 0$); when $\sqrt{\omega_x^2 + \omega_y^2}$ is dominant it is blue or red depending on the sign of ω_x

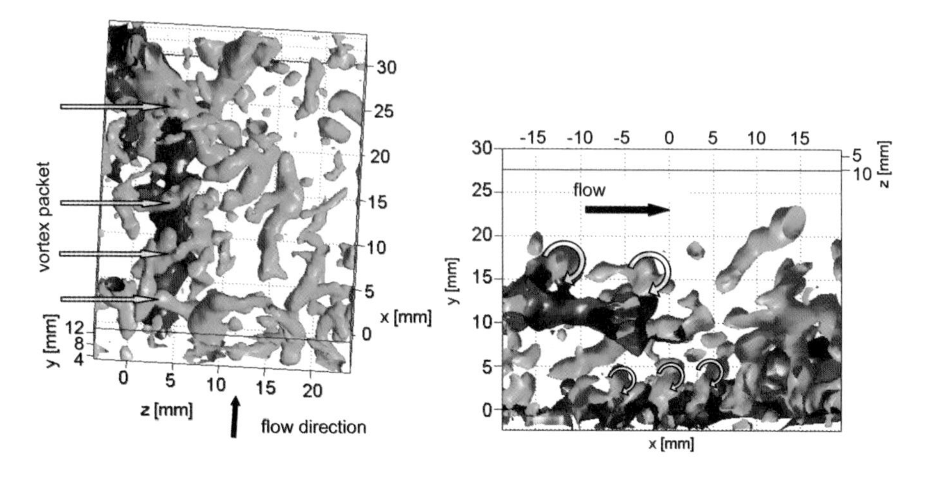

Fig. 13. Instantaneous vortex structures (Q-criterion) in *green* and low-speed regions in *blue* ($u < 85\,\%$ of local mean) in a volume parallel to the wall (*left*) and normal to the wall (*right*)

volume was 26.6 voxels/mm resulting in a vector spacing of 0.47 mm (applying 50^3 crosscorrelation volumes at 75 % overlap).

Two examples of the measured spatial organization of the vortex structures and low-speed regions are presented in Fig. 13. Vortical structures are identified by means of the Q-criterion [21], whereas low-speed streaks are defined with a threshold of $u < 85\,\%$ of local mean velocity. Figure 13, left, shows a measurement volume with the largest dimensions parallel to the wall where the x-axis corresponds to the streamwise direction, y is the distance from the wall and z is the spanwise direction. A range of vortex structures can be observed such as streamwise vortices (e.g., $5 < x < 20\,\text{mm}$, $z = 15\,\text{mm}$) and arch cane or hairpin vortices, which can be both symmetrical (e.g., $x = 0\,\text{mm}$, $10 < z < 17\,\text{mm}$) or asymmetrical (e.g., $x = 17\,\text{mm}$, $z = 12\,\text{mm}$). Furthermore, a series of four hairpin-like vortices aligned in the streamwise direction with a low-speed region between their legs is observed at $z = 4\,\text{mm}$, which may be recognized as a hairpin packet as introduced by [22]. The length of this packet is 25 mm or 1.3δ and its width is approximately 5 mm, corresponding to 100 viscous length scales. More hairpin packets are shown in Fig. 13, right, where the measurement volume depth is oriented in the spanwise direction (z) such that the whole boundary-layer height is included in the volume. Also in this case, a packet of three hairpin vortices is observed close to the wall. Arrows indicate the orientation of vorticity in the hairpin heads, which are separated in the streamwise direction by approximately 100 viscous length scales. Downstream of the packet and shifted in the z-direction more hairpin structures are detected, however, it is not evident whether they also belong to or interact with the packet upstream. Further from the wall at $y = 15\,\text{mm}$

two hairpins (indicated by the arrows) seem to form another packet. Away from the wall both the size of the individual vortices and their spacing in the streamwise direction increase as expected.

The statistical relevance of each of the above structures remains to be explored, but the presented results show that these larger structures within the boundary layer can be observed directly without looking at their signatures as in planar and point methods, which is an important advantage of 3D PIV methods in turbulence research.

4.2 Time-Resolved Measurements of a Turbulent Boundary Layer and Spot in Air

In a recent feasibility study [23] tomographic PIV has been applied to time-resolved PIV recordings of a turbulent spot and a turbulent boundary layer in air. The turbulent spot is generated in a laminar boundary layer flow over a flat glass plate with elliptical leading edge in the 1-m wind tunnel of DLR Göttingen at a free-stream velocity of 7 m/s with zero pressure gradient (Fig. 14, right). With a short 1-ms local flow injection at $(\mathrm{Re}_x)^{0.5} = 300$ an intense initial disturbance is introduced into the laminar boundary layer resulting in the growth of a turbulent spot, which is measured downstream at $(\mathrm{Re}_x)^{0.5} = 450$ in a 34(L) $\times$ 30(H) $\times$ 19(D) mm^3 volume. The turbulent boundary layer is generated by a series of tripping wires

The experimental arrangement and procedure are similar to the experiments presented above with two important modifications. First, a high repetition rate PIV hardware is used, which consists of four Photron APX-RS CMOS cameras and a high repetition rate Nd:YAG laser from Lee Laser Inc with a pulse energy of 21 mJ at 5 kHz. The high-speed system is capable of recording 800 $\times$ 768 pixel images at 5 kHz without frame straddling. Secondly, the light scattered by the 1.5-μm oil particles is increased by adding two coated and highly reflective dielectric mirrors to the laser optics reflecting the light back and forth through the test section (see sketch in Fig. 14, left), which effectively increased the laser pulse energy from 21 mJ to an equivalent of 150 mJ. The mirrors are placed on both spanwise sides outside the free stream so as not to disturb the flow. Optimization of the particle illumination is crucial under the present conditions, since laser power and camera sensitivity are limited, while the illuminated region is thick.

The 4D data with approximately 48 000 instantaneous velocity vectors for each volume measured with 5 kHz and 4 kHz provides all main flow structure information, which has often been desired in former investigations of wall-bounded turbulent flows. The number of spurious vectors is less than 3 % for the considered cases.

Four instantaneous velocity fluctuation vector volumes separated in time by 500 μs each are represented in Fig. 15 by 3D isovorticity contour surfaces and two 3C-velocity vector fields in xy-planes with 2.5 mm distance. The shown vectors are with respect to the nonconverged average velocity profile

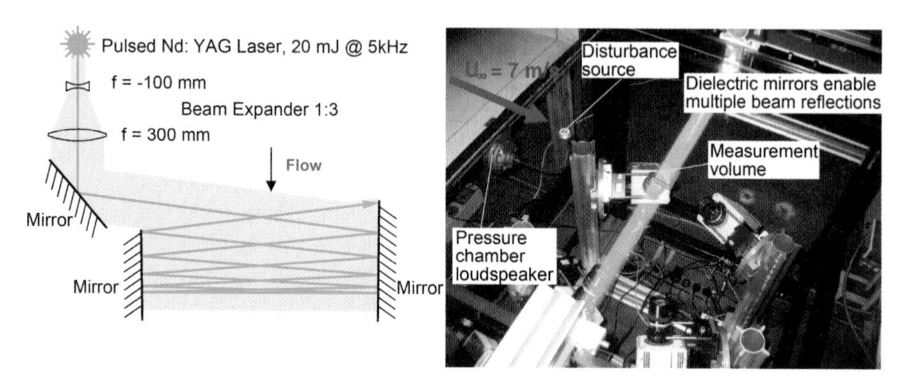

Fig. 14. Sketch of the optical setup that enables the optimal usage of the laser pulse energy of only 21 mJ per pulse for illumination of the PIV measurement volume (*left*). Experimental arrangement (*right*). The 4 high-speed cameras are located underneath the glass plate

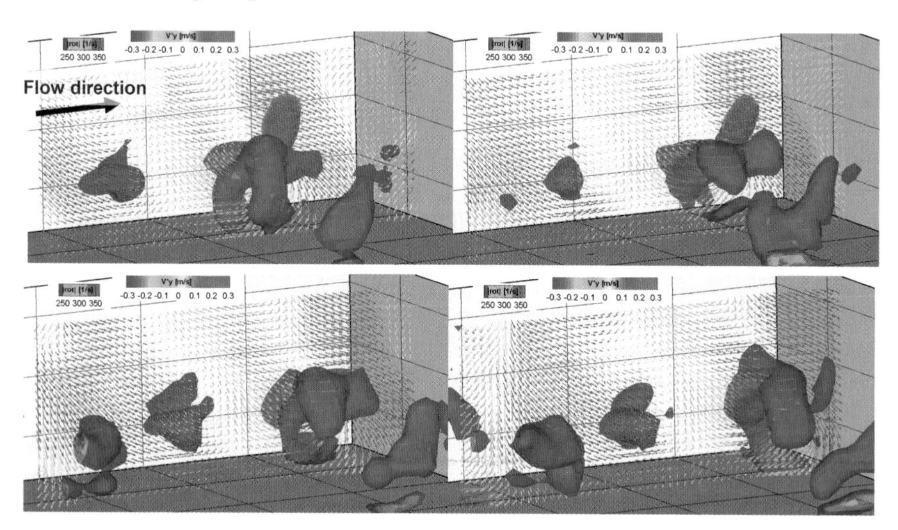

Fig. 15. Time series of instantaneous isovorticity surfaces and two xy-planes of 3C fluctuation velocity vector fields with $t = 500\,\mu s$ shift each in a tripped turbulent boundary layer

over only 70 volumes. Arch or hairpin structures with negative relative velocities u' are found with a strong Q2 event ($u' < 0$; $v' > 0$) region located directly upstream of this vortex structure and a Q4 event ($u' > 0$; $v' < 0$) region located directly downstream. Following the flow development in time the Q4 event downstream of the mentioned arch turns into a Q3 event ($u' < 0$; $v' < 0$) close to the wall, which guide the flow in between the arch into an upstream Q2 event. This Q2 event influences the upstream flow so that a deflected Q4 event leads to another young arch structure following the path of the downstream one. Here, almost the same velocity distribution can be found up- and downstream of this structure.

In Fig. 16 a series of three velocity vector volumes with a time difference of 400 µs in between demonstrates the explosive growth of a Q2 event and related arch or hairpin-like structure at the trailing edge (TE) of the turbulent spot. The preprocess for the rapid growth of this new turbulence producing structure is a deflected Q4 event, which interacts with the downstream flow structures of the spot with much lower velocities. This process is visible in the related part of the vector volume (not shown here). This blockage for the Q4 flow results in a deflection and change of the relative flow direction. The Q4 event with high velocities transforms into a slightly spanwise directed Q3 event, which then hits the near-wall region. At this position only a Q2 event can help to "satisfy the continuity equation". The fluid rapidly shoots upwards against the relative flow velocity while convecting downstream and the precondition for the start of a new similar process is given. The effect of the rapid growth of Q2 events has also been shown by *Schröder* and *Kompenhans* [24] by means of multiplane stereo-PIV, but the number of time steps per measurement plane was limited to two, so that the whole complex process could not be fully described nor understood. This scenario has been recognized also at other TEs of spots in the tomographic PIV dataset and can be found in a more complex distribution also inside the center of the spot structure. Of course, the present observations are based only on a few velocity vector volumes, but the similarities found in this investigation can guide future work.

5 Summary

Tomographic PIV is a recently developed technique for 3D velocity measurements, in which the three-dimensional particle distribution within the measurement volume is reconstructed by optical tomography from particle image recordings taken simultaneously from several viewing directions. The optical arrangement for illumination and recording resembles that of stereoscopic PIV and 3D-PTV. Velocity information results from three-dimensional particle pattern crosscorrelation of two volume reconstructions obtained from subsequent exposures of the particle images. The measurement chain is therefore fully digital with the advantage of being instantaneous within a volume,

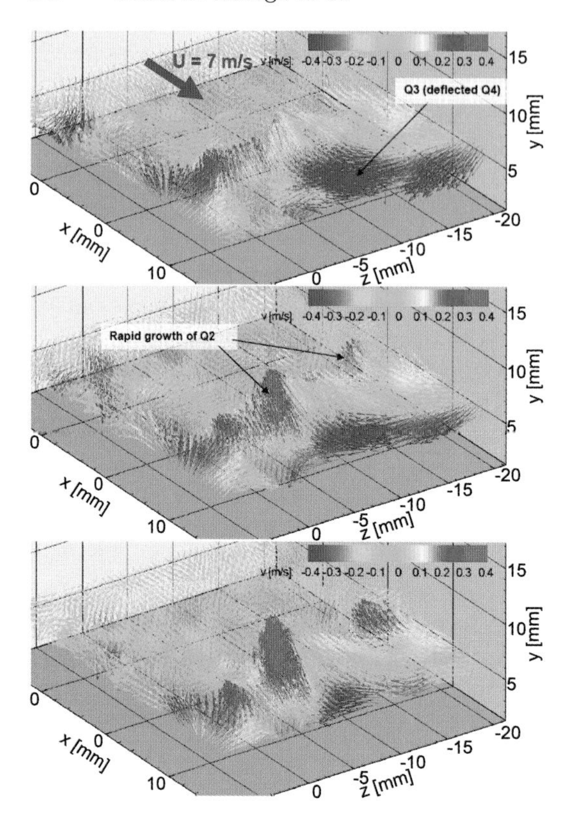

Fig. 16. Three instantaneous 3D-3C-velocity vector fields at $y = 5.6$, 6.1 and 6.6 mm of the same turbulent spot at $t = 87.4$ (*top*), 87.8 (*middle*) and 88.2 ms (*bottom*) after an initial disturbance showing a rapid growth of Q2 events or hairpin-like vortices at the trailing edge ($u_{\mathrm{ref}} = u - 6.6\,\mathrm{m/s}$). The colors represent the wall-normal velocity component

which makes it suitable for the analysis of flows irrespective of the flow speed. The parameters dominating the accuracy of the reconstruction procedure are the particle seeding density and the number of viewing cameras. However, the accuracy of the velocity measurement by crosscorrelation has been demonstrated not to be sensitive to the seeding density, which makes the technique particularly robust.

Using high repetition rate PIV equipment the requirements for time-resolved experiments in air flows at a moderate speed can be matched. A first application of this type was presented in Sect. 4.2 showing a turbulent boundary layer in air measured at 5 kHz. A specific illumination arrangement with multiple reflections of the light beam was needed to increase the local light intensity by a factor 7.

The chapter presents four applications of the technique to turbulent wake flows and boundary layers in both air and water describing the experimental

Table 1. Applications overview with some experimental parameters

	Cylinder wake in air	Time-resolved cylinder wake in water	Turbulent boundary layer in air	Time-resolved boundary layer in air
Image size (pixels)	1376×1040	2048×2048	1376×1040	800×768
Recording rate (Hz)	1	10	2	5000
Measurement volume, $L \times H \times D$ (mm^3)	$37 \times 36 \times 8$	$88 \times 59 \times 16$	$33 \times 26 \times 11$	$34 \times 30 \times 19$
Spatial resolution (voxels/mm)	18.2	23.6	26.6	24
Particle concentration (particles/mm^3) (total number)	2.1 (23 000)	1.2 (98 000)	3.3 (32 000)	0.94 (18 000)
Number of vectors* (total)	$77 \times 79 \times 15$ (91 000)	$174 \times 117 \times 32$ (651 000)	$69 \times 54 \times 25$ (93 000)	$46 \times 41 \times 24$ (45 000)

* using 75 % overlap between interrogation volumes

arrangements and briefly discussing the results. All applications use a four-camera system typically returning 50k to 100k velocity vectors using $1k \times 1k$ cameras or up to $650k$ vectors using $2k \times 2k$ cameras. An overview of the measurement volumes, vector yields and related experimental parameters is given in Table 1. The accuracy of the velocity vector is close to that of standard PIV as assessed through synthetic experiments and by direct comparison with stereo-PIV in a real experiment.

The 3D flow organization of a Kármán wake was captured with details describing both the vortex shedding and the secondary vortex structures in the cylinder wake. A complete 3D description of hairpins and arch vortices was possible from the applications to boundary layers. Also, the interaction between low-speed regions and hairpin packets in the turbulent boundary layer was visualized. The evolution and interaction of these 3D structures is still not fully understood and therefore the application of 3D PIV techniques to those fluid-dynamic problems is expected to give important new insight in the future.

One of the current limitations of the technique is the processing time required for the reconstruction and interrogation (approximately 1 h for 10^8 voxels on a Pentium 4 PC). Therefore, further developments of the tomographic-PIV technique are directed towards the reduction of the computational load by advanced algorithms for reconstruction and 3D correlation with sparse and adaptive correlation approach. Finally, the implementation of a volume self-calibration procedure will result in the improvement of the calibration accuracy.

References

[1] M. P. Arroyo, C. A. Greated: Stereoscopic particle image velocimetry, Meas. Sci. Technol. **2**, 1181–1186 (1991)
[2] C. J. Kähler, J. Kompenhans: Fundamentals of multiple plane stereo particle image velocimetry, Exp. Fluids, Suppl. pp. S70–S77 (2000)
[3] Ch. Brücker: Digital-particle-image-velocimetry (DPIV) in a scanning light-sheet: 3D starting flow around a short cylinder, Exp. Fluids **19**, 255–263 (1995)
[4] K. D. Hinsch: Holographic particle image velocimetry, Meas. Sci. Technol. **13**, R61–R72 (2002)
[5] H. G. Maas, A. Gruen, D. Papantoniou: Particle tracking velocimetry in three-dimensional flows, Exp. Fluids **15**, 133–146 (1993)
[6] G. E. Elsinga, F. Scarano, B. Wienele, B. W. Van Oudheusden: Tomographic particle image velocimetry, Exp. Fluids **41**, 933–947 (2006)
[7] G. T. Herman, A. Lent: Iterative reconstruction algorithms, Comput. Biol. Med. **6**, 273–294 (1976)
[8] F. Scarano, M. L. Riethmuller: Advances in iterative multigrid PIV image processing, Exp. Fluids, Suppl pp. S51–S60 (2000)
[9] R. Y. Tsai: An efficient and accurate camera calibration technique for 3D machine vision, in *Proc. IEEE Conf. on Computer Vision and Pattern Recognition* (1986) pp. 364–374
[10] S. M. Soloff, R. J. Adrian, Z.-C. Liu: Distortion compensation for generalized stereoscopic particle image velocimetry, Meas. Sci. Technol. **8**, 1441–1454 (1997)
[11] B. Wieneke: Stereo-PIV using self-calibration on particle images, Exp. Fluids **39**, 267–280 (2005)
[12] G. E. Elsinga, B. W. Van Oudheusden, F. Scarano: Experimental assessment of tomographic-PIV accuracy, in *13th Int. Symp. on Applications of Laser Techniques to Fluid Mech* (2006) paper 20.5
[13] F. Pereira, H. Stüer, E. C. Graff, M. Gharib: Two-frame 3D particle tracking, Meas. Sci. Technol. **17**, 1680–1692 (2006)
[14] F. Scarano, G. E. Elsinga, B. W. Van Oudheusden: Investigation of 3-D coherent structures in the turbulent cylinder wake using tomo-PIV, in *13th Int. Symp. on Applications of Laser Techniques to Fluid Mech* (2006) paper 20.4
[15] J. Westerweel, F. Scarano: A universal detection criterion for the median test, Exp. Fluids **39**, 1096–1100 (2005)
[16] C. H. K. Williamson: Vortex dynamics in the cylinder wake, Annu. Rev. Fluid Mech. **28**, 477–539 (1996)
[17] J. F. Huang, Y. Zhou, T. Zhou: Three-dimensional wake structure measurement using a modified PIV technique, Exp. Fluids **40**, 884–896 (2006)

[18] B. Wieneke, S. Taylor: Fat-sheet PIV with computation of full 3D-strain tensor using tomographic reconstruction, in *13th Int. Symp. on Applications of Laser Techniques to Fluid Mech* (2006) paper 13.1
[19] D. Michaelis, C. Poelma, F. Scarano, J. Westerweel, B. Wieneke: A 3D time-resolved cylinder wake survey by tomographic PIV, in *12th Int. Symp. on Flow Visualization* (2006) paper 12.1
[20] G. E. Elsinga, D. J. Kuik, B. W. Van Oudheusden, F. Scarano: Investigation of the three-dimensional coherent structures in a turbulent boundary layer with tomographic-PIV, in *45th AIAA Aerospace Sciences Meeting, Reno, NV, USA, AIAA-2007-1305* (2007)
[21] J. C. R. Hunt, A. Wray, P. Moin: *Eddies, stream, and convergence zones in turbulent flows*, Center for turbulence research report, CTR-S88 (1988)
[22] R. J. Adrian, C. D. Meinhart, C. D. Tomkins: Vortex organization in the outer region of the turbulent boundary layer, J. Fluid Mech. **422**, 1–54 (2000)
[23] A. Schröder, R. Geisler, G. E. Elsinga, F. Scarano, U. Dierksheide: Investigation of a turbulent spot using time-resolved tomographic PIV, in *13th Int Symp on Applications of Laser Techniques to Fluid Mech* (2006) paper 1.4
[24] A. Schröder, J. Kompenhans: Investigation of a turbulent spot using multi-plane stereo PIV, Exp. Fluids **36**, 82–90 (2004)

Index

Recent Developments of PIV towards 3D Measurements

M. Pilar Arroyo[1] and Klaus D. Hinsch[2]

[1] Aragón Engineering Research Institute (I3A), University of Zaragoza, Spain
arroyo@unizar.es
[2] Applied Optics, Institute of Physics, Carl von Ossietzky University Oldenburg, Germany
klaus.hinsch@uni-oldenburg.de

Abstract. This chapter reviews the different techniques that have been proposed in the last few years for turning PIV into a 3D velocimetry technique. Any technique capable of simultaneously measuring more than one plane is included. In expanding normal-viewing PIV depth increases from dual-plane PIV, multiple-plane PIV in its version of digital image plane holography to adjustable-depth volume PIV methods like defocus-evaluating PIV, tomographic PIV, and off-axis holography. Other volume holographic setups utilize reusable real-time recording material (polarization multiplexing with bacteriorhodopsin), explore digital in-line holography and promise extensions to even deeper volumes (light-in-flight holography). The principles, the present state-of-the-art and some ideas on future developments are presented.

1 Introduction

Most interesting problems in fluid dynamics involve complex three-dimensional nonstationary flows, the velocity being the essential characteristic quantity. The experimental study of such flows requires measurement of the instantaneous three-dimensional (3D), three-component (3C) velocity field. The analysis of this 3D–3C velocity field – yielding also the corresponding stress tensor and vorticity vector field – is a key element in learning about the topology of unsteady flow structures.

Particle image velocimetry (PIV) is a well-established 2D fluid velocimetry technique both for 2C and 3C measurements. The measured 2D region (fluid plane) is fixed by illumination with a lightsheet. 2C measurements involve normal viewing of the fluid plane, while 3C measurements require two different (stereoscopic) viewing angles. PIV as a 3D technique, however, is not well established yet and various approaches are still being investigated. Here, we will review systems suitable for simultaneously measuring in more than one plane.

We will not discuss scanning lightsheet systems – though they work as 3D techniques and have a long history of development [1]. However, they do not yield truly simultaneous data. The time needed for the successive scan of the depth sets upper limits to the dynamics of the flow and focus requirements

A. Schroeder, C. E. Willert (Eds.): Particle Image Velocimetry,
Topics Appl. Physics **112**, 127–154 (2008)

limit the depth of field or the available image light. Yet, with continuing advances in high-speed cameras and sophisticated triggering schemes these methods improve in versatility and provide high spatial resolution at several hundred Hz repetition rate, typically [2].

In the course of our analysis we will gradually extend the volume under consideration. Dual-plane PIV is a first step towards 3D measurements. It maintains the side-view characteristic of PIV, but records data from two neighboring planes on separate cameras that are discriminated by polarization or wavelength. The implementation of this system is quite straightforward, but can not be extended easily to more than two planes or even into a volume.

Multiple-plane PIV systems are the next step forward. Most of these systems were developed to improve the performance of some thick lightsheet techniques. A novel proposal is a multiplexed digital image plane hologram (DIPH) that uses multiple lightsheets of different optical path lengths and corresponding multiple reference beams, each one coherent with only one lightsheet. In this case, the 3C velocity field is obtained by combining a PIV analysis with an interferometric analysis with the advantage of using only one common view.

Adjustable-depth volume PIV techniques are fully 3D techniques that use expanded beam illumination, while still maintaining normal viewing. Illumination sets the depth of the recorded 3D region, while the field area is limited by the sensor size and imaging ratio. Defocus-evaluating PIV and tomographic PIV are fully digital techniques that use stereoscopic or multi-angle viewing for the 3C-3D measurements. The latest advances in classical off-axis holographic PIV are also included here.

Finally, full-flow-depth volume techniques are covered where the fluid is viewed in forward or backward scattering. In this case, the field area of the 3D region is limited by the illumination beam cross section or the sensor size, while there is no control over the recorded volume depth, which can only be limited with localized seeding. In-line holography – especially in its digital version – and light-in-flight holography are such 3D techniques, however, with restricted resolution in the depth component. They can be expanded for 3C measurements by illuminating the flow with two perpendicular beams (in-line holography) or stereoscopic viewing (light-in-flight holography). Polarization holography using bacteriorhodopsin (BR) has improved the lengthy process of silver-halide holography, while preserving its spatial resolution.

2 Dual-Plane PIV

The basic principle of dual-plane PIV is shown in Fig. 1. Two lightsheets illuminate the flow and both fluid planes are separately recorded by two cameras. The lightsheet separation can range from almost zero up to the flow depth. The camera positions are adjusted such that the fluid planes are

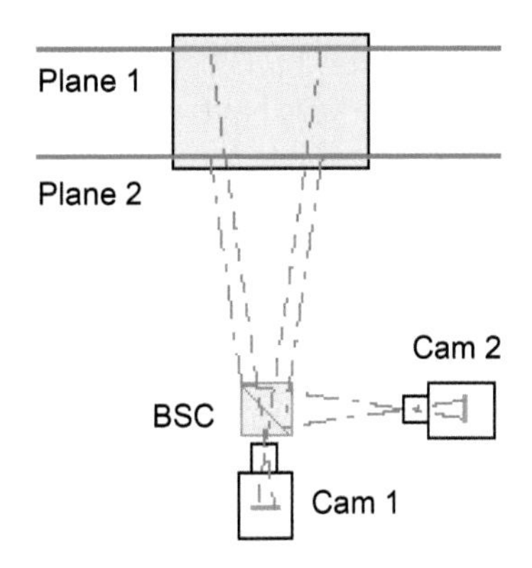

Fig. 1. Dual-plane PIV – monoscopic version

recorded in focus and with the same magnification. We show the monoscopic version, usually a twofold setup is used for stereo applications.

The recordings can be separated either by light polarization [3–5] or wavelength (color) [6]. A beamsplitter cube BSC is used for a common view of both planes through the same window [3,4]. Alternatively, each camera could be placed to look from opposite windows [6]. Additional polarization or color filters in front of each camera improve the separation. When light polarization is used, there still may be some crosstalk due to a changed polarization of the light scattered by the particles. Generally, however, this crosstalk is not very pronounced. Furthermore, its effect is even more negligible when the planes are not close together since the wrong particles will not be in focus.

Very closely spaced planes are used in determining the stress tensor and vorticity vector [3–6]. In this case, the 3C data are essential. This requires stereo recording of each plane and thus four cameras need to be properly placed. They can be arranged such that both stereo pairs have the same camera angle and are symmetrical relative to the normal viewing direction [3,4]. Alternatively, they are placed at slightly different geometries [6]. Furthermore, a system with only three cameras has also been used [5] where the out-of-plane velocity component is only measured for one plane. From this, the value for the second plane is calculated utilizing the continuity equation.

Widely spaced planes can be used for spatial correlation calculations. Much useful information can thus be obtained – even with only 2C data. For 3C data four cameras are required since the continuity equation cannot be applied.

Both lightsheets in the polarization-based system can be obtained from just one double-pulse Nd:YAG laser [5] although most of the experiments reported in the literature use two double-pulse lasers [3, 4, 6]. The main advantages in the two-laser systems are higher energy per lightsheet and time flexibility that allows, e.g., to measure acceleration – a topic that is beyond the scope of this chapter.

In the color-based system, one double-pulse Nd:YAG laser provides green lightsheets and another double-pulse Nd:YAG laser pumping two dye lasers provides the red lightsheets [6–8]. While the polarization-based system requires polarization-preserving tracer particles, the color-based system works with any tracer particles. However, lower energies are available for the color-based system since the emitted red light energy is about 10 times lower than the pumping-beam energy.

The dual-plane PIV is quite well developed and has been applied in several turbulent-flow investigations [7–11]. Probably, there will be no significant further development. Although the color-based system could be extended to more planes, the need of a separate laser source for each plane makes the system expensive and complicated. While the polarization-based system is limited to two planes only, it is easier to implement and renders useful information – especially for two closely spaced planes. Here, the main challenge is to know the spacing with high accuracy because spatial derivatives need to be calculated. *Ganapathisubramani* et al. [5] claim an accuracy of 0.1 mm for a spacing of 1.3 mm in lightsheets of about 0.4 mm in thickness (8 % error) while *Mullin* and *Dahm* [6] claim an accuracy of 0.025 mm for a spacing of 0.4 mm (6 % error).

3 Multiple-Plane PIV

The main idea behind multiple-plane techniques is to cover a 3D region with a number of evenly spaced lightsheets. There are several advantages in this setup as compared with the illumination of the whole 3D region. First, the energy density is higher in the multiple-plane illumination than in the volume illumination; the particle images reconstructed in the analysis process will have an improved signal-to-noise ratio (SNR). SNR will also be better because the number of out-of-focus particles will be smaller. Given that any recording medium has a limit on the number of particles that can be recorded the multiple-plane techniques will work at a higher seeding concentration than the volume techniques. The main challenge, however, is the need to accurately know all lightsheet positions.

3.1 Generation of Lightsheets

Figure 2 shows two systems described in the literature for the production of the set of lightsheets. The first system (Fig. 2a) uses polarizing beamsplitter cubes PBS for the splitting and half-wave plates HP to control the splitting ratio [12]. The main advantages are that the intensity of each lightsheet can be well controlled by the half-wave plates and that all lightsheets have the same vertical polarization. The main drawback is the need for a beamsplitter and a half-wave plate for each lightsheet. Then, there are limitations on the sheet spacing imposed by the physical size of the cubes and plates. A much simpler and more flexible solution is the use of a multibeamsplitter plate MBS (Fig. 2b) that consists of a glass plate with one face perfectly

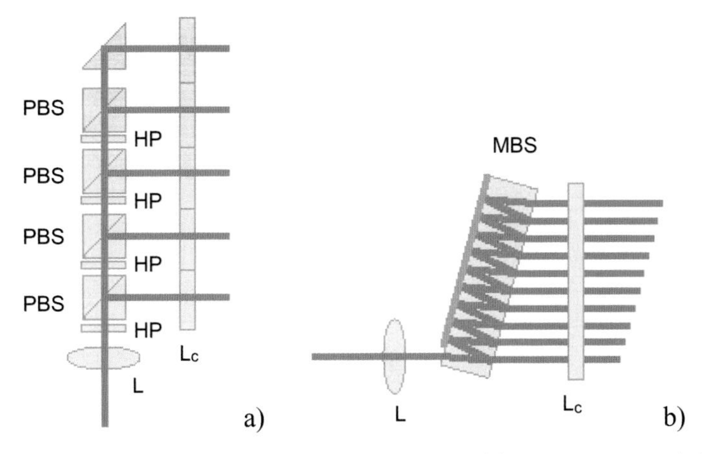

Fig. 2. Production of multiple lightsheets: (**a**) Fixed spacing; (**b**) variable spacing

reflecting (except for an antireflecting slit on one side) and the other face partially transmitting [13]. The number of lightsheets and their spacing can be easily controlled by changing the plate angle – however, at the expense of nonuniform splitting of the energy that depends on the number of lightsheets. A system that is easier to change and producing the same multiple-beam splitting as the MBS consists of one mirror and one partially transmitting plate. This system may be bigger in size, is flexible as to lightsheet spacing, and allows easy changes in the relative optical path lengths – a feature important in holographic use. Furthermore, the splitting ratio can be adjusted by exchange of the transmitting plate.

The multiple-plane illumination can, of course, be used with any of the recording techniques that will be described in Sect. 4. Up to now some experiments using the defocusing principle have been reported [12]. However, the multiple-plane illumination is essential in a quasi-3D technique called digital image plane holography (DIPH) that will be explained later in this section.

3.2 Holographic Recording of Particles

When a set of lightsheets has been prepared to illuminate the particles in the flow the essential task is to acquire particle images of sufficient quality over the whole depth covered by the sheets. In classical recording this is limited by the depth of field of the imaging device – the typical 3D problem! Here, holography provides an essential approach that is free of such constraints. Let us therefore briefly recall the features of holographic particle recording that will also be used in later sections.

Holography consists of two steps: the recording of the laser light-field scattered by the particles and the reconstruction of this field from the hologram. Usually, the hologram is obtained by photographic recording of a superposition of the object light with a reference wave. Thus, the recorded interference pattern (the hologram) stores not only the amplitude of the object wave but also its phase. In the reconstruction the hologram is illuminated with the reference wave and diffraction reproduces the original object wave that can be used for further processing. In a single instant of time the complete information about the particle field is stored to be evaluated later from an analysis of the reconstructed image. When the hologram is illuminated with the phase conjugate of the reference wave even a real image of the object can be reconstructed, i.e., in space there originates a 3D light field that reproduces particle images at their former locations – available for PIV interrogation. We will learn in Sects. 4 and 5 how this has been implemented for holographic particle image velocimetry (HPIV).

To avoid crosstalk between direct and conjugate images, as well as the superposition with the reconstructing wave the reference beam in holography is mostly introduced at a pronounced angle with the object light (off-axis holography). In weakly scattering objects like particle fields it is possible to avoid any extra reference light and use the unaffected background light for the reference wave (in-line holography). This allows very simple and stable setups with low coherence requirements but suffers from additional noise and little flexibility. Yet, we will see that there are situations in particle velocimetry where in-line holography is the configuration of choice.

As in ordinary imaging the resolution in holographic imaging is governed by the effective aperture angle that collects object light that now is determined by the size of the hologram and its distance from the object. Here, the classical diffraction limits hold, which once again results in a much poorer resolution in the depth direction than in the transverse direction.

It should be noted that the successful recording of a hologram requires coherent superposition of object and reference waves. Even laser light has only a finite coherence length and the path-length difference between these two waves must be kept shorter than this value to ensure interference. Several techniques treated later in this chapter make use of this feature.

3.3 Digital Image Plane Holography (DIPH)

DIPH is a digital holographic technique [14–16] that makes use of intensity and phase in the reconstructed object light. In DIPH as a multiplane technique (Fig. 3), the fluid is illuminated with several lightsheets according to Fig. 2b. The volume occupied by the lightsheets is imaged onto a digital camera like in PIV; the f-number, however, is higher and typically more than 11. The optical-fiber-guided reference beams are slightly off-axis divergent beams, which are sent to the sensor via a beamsplitter cube. Furthermore, we use short-coherent light and adjust the optical path lengths such that each reference wave interferes on the sensor with object light from within only one of the lightsheets. The coherence restriction ensures that many independent holograms (one per fluid plane) are superimposed in the recording. Crosstalk

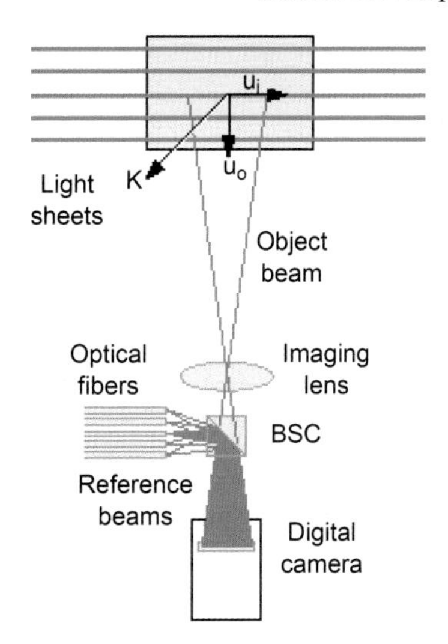

Fig. 3. Multiple-plane digital image plane holography (DIPH) recording setup

is avoided by also introducing a proper optical path length difference between consecutive lightsheets.

As in digital holograms, the object complex wave can be calculated at any plane through numerical modeling of the light propagation. For the lens aperture plane, the numerical reconstruction turns into a Fourier transform. In fact, the recorded holograms are not only image plane holograms of the fluid planes but also lensless Fourier holograms of the imaging-lens aperture. Thus, this Fourier transform (Fig. 4a) reconstructs both the real and the virtual images of the lens aperture. The two images are symmetric with respect to the center, its distance to this center being proportional to the relative position between the reference beam focus and the lens-aperture center. In the multiple-plane setup, there will be as many pairs of aperture images as separate reference beams used to record the hologram (Fig. 4a).

The complex amplitude of each object wave at the sensor plane is reconstructed from the corresponding real (or virtual) aperture image by an inverse Fourier transformation, after displacing the aperture image to the Fourier-plane center [14]. Alternatively, by numerical modeling of light propagation in space we could also calculate the complex light distribution in any plane [16]. The squared modulus of this complex object wave gives the light intensity in the sheet that has the same properties as a PIV image (Fig. 4b). Thus, a standard PIV analysis of two such consecutive images yields the in-plane velocity field (Fig. 4c).

Further, we can make use of the object light phase. Due to the random particle positions the phase map in a single image is a random function. However, the phase difference obtained from the subtraction of two phase maps obtained at consecutive instants of time is proportional to the particle displacement. Such phase differences are obtained modulo 2π and the

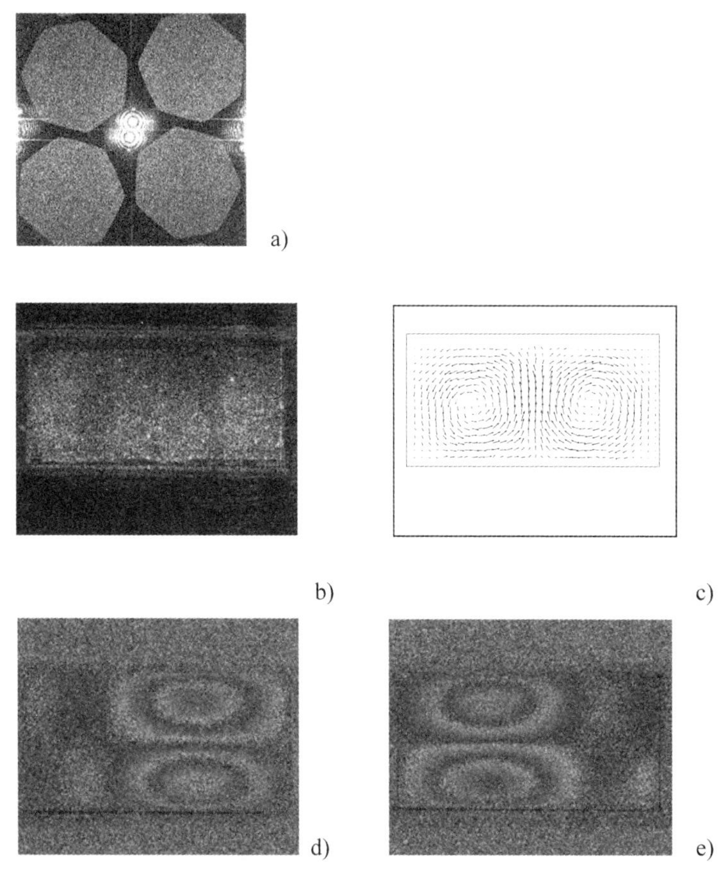

Fig. 4. Two-plane DIPH analysis: (**a**) Reconstructed intensity in the lens-aperture plane showing the real and virtual aperture images; (**b**) reconstructed intensity in the image plane for one fluid plane, as obtained from the *upper left* aperture image and (**c**) the corresponding data after PIV analysis of two such intensity fields; (**d**) phase difference map for one fluid plane, obtained from the *upper left* aperture image; (**e**) phase difference map for the second plane, obtained from the *upper right* aperture image

corresponding maps encode these data as gray levels between 0 and 255 (Figs. 4d,e).

To turn phase difference into displacement we must take into account illumination and observation directions, $\boldsymbol{u}_{\mathrm{i}}$ and $\boldsymbol{u}_{\mathrm{o}}$. They are represented by a sensitivity vector $\boldsymbol{K}$ such that the phase maps represent isoline contours of the displacement component parallel to $\boldsymbol{K}$ (Fig. 3). Typically, contour spacing corresponds to 300–400 nm. The sensitivity vector is never an in-plane vector, which makes the information obtained from the phase analysis complementary to the information obtained from the intensity analysis. In

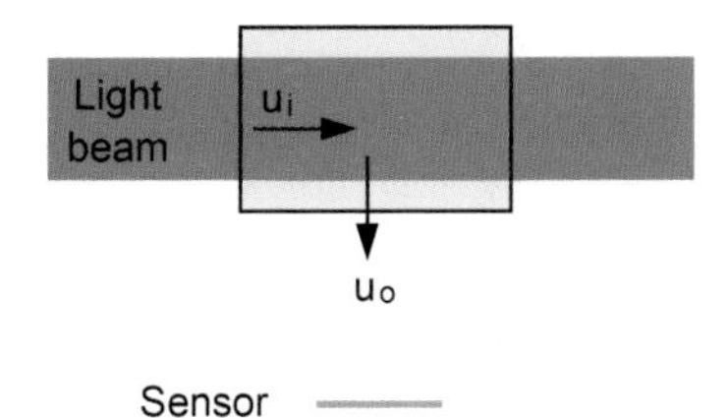

Fig. 5. Schematic of setup for adjustable-depth volume PIV techniques

this sense DIPH is a 3C technique, where all components are obtained under the same illumination and viewing directions. This makes DIPH suitable for confined flows where stereo viewing is not possible or for liquid flows where stereo viewing is not easy to implement [14–17]. Since the interferometric sensitivity is much higher than the PIV sensitivity different time intervals are needed, requiring three laser pulses. Let us point out that the phase analysis has no depth selection at all, which requires multiplexed hologram recording in multiple-plane illumination to separate each sheet individually.

So far, DIPH as a 3D velocimetry technique has been demonstrated only in a 2-plane configuration, although its extension for recording multiple-planes is quite straightforward. The success of DIPH depends on laser parameters like beam shape, divergence and coherence length that affect the lightsheet properties – especially on sheet thickness and local direction of illumination. Here, rather small changes have a noticeable effect on SNR in both the intensity and the phase fields. PIV data from DIPH recordings required bigger interrogation cells than in equivalent standard PIV images. This might be due to different depth of focus in PIV and DIPH. Furthermore, much better overlapping is required for DIPH than for PIV.

In spite of all these restrictions successful DIPH recordings have been made with various commercial PIV lasers. Due to the complexity of DIPH, however, it is expected to be used only when stereo viewing is not possible or when multiple-plane recording is essential.

4 Adjustable-Depth Volume PIV

In the adjustable-depth volume PIV techniques the cross-sectional area of the illumination is increased in depth while maintaining normal viewing, i.e., side scattering by the particles. The challenge now is to discriminate depth positions since the sensor is 2D and more or less parallel to the sheet (Fig. 5).

4.1 Defocus-Evaluating PIV

A recording system of large numerical aperture produces defocused particle images whose size or shape depend on depth position – which may be used to determine the depth coordinate. Special aperture masks have been employed to encode particle depth. Particle image shape is used by *Hain* and

Kähler [18]. An astigmatic imaging lens produces ellipsoidal particle images whose elongation and orientation depend on the separation of the particle from the plane of best focus. Particle image size is used by *Liberzon* et al. [12] in a three-plane setup as a way to identify the particles from each plane and thus produce three separated PIV images. Two cameras in a stereoscopic setup are used to measure the 3C velocity fields corresponding to the three planes. In a recent study, anamorphic imaging and imaging at several planes through a distorted grating were investigated, yielding depth resolution of less than 10 μm – the best value yet achieved [19].

In another approach [20–22], three images from each particle are produced, the depth position being encoded in the distance between these images. Originally, a single camera with a modified three-hole aperture was used [23]. Lately, three individual cameras are used instead, allowing the large three-hole separation required for improved accuracy. The triangle-image of each particle is obtained by combining the three camera images into one 3×8 bit image suitable for processing. Processing includes the detection of all particle images and their fit to 2D Gaussian functions, the identification of triplets of detected images forming equilateral triangles, and a 3D cross-correlation displacement calculation. Such a system has been described [20] for observing a volume of 300 mm × 300 mm × 300 mm. It uses a mean observation distance of 1265 mm and a separation between apertures of 76.4 mm. Although the estimated accuracies in object space are 4.6 μm for the in-plane components (based on assuming an accuracy of 0.01 pixel in image space) and 147.2 μm for the out-of-plane component, the actually measured errors using an experimental calibration procedure are 0.025 pixel for the in-plane components and 15 times bigger for the out-of-plane component. Some preliminary measurements in a two-blade model boat propeller demonstrate the functionality of the system. A more detailed theoretical analysis is described by *Pereira* and *Gharib* [21] and *Kajitani* and *Dabiri* [22]. Some simulations using synthetic images show that particle densities should be lower than 0.01 particles per pixel.

4.2 Tomographic PIV

In tomographic PIV (tomo-PIV) the illuminated fluid volume is viewed from several directions by digital cameras. The method is closely related to the long and successful development of 3D particle tracking velocimetry (PTV) that aimed at the coverage of the third dimension by combining the output from at least three cameras and exploring approaches from photogrammetry to extract particle trajectories; the state-of-the-art is illustrated in [24]. The problem of depth identification with a single camera has also been tackled repeatedly by color coding the depth position of the tracked particles – a recent study employs acousto-optical color selection from a white-light laser [25].

In the most advanced setups for tomo-PIV [26] three or four cameras record the fluid volume at large f-number and complying with the

Scheimpflug condition so that the entire volume is in focus. The cameras image 2D projections of the 3D particle distribution in the fluid. Recovering a 3D distribution from its 2D projections is the topic of tomography. The special feature of this technique is to apply tomographic algorithms for recovering the 3D intensity distribution – different from an already established photogrammetric technique in PTV [27] that relies on particle identification. Yet, the basic idea is essentially the same: the position of an intensity peak (particle) in 3D space is obtained by triangulation of the lines of sight originating from peak intensities in the projections. A common problem is the appearance of "ghost particles" [28], i.e., peak intensities produced by the triangulation procedure at positions where there were no particles in the object. Common to both techniques is also the need of an accurate calibration of the setup to establish a relation between image and object coordinates. Different from the 3D PTV that works at low seeding density (0.005 particles per pixel) tomo-PIV can handle higher seeding densities (up to 0.1 particles per pixel).

The image-to-object coordinate relation is calculated with 0.1 pixel accuracy using calibration procedures from stereo-PIV. The tomographic reconstruction utilizes the multiplicative algebraic reconstruction technique (MART) producing distinct particles without many artifacts. It starts with a guess for the 3D intensity distribution and iteratively improves the image. The 3C displacement in any interrogation volume is obtained by 3D cross-correlation of consecutive images.

A numerical study of tomo-PIV has shown that at least three cameras are needed [26]. Operation parameters are: optimum angle about 30°, seeding density up to 0.075 particles per pixel, and calibration errors must be less than 0.4 pixels. Random noise up to 25 % of peak intensity has only a small effect after preprocessing the images. All of the illuminated volume should be reconstructed because ghost particles outside the volume degrade the reconstruction much more rapidly than ghost particles inside the volume.

In an experimental validation the wake behind circular cylinders in a low-speed wind tunnel has been studied [28, 29]. Four cameras at f-number 8 were placed at the corners of a square making angles of about 20° with the cylinder axis. Measurements were performed at seeding densities ranging from 0.01 to 0.1 particles per pixel for a fixed section of 36.5 mm × 35.8 mm and lightsheet thickness from 2 to 12 mm. The measured mean and RMS displacements agree with stereo-PIV measurements within 0.3 and 0.18 pixel, respectively, for a flow characterized by a free-flow displacement of 3.2 pixels.

Tomo-PIV has also been used as an alternative to dual-plane PIV for the calculation of the 3D strain tensor, using a 3.7-mm thick lightsheet and evaluating each half as separated planes [30]. Measurements were also made on a turbulent cylinder wake and on a turbulent spot [31, 32].

Tomo-PIV is very attractive because the optical configuration is very similar to stereo-PIV – just with novel software. Research is still needed to improve analysis speed and maybe also accuracy. There are good chances that tomo-PIV may become a reference technique for 3D measurements in the same way as stereo-PIV is for 3C measurements.

4.3 Off-Axis Holography

Optical off-axis particle holography as introduced in Sect. 3.2 is a straightforward approach to solve the depth-of-focus problem PIV faces in a thick lightsheet. The 90°-scattered light usually imaged by a CCD camera is now collected on a holographic plate and superimposed with an obliquely incident reference beam. When this is a collimated planar wave a real-image particle field can be produced upon reconstruction by illuminating the plate backwards. This technique has been the basis for several quite sophisticated holographic setups rendering impressive 3D velocity fields [33–35]. The instrumental inputs as well as the computational effort, however, are quite large for such devices. In addition, due to the photographic processing they lack real-time capability. With the exception of bacteriorhodopsin, which will be covered in Sect. 5.2, alternative recording media with the feature of instant response – like photorefractive crystals – have not yet been found suitable for the task of holographic particle velocimetry. Thus, recent work in this area has been concentrated rather on issues in understanding the holographic imaging to improve the quality of the displacement data than on new concepts.

The essential task in holographic particle velocimetry is to determine the 3D displacement of single particles or clusters of particles from the reconstructed light fields. A real particle image reconstructed from a hologram, however, will rarely be a faithful replica of the original three-dimensional particle. The holographic plate intercepts only part of the rather complicated angular distribution of scattered light from the particle (Mie scattering). Furthermore, its limited dynamic range weights input light by its intensity. Thus, the reconstructed wave already differs from the wave originally incident. Furthermore, this wave backpropagates through space onto a location that was originally occupied by a particle. It has been studied in detail how all of this modifies the reconstructed wave to introduce aberrations in the position of the former particle [36].

The thicker the lightsheet the more light in out-of-focus particle images competes with the useful particle-image light. This results in an intrinsic speckle background that limits the performance of the system. Theoretical studies have established relations between the information capacity of particle holography and essential parameters like particle number density and thickness of the sheet [37] that assist in system design and favor large-aperture holography. The precise knowledge of all influential factors has guided the development of strategies and algorithms for a precise determination of particle position [38].

The final objective in 3D velocimetry is to obtain a series of holograms during the dynamic development of a flow (four-dimensional flow measurement). In the case of photographic recording this requires either the rapid

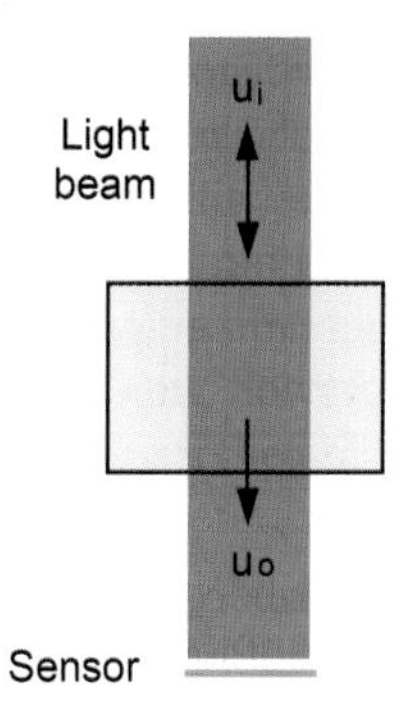

Fig. 6. Schematic for full-flow-depth volume techniques

exchange of the holographic carrier (spatial multiplexing) or switching between a set of reference beam directions (angular multiplexing) – both solutions yet requiring a large instrumental effort. In a µPIV configuration the latter scheme has been studied in detail recording several tens of image pair holograms on a photopolymer that was slightly rotated between the pair of exposures [39]. In another study the temporal measurement capability of HPIV has been illustrated by recording a series of individual holograms phase-locked to a synchronized flow [38].

Once good-quality particle holograms have been obtained it requires much computational effort to extract the displacement data – a task that has been dealt with extensively in the past. In extension of the optical processing schemes of early PIV centering on Young's fringes similar concepts have been developed for the 3D case – compare the detailed review of this method called object conjugate reconstruction in [35]. Meanwhile, this approach has been supplemented by computer assistance to provide what is called the digital shearing method [40].

5 Full-Flow-Depth Volume Techniques

The main advantage of these techniques, which are also holographic techniques, is the use of forward or backward light scattering (Fig. 6), which is several times stronger than side scattering and thus compensates for the low sensitivity of holographic materials. Particle holography with forward scattering on photographic film has been used for a long time. Recently, studies were made using the instant recording material bacteriorhodopsin and digital sensors (digital in-line holography). Backward particle scattering is a key factor in light-in-flight holography to improve the signal-to-noise ratio.

5.1 Optical Forward Scattering Holography

In-Line holography is the simplest holographic setup (Fig. 6 for $\boldsymbol{u}_\mathrm{i} = \boldsymbol{u}_\mathrm{o}$) for forward scattering holography. Due to its simplicity and low coherence requirements in-line holography was proposed for velocimetry already in the 1990s. However, several important drawbacks had to be tackled. First, there is an upper limit in particle density to leave enough nonscattered light for a good reference beam. This limit affects spatial resolution and often ranges inferior to standard PIV. It can be overcome by using an independent reference beam while maintaining forward scattering (Fig. 7) and blocking the direct light. In this case, we lose the advantage of relaxed coherence requirements. Furthermore, speckle noise from the interference of the many scattered waves all along the traveling beam also increases with higher coherence. Thus, noise grows with the depth of the flow and the only solution now is to use the thick lightsheet techniques described in Sect. 4. The main problem in 3D measurements, however, is the long axial depth of the reconstructed particle images, which is several times bigger than its transversal diameter. Thus, axial accuracy is much poorer than transversal accuracy. This is especially pronounced in forward scattering since the effective aperture is small due to the narrow lobe of scattered light – even more for bigger particles. This depth-resolution problem has been addressed by simultaneously recording two holograms with perpendicular viewing directions [41]. Some impressive measurements have been reported from optical forward scattering holography with an off-axis reference beam (Fig. 7b), which is free of the noise from the direct beam and the unfocused twin image of the pure in-line setup [42,43]. In a simple mirror system both fluid views are joined in one hologram [44].

5.2 Polarization Multiplexed Holography with Bacteriorhodopsin

Bacteriorhodopsin (BR) is a photochromic material whose use as a reusable, high-resolution (5000 lines/mm), real-time recording material in velocimetry was proposed in 2002 [45]. Its main weakness, however, is the photoinduced erasure during readout. BR can only be used for temporary storage of holographic information since the produced transparency decays at room temperature within a few minutes. Strategies for hologram recording and data retrieval have been proposed [46]. In practice, an off-axis reference beam and a forward-scattering setup have proven useful to cope with the material's low sensitivity (a few mJ/cm^2 at 532 nm) [47–49] – after initial tests with a side-scattering configuration [50]. Data retrieval uses optical reconstruction of the real-image particle field at the recording wavelength. The particle image field is digitally recorded with a plane-by-plane strategy. A liquid-crystal shutter synchronized with the acquisition camera is used to reduce the photoinduced erasure of the hologram during reconstruction. Yet, this erasure still limits the amount of data that can be retrieved. Cooling decreases erasure, but the cyclic warming for recording and cooling for reconstruction tends to damage the material [49]. Higher speed and number of pixels in acquisition cameras will improve data collection.

A nice feature of BR is its sensitivity to light polarization. It has been observed that the most efficient way to multiplex two holograms for their

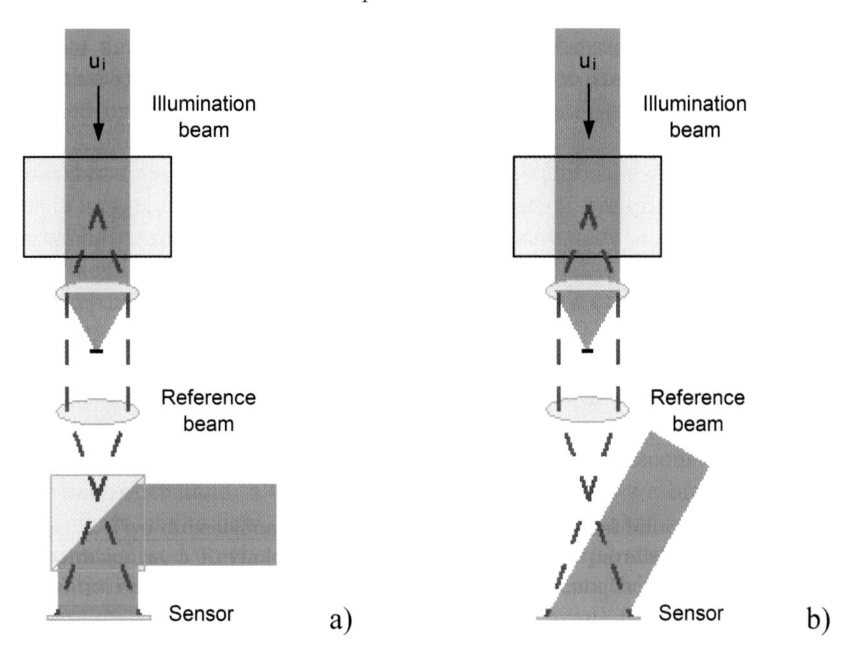

Fig. 7. Schematic of setup for forward-scattering holography with independent reference beam: (**a**) In-line beam; (**b**) off-axis beam

independent reconstruction is by using beams of opposite circular polarization [51]. By changing polarization between the first and second exposure, each object wave will be reconstructed independently by a reconstruction beam of the same polarization as the reference beam in the recording – and this with maximum efficiency and differing polarization. Such polarization multiplexing resolves the directional ambiguity in double-exposure holograms [47–50], is easier to set up and produces more accurate data than the traditional angular multiplexing because it avoids a change in reference-beam direction. BR has proven useful in taking and acquiring the information of several holograms per hour. However, time-series recording with spatial/angular multiplexing will not be feasible until the photoinduced erasure can be lowered significantly.

While there are clear advantages in using a real-time recording material, the experimental implementation for holography with BR is still very sophisticated and far from routine.

5.3 Digital In-Line Holography

The latest proposal for in-line holography is the use of digital (CCD/CMOS) sensors. Digital in-line holography (DiH) as a velocimetry technique is not even a decade old. The drawback of the low spatial resolution (about 100 lines/mm) and low capacity information (nowadays up to 4000 $\times$ 4000 pixels) of sensors as compared with film (3000 lines/mm in $100 \times 125\,\text{mm}^2$) is counteracted by the advantages of very fast digital recording (time recordings are possible), image processing capabilities (reduction of noise and aberrations) and full complex-wave information (phase data can improve depth resolution). It is expected that a special concept of undersampling may even overcome the resolution limitations [52].

Extensive research has been carried out to demonstrate the feasibility of digital in-line holography as a 3D velocimetry technique. The first step in DiH is recording the digital hologram. Some image preprocessing of the hologram data might be carried out as a second step to remove background noise for cleaner and higher-contrast holograms. The original object wave at any desired position/plane in space is numerically reconstructed from these holograms and the whole volume of interest is assembled. In the next step, particle images are identified in the intensity field to extract both particle 3D position and size. The 3C velocity field is obtained either by tracking of particle positions or by 3D correlation of interrogation volumes.

The most widely used optical setup for recording is the simple arrangement of Fig. 6 although the independent in-line reference beam setup (Fig. 7a) has also been investigated [53, 54]. The independent off-axis reference beam setup (Fig. 7b) has not been considered yet due to its higher spatial resolution requirements. However, an implementation with a divergent reference beam and an imaging lens as in DIPH or a small-volume flow a long distance from the sensor as in lensless Fourier holography [55] should be feasible. The main advantage of this setup is noise reduction due to the spatial separation of real and virtual images. The main drawbacks are low object-light intensity and poor depth resolution. Another alternative could be the use of a parallel phase-shifting holography setup [56].

One of the most important steps in DiH is the numerical reconstruction of the object wave from the hologram data. This is always based on the Rayleigh–Sommerfeld diffraction theory where the hologram is treated as a complex grating. Under typical conditions of small viewing angles, the amplitude of the wave emerging after illumination with a reconstructing light beam is expressed as a convolution with a diffraction kernel. The convolution can be calculated by using FFT algorithms [53, 54, 57–59]; two FFTs are required for each distance in depth since the Fourier transform of the diffraction kernel is an analytical function. Most papers, however, use the Fresnel approximation, valid only when transverse dimensions are very small compared to distance. In this case, the convolution turns into a Fourier transform, allowing a faster computation since only one FFT per distance is needed. However, the reconstructed image scales with the distance [60, 61]. Since this is a big drawback for volume reconstruction, the convolution approach is still used with the Fresnel approximation [62–65]. Other alternative ways of calculating the reconstructed light field within the regime of Fresnel approximation are wavelet transforms [66–69] or fractional-order Fourier transforms [70], which also address the issue of axial particle localization.

A lot of effort has been dedicated to successful particle identification and localization [66, 71], an issue closely related to the numerical reconstruction. In general, particle identification is based on finding local intensity peaks yielding the 3D positions. The discrimination of particles from speckle noise is a big challenge that increases with particle density [68, 72]. Another challenge is to obtain accurate values for the 3D position, especially its axial value [54, 60, 63, 64]. Numerical simulations together with experimental verification in simple objects have been carried out to test different proposals. It was shown that for opaque particles evaluation of the phase field around the local intensity peaks gives a fourfold increase in axial accuracy relative to pure intensity evaluation [60]. Similar improvements are also obtained by applying filters either in the recording [54] or in the reconstruction [65]. The recording of two holograms with perpendicular viewing has also been tested in digital holography [62].

Some 3D velocity measurements in flows [57, 62, 73] have already been reported, although DiH is still far from being a routine technique. All take time-resolved measurements using CW-laser illumination. 3D velocity vector maps obtained in a laminar jet and in the flow past a cylinder are shown to quantify the accuracy of a DiH system [57]. A 2.5 % error is claimed using seeding densities of 12 particles/mm^3, recording volumes of 8 mm in width and 25.4 mm in depth, and a 1280×1024 6.7 µm pixel camera, corresponding to 0.004 and 0.014 particles per pixel. In [62] a $15 \times 15 \times 64\,\text{mm}^3$ volume has been recorded with a seeding density of 4 particles/mm^3 and a 2048×2048 7.4 µm pixel camera (0.014 particles per pixel) running at 15 frames/s. An axial resolution of 0.5 mm in each view and an accuracy of about a pixel in the measured 3D displacements are reported. The measurements illustrated the velocity field and particle entrainment produced by a freely swimming 1 mm size animal. In [73] a $9 \times 9 \times 25\,\text{mm}^3$ volume was recorded with a 1008×1018 9 µm pixel camera running at 30 frames/s. $32 \times 32 \times 21$ vectors were obtained in a volume of $9 \times 9 \times 9\,\text{mm}^3$ by 3D crosscorrelation on $64 \times 64 \times 40$ pixel interrogation volumes from which 3D streamlines and the isovorticity maps were calculated.

At present, DiH is working more as a PTV than as a PIV technique because it relies on particle detection. Thus, only holograms with up to 0.01 particles per pixel have been analyzed successfully. Although there are several reasons that prevent DiH from working at higher seeding densities, we believe that the main problem is the coherent speckle noise. Part of this noise originates from the virtual images, but could be removed by using the off-axis feature of the independent-reference beam setup. However, the considerable speckle noise from the interference of the scattered waves from all particles stacked in depth will still be there. This could be the main factor limiting the seeding densities to levels lower than in tomo-PIV that is free of interference. If so, DiH could improve by using the independent-reference beam setup with short coherence length lasers like pulsed Nd:YAG lasers (coherence length about 1 cm). By using backward scattering, only the scattered waves from particles contained in a 5-mm thick slice will interfere with each other (cf. Sect. 5.4). Thus, the volume depth is restricted while still using volume illumination.

5.4 Light-in-Flight Holography

We have seen that in full-flow-depth particle holography an essential advantage of lightsheet illumination is lost, i.e., the suppression of scattered light from the majority of tracer particles that are not contained in the volume under investigation. This results in a considerable noise background when the depth of the volume or the particle number density increase – compare the theoretical studies on the information capacity in particle holography [37]. Thus, concessions must be made in either spatial resolution or total depth. This problem is tackled in light-in-flight holography of particle fields where coherence properties are utilized such that the complete flow field is recorded instantaneously while particle images are later reconstructed within separate shells in depth without disturbance by the rest of the field. The method has been termed light-in-flight holography PIV (LiFH-PIV).

Let us briefly recall the use of light sources with a coherence length of only a few millimeters for applications in flow investigations. The idea is the following: object wave and reference wave can interfere at the holographic plate only when the difference in path these waves have traveled is less than the coherence length of the light – a feature that has already been mentioned in Sect. 3.1 to separate multiple lightsheet data. As shown in the schematic of Fig. 8, reference light incident from the left has to travel a longer path to the right side of the holographic plate than to the left. Object light scattered from particles is recorded only if this path length differs by no more than the coherence length L from that of the corresponding reference light. Thus, with proper alignment, particles from a shell in the middle of the observed field are recorded in a small region in the middle of the plate, particles from a front shell on the left and from a rear shell on the right. For the reconstruction of a real image a conjugate reference wave (i.e., inversely traveling wave) is needed to illuminate the hologram. As mentioned earlier, this is achieved with a single planar wave for recording and reconstruction, when the hologram is turned by 180° after development. Upon reconstruction from a thin slit-aperture of the hologram, only a shell with a depth of roughly half a coherence length shows up – in the case of an extended aperture this grows by about half its diameter D and is referred to as the coherence depth l_D. The background of defocused holographic particle images is therefore reduced considerably.

This method was originally applied by using a double-pulsed ruby laser – a light source of high energy, yet of many disadvantages. Meanwhile, the technique has been implemented successfully for illumination and reconstruction in Nd:YAG-laser light (coherence length some 7 mm) and found a first application in a wind-tunnel study [74, 75].

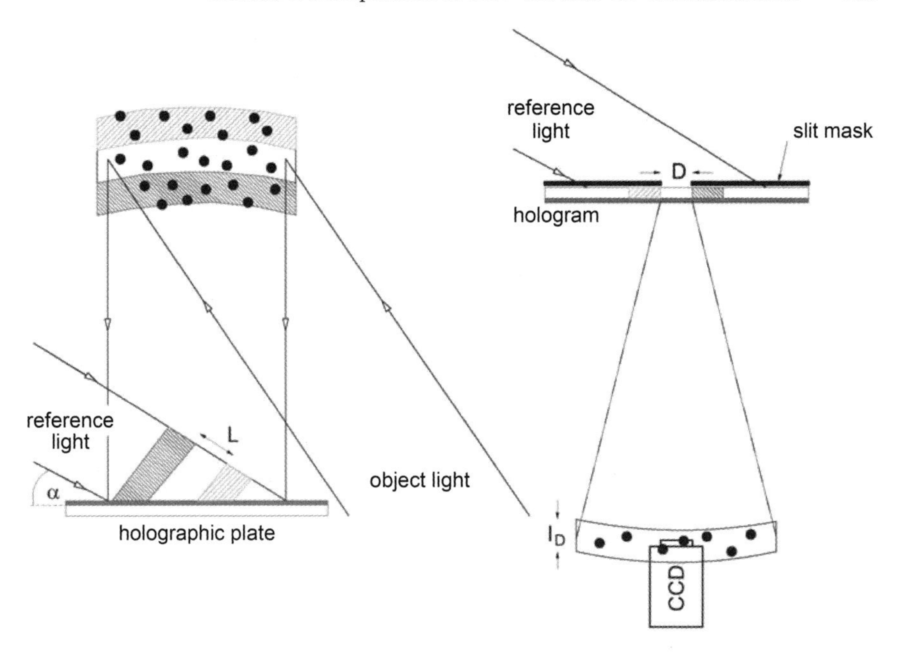

Fig. 8. Optical setup for recording and reconstruction in light-in-flight holographic PIV

The preferred setup utilizes backscatter illumination to enhance the path length traveled by the light and applies angular multiplexing of the reference beams to separate the two recordings from a single holographic plate. The reconstruction is usually performed in a separate setup with a Nd:YAG laser in continuous-wave operation. Alignment markers or even interferometric control of the reference waves allow for optimum adjustment of the reconstruction geometry.

The reconstructed particle image fields are scanned by a CCD array on a three-dimensional traversing stage. Complete images of the light intensity are combined from the many patches thus obtained. The accumulation of each particle field still is quite a time-consuming task. For the evaluation, a straightforward three-dimensional correlation has been applied that makes use of many of the modern algorithms and ideas that have been developed for regular 2D PIV.

When two reference waves from different directions are used in the angular multiplexing method the same shell in depth for the first and second exposures is usually reconstructed from apertures at different locations on the hologram because path-length conditions are dependent on the angle of incidence. A particle-field image in 3D real space that gets its light from the hologram aperture, however, looks different when it is produced from a different projection direction. Thus, three-dimensional crosscorrelation finds two different light distributions in space and does not work very well. A special geometry that makes use of azimuth as well as height angle was therefore invented to allow for both reconstructions to be made from the same small circular aperture.

The basic idea of light-in-flight sounds convincing – background noise from out-of-focus particles is suppressed and the signal-to-noise ratio, SNR, improved. However, this is done at the expense of a smaller effective aperture because only part of the whole hologram is used for each shell in depth. Consequently, the resolution as well as the overall intensity is affected and it has to be investigated whether and under which conditions the overall performance is still superior to a classical HPIV setup. The Nd:YAG system is well suited for comparative studies because a simple switch allows + to change from low to high coherence conditions. To quantify the performance SNR values have been derived from particle-field images and an improvement could be shown [76]. However, it is very difficult to discriminate signal and noise in images that contain all grades from a well-focused particle to a speckle background. Therefore, in an investigation underway the performance is judged by the rates for valid vectors that have been used extensively also in ordinary PIV work to assess its performance. Once again, it is found that LiFH-PIV performs better for high particle concentrations and deep volumes.

6 Conclusions

The development of PIV as a 3D velocimetry technique is still an important ongoing research subject. Different approaches have been tested over the last few years. Table 1 summarizes the main features of the different approaches. Table 2 presents some quantitative data on recording and analysis parameters, showing the present state-of-the-art for each technique. The simultaneous 3C velocity measurement in two fluid planes is already well solved with dual-plane PIV for flows with plenty of optical access – needed for the stereo viewing. Although DIPH can do this task in confined flows and even in more than two fluid planes, DIPH is still far from a widely accepted technique. This is due to the complexity in both the concept of digital image plane holography and its experimental implementation. Although holography seems the natural option for measurements in a volume, most of the developed techniques are quite complex for a widespread use. Tomo-PIV might establish itself as the reference technique for the measurement of 3D-3C velocity fields in open flows with plenty of optical access. For confined flows and/or small-size flows, digital holography could still be the best option. Straight advances in digital holography are directly linked with technological advances in digital sensors, such as more and smaller-size pixels and higher frame rates.

Table 1. Main features of 3D techniques

		Dual-plane PIV	Digital image plane holography	Defocus-evaluating PIV	Tomo-PIV	Side scattering holography	Forward scattering holography	Polarization multiplexed holography	Digital inline holography	Light-in-flight holography
Recording media		CCD	CCD	CCD	CCD	Film	Film	BR	CCD	Film
Encoding for z		Polarization	Coherence	Defocus	Tomo	Depth of focus	Depth of focus	Depth of focus	Depth of focus	Depth of focus
Encoding for Vz		Stereo	Interferometry	Defocus	Tomo	Depth of focus	Two perpendi-cular illuminations	Stereo	Two illuminations or complex wave	Stereo
Required optical access for observation	3D	Narrow	Narrow	Narrow	Wide	Narrow	Narrow	Narrow	Narrow	Narrow
	3D-3C	Wide	Narrow	Narrow	Wide	Narrow	Wide	Wide	Wide	Wide
Particle concentration		High	High	Low	Medium	Medium	Low	Low	Very low	Medium
Exp. Complexity		Low	Medium	Low	Low	High	High	High	Low	High
Required sensors / laser pulses for recording	3D	2 / 2 or 4	1 / 2	1 or 2 / 2	4 / 2	1 / 2	1 / 2	1 / 2	1 /2	1 / 2
	3C	+2 / -	+1 / +1	+1 / -	- / -	- / -	+1 / -	- / -	+1 / -	- / -
Required laser energy		Medium	Medium	Medium	High	Very high	High	Very high	Low	Very high

Table 2. Selected quantitative data for each technique as from published material

		Dual-plane PIV	Digital image plane holography	Defocus-evaluating PIV	Tomo-PIV	Side-scattering holography	Forward scattering holography	Polarization multiplexed holography	Digital inline holography	Light-in-flight holography
		Ref 9	Ref 15	Ref 20	Ref 26	Ref 33	Ref 42	Ref 48	Ref. 57	Ref 75
Sensor parameters	Recording	4 CCD	1 CCD	3 CCD	4 CCD	75 mm film	70 mm Holotest 10E75 film	100 mm BR film	1 CCD	Slavich film
	Analysis (pixels)	1.25kx1kx4	1.25kx1k	1kx1kx3	1.25kx1kx4	25kx20kx0.5k	10kx10kx10k	1 CCD1	1.25kx1k	3.5kx2.7kx0.6k
	Pixel size in object space (μm^3)	52x34x600	22x22x-	300x300x-	55x55x55	2x2x100	4.9 x 4.9	6.45x6.45x200	6.7x6.7x6.7	6.7x6.7x49
Field of view (mm^2)		67x35	28.5x22.9	300x300	50x50	50x40	57x57	32x32	6.86x6.86	24.0x18.8
Volume depth (mm)		1.2	3	300	8	50	45	32	25.4	29.1
Particle parameters	Diameter (μm)	2	5	1.5	1	5	20	10	7	1
	C (part/mm^3)			2.6x10^4	2	30	4-8		12	12
Computed vectors	Mesh	124x64x2	32x16x2	-	64x64x30	25x20x25	136x130x128		9x9x23	
	Total / % overlap	15872 / 50	1024 / 0	7000	122880 / 75	92000	2.26x10^6 / 65	100000	1863 / -	124800 / 75
Accuracy	Inplane (μm)		5	42	5.5	1-2	2		2.5%	
	Out-of-plane (μm)		0.05	658	8.8	18	2		20	
Laser parameters	Wavelength (nm)	532	633	532	532	532	694	532	514	532
	Energy (mJ/pulse)	250	0.085	125	400	400	25	300		1500
	Number of pulses	4	3	2	2	2	2	2		2
Maximum displacement (pixels)		8	5	8	3.2		26			

However, further advances could also come from the application of concepts of noise reduction developed in photographic film recording like light in-flight holography and the separation of virtual and real images.

References

[1] C. Brücker: 3-D scanning-particle-image-velocimetry: Technique and application to a spherical cap wake flow, Appl. Sci. Res. **56**, 157–179 (1996)

[2] T. Hori, J. Sakakibara: High-speed scanning stereoscopic PIV for 3D vorticity measurement in liquids, Meas. Sci. Technol. **15**, 1067–1078 (2004)

[3] C. J. Kähler, J. Kompenhans: Fundamentals of multiple plane stereo particle image velocimetry, Exp. Fluids, Suppl. pp. S70–S77 (2000)

[4] H. Hu, T. Saga, T. Kobayashi, N. Taniguchi, M. Yasuki: Dual-plane stereoscopic particle image velocimetry: System set-up and its application on a lobed jet mixing flow, Exp. Fluids **31**, 277–293 (2001)

[5] B. Ganapathisubramani, E. K. Longmire, I. Marusic, S. Pothos: Dual-plane PIV technique to determine the complete velocity gradient tensor in a turbulent boundary layer, Exp. Fluids **39**, 222–231 (2005)

[6] J. A. Mullin, W. J. A. Dahm: Dual-plane stereo particle image velocimetry (DSPIV) for measuring velocity gradient fields at intermediate and small scales of turbulent flows, Exp. Fluids **38**, 185–196 (2005)

[7] J. A. Mullin, W. J. A. Dahm: Dual-plane stereo particle image velocimetry measurements of velocity gradient tensor fields in turbulent shear flow I. Accuracy assessments, Phys. Fluids **18**, 035101 (2006)

[8] J. A. Mullin, W. J. A. Dahm: Dual-plane stereo particle image velocimetry measurements of velocity gradient tensor fields in turbulent shear flow II. Experimental results, Phys. Fluids **18**, 035102 (2006)

[9] C. J. Kähler: Investigation of the spatio-temporal flow structure in the buffer region of a turbulent boundary layer by means of multiplane stereoPIV, Exp. Fluids **36**, 114–130 (2004)

[10] A. Schröeder, J. Kompenhans: Investigation of a turbulent spot using multiplane stereo particle image velocimetry, Exp. Fluids **36**, 82–90 (2004)

[11] N. Saikrishnan, I. Marusic, E. K. Longmire: Assessment of dual plane PIV measurements in wall turbulence using DNS data, Exp. Fluids **41**, 265–278 (2006)

[12] A. Liberzon, R. Gurka, G. Hetsroni: XPIV-multi-plane stereoscopic particle image velocimetry, Exp. Fluids **36**, 355–362 (2004)

[13] M. P. Arroyo, K. von Ellenrieder, J. Lobera, J. Soria: Measuring 3-C velocity fields in a 3-D flow domain by holographic PIV and holographic interferometry, in *4th Int. Symp. on Particle Image Velocimetry* (2001)

[14] J. Lobera, N. Andrés, M. P. Arroyo: Digital image plane holography as a three-dimensional flow velocimetry technique, SPIE **4933**, 279–284 (2003)

[15] J. Lobera, N. Andrés, M. P. Arroyo: From ESPI to digital image plane holography (DIPH): Requirements, possibilities and limitations for velocity measurements in a 3-D volume, in M. Stanislas, J. Westerweel, J. Kompenhans (Eds.): *Particle Image Velocimetry: Recent Improvements* (Springer, Berlin, Heidelberg 2004) pp. 363–372

[16] J. Lobera, N. Andrés, M. P. Arroyo: Digital speckle pattern velocimetry as a holographic velocimetry technique, Meas. Sci. Technol. **15**, 718–724 (2004)
[17] M. P. Arroyo, J. Lobera, S. Recuero, J. Woisetschläger: Digital image plane holography for three-component velocity measurements in turbomachinery flows, in *13th Int. Sym. on Applications of Laser Techniques to Fluid Mechanics* (2006)
[18] R. Hain, C. J. Kähler: 3D3C time-resolved measurements with a single camera using optical aberrations, in *13th Int. Symp. on Applications of Laser Techniques to Fluid Mechanics* (2006)
[19] N. Angarita-Jaimes, E. McGhee, M. Chennaoui, H. I. Campbell, S. Zhang, C. E. Towers, A. H. Greenaway, D. P. Towers: Wavefront sensing for single view three-component three-dimensional flow velocimetry, Exp. Fluids **41**, 881–891 (2006)
[20] F. Pereira, M. Gharib, D. Dabiri, M. Modarress: Defocusing PIV: A three component 3-D PIV measurement technique. Application to bubbly flows, Exp. Fluids **29**, S78–S84 (2000)
[21] F. Pereira, M. Gharib: Defocusing digital particle image velocimetry and the three-dimensional characterization of two-phase flows, Meas. Sci. Technol. **13**, 683–69 (2002)
[22] L. Kajitani, D. Dabiri: A full three-dimensional characterization of defocusing digital particle image velocimetry, Meas. Sci. Technol. **16**, 790–804 (2005)
[23] C. E. Willert, M. Gharib: Three-dimensional particle imaging with a single camera, Exp. Fluids **12**, 353–358 (1992)
[24] J. Willneff, A. Grün: A new spatio-temporal matching algorithm for 3D-particle tracking velocimetry, in *9th Intl. Symp. Transport Phenomena and Dynamics of Rotating Machinery* (2002)
[25] B. Ruck: Color-coded tomography, in *7th Int. Symp. on Fluid Control, Measurement and Visualization* (2003)
[26] G. E. Elsinga, F. Scarano, B. Wienecke, B. W. Oudheusden: Tomographic particle image velocimetry, Exp. Fluids **41**, 933–947 (2006)
[27] H. G. Maas, A. Gruen, D. Papantoniou: Particle tracking velocimetry in three-dimensional flows. Part 1. Photogrammetric determination of particle coordinates, Exp. Fluids **15**, 133–146 (1993)
[28] G. E. Elsinga, B. W. van Oudheusden, F. Scarano: Experimental assessment of tomographic-PIV accuracy, in *13th Int. Symp. on Applications of Laser Techniques to Fluid Mechanics* (2006)
[29] G. E. Elsinga, B. Wienecke, F. Scarano, B. W. van Oudheusden: Assessment of tomo-PIV for three-dimensional flows, in *6th Int. Symp. on Particle Image Velocimetry* (2005)
[30] B. Wieneke, S. Taylor: Fat-sheet PIV with computation of full 3D-strain tensor using tomographic reconstruction, in *13th Int. Symp. on Applications of Laser Techniques to Fluid Mechanics* (2006)
[31] F. Scarano, G. E. Elsinga, E. Bocci, B. W. van Oudheusden: Investigation of 3-D coherent structures in the turbulent cylinder wake using tomo-PIV, in *13th Int. Symp. on Applications of Laser Techniques to Fluid Mechanics* (2006)
[32] A. Schröder, R. Geisler, G. E. Elsinga, F. Scarano, U. Dierksheide: Investigation of a turbulent spot using time-resolved tomographic PIV, in *13th Int. Symp. on Applications of Laser Techniques to Fluid Mechanics* (2006)

[33] Y. Pu, H. Meng: An advanced off-axis holographic particle image velocimetry system, Exp. Fluids **29**, 184–197 (2000)
[34] Y. Pu, X. Song, H. Meng: Off-axis holographic particle image velocimetry for diagnosing particulate flows, Exp. Fluids **29**, S117–S128 (2000)
[35] K. D. Hinsch: Holographic particle image velocimetry, Meas. Sci. Technol. **13**, R61–R72 (2002)
[36] Y. Pu, H. Meng: Intrinsic aberrations due to Mie scattering in particle holography, J. Opt. Soc. Am. A **20**, 1920–1932 (2003)
[37] Y. Pu, H. Meng: Intrinsic speckle noise in off-axis particle holography, J. Opt. Soc. Am. A **21**, 1221–1230 (2004)
[38] Y. Pu, H. Meng: Four-dimensional dynamic flow measurement by holographic particle image velocimetry, Appl. Opt. **44**, 7697–7708 (2005)
[39] C. T. Yang, H. S. Chuang: Measurement of a microchamber flow by using a hybrid multiplexing holographic velocimetry, Exp. Fluids **39**, 385–396 (2005)
[40] H. Yang, N. Halliwell, J. M. Coupland: Application of the digital shearing method to extract three-component velocity in holographic particle image velocimetry, Meas. Sci. Technol. **15**, 694–698 (2004)
[41] J. Zhang, B. Tao, J. Katz: Turbulent flow measurement in a square duct with hybrid holographic PIV, Exp. Fluids **23**, 373–381 (1997)
[42] B. Tao, J. Katz, C. Meneveau: Statistical geometry of subgrid-scale stresses determined from holographic particle image velocimetry measurements, J. Fluid Mech. **457**, 35–78 (2002)
[43] A. Svizher, J. Cohen: Holographic particle image velocimetry system for measurements of hairpin vortices in air channel flows, Exp. Fluids **40**, 708–722 (2006)
[44] J. Sheng, E. Malkiel, J. Katz: Single beam two-views holographic particle image velocimetry, Appl. Opt. **42**, 235–250 (2003)
[45] D. H. Barnhart, N. Hampp, N. A. Halliwell, J. M. Coupland: Digital holographic velocimetry with bacteriorhodopsin (BR) for real-time recording and numeric reconstruction, in *11th Int. Symp. on Applications of Laser Techniques to Fluid Mechanics* (2002)
[46] D. H. Barnhart, W. D. Koek, T. Juchem, N. Hampp, J. M. Coupland, N. A. Halliwell: Bacteriorhodopsin as a high-resolution, high-capacity buffer for digital holographic measurements, Meas. Sci. Technol. **15**, 639–646 (2004)
[47] V. S. S. Chan, W. D. Koek, D. Barnhart, C. Poelma, T. A. Ooms, N. Bhattacharya, J. J. M. Braat, J. Westerweel: HPIV using polarization multiplexing holography in bacteriorhodopsin (BR), in *12th Int. Symp. on Applications of Laser Techniques to Fluid Mechanics* (2004)
[48] T. Ooms, J. Braat, J. Westerweel: Optimizing a holographic PIV system using a bacteriorhodopsin film, in *13th Int. Symp. on Applications of Laser Techniques to Fluid Mechanics* (2006)
[49] W. D. Koek: *Holographic Particle Image Velocimetry using Bacteriorhodopsin*, PhD dissertation, Delft Univ. (2006)
[50] V. S. S. Chan, W. D. Koek, D. H. Barnhart, N. Bhattacharya, J. J. M. Braat, J. Westerweel: Application of holography to fluid flow measurements using bacteriorhodopsin (br), Meas. Sci. Technol. **15**, 647–655 (2004)
[51] W. D. Koek, N. Bhattacharya, J. M. Braat, V. S. S. Chan, J. Westerweel: Holographic simultaneous readout polarization multiplexing based on photoinduced anisotropy in bacteriorhodopsin, Opt. Lett. **29**, 101–103 (2004)

[52] J. M. Coupland: Holographic particle image velocimetry: Signal recovery from under-sampled CCD data, Meas. Sci. Technol. **15**, 711–717 (2004)
[53] W. D. Koek, N. Bhattacharya, J. M. M. Braat, T. A. Ooms, J. Westerweel: Influence of virtual images on the signal-to-noise ratio in digital in-line particle holography, Opt. Exp. **13**, 2578–2589 (2005)
[54] T. Ooms, W. Koek, J. Braat, J. Westerweel: Optimizing Fourier filtering for digital holographic particle image velocimetry, Meas. Sci. Technol. **17**, 304–312 (2006)
[55] U. Schnars, W. P. O. Jüptner: Digital recording and numerical reconstruction of holograms, Meas. Sci. Technol. **13**, R85–R101 (2002)
[56] Y. Awatsuji, A. Fujii, T. Kubota, O. Matoba: Parallel three-step phase-shifting digital holography, Appl. Opt. **45**, 2995–3002 (2006)
[57] H. Meng, G. Pan, Y. Pu, S. H. Woodward: Holographic particle image velocimetry: From film to digital recording, Meas. Sci. Technol. **15**, 673–685 (2004)
[58] W. Xu, M. H. Jericho, I. A. Meinertzhagen, H. J. Kreuzer: Digital in-line holography of microspheres, Appl. Opt. **41**, 5367–5375 (2002)
[59] W. Xu, M. H. Jericho, H. J. Kreuzer, I. A. Meinertzhagen: Tracking particles in four dimensions with in-line holographic microscopy, Opt. Lett. **28**, 164–166 (2003)
[60] G. Pan, H. Meng: Digital holography of particle fields: reconstruction by the use of complex amplitude, Appl. Opt. **42**, 827–833 (2003)
[61] S. Kim, S. J. Lee: Digital holographic PTV measurements of a vertical jet flow, in *6th Int. Symp. on Particle Image Velocimetry* (2005)
[62] E. Malkiel, J. Sheng, J. Katz, J. R. Strickler: The three-dimensional flow field generated by a feeding calanoid copepod measured using digital holography, J. Exp. Bio. **206**, 3657–3666 (2003)
[63] C. Fournier, C. Ducottet, T. Fournel: Digital in-line holography: influence of the reconstruction function on the axial profile of a reconstructed particle image, Meas. Sci. Technol. **15**, 686–693 (2004)
[64] W. Yang, A. B. Kostinski, R. A. Shaw: Depth-of-focus reduction for digital in-line holography of particle fields, Opt. Lett. **30**, 1303–1305 (2005)
[65] S. Satake, T. Kunugi, K. Sato, T. Ito, H. Kanamori, J. Taniguchi: Measurements of 3D flow in a micro-pipe via micro digital holographic particle tracking velocimetry, Meas. Sci. Technol. **17**, 1647–1651 (2006)
[66] C. Buraga-Lefevre, S. Coëtmellec, D. Lebrun, C. Özkul: Application of wavelet transform to hologram analysis: Three-dimensional location of particles, Opt. Lasers Eng. **33**, 409–421 (2000)
[67] S. Coëtmellec, C. Buraga-Lefevre, D. Lebrun, C. Özkul: Application of in-line digital holography to multiple plane velocimetry, Meas. Sci. Technol. **12**, 1392–1397 (2001)
[68] M. Malek, D. Allano, S. Coëtmellec, D. Lebrun: Digital in-line holography: Influence of the shadow density on particle field extraction, Opt. Exp. **12**, 2270–2279 (2004)
[69] M. Malek, D. Allano, S. Coëtmellec, C. Özkul, D. Lebrun: Digital in-line holography for three-dimensional-two-components particle tracking velocimetry, Meas. Sci. Technol. **15**, 699–705 (2004)

[70] S. Coëtmellec, D. Lebrun, C. Özkul: Application of the two-dimensional fractional-order fourier transformation to particle field digital holography, J. Opt. Soc. Am. A **19**, 1537–1546 (2002)
[71] S. Murata, N. Yasuda: Potential of digital holography in particle measurement, Opt. Las. Technol. **32**, 567–574 (2000)
[72] W. Yang, A. B. Kostinski, R. A. Shaw: Phase signature for particle detection with digital in-line holography, Opt. Lett. **31**, 1399–1401 (2006)
[73] G. Shen, R. Wei: Digital holography particle image velocimetry or the 3Dt-3C flows, in *5th International Symposium on Particle Image Velocimetry. Busan. Korea, September 22–25* (2003)
[74] S. F. Hermann, K. D. Hisch: Light-in-flight holographic particle image velocimetry for wind-tunnel application, Meas. Sci. Technol. **15**, 613–621 (2004)
[75] S. F. Hermann, K. D. Hinsch: Advances in light-on-flight HPIV for the study of wind tunnel flows, in M. Stanislas, J. Westerweel, J. Kompenhans (Eds.): *Particle Image Velocimetry: Recent Improvements* (Springer, Berlin, Heidelberg 2004) pp. 317–331
[76] K. D. Hinsch, S. F. Hermann: Signal quality improvements by short-coherence holographic particle image velocimetry, Meas. Sci. Technol. **15**, 622–630 (2004)

Index

Digital In-Line Holography System for 3D-3C Particle Tracking Velocimetry

Mokrane Malek[1], Denis Lebrun[2], and Daniel Allano[2]

[1] UMR A 1114 Climate, Soil, Environment,
University of Avignon,
33, rue Louis Pasteur,
84000 Avignon, France
mokrane.malek@univ-avignon.fr

[2] UMR 6614 CORIA,
Université de Rouen,
76801 Saint Etienne du Rouvray, France
lebrun@coria.fr

Abstract. Digital in-line holography is a suitable method for measuring three dimensional (3D) velocity fields. Such a system records directly on a charge-coupled device (CCD) camera a couple of diffraction patterns produced by small particles illuminated by a modulated laser diode. The numerical reconstruction is based on the wavelet transformation method. A 3D particle field is reconstructed by computing the wavelet components for different scale parameters. The scale parameter is directly related to the axial distance between a given particle and the CCD camera. The particle images are identified and localized by analyzing the maximum of the wavelet transform modulus (WTMM) and the equivalent diameter of the particle image (D_{eq}). Afterwards, a 3D point-matching (PM) algorithm is applied to the pair of sets containing the 3D particle locations. In the PM algorithm, the displacement of the particles is modeled by an affine transformation. This affine transformation is based on the use of the dual number quaternions. Afterwards, the velocity-field extraction is performed. This system is tested with simulated particle field displacements and the feasibility is checked with an experimental displacement.

1 Introduction

Recently, we have demonstrated that digital in-line holography can be used to determine 2D velocity fields in several slices of a sample volume [1]. In-line holograms record on a CCD camera the far-field diffraction patterns of particles so that a large field can be stored. Then, from the reconstructed sample volume, the 3D location and size of each particles can be determined [2]. This technique constitutes an alternative to classical particle-field holography [3–9].

Generally, the diffraction pattern is analyzed by means of space-frequency operators. For example, *Onural* [10] showed that the diffraction process can be seen as a convolution operation between the amplitude transmission in the space plane and a family of wavelet functions. In a similar way, the holographic reconstruction process can be seen as a wavelet transform (WT)

A. Schroeder, C. E. Willert (Eds.): Particle Image Velocimetry,
Topics Appl. Physics **112**, 155–170 (2008)

of the hologram transmission function. More recently, it has been shown that the fractional fourier transformation (FRFT) was well adapted to carry out a digital reconstruction of near-field holograms [11, 12].

In this chapter, we show that in-line holography can be extended to determine 3D velocity fields in a flow that contains particle tracers. By this technique, we record directly on a charge coupled device (CCD) camera a couple of diffraction patterns produced from small particles illuminated by a modulated laser diode. The numerical reconstruction is achieved by applying the wavelet transformation method. The determination of the velocity vector is based on two main steps. First, the 3D coordinates of each particle image within a predefined reconstructed volume are extracted. In the second step, a particle tracking velocimetry (PTV) based on a point-matching (PM) algorithm is applied to find the best correspondence between the two sets of particle positions.

In Sect. 2, the theoretical background used for digital in-line reconstruction of holograms is recalled. In Sect. 3, we present a description of the proposed method for the estimation of 3D coordinates of each particle by the analysis of the variations of WTMM and the D_{eq} of the reconstructed image. Then, the main objective of this section is devoted to the presentation of the algorithm used to find the best correspondence between particle images. In Sect. 4, this method is applied on simulated holograms. Then, experimental results about 3D displacements of particle fields attest that such an approach can provide an interesting tool for diagnostics in flows. Finally, the advantage and the limitation of the presented HPTV method are discussed.

2 Theoretical Background

Let us recall the digital reconstruction process proposed by authors of [9]. Consider an opaque object $O(\xi, \eta)$ of diameter d illuminated by a monochromatic plane wave (see Fig. 1). The intensity distribution on a plane (x, y) located at a distance z_0 from this object can be approximated by the following convolution operation:

$$I_{z_0}(x, y) = 1 - O(x, y) ** \frac{2}{\lambda z_0} \sin\left[\frac{\pi(x^2 + y^2)}{\lambda z_0}\right]. \tag{1}$$

Note here that the intermodulation term has been neglected. This simplification is valid in the general case of far-field in-line holography (i.e., $\pi d^2/\lambda z_0 \ll 1$). The expression given by (1) can be rewritten as a WT of the amplitude distribution in the object plane:

$$I_{z_0}(x, y) = 1 - \frac{2}{\pi}\mathrm{WT}_O(a_0, x, y). \tag{2}$$

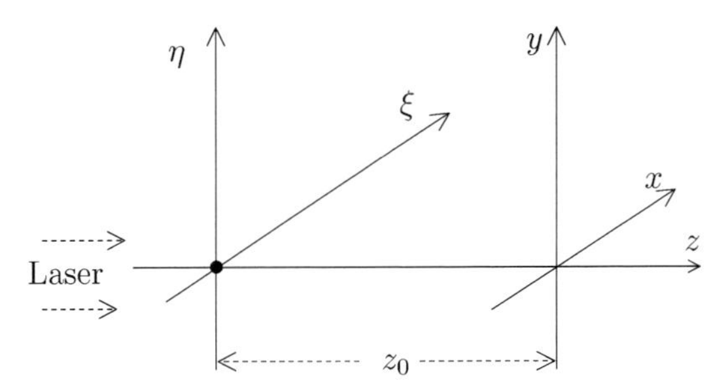

Fig. 1. Hologram recording in the Gabor configuration

The daughter wavelet functions are defined for the scale parameter a by:

$$\Psi_a = \frac{1}{a^2} \sin\left(\frac{x^2 + y^2}{a^2}\right) . \tag{3}$$

The scale parameter a_0 of the wavelet is related to the distance z_0 as follows:

$$a_0 = \sqrt{\frac{\lambda z_0}{\pi}} . \tag{4}$$

In the same way, this approach can be used for hologram reconstruction. The reconstructed image at a given distance z_r can be seen as the WT of the intensity distribution recorded by the photosensitive plane. When $a_r = a_0 = a$ (i.e., the interrogation plane located at a distance z_r corresponds to the object plane located at a distance z_0), it can be shown that by dropping a multiplicative constant term:

$$\mathrm{WT}_{I_z}(a, x, y) = 1 - O(x, y) - \frac{1}{2\lambda z} O(x, y) ** \sin\left[\frac{\pi(x^2 + y^2)}{2\lambda z}\right] . \tag{5}$$

The object function $O(x, y)$ can be easily reconstructed. This method can be extended to the case of several particles of different diameters provided that the far-field condition is maintained.

In fact, the function given by (3) is not really a wavelet and must be modified in order to check the admissibility conditions of a wavelet function (zero mean and localization). It is shown [9] that the following function can be used:

$$\Psi_{\mathrm{Ga}}(x, y) = \frac{1}{a^2}\left[\sin\left(\frac{x^2 + y^2}{a^2}\right) - M_\Psi(\sigma)\right] \exp\left(-\frac{x^2 + y^2}{\sigma^2 a^2}\right) , \tag{6}$$

where σ is a parameter that depends on the recording system (CCD camera) specifications. Its value is well discussed in [13]. The parameter M_Ψ is adjusted in order to have a zero mean value of $\Psi_{\mathrm{Ga}}(x, y)$.

3 3D Velocity Field Extraction and Data Postprocessing

In fluid flows studies, the analysis of the velocity in multiple planes or in a sample volume is of great interest. In our previous work we have demonstrated the feasibility of investigating several parallel planes by using digital in-line holography [1]. In this section we generalize this method to extract a 3D velocity field in a sample volume. The particle field to be studied is twice illuminated by a collimated laser beam emanating from a low-power modulated laser diode. When the object field to be studied is not too large, the diffraction patterns of particles can be directly recorded by means of a CCD camera. However, it should be admitted that generally an optical imaging setup with magnification under unity must be used so that larger image fields are both reported near the CCD camera. Compared to the equipment required in PIV experiments, this method has several advantages. Firstly, a low-power laser source is required and the same laser cavity is used to generate the two pulses. Such a source has been used successfully in a three-dimensional two-component (3D-2C) particle tracking velocimetry with a real flow [1]. Secondly, the in-line configuration is appreciated when optical access is limited.

In digital in-line holography, out-of-focus particle images give rise to high-contrast circular rings. Consequently, classical correlation-based algorithms (that are commonly used in PIV experiments) are not suitable for finding a local displacement of holographic particle images. Indeed, the displacement of out-of-focus images has the same influence on the correlation map as the infocus image displacements. Therefore, the mean velocity cannot be directly evaluated by using a statistical operator based on the gray-level analysis of the reconstructed images.

Our approach for velocity field analysis is based on two main steps. First, The 3D particle coordinates of each particle identified within a predefined reconstructed slice are extracted. The method used to validate a bright spot as a particle image is developed in the next section. In the second step, the point-matching algorithm is applied to find the best correspondence between the two sets of particle positions. This result is used to extract the 3D velocity field.

3.1 Extraction of 3D Particle Images

The computation of the velocity field from a couple of images gives rise to the need of a robust and accurate method for extracting particle coordinates from the reconstructed 3D images.

First, let us consider a space region corresponding to a set of reconstructed z-planes $WT_{I_z}(a, x, y)$ such that the inequalities $z_\mathrm{r} - \delta z \leq z \leq z_\mathrm{r} + \delta z$ are checked. This region defines a sample volume built from several reconstructed planes and whose thickness is equal to $2\delta z$ (typically a few millimeters).

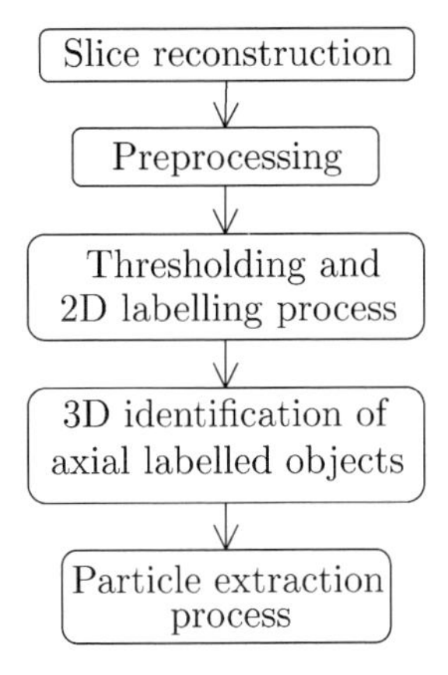

Fig. 2. General process for determining the 3D particles's locations

The general process to determine the 3D particle locations within this volume is outlined in Fig. 2.

Before selecting the supposed particle images by a thresholding operation, a background substraction is performed. The background is estimated by using a morphological opening operator. The size of the structuring element of this operator must be much higher than the particle-image diameters. The objective of this operation is to improve the signal-to-noise ratio by enhancing the contrast of the particle images in each plane. This operation is followed by a thresholding and a 2D labeling operations. Then, the x–y positions of labeled objects in each plane are available.

In the fourth step, a 3D labeled object is built by clustering the 2D labeled objects located on the same z-axis. During this operation, only the 3D labeled objects that are contained at least in three consecutive planes are considered as particle images. The use of this rejection principle is based on the great depth of field property of in-line holographic images. Therefore, a given labelled particle image is inevitably detected in several adjacent z-planes (see Fig. 3).

Finally, the result of the above step is used for recognizing the particles among all the three-dimensional labeled objects in the slice by using both the wavelet transform maximum modulus (WTMM) and the equivalent diameter (D_{eq}) variations versus the reconstruction distance z_{r}.

The WTMM method has been proposed in [14] to estimate the axial location of the particle images. The best focus plane is supposed to be the axial distance that gives the maximum of $\mathrm{WTMM}(z_{\mathrm{r}})$. The use of the $D_{\mathrm{eq}}(z_{\mathrm{r}})$ has been proposed in [15] to evaluate particle location. It can be shown that the axial position of the particle corresponds to a minimum of D_{eq} (see Fig. 4). Let us consider Z_1 and Z_2 as the axial coordinates corresponding, respectively, to the maximum of $\mathrm{WTMM}(z_{\mathrm{r}})$ and the minimum of $D_{\mathrm{eq}}(z_{\mathrm{r}})$. The joint exploitation of these extrema efficiently eliminates ambiguities between the 3D particle images and other 3D objects that are present in several planes.

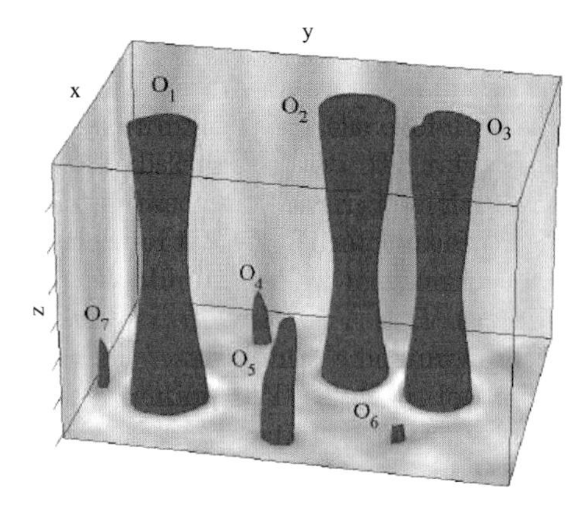

Fig. 3. An isosurface computed from the constructed sample volume containing 3 particle images (3D objects denoted O_1, O_2 and O_3)

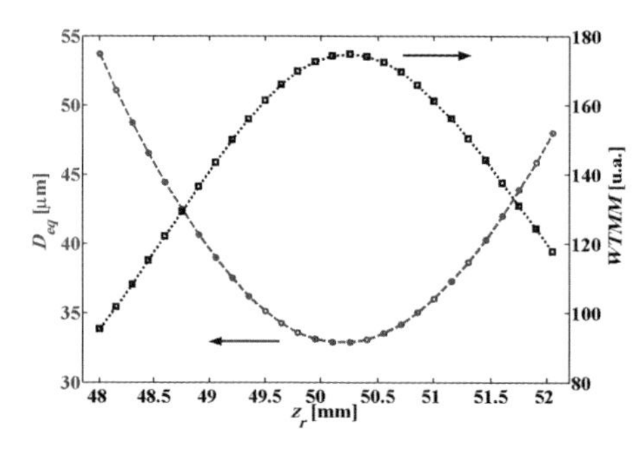

Fig. 4. Variations in WTMM and D_{eq} versus z_{r}

In the actual sample volume, only the 3D labeled objects that present the parabolic shapes of the WTMM and D_{eq} curves (as shown in Fig. 4) and that check $Z_1 \approx Z_2$ are considered as 3D particle images. In the example illustrated in Fig. 3, only 3D objects denoted O_1, O_2 and O_3 are considered as particle images.

3.2 Computation of the 3D Velocity Field

3.2.1 Dual-Number Quaternion for 3D Pose Estimation

In this section we recall the properties of dual-number quaternions as proposed in [16]. Quaternions are an extension of complex numbers to R^4 that consist of a 3×1 vector and a scalar component. Formally, a quaternion $\boldsymbol{q}$ can be defined as

$$\boldsymbol{q} = \begin{bmatrix} q_1 \\ q_2 \\ q_3 \\ q_4 \end{bmatrix} = \begin{bmatrix} \boldsymbol{q} \\ q_4 \end{bmatrix} . \tag{7}$$

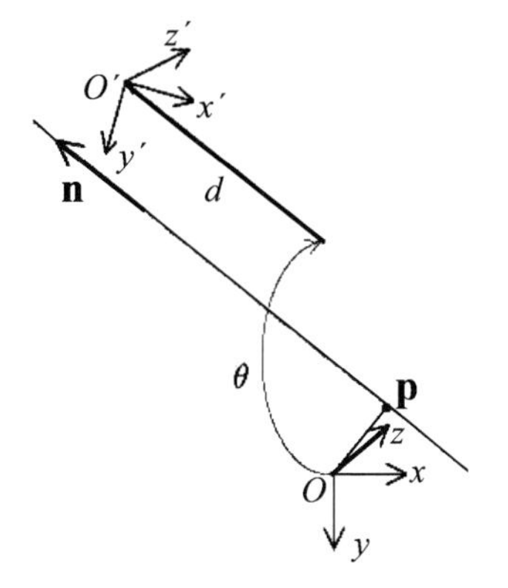

Fig. 5. Rotation and translation by using dual-number quaternion

In this chapter, each quaternion is represented by a bold-face italic character (for example $\boldsymbol{q}$) and all 3×1 vector by a standard bold-face character (for example $\mathbf{q}$).

Quaternions are limited to representing 3D rotation. In this case, the components of a quaternion are called the Euler symmetric parameters and we can rewrite the expression of the quaternion as

$$\boldsymbol{q} = \begin{bmatrix} \sin(\theta/2)\mathbf{n} \\ \cos(\theta/2) \end{bmatrix} . \tag{8}$$

The unit vector $\mathbf{n}$ is the vector about which the coordinate system has rotated and θ is the amount of rotation about $\mathbf{n}$.

By using $\mathbf{q}$ and q_4 the rotation matrix is expressed as

$$\mathbf{R} = (q_4^2 - \mathbf{q}^T\mathbf{q})\mathbf{I} + \mathbf{q}\mathbf{q}^T + 2q_4\mathbf{K}(\mathbf{q}) , \tag{9}$$

where

$$\boldsymbol{K}(\mathbf{q}) = \begin{bmatrix} 0 & -q_3 & q_2 \\ q_3 & 0 & -q_1 \\ -q_2 & q_1 & 0 \end{bmatrix} . \tag{10}$$

$\boldsymbol{I}$ is 4×4 identity matrix.

To represent both rotation and translation at the same time we can use the dual-number quaternions. These numbers consist of two parts,

$$\widehat{\boldsymbol{q}} = \boldsymbol{r} + \epsilon , \boldsymbol{s} , \tag{11}$$

where $\boldsymbol{r}$ and $\boldsymbol{s}$ are quaternions and are called the real part and dual part of $\widehat{\boldsymbol{q}}$ and $\epsilon^2 = 0$.

By this representation, a given 3D transformation can be described by first translating the original coordinate frame along the direction of unit vector $\mathbf{n}$ by the distance d and then by rotating it by an angle θ around the line having $\boldsymbol{n}$ as unit vector and passing through a point $\mathbf{p}$ (see Fig. 5). The quaternions

$\boldsymbol{r}$ and $\boldsymbol{s}$ are related to the screw parameters ($\mathbf{n}$, θ and $\mathbf{p}$) by the following equations:

$$\boldsymbol{r} = \begin{bmatrix} \sin(\theta/2)\mathbf{n} \\ \cos(\theta/2) \end{bmatrix} \tag{12}$$

and

$$\boldsymbol{s} = \begin{bmatrix} \frac{d}{2}\cos(\theta/2)\mathbf{n} + \sin(\theta/2)(\mathbf{p} \times \mathbf{n}) \\ \cos(\theta/2) \end{bmatrix} . \tag{13}$$

Note that $\boldsymbol{r}$ and $\boldsymbol{s}$ must satisfy the following constraints:

$$\boldsymbol{r}^T\boldsymbol{r} = 1 \tag{14}$$

and

$$\boldsymbol{r}^T\boldsymbol{s} = 0 . \tag{15}$$

From the dual-quaternion representation of the 3D transform we can compute the homogeneous transform. The rotation matrix $\mathbf{R}$ can be deduced from the following expression:

$$\boldsymbol{R} = \begin{bmatrix} \mathbf{R} & 0 \\ \mathbf{0}^T & 1 \end{bmatrix} = \boldsymbol{W}(\boldsymbol{r})^T\boldsymbol{Q}(\boldsymbol{r}) , \tag{16}$$

where $\boldsymbol{W}(\boldsymbol{r})$ and $\boldsymbol{Q}(\boldsymbol{r})$ are defined as

$$\boldsymbol{W}(\boldsymbol{r}) = \begin{bmatrix} r_4\boldsymbol{I} + \boldsymbol{K}(\boldsymbol{r}) & \boldsymbol{r} \\ -\boldsymbol{r}^T & r_4 \end{bmatrix} \tag{17}$$

and

$$\boldsymbol{Q}(\boldsymbol{r}) = \begin{bmatrix} r_4\boldsymbol{I} - \boldsymbol{K}(\boldsymbol{r}) & \boldsymbol{r} \\ -\boldsymbol{r}^T & r_4 \end{bmatrix} . \tag{18}$$

The $\mathbf{0}^T$ is a 1×3 vector where all elements are equal to zero: $\mathbf{0}^T = [000]$.
The matrix $\boldsymbol{K}(r)$ is defined by (10).
The translation vector ($\mathbf{t}$) is related to the translation quaternion ($\boldsymbol{t}$) by

$$\mathbf{t} = \begin{bmatrix} \boldsymbol{t} \\ 0 \end{bmatrix} = \boldsymbol{W}(\boldsymbol{r})^T\boldsymbol{s} . \tag{19}$$

The full 3D affine transformation is represented here by $T_{[\boldsymbol{r},\boldsymbol{s}]}$ which consists of a rotation $\boldsymbol{R}$ followed by a translation $\boldsymbol{t}$.

3.2.2 Three-Dimensional Point-Matching Algorithm

In this section we present the formulation of our optimization problem by using dual-number quaternions for reaching the best correspondence between particle images of the first and the second exposure.

Firstly, both of the reconstructed sample volume are divided into 3D interrogation cells (IC). Let us consider that each IC consists of two frames X (taken at $t = t_0$), and Y (taken at $t = t_0 + \Delta t$), obtained from a double-exposure HPTV experiment. Such frames contain the two lists $\{\boldsymbol{X}_j, j = 1, \ldots, J\}$, and $\{\boldsymbol{Y}_k, k = 1, \ldots, K\}$ of the 3D particles' coordinates extracted as explained in Sect. 3.1. X is supposed to be related to Y by an affine transformation $T_{[\boldsymbol{r},\boldsymbol{s}]}$ defined below. The correspondence between particle j in the first set and particle k in the second set is defined by the correspondence variables m_{kj} that constitute the match-matrix $\boldsymbol{M}$ where

$$m_{kj} = \begin{cases} 1, & \text{if point } Y_k \text{ corresponds to point } X_j; \\ 0, & \text{otherwise.} \end{cases} \tag{20}$$

m_{kj} must satisfy the following equality constraints [17]:

$$\begin{aligned} m_{k(J+1)} + \sum_{j=1}^{j=J} m_{kj} &= 1\,, \\ m_{(K+1)j} + \sum_{k=1}^{k=K} m_{kj} &= 1\ . \end{aligned} \tag{21}$$

The elements $m_{k(J+1)}$ and $m_{(K+1)j}$ are added to interpret the missing particles. These variables are set to one if an error occurs during the extraction of positions from the reconstructed image or if a given particle image leaves the IC or enters in the IC.

By estimating the matrix $\boldsymbol{M}$ and the transformation parameters $\boldsymbol{r}$ and $\boldsymbol{s}$, the velocity vectors can be deduced in the considered IC.

Now, the problem is to estimate the optimal value of the $\boldsymbol{M}$, $\boldsymbol{r}$ and $\boldsymbol{s}$. The method used here consists in minimizing a cost function $E(\boldsymbol{M}, \boldsymbol{r}, \boldsymbol{s})$. According to reference [18], the form of $\mathrm{E}(\boldsymbol{M}, \boldsymbol{r}, \boldsymbol{s})$ used in this work is:

$$\begin{aligned} E(\boldsymbol{M}, \boldsymbol{r}, \boldsymbol{s}) = \sum_{k=1}^{k=K} \sum_{j=1}^{j=J} m_{kj} \|\boldsymbol{Y}_k - T_{[\boldsymbol{r},\boldsymbol{s}]}(\boldsymbol{X}_j)\|^2 \\ + \alpha \left[\sum_{j=1}^{j=J} m_{(K+1)j} + \sum_{k=1}^{k=K} m_{k(J+1)} \right] . \end{aligned} \tag{22}$$

By using (16) and (19) the expression of $T_{[\boldsymbol{r},\boldsymbol{s}]}(\boldsymbol{X}_j)$ is:

$$T_{[\boldsymbol{r},\boldsymbol{s}]}(\boldsymbol{X}_j) = \boldsymbol{W}(\boldsymbol{r})^T \boldsymbol{s} + \boldsymbol{W}(\boldsymbol{r})^T \boldsymbol{Q}(\boldsymbol{r}) \boldsymbol{X}_j\,. \tag{23}$$

Note that here we use the quaternion representation of particle positions:

$$\boldsymbol{Y}_k = \frac{1}{2} \begin{bmatrix} \mathbf{Y}_k \\ 0 \end{bmatrix} \tag{24}$$

and

$$\mathbf{X}_j = \frac{1}{2}\begin{bmatrix} \boldsymbol{X}_j \\ 0 \end{bmatrix} . \tag{25}$$

The cost function $E(\boldsymbol{M}, \boldsymbol{r}, \boldsymbol{s})$ evaluates the degree of similarity between two sets of particle coordinates X and Y. The first term of this function is the sum of the assignment costs and the second represents the cost that incurred if a particle in one frame has no partner in the other frame. The α parameter is the cost of a no match and must satisfy the following inequality (see Sect. 2.1.3 of [18]):

$$\|\boldsymbol{Y}_k - T_{[\boldsymbol{r},\boldsymbol{s}]}(\boldsymbol{X}_j)\|^2 < \alpha . \tag{26}$$

By introducing the expression of $T_{[\boldsymbol{r},\boldsymbol{s}]}(\boldsymbol{X}_j)$ in (22) and using the properties $\boldsymbol{Q}(\boldsymbol{a})\boldsymbol{b} = \boldsymbol{W}(\boldsymbol{b})\boldsymbol{a}$ and $\boldsymbol{W}(\boldsymbol{a})^T\boldsymbol{W}(\boldsymbol{a}) = \boldsymbol{W}(\boldsymbol{a})\boldsymbol{W}(\boldsymbol{a})^T = \boldsymbol{a}^T\boldsymbol{a}\boldsymbol{I}$, the cost function can be written as a quadratic function of $\boldsymbol{r}$ and $\boldsymbol{s}$,

$$E(\boldsymbol{M}, \boldsymbol{r}, \boldsymbol{s}) = \boldsymbol{r}^T\boldsymbol{C}_1\boldsymbol{r} + \boldsymbol{s}^T\boldsymbol{C}_2\boldsymbol{s} + \boldsymbol{s}^T\boldsymbol{C}_3\boldsymbol{r} + \boldsymbol{C}_4 , \tag{27}$$

where

$$\begin{aligned}
\boldsymbol{C}_1 &= -2\sum_{k=1}^{k=K}\sum_{j=1}^{j=J} m_{kj}\boldsymbol{Q}^T(\boldsymbol{Y}_k)\boldsymbol{W}(\boldsymbol{X}_j) , \\
\boldsymbol{C}_2 &= \Big(\sum_{k=1}^{k=K}\sum_{j=1}^{j=J} m_{kj}\Big)\boldsymbol{I} , \\
\boldsymbol{C}_3 &= 2\sum_{k=1}^{k=K}\sum_{j=1}^{j=J} m_{kj}[\boldsymbol{W}(\boldsymbol{X}_j) - \boldsymbol{Q}(\boldsymbol{Y}_k)] , \\
\boldsymbol{C}_4 &= 2\sum_{k=1}^{k=K}\sum_{j=1}^{j=J} m_{kj}[\boldsymbol{X}_j^T\boldsymbol{X}_j + \boldsymbol{Y}_k^T\boldsymbol{Y}_k] \quad .
\end{aligned} \tag{28}$$

With this new representation, the cost function is minimized by the couple $(\widehat{\boldsymbol{r}}, \widehat{\boldsymbol{s}})$ where $\widehat{\boldsymbol{r}}$ is the eigenvector corresponding to the largest positive eigenvalue of matrix $\boldsymbol{A}$. $\boldsymbol{A}$ is defined by the following expression:

$$\boldsymbol{A} = \frac{1}{2}\boldsymbol{C}_3^T(\boldsymbol{C}_2 + \boldsymbol{C}_2^T)^{-1}\boldsymbol{C}_3 - \frac{1}{2}(\boldsymbol{C}_1 + \boldsymbol{C}_1^T) . \tag{29}$$

The knowledge of $\widehat{\boldsymbol{r}}$ enables us to calculate $\widehat{\boldsymbol{s}}$ by

$$\widehat{\boldsymbol{s}} = -(\boldsymbol{C}_2 + \boldsymbol{C}_2^T)^{-1}\boldsymbol{C}_3\widehat{\boldsymbol{r}} . \tag{30}$$

In order to find the optimal values for the correspondence matrix and the pose parameters, the cost function $E(\boldsymbol{M}, \boldsymbol{r}, \boldsymbol{s})$ is minimized with respect to its arguments. These optimal solutions are computed by using a 3D PM algorithm originally proposed by *Gold* et al. [17]. This algorithm has been

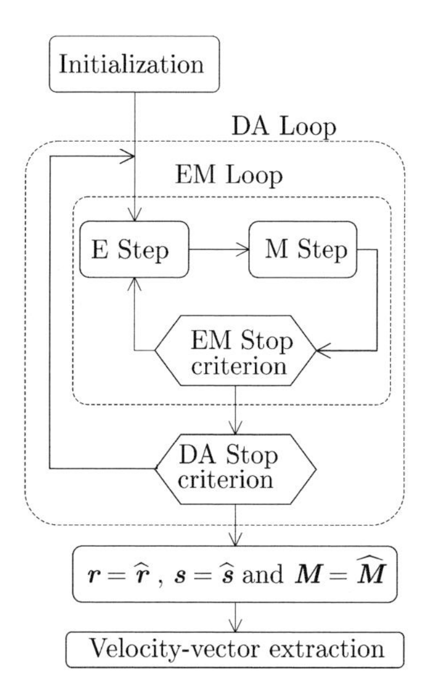

Fig. 6. Point-matching algorithm for 3D velocity-field extraction

extended by *Stellmacher* and *Obermayer* [18] for the development of a new particle tracking algorithm and by *Krepki* et al. [19] for the interrogation of 3D holographic PIV data.

As illustrated by Fig. 6, the solution of our optimization problem is realized by an expectation-minimization (EM) algorithm, which is combined with a deterministic annealing (DA) scheme. In the EM algorithm, we proceed in two steps. In the first step (M step) we estimate the correspondence matrix using the softassign algorithm [17]. This algorithm uses an iterative normalization along columns and rows with the exception of the slack variables. In the second step (E step) we estimate the pose parameters ($\boldsymbol{r}$, $\boldsymbol{s}$) using the dual-number quaternion method presented in Sect. 3.2.1. The whole EM algorithm is enclosed by a DA that is used to ensure convergence toward a global minimum of the cost function $E(\boldsymbol{M}, \boldsymbol{r}, \boldsymbol{s})$. The convergence of the DA scheme is governed by the inverse-temperature parameter β. This parameter can be selected between $\beta = \beta_{\text{begin}}$ (theoretically tends to zero) until $\beta = \beta_{\text{end}}$ (theoretically tends to $+\infty$) with an increment factor β_{fac}. In our case we have chosen $\beta_{\text{begin}} = 0.5$, $\beta_{\text{end}} = 1000$ and $\beta_{\text{fac}} = 1.5$. Finally, the velocity field is extracted by exploiting the correspondence matrix obtained. To improve this result we can use an adequate postprocessing as filtering and interpolation.

4 Simulations

In this section, we show the potentiality of the proposed method to evaluate the displacement according z-axis. We have simulated two intensity distribu-

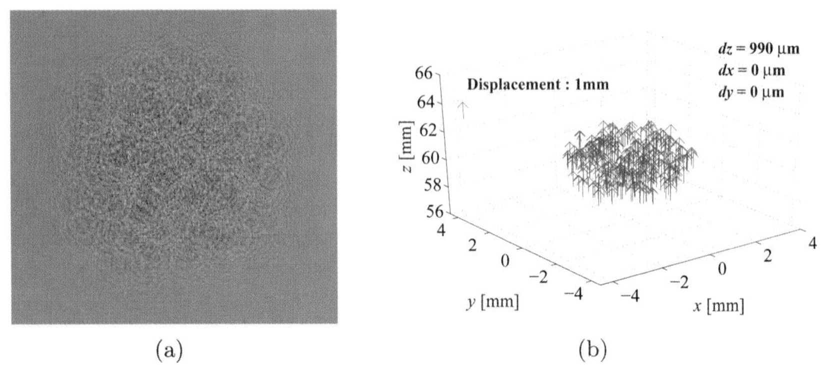

(a) (b)

Fig. 7. 3D-3C HPTV: (**a**) Intensity distribution in one of the diffraction patterns and (**b**) reconstruction of the 3D velocity vector field

tions $I_1(x, y)$ and $I_2(x, y)$. These distributions are supposed to represent the diffraction patterns of the particle field at two different times. The concentration number in this simulation is set to 5 particles per mm^3 and the diameter of each particle is fixed to 20 μm. The particles are distributed uniformly in a spherical region with radius equal to 2.3 mm and center coordinate at ($x = 0$, $y = 0$, $z = 59.6$ mm). We have simulated a z-displacement by moving the particle field along the z-axis by 1 mm. Then, the pair of the reconstructed sample volumes is processed according to the procedure discussed in Sects. 3.1 and 3.2, respectively.

The result of this simulation is presented in Fig. 7. In this figure, dx, dy and dz are the average value of the measured displacement in x-, y-, and z-directions. As we can see, we have recovered the displacement vectors in the simulated sample volume. In this simulation, few false vectors have been produced. These vectors have been reduced by adjusting the parameters of the PM algorithm as the convergence criteria of the softassign algorithm, the EM algorithm and the DA.

The errors observed on the calculation of particle displacements are mainly due to the localization uncertainties during the extraction process. In fact, in a previous study we have established that the reliability of the extraction process depends on the signal-to-noise ratio (SNR) of the reconstructed particle fields [15]. We have shown that the particle number density should be drastically reduced for successful diagnostics in thick volume. Recall that the results presented in this simulation are obtained by a particle number density of 5 particles per mm^3. By using this low value, the error obtained on the measured z-displacement is 1 %.

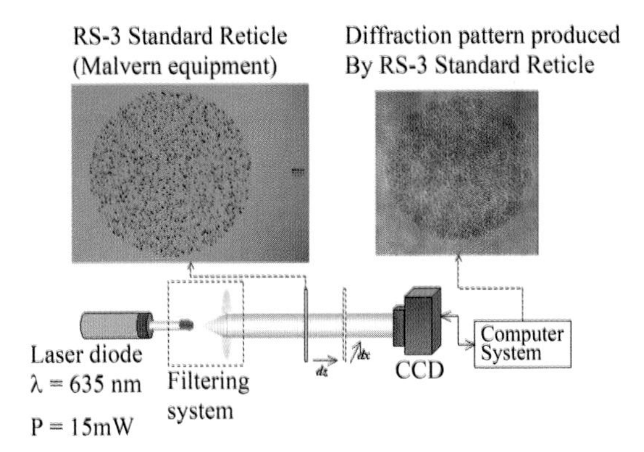

Fig. 8. Experimental setup for the 3D HPTV (z-axis and x-axis displacement: $d_z = 1\,\text{mm}$ and $d_x = 50\,\mu\text{m}$)

5 Experimental Results

In order to test our 3D HPTV system, we have used an RS-3 standard reticle (Malvern Equipment) that is approximately parallel to the recording plane. This reticle is an optical glass plate with a pattern of small opaque disks ($10\,\mu\text{m} < d < 90\,\mu\text{m}$) photographically deposited on the surface. This reticle is illuminated by a collimated laser beam generated by a low-power modulated laser diode ($P = 15\,\text{mW}$ and $\lambda = 635\,\text{nm}$).

The diffraction patterns corresponding to each exposure are both recorded by means of a CCD camera and digitized by a computer vision system as shown in Fig. 8. The size of the registered image is 1008×1018 pixels and the pixel size is equal to $9\,\mu\text{m}$.

At first, we have recorded one diffraction pattern generated by the reticle at a given position. Then, we have moved it by $1000\,\mu\text{m}$ in the z-direction and by $50\,\mu\text{m}$ in the x-direction. At this new position, a second diffraction pattern has been recorded.

Figure 9a shows an example of the diffraction pattern corresponding to one of the exposure. The information contained in this diffraction pattern leads to a satisfactory digital reconstruction. We have applied the process described in Sect. 3.1 to reconstruct the volume containing the particle field and to extract the 3D particle locations corresponding to each exposure. In this experiment, we have used the nearest-neighbor algorithm to reduce the level of miscalculation during the PM procedure. The reconstructed velocity vector map is presented in Fig. 9b. This result is obtained by applying a filtering process based on the analysis of the velocity histogram.

As shown by this figure, the average value of the measured displacement along the z-axis is $900\,\mu\text{m}$ (i.e., error of $10\,\%$). Here, the error is larger than in the simulation case. We believe that the experimental configuration causes a supplementary error. In fact, in this experiment, the displacements are realized manually. Thus, a systematic error is probably introduced into the

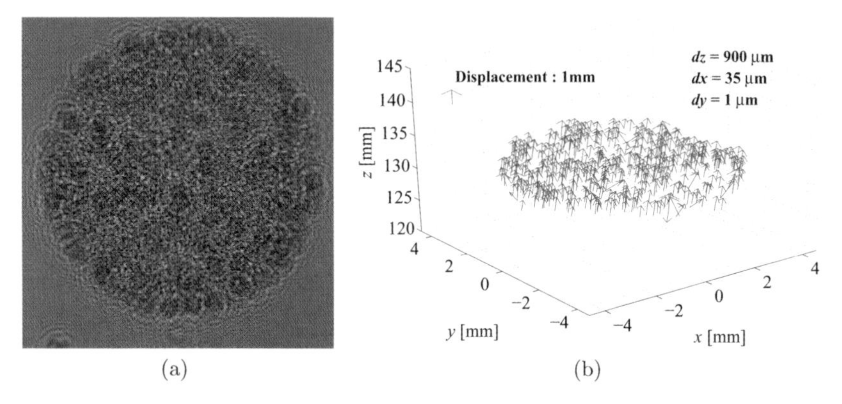

Fig. 9. Experimental result of 3D-3C HPTV: (**a**) Intensity distribution in one of the diffraction patterns and (**b**) reconstructed of 3D velocity vector

measurements. A second explanation concerning this difference is that the recording beam is not perfectly collimated. Consequently, a magnification factor less than unity could be introduced if a convergent beam was used.

Finally, some false vectors can be observed in Fig. 9b. These nonexpected vectors may significantly reduce the average value of the measured displacement.

These remarks are also valid for the difference observed in x- and y-directions.

According to the conclusion of [9], the best resolution that can be expected for the determination of the z-coordinate is about 50 µm when a plane wave illuminating a few particles is considered. More generally, when a higher particle concentration is considered, the accuracy in z can exceed more than several hundreds of micrometers.

Moreover, it must be noted that a uniform displacement has been used for testing our algorithm. Consequently, large IC ($128 \times 128 \times 128$) have been used. When small scales are investigated, for example in more complex two-plane flows, small IC are recommended. This point can be connected to the general problem of the spatial resolution in PIV experiment. Here, the discussion is extended in the 3D case.

6 Conclusion

In this study, it has been shown that the use of the digital in-line holography provides an appropriate tool to determine 3D velocity fields. This approach cannot be directly compared to other methods like particle image velocimetry (PIV) or laser Doppler velocimetry (LDV) where high-density seeding is recommended. Here, considering the limitations imposed by in-line holography, it is necessary to work with a low concentration of particles involved. Then 3D particle tracking velocimetry methods, based on the use of the point-matching algorithm, are more suited to this situation than correlation-based methods. Although, the velocity vector fields are not statistically as reliable as those obtained by a PIV system, additional useful information concerning instantaneous velocity in a given 3D flow can be brought. Our experimental results show the feasibility and possibilities offered by this method.

References

[1] M. Malek, D. Allano, S. Coëtmellec, C. Özkul, D. Lebrun: Digital in-line holography for three-dimensional-two-components particle tracking velocimetry, Meas. Sci. Technol. **15**, 699–705 (2004)
[2] G. A. Tyler, B. J. Thompson: Fraunhofer holography applied to particle size analysis a reassessment, Opt. Acta **23**, 685–700 (1976)
[3] T. M. Kreis, W. Jüptner: Suppression of the DC term in digital holography, Opt. Eng. **36**, 2357–2360 (1997)
[4] M. K. Kim: Tomographic three-dimensional imaging of a biological specimen using wavelength-scanning digital interference holography, Opt. Exp. **7**, 305–310 (2000)
[5] O. Schnars, W. Juptner: Direct recording of holograms by a CCD target and numerical reconstruction, Appl. Opt. **33**, 179–181 (1994)
[6] I. Yamaguchi, T. Zhang: Phase-shifting digital holography, Opt. Lett. **22**, 1268–1270 (1997)
[7] W. Xu, M. H. Jericho, H. J. Kreuzer, I. A. Meinertzhagen: Tracking particles in four dimensions with in-line holographic microscopy, Opt. Lett. **28**, 164–166 (2003)
[8] G. Pan, H. Meng: Tracking particles in four dimensions with in-line holographic microscopy, Appl. Opt. **42**, 827–833 (2003)
[9] C. Buraga-Lefebvre, S. Coëtmellec, D. Lebrun, C. Özkul: Application of wavelet transform to hologram analysis: Three-dimensional location of particles, Opt. Laser Eng. **33**, 409–421 (2000)
[10] L. Onural: Diffraction from a wavelet point of view, Opt. Lett. **18**, 846–848 (1993)
[11] S. Coëtmellec, D. Lebrun, C. Özkul: Characterization of diffraction patterns directly from in-line holograms with the fractional fourier transform, App. Opt. **41**, 312–319 (2002)
[12] S. Coëtmellec, D. Lebrun, C. Özkul: Application of the two-dimensional fractional-orders fourier transformation to particle field digital holography, J. Opt. Soc. Am. A **19**, 1537–1546 (2002)
[13] M. Malek, S. Coëtmellec, D. Lebrun, D. Allano: Formulation of in-line holography process by a linear shift invariant system, Opt. Commun. **223**, 263–271 (2003)
[14] S. Belaïd, D. Lebrun, C. Özkul: Application of two dimensional wavelet transform to hologram analysis: Visualization of glass fibers in a turbulent flame, Opt. Eng. **36**, 1947–1951 (1997)
[15] M. Malek, D. Allano, S. Coëtmellec, D. Lebrun: Digital in-line holography influence of the shodow density on particle field extraction, Opt. Exp. **12**, 2270–2279 (2004)
[16] M. W. Walker, L. Shao, R. A. Volz: Estimating 3-d location parameters using dual number quaternions, CVGIP:Image Understanding **54**, 358–367 (2004)

[17] S. Gold, C. P. Lu, A. Rangaragan, S. Pappu, E. Mjolsness: New algorithms for 2d and 3d point matching. pose estimation and correspondence, Pattern Recog. **31**, 1019–1031 (1998)
[18] M. Stellmacher, K. Obermayer: New particle tracking algorithm based on deterministic annealing and alternative distance measures, Exp. Fluids **28**, 506–518 (2000)
[19] B. Krepki, Y. Pu, H. Meng, K. Obermayer: A new algorithm for the interrogation of 3d holographic PTV data based on deterministic annealing and expectation minimization optimization, Exp. Fluids (Suppl.) **29**, S099–S0107 (2000)

Index

Holographic PIV System Using a Bacteriorhodopsin (BR) Film

Thomas Ooms[1], Victor Chan[1], Jerry Westerweel[1], Wouter Koek[2], Nandini Bhattacharya[2], and Joseph Braat[2]

[1] Faculty of Mech. Engineering, Delft University of Technology, Delft, The Netherlands,
t.a.ooms@wbmt.tudelft.nl
[2] Faculty of Applied Sciences, Delft University of Technology, Delft, The Netherlands

Abstract. We present a holographic particle image velocimetry system that uses a reusable holographic material as the recording medium. The measurement system records double-exposure holograms in a film containing the photochromic protein bacteriorhodopsin (BR). The two constituent particle holograms are recorded in a single film using polarization multiplexing. We describe the system in detail and present three-dimensional measurements that determine the system's accuracy and demonstrate its capabilities.

1 Introduction

Particle image velocimetry (PIV) is a common and valuable technique for the study of fluid flows [1]. In its basic form PIV yields planar two-component flow-field information. Although special stereoscopic or scanning implementations have been developed [2, 3], it has been impossible to perform truly instantaneous three-dimensional three-component flow measurements with these PIV methods.

The need for instantaneous three-dimensional three-component flow measurements for both engineering and fundamental flow studies suggests the use of holographic imaging systems, and much effort is currently focused on the development of such systems. Although several successful holographic particle image velocimetry (HPIV) systems have been demonstrated, they all use silver-halide film as the holographic medium [4–7]. The chemical development associated with such films introduces distortion to the recorded hologram, requires a skilled person for the development, and is time consuming. Digital HPIV forms an alternative to film-based HPIV [8]. However, the space-bandwidth of current state-of-the-art CCDs is very small when compared to holographic film, thereby limiting the number of flow vectors that can be obtained from one recording. *Barnhart* et al. [9] realized that a practical HPIV system must use a holographic film that does not require development, and they suggested the use of bacteriorhodopsin (BR).

In this chapter we present the first operational HPIV system to employ BR as the recording material. Furthermore, the presented HPIV system is the

A. Schroeder, C. E. Willert (Eds.): Particle Image Velocimetry,
Topics Appl. Physics **112**, 171–189 (2008)

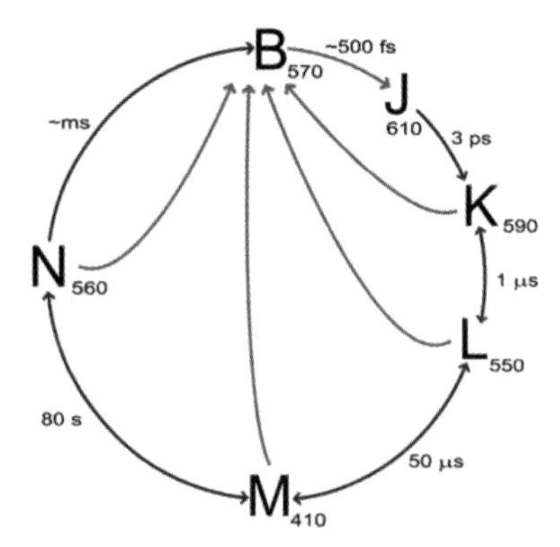

Fig. 1. Photocycle for D96N BR. *Black arrows* indicate a thermal transition. *Gray arrows* indicate photoinduced transitions. The absorption peak for each photointermediate is shown

first to use polarization multiplexing for the independent recording of the two particle holograms. We will show results to demonstrate the system's capabilities for the study of three-dimensional three-component flow fields.

2 BR Films

In nature, the photochromic protein BR plays a crucial role in the photosynthetic system of the salt-marsh bacterium Halobacterium salinarium [10]. BRs are found in the bacterial cell membrane in the form of two-dimensional crystalline patches. These patches have a diameter of up to 5 μm and are called purple membrane (PM). Upon absorption of light the BR molecule transports a proton from the cell's cytoplasm to the outer medium. The energy that is stored in this proton gradient is used by a membrane-bound enzyme (ATPase) to regenerate adenosine triphosphate (ATP) from adenosine diphosphate (ADP). Like all other living cells, the halobacterium uses ATP as the energy source to drive its cellular molecular processes. Effectively, the bacterium uses BR to convert sunlight directly into chemical energy.

Prior to absorption of a photon the BR molecule is in the ground state, called B. The absorption of a photon triggers a sequence of molecular configurational changes resulting in the pumping of a proton, and finally leading the molecule back into the B state. Due to their different molecular configurations the various intermediates exhibit different absorption spectra.

Figure 1 displays the photocycle of BR, and shows the absorption maxima of the various intermediates. As shown in Fig. 2, the BR photocycle is often simplified by adopting a two-state model using B and M. Not only is M the longest-lived intermediate state, its shift of the absorption peak over 160 nm is much greater than that of any other state. The ability to use light to induce a large modulation of the optical density makes the protein a good

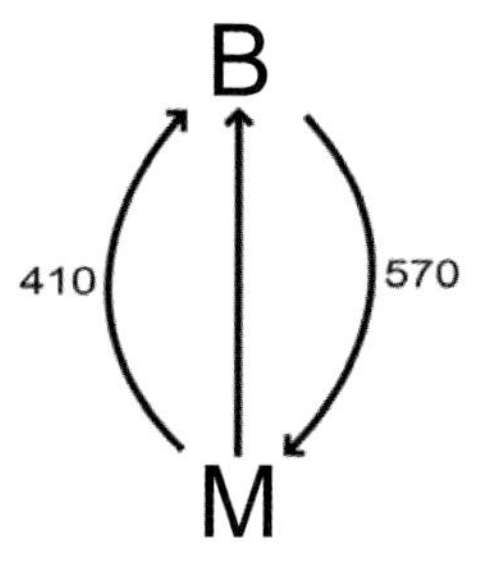

Fig. 2. Simplified two-state model of the BR photocycle

material for optical recording. Since the configurational changes also lead to a change in refractive index of the order of 10^{-3}, BR can be used for combined phase-amplitude recordings. The protein's size, that is of the order of a few nanometers, combined with the dense packing of BR within PM, allows for high-resolution recording ($> 5000\,\mathrm{lp/mm}$). The above-mentioned features make BR very attractive as a reversible holographic recording medium, and as such BR has been integrated in an interferometric measurement system [11].

When illuminating the protein with linearly polarized light, for the probability P to initiate a photoinduced transition can be written:

$$P(\varphi) \propto \cos^2\varphi + k_1 \sin^2\varphi\,, \tag{1}$$

where φ is the angle between the molecule's long axis and the plane of polarization of the exciting light, and the dichroic ratio $k_1 = 0.04$. Illuminating a normally isotropic collection of BR molecules with linearly polarized light will thus result in a photoinduced anisotropic distribution of molecules in the B and the M state. *Kakichashvili* [12] was the first to realize the extra possibilities that materials with photoinduced anisotropy offer for holographic recording. He showed the feasibility of recording a hologram in polarization-sensitive materials using an object and reference wave with orthogonal polarizations, both linear and circular, by recording a varying state of polarization rather than recording a varying state of intensity, as is done in traditional holography.

Through many millions of years of evolution BR has been optimized for light absorption and energy conversion. Unfortunately, the requirements for an intracellular energy converter tend to differ from those for a holographic recording material. For example, in wild-type (WT) BR, the form as it is found in nature, the half-lifetime of the M state is only 15 ms. This rapid loss of information makes the medium too volatile for many practical applications. However, several genetic variants exist that have different properties from WT BR. The mutant used in our experiments is D96N. Obtained through random mutagenesis this variant has an M half-lifetime of 100 s, thereby providing more time to extract the information contained in a hologram.

Table 1. The polarization of the reconstructed image is shown for different combinations of the reference beam, object beam, and reconstruction beam polarization (L = LHCP, R = RHCP, – indicates no image is reconstructed)

Reference beam	Object beam	Reconstruction beam	Reconstructed image
L	R	L	R
L	R	R	–
R	L	L	–
R	L	R	L

For use in a technical application it is practical when the BR is contained in a film. Fortunately, upon extraction of PM from the bacterial cell the protein BR maintains its functionality. The extracted PM can be embedded in a film matrix that is subsequently sandwiched between two glass plates. The obtained film has a homogeneous optical density and a uniform thickness. Since the angular distribution of BR is isotropic, the film is well suited for the recording of polarization holograms. The film used in this experiment has an aperture of 100×100 mm, a thickness of 30 μm, and optical density of 1.5 at 570 nm, and is commercially available from Munich Innovative Biomaterials (MIB) GmbH.

3 Polarization Multiplexing

Polarization holograms have the unique ability to modify the polarization of the reconstruction beam resulting in a reconstruction of the original object polarization. Polarization holograms formed by two orthogonal circularly polarized waves also have the ability of reconstructing an image in the $+1$ order only and not in the -1 order and *vice versa* [13]. *Koek* et al. demonstrated the crosstalk-free polarization multiplexing of two images in BR, using either a sequential or a simultaneous reconstruction [14, 15]. As is illustrated in Table 1, the technique of sequential readout polarization multiplexing involves recording the first image using left-hand circular polarization (LHCP) for the reference beam, and right-hand circular polarization (RHCP) for the object beam.

The second image is then recorded using RHCP for the reference beam and LHCP for the object beam. If the hologram is reconstructed with LHCP the first recorded image will be reconstructed having RHCP, whereas the second image is fully suppressed. Alternatively, if the hologram is reconstructed using RHCP the second image is reconstructed having LHCP, while the first image remains absent. Thus, by choosing the proper polarization for the reconstruction beam the two holographically stored images can be retrieved independently. This is the type of multiplexing that is used for the experiments presented in this chapter.

The ability to use polarization multiplexing gives polarization-sensitive materials (such as BR) a distinct advantage for use in HPIV over nonpolarization sensitive materials (such as silver halide, or photopolymer films). HPIV systems based on silver-halide films normally use angular multiplexing to store the two constituent holograms. The need for two reference beam paths obviously makes the system more complex and subject to alignment errors. Polarization multiplexing allows both pulses to follow a common beam path, thereby keeping the system simple.

4 Holographic Imaging System and Data Processing

The potential of our holographic imaging system is investigated by performing three experiments; a calibration measurement, a jet-flow measurement and a vortex-ring flow measurement. The first two experiments (calibration and jet flow) are performed with a similar imaging system configuration and a similar data-processing method. These are described in Sect. 4.1. The third experiment (vortex-ring flow) is performed with an improved imaging system and an improved data-processing method, as described in Sect. 4.2.

4.1 System Configuration of Calibration Measurement and Jet-Flow Measurement

Our HPIV system based on polarization multiplexing in BR will be demonstrated by means of experiments on a water jet seeded with 100 μm solid glass spheres. As is illustrated in Fig. 3 the jet is injected into a tank of clear water through a syringe nozzle having an exit diameter of 2 mm. With a tank width of 50 mm the wall boundaries should have little influence on the near field of the flow.

In contrast to the neutrally buoyant hollow glass spheres that are frequently used in PIV, the solid glass spheres used in this experiment are obviously heavier than water, thereby influencing the fluid flow. However, the choice of particle ensures maximum efficiency of the object illumination beam. Rather than scattering in all directions as is the case with hollow glass spheres, a solid glass sphere acts as a ball lens directing almost all the incident light into a more or less uniform cone with NA = 0.15. This value of the NA was determined by performing a three-dimensional scan of the holographic reconstruction of such a particle.

In the remainder of this section we will discuss the three main steps that are involved in performing a holographic measurement; recording a hologram, reconstructing the hologram, and finally analyzing the data. In the next section we will present the experimental results.

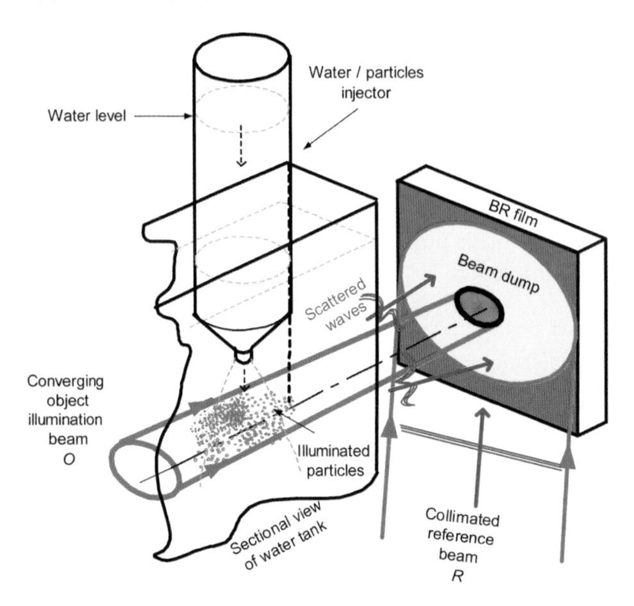

Fig. 3. Schematic drawing of analyzing jet flow by holographically recording the particle field in BR

4.1.1 Recording the Hologram

In order to detect the velocity of the particles, their location must be recorded twice with a known time separation between the two recordings. By means of polarization multiplexing the spatial distribution of the particle field at the two moments is recorded into a single BR film using the experimental setup shown in Figs. 3 and 4. Prior to each recording the BR film is exposed to light from a UV LED array (410 nm). This erases any prior information in the film by forcing all molecules into the B state (Fig. 2). The holograms are recorded using a pulsed frequency-doubled Nd:YAG laser that outputs a maximum of 320 mJ per 7 ns pulse at 532 nm into a beam with 8 mm beam diameter. Using a half-wave plate and a Glan laser polarizer the light from the pulsed Nd:YAG laser is split into an object illumination wave O, and a reference wave R. The quarter-wave plates are adjusted such that when the Pockels cells are inactive the reference beam has LHCP at the location of the BR film, and the object beam has RHCP at the location of the BR film. Upon activation, the Pockels cells function as half-wave plates allowing to record the second hologram using RHCP for the reference beam and LHCP for the object beam. After passing through the Pockels cell wavefront R is expanded to 65 mm and collimated by lenses L1 and L2, after which it falls onto the BR film at an incidence angle of approximately 40°. Using lenses L3 and L4 wavefront O is expanded to 17 mm and made slightly convergent (NA = 0.005).

The convergent object illumination beam enters the tank of clear water, in which the injected particles close to the injection nozzle will refract the

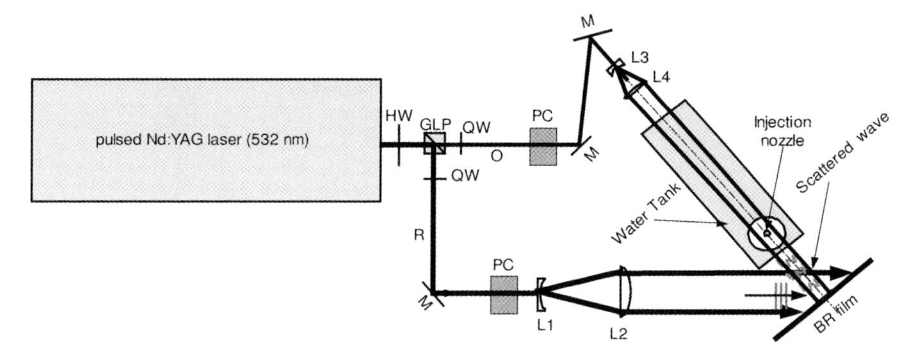

Fig. 4. Schematic for recording a double-exposed particle hologram. HW: $\lambda/2$ plate; GLP: Glan laser polarizer; QW: quarter-wave plate; O: object wavefront; R: reference wavefront; M: mirror; PC: Pockels cell; Lx: lens

incident light. The resulting divergent waves have NA = 0.15 inside the water tank and NA = 0.20 after exiting the tank. The NA of the holographic imaging system is sufficient to capture all the refracted light. The BR film is partially covered with a circular metallic film beam dump (diameter = 8 mm) to protect it from permanent damage by the unrefracted part of the object illumination beam. The beam dump also prevents the presence of the unrefracted beam in the reconstruction that leads to a better particle detectability. The refracted part of the light will fall onto the BR film and interfere with the reference beam to form a holographic recording of the particle field. At the location of the BR film wavefront R has a fluence of $5\,\mathrm{mJ/cm^2}$. Wavefront O has a fluence of $80\,\mathrm{mJ/cm^2}$ at the location of the injection nozzle, obviously only a fraction of this energy will contribute to the formation of the hologram. The system is controlled by dedicated timing electronics that synchronize the firing of the two laser cavities and the switching of the polarization between the two recordings.

Because the object beam is convergent, the beam dump on the BR film can be smaller than the cross-sectional area of the illuminated particle field. With this configuration more refracted light is captured on the holographic film than with a collimated object beam (which would require a beam dump that is as large as the cross-sectional area of the illuminated object). Because more scattered light is recorded on the film, more light is diffracted during reconstruction that leads to a higher reconstructed particle image intensity. Another elegant solution for this problem is using a hybrid geometry where the zeroth-order beam is filtered out by means of a Fourier filter [16]. This geometry is applied in the experiment of the measurement of the vortex ring (Sect. 4.2.1).

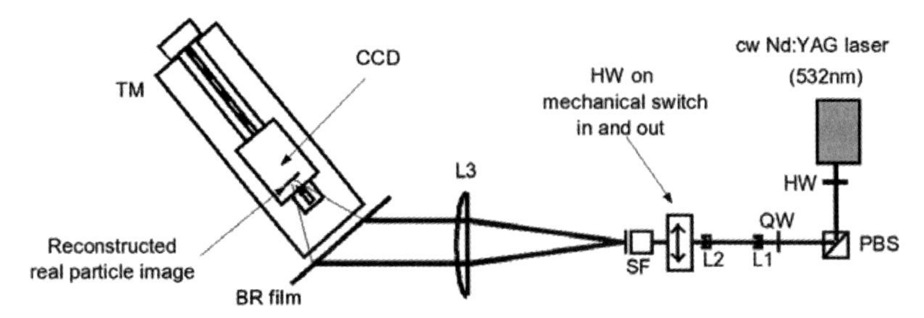

Fig. 5. Schematic for sequentially reconstructing two polarization multiplexed particle holograms. HW: $\lambda/2$ plate; PBS: polarizing beam splitter; QW: quarter-wave plate; Lx: lens; SF: spatial filter; TM: motorized translation mechanism

4.1.2 Reconstructing the Hologram

After the two particle holograms have been recorded, the information contained in the BR film must be extracted. This is done using the experimental setup in Fig. 5. Light from a CW frequency-doubled Nd:YAG laser (532 nm) is used for the reconstruction. Using a half-wave plate and polarizing beam-splitter the beam intensity can be adjusted. A quarter-wave plate converts the linearly polarized light to RHCP. By means of a half-wave plate that is mounted on a mechanical switch the beam can switch between RHCP and LHCP with speeds up to 10 Hz. This switching is used to alternate between the polarization multiplexed images. Lenses L1 and L2 expand the beam to fit the aperture of the spatial filter. Using the spatial filter and lens L3 the beam is expanded and collimated. At the location of the BR film the reconstruction beam has 65 mm diameter and an intensity of $130\,\mu\text{W/cm}^2$.

The reconstruction beam illuminates the BR film in a phase-conjugate manner. The holographic grating inside the BR film will diffract light to form real images of the recorded particle fields. A CCD camera (1376×1040 pixels, pixel size 6.45 µm) that is mounted on a motorized translation stage is used to digitize the three-dimensional intensity distribution. The three-dimensional field is digitized as 100 parallel planes. The frame capturing of the CCD camera, the motion of the translation stage, and the mechanical movement of the half-wave plate have been synchronized. As a result, each captured odd frame contains the intensity distribution of the first recorded particle field, whereas each captured even frame contains information concerning the second recorded particle field. This leads to 200 digitized frames per hologram. Currently, the various operations during reconstruction lead to a readout rate of about 2 frames per second, which leads to a total readout time of about 100 s. The translation mechanism moves the camera 25 µm in between each frame. Therefore, the z-resolution for each particle field is currently 50 µm in air. Because the particle field is recorded in water and reconstructed in air a longitudinal magnification is introduced during the reconstruction. This is illustrated in Fig. 6.

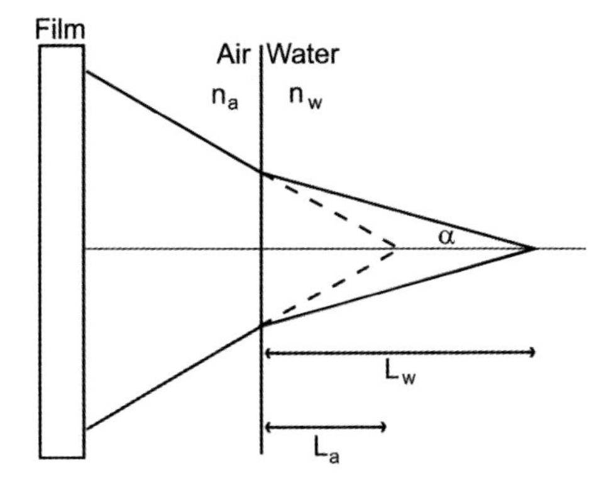

Fig. 6. Longitudinal magnification due to change in refractive index. *Solid lines* indicate the ray path during the recording (particle in water, BR film in air). The *dashed line* represents the ray path when during the reconstruction the air/water interface has disappeared

A particle is located at a distance L_w behind the air/water interface. Rays that come from the particle at an angle α with respect to the optical axis refract at the air/water interface and are recorded in the BR film. During the reconstruction, the light does not refract and the rays focus at a distance L_a behind the location where the air/water interface was at the time of recording. The distance that was traveled through water at the time of the recording is thus magnified by a factor $M = L_\mathrm{a}/L_\mathrm{w}$. It is easily shown that this magnification may be written:

$$M = \frac{n_\mathrm{a}\sqrt{1-\left(\frac{n_\mathrm{w}}{n_\mathrm{a}}\sin\alpha\right)^2}}{n_\mathrm{w}\cos\alpha}, \tag{2}$$

where n_a and n_w are the refractive index in air (1) and water (1.33), respectively. For the particles that are used in this experiment (NA = 0.15) we find for the outer rays that M = 0.745. The dependency of the longitudinal magnification on the NA implies that the image will be affected by spherical aberration. Using ray tracing it can be shown that a point-source-like particle that is 20 mm into the tank and refracts light with NA = 0.15, will be reconstructed in air having a geometrical spot size of 16 µm. We did not take measures to minimize the induced spherical aberration. However, measurements that have higher NA, require a better z-resolution, or require higher particle-seeding densities, elimination of spherical aberration may be necessary. This can be achieved by introducing an equal amount of water between the BR film and the CCD camera during the reconstruction as was present between the particle of interest and the BR film during the recording. A ray-tracing program has shown that the spherical aberrations that are caused by refraction from the air/water interfaces can be completely eliminated with this approach. Practically, this approach can be realized by placing an expandable water tank between the hologram and the CCD camera during reconstruction, with a fixed glass window on one side and a glass window on the other side of the tank that moves synchronized with the camera. (The tank would resemble an accordion musical instrument.) This approach removes the need to place the CCD chip inside a water-filled tank during reconstruction to eliminate the spherical aberrations.

When reconstruction is performed *with* an expandable water tank, the discussed spherical aberrations are absent and the inplane resolution (x–y) is only determined by the pixel size. When reconstruction is performed *without* an expandable water tank, spherical aberrations lead to a transverse particle-image diameter of 16 µm. This implies a small reduction of the accuracy of the transverse particle-image position. This reduction, however, is acceptable in the current experiments. In both cases there is no transverse magnification.

Being limited by computer memory we could capture 280 frames per measurement. Therefore, our reconstruction volume is currently limited to $7.0 \times 9.2 \times 7.0\,\mathrm{mm}^3$ in air. Due to the volatile nature of BR there is a slight decrease in image intensity and contrast during the reconstruction. All images are written to the computer's hard disk for subsequent analysis.

4.1.3 Data Extraction

The two sets of three-dimensional intensity distributions are used to create two three-dimensional particle location maps. Image processing is performed on each image to enhance the detectability of the reconstructed particles. This is achieved by using a phase-preserving denoizing image-processing routine outlined by *Kovesi* [17]. Next, we perform particle detection by crosscorrelating each image with a 5×5 Gaussian particle mask. After this correlation the software detects the local maxima in the correlation plane, corresponding to the particle centers. These two-dimensional particle maps are stacked to form a volumetric map containing particle streaks. One such streak obviously corresponds to a particle coming in, and going out of, focus. Within each streak every slice is given a value based on the unfiltered and unprocessed intensity value of the corresponding particle image. We found that the maximum intensity value serves as a reliable indicator for the particle's longitudinal position (see Fig. 7).

After the two three-dimensional particle location maps are retrieved, the particle displacement is determined using a nearest-neighbor search routine operating on the two datasets. A particle from the first dataset is thus assumed to have moved to the nearest particle in the other dataset. Obviously this is only valid when the mean particle displacement is smaller than the mean distance between two adjacent particles. Finally, the displacement in the z-direction is compensated for the induced magnification using (2). All image processing and data extraction is performed on a personal computer

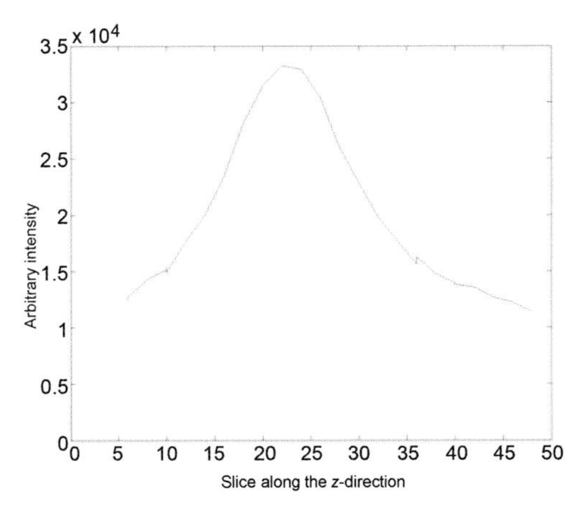

Fig. 7. Particle intensity as a function of position in reconstructed volume

using MATLAB. It typically takes 3–4 h to fully analyze 2 sets of 140 images, in total containing several hundreds of particles.

4.2 Configuration of Vortex-Ring-Flow Measurements

The configuration of the vortex-ring experiment is similar to the earlier described configuration, with the exception of some changes. The changes are described here.

4.2.1 Recording

As shown in Fig. 8, the object beam is expanded and collimated by lenses L3 and L4. The collimated object beam reaches a water-filled glass tank with a transverse beam-diameter of 32 mm and a typical fluence of 29 mJ/cm^2. The incoming light partially scatters from tracer particles and passes through a highpass optical Fourier filter that consists of a lens, L5, a beamstop and a lens, L6. L5 and L6 have a focal length, f, of 200 mm and a beamstop of 6 mm yields a Fourier filter spatial cutoff frequency of 28 mm^{-1}. The Fourier filter generates a real image (RI) of the tracer particles at distance f behind L6, which is recorded (380 mm further) on a BR film. The reference beam is expanded and collimated by lenses L1 and L2 and reaches the BR film with a fluence of 0.53 mJ/cm^2. The fluence values correspond to a beam-energy ratio between the object beam and the reference beam of 9 : 2 respectively. (Note: This ratio refers to the object beam fluence *before* it reaches the object. The scattered object light that passes the spatial Fourier filter and reaches the holographic film is significantly weaker.) These values are a result of experimental maximization of the reconstructed signal intensity.

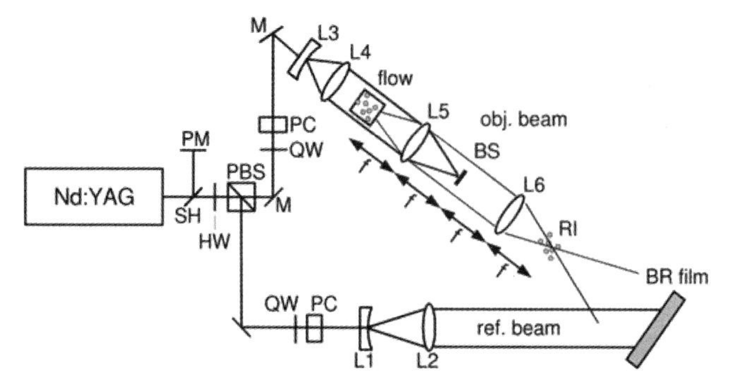

Fig. 8. Experimental configuration that is used for vortex-ring-flow measurements. BS: beamstop; RI: real image

The reference beam reaches the film at an angle of 26° from the film-normal, while the film-normal is positioned parallel to the optical axis of the object beam.

4.2.2 Flow

The flow around a vortex ring in water is measured. The tank size is increased to 60 mm (along the optical axis) ×135 mm (vertical) ×200 mm (horizontal, perpendicular to the optical axis). A tube filled with strongly diluted ink is placed just above the water surface (∼5 mm). Labview software controls the release of one droplet that forms a downward-traveling vortex ring. The double laser pulse is fired 1–2 s after the release of the droplet. The vortex ring is then in the illuminated part of the tank. The horizontal size of the vortex ring grows towards a typical size of 2 cm near the bottom of the tank.

The tracer particles in this experiment are hydrogen bubbles with a typical diameter of 20–30 µm. The hydrogen bubbles are generated by electrolysis of water. The typical traveling velocity of the vortex ring is five times larger than the maximum observed rise velocity of the hydrogen bubbles. This suggests that the hydrogen bubbles are reasonably suitable tracer particles in this experiment. Although the non-neutral buoyancy of the hydrogen bubbles is clearly a disadvantage, the strong scattering behavior makes their use attractive.

4.2.3 Reconstruction

The intensity of the reconstruction beam is, in the second experiment, 80 mW/cm^2. This value is experimentally determined by considering two opposing effects: If the reconstruction-beam intensity is lower, the readout process becomes so slow that thermal-induced erasure of the hologram prevents successful readout of all frames. However, a higher reconstruction-beam

Fig. 9. Vortex rings are generated in a water-filled tank. For visualization, three vortex rings are generated in this photograph

intensity leads to excessive photoinduced erasure during overhead readout operations (i.e., stage motion, polarization switching).

4.2.4 Data Analysis

Data analysis is performed by MATLAB 7 on a conventional desktop computer (CPU: AMD 2.5 GHz, RAM: 1.8 GB, OS: Linux). Both reconstructions are each split into interrogation volumes. The size of an interrogation volume varies from 8 to 128 pixels in the transverse direction and is fixed at 8 pixels in the longitudinal direction. The volume overlap between adjacent interrogation volumes is generally 50 %. Interrogation volumes are processed comparably to conventional 2D digital PIV: a three-dimensional crosscorrelation is performed with an FFT algorithm according to this formula: the average intensity of the two interrogation volumes is subtracted and both volumes are zero-padded. A three-dimensional crosscorrelation is performed with an FFT algorithm: the FFT of interrogation volume 1 is multiplied by the complex conjugate of the FFT of interrogation volume 2. The absolute value of the inverse FFT of the product is calculated and the result then divided by the square root of the autocovariance of interrogation volumes 1 and 2 to obtain a 3D volume of crosscorrelation coefficients:

$$
\begin{aligned}
r_{12}(\boldsymbol{x}) &= \frac{\langle \tilde{I}_1(\boldsymbol{\xi}) \cdot \tilde{I}_2(\boldsymbol{\xi}+\boldsymbol{x}) \rangle}{\sqrt{\langle \tilde{I}_1(\boldsymbol{\xi}) \cdot \tilde{I}_1(\boldsymbol{\xi}) \rangle \cdot \langle \tilde{I}_2(\boldsymbol{\xi}) \cdot \tilde{I}_2(\boldsymbol{\xi}) \rangle}}\,, \\
&= \frac{\left| \mathrm{IFFT}\big(\mathrm{FFT}(\tilde{I}_1) \cdot \mathrm{FFT}^*(\tilde{I}_2)\big) \right|}{\sqrt{\langle \tilde{I}_1 \cdot \tilde{I}_1 \rangle \cdot \langle \tilde{I}_2 \cdot \tilde{I}_2 \rangle}}\,,
\end{aligned} \tag{3}
$$

where $r_{12}(\boldsymbol{x})$ is the crosscorrelation coefficient, $\langle\,\rangle$ is the mean, $\tilde{I}_1$ and $\tilde{I}_2$ are the intensity data of interrogation volumes 1 and 2, respectively, after subtraction of the mean intensity and after zero-padding, $\boldsymbol{x}$ and $\boldsymbol{\xi}$ are three-dimensional spatial coordinates and * is the complex conjugate.

The integer-pixel position of the global maximum of the correlation cube $(x_{\text{peak}}, y_{\text{peak}}, z_{\text{peak}})$ is found and the correlation cube is then divided by a 3-dimensional kernel to compensate for the fact that each point in the correlation volume is the result of a different volume overlap between interrogation volume 1 and 2. Then, a subpixel estimate of the z-position of the correlation peak is determined by making a 7-point least-square Gaussian fit on the points between $(x_{\text{peak}}, y_{\text{peak}}, z_{\text{peak}} - 3)$ and $(x_{\text{peak}}, y_{\text{peak}}, z_{\text{peak}} + 3)$. No subpixel estimate is made of the x- and y-positions of the correlation peak because the accuracy of the integer-pixel position is currently sufficient.

5 Measurements

A series of measurements were performed using the described system. First, we calibrated the system's accuracy. Next, we used the holographic camera to analyze an actual flow.

5.1 Accuracy Test/Calibration

For testing the system's accuracy we prefer to induce a known physical shift to a particle field and verify the system's capability to retrieve the associated displacement vector based on two holographic recordings. For this purpose we obtained a tank with tracer particles in resin. Unfortunately, the resin depolarized the light, thereby introducing unwanted crosstalk into our polarization multiplexed holograms. An alternative approach was to stick particles onto a microscope slide, after which the microscope slide was mounted on a translation stage. However, we had practical problems in sticking the 100-µm glass particles to the microscope slide while maintaining a good optical quality. Based on before-mentioned experiences we decided to use an artificial or null shift rather than an induced physical shift of the particle field.

First, we recorded a hologram of the jet flow coming out of a syringe nozzle using one laser only. Using this single three-dimensional intensity distribution we created a second stack of images with an artificially induced known shift. The intensity of the second stack of images was altered to simulate a slight difference in diffraction efficiency between two actual reconstructed holograms. For multiple artificially induced shifts, in total having analyzed thousands of particle images, the software retrieved the actual displacement to within one pixel accuracy (x, $y = 6.45\,\mu\text{m}$, $z = 67\,\mu\text{m}$) for more than 95 % of the identified particles.

Next, we recorded a hologram of the jet flow coming out of a syringe nozzle using two orthogonally polarized laser pulses with a time separation of 10 µs. Given the speed of the jet this time delay was short enough such that only subpixel particle displacements should occur. Data extraction yielded 677 vectors, of which 647 were $[0, 0, 0]$. Again the system provided single-pixel accuracy for more than 95 % of the vectors.

Fig. 10. Intensity distribution of the reconstructed particle field in a single plane. The meandering contour of the jet is clearly visible

5.2 Jet Flow

The holographic system was used to analyze the jet flow coming out of a syringe nozzle. This experiment was repeated 30 times on a single day. We will present one typical result. Two particle holograms were recorded with a time separation of 400 μs between the two consecutive pulses. Figure 10 shows one of the 280 unfiltered images that were captured during the reconstruction. The particles are clearly visible. Also, notice that the jet is not symmetric but has a meandering flow profile. This effect may be due to the nonideal tracer particles. After processing all the 280 images the jet's velocity profile is known. Figure 11 shows the resulting three-dimensional velocity-vector plot containing 1160 raw vectors. This vector plot shows the same meandering flow profile as was seen in Fig. 10. As expected, the main flow is in the downward ($-y$) direction, with the highest velocities near the center of the jet. In Fig. 12 the mean downward velocity is plotted over the footprint of the jet. Again the highest velocities are found near the center of the jet.

5.3 Vortex Ring Flow

The resulting velocity field is shown in Fig. 13. The velocity field is here measured in only one plane. The main reason for this is the planar distribution of the hydrogen-bubble tracer particles. The time delay between the two recordings is 3 ms. In the correlation analysis, the transverse size of the interrogation volume is 64 pixels (= 0.4 mm), the longitudinal size is 8 pixels (= 1.6 mm) and the interrogation volume overlap is 50 %. Only the transverse component of the vectors is shown and all 1120 vectors are located in one transverse plane. A vertical vector (11 mm/s) is subtracted to compensate for the rising motion of the hydrogen bubbles.

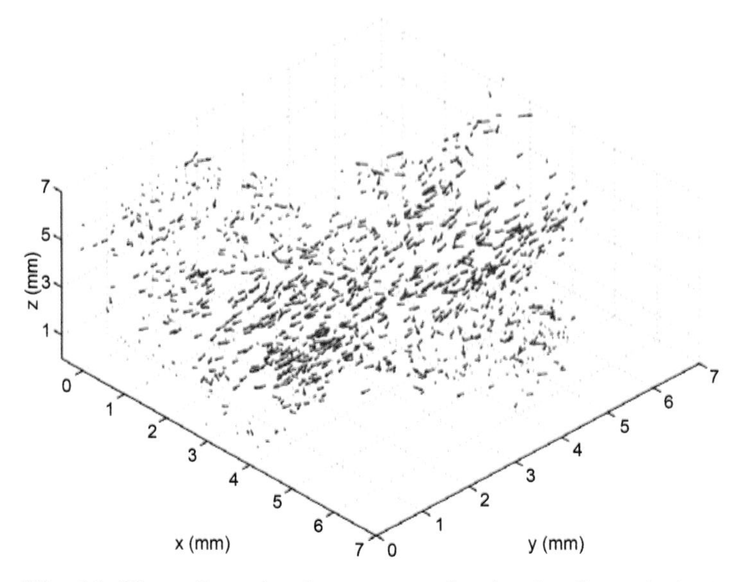

Fig. 11. Three-dimensional raw vector plot showing flow velocity inside a particle-loaded jet

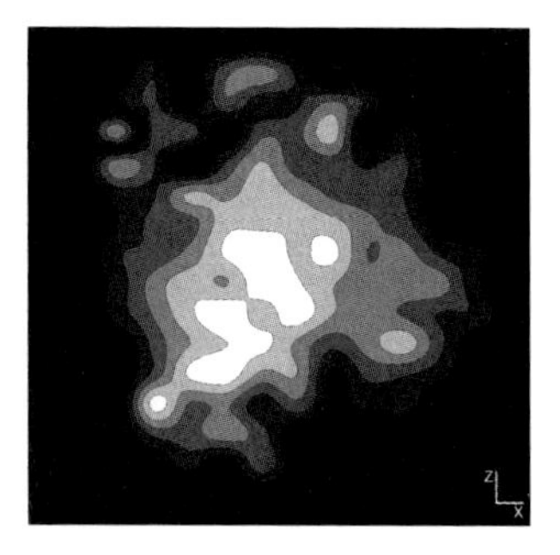

Fig. 12. Mean downward velocity over the jet's footprint. White areas correspond to a mean downward velocity of 15 cm/s

6 Future Outlook

The present system clearly demonstrates the potential of using BR films and polarization multiplexing in velocimetry measurements. As with most systems several improvements are possible. We may correct for the introduced spherical aberration in the future. As stated before, this can be achieved using an expandable water tank.

Alternative to obtaining the three-dimensional intensity distribution by scanning the reconstruction volume using the CCD camera, we will soon look into performing a single wavefront scan as described by *Barnhart* et al. [18].

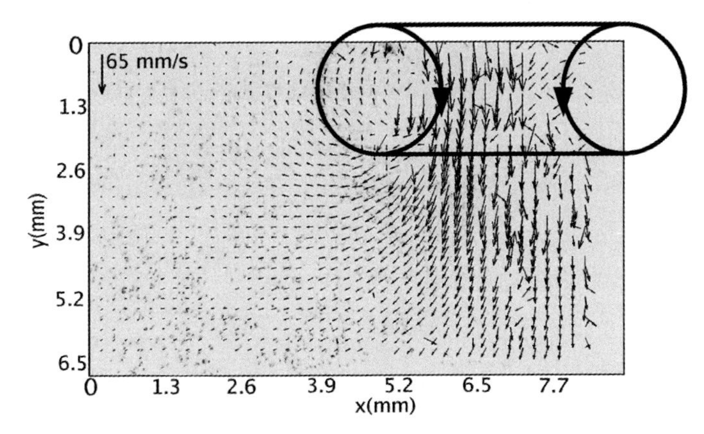

Fig. 13. Vortex-ring-flow measurement result. The illustrated plane is a slice through the center of the vortex ring

Using this digitized wavefront all particle locations may be found be digital holographic reconstruction. This method will also eliminate the inherent spherical aberration because the air–water transition can be included in the numerical model.

7 Conclusion

We have presented the first ever holographic particle image velocimetry system that uses a reusable holographic material as the recording medium. The measurement system records double-exposure holograms in a film containing the photochromic protein BR. The two constituent particle holograms are recorded in a single film using polarization multiplexing. To our best knowledge it is the first time that this type of multiplexing has been used in (particle) velocimetry measurements. The entire system delivers single pixel accuracy for more than 95 % of the generated displacement vectors.

The current system records, reconstructs, digitizes and erases a particle field hologram in less than 3 min. This allows for large-scale repetitive measurements (more than a hundred holograms on a single day), thereby enabling the acquisition of three-dimensional flow statistics. Such an experiment would be extremely laborious for a silver-halide-based HPIV system.

Although the system is still a laboratory setup, we have shown that it is possible to create an easy to operate, fully reusable HPIV camera. We believe that such a system will prove to be an extremely valuable tool for both fundamental and engineering flow study in the near future.

Acknowledgements

This work has been funded by the Foundation for Fundamental Research on Matter (FOM), the Netherlands. The authors thank Christian Poelma for his programming efforts. Furthermore, we thank Prof. Norbert Hampp (Philipps-Universität Marburg) and Donald Barnhart for their support on BR-related issues.

References

[1] C. E. Willert, M. Gharib: Digital particle image velocimetry, Exp. Fluids **10**, 181–193 (1991)

[2] A. K. Prasad, R. J. Adrian: Stereoscopic particle image velocimetry applied to liquid flows, Exp. Fluids **15**, 49–60 (1993)

[3] D. Rockwell, et al.: High image-density particle image velocimetry using laser scanning techniques, Exp. Fluids **14**, 181–192 (1993)

[4] D. H. Barnhart, R. J. Adrian, G. C. Papen: Phase conjugate holographic system for high resolution particle image velocimetry, Appl. Opt. **33**, 7159–7170 (1994)

[5] J. Zhang, B. Tao, J. Katz: Turbulent flow measurement in a square duct with hybrid holographic PIV, Exp. Fluids **23**, 373–381 (1997)

[6] Y. Pu, H. Meng: An advanced off-axis holographic particle image velocimetry (HPIV) system, Exp. Fluids **29**, 184–197 (2000)

[7] S. Herrmann, H. Hinrichs, K. D. Hinsch, C. Surmann: Coherence concepts in holographic particle image velocimetry, Exp. Fluids **29**, S108–S116 (2000)

[8] S. Murata, N. Yasuda: Potential of digital holography in particle measurement, Opt. Laser Technol. **32**, 567–574 (2000)

[9] D. H. Barnhart, N. Hampp, N. A. Halliwell, J. M. Coupland: Digital holographic velocimetry with bacteriorhodopsin (BR) for real-time recording and numeric reconstruction, in *11th Int. Symp, on Applications of Laser Techniques to Fluid Mechanics* (2002)

[10] N. Hampp: Bacteriorhodopsin as a photochromic retinal protein for optical memories, Chem. Rev. **100**, 1755–1776 (2000)

[11] N. Hampp, T. Juchem: Fringemaker – the first technical system based on bacteriorhodopsin, in A. Dér, L. Keszthelyi (Eds.): *Bioelectronic Applications of Photochromic Pigments* (IOS Press 2001) pp. 44–53

[12] S. D. Kakichashvili: Polarization recording of holograms, Opt. Spectrosc. (USSR) **33**, 171–173 (1972)

[13] L. Nikolova, T. Todorov: Diffraction efficiency and selectivity of polarization holographic recording, Opt. Acta **31**, 579–588 (1984)

[14] W. D. Koek, V. S. S. Chan, N. Bhattacharya, J. J. M. Braat, J. Westerweel: Polarization multiplexing based on photo-induced anisotropies in bacteriorhodopsin, in J. Coupland (Ed.): *Proc. Int. Workshop on Holographic Metrology in Fluid Mechanics* (Loughborough University Press, Loughborough 2003) pp. 123–129

[15] W. D. Koek, N. Bhattacharya, J. J. M. Braat, V. S. S. Chan, J. Westerweel: Holographic simultaneous readout polarization multiplexing based on photoinduced anisotropy in bacteriorhodopsin, Opt. Lett. **29**, 101–103 (2004)

[16] F. Liu, F. Hussain: Holographic particle velocimeter using forward scattering with filtering, Opt. Lett. **23**, 132–134 (1998)
[17] P. Kovesi: Phase preserving denoising of images, in *The Australian Pattern Recognition Society Conference: DICTA'99* (Australian Pattern Recognition Society 1999) pp. 212–217
[18] D. H. Barnhart, W. D. Koek, T. Juchem, N. Hampp, J. M. Coupland, N. A. Halliwell: Bacteriorhodopsin as a high-resolution, high-capacity buffer for digital holographic measurements, Meas. Sci. Technol. **15**, 639–646 (2004)

Index

Assessment of Different SPIV Processing Methods for an Application to Near-Wall Turbulence

Jie Lin, Jean-Marc Foucaut, Jean-Philippe Laval, Nicolas Pérenne, and Michel Stanislas

Laboratoire de Mécanique de Lille (LML, UMR 8107),
Boul. P. Langevin, 59655 Villeneuve d'Ascq, France
michel.stanislas@ec-lille.fr

Abstract. An experiment has been performed in a large wind tunnel with the objectives to record 2D3C velocity fields of a fully developed turbulent boundary layer along a flat plate by means of stereoscopic PIV (SPIV) and to study the characteristics of this turbulence. The present study starts from determining the suitable method to process the database that was recorded with the stereoscopic PIV system. It suggests that the Soloff method with 3 calibration planes and integer shift is the best choice. Then, by using this method the analysis of the mean streamwise velocity, velocity fluctuations, Reynolds shear stress, spectrum, probability density function (PDF) as well as skewness and flatness, was performed and compared with values from hot-wire anemometry (HWA) and direct numerical simulation (DNS). The comparison indicates that SPIV is a well-qualified method to investigate near-wall turbulence.

1 Introduction

PIV is a quantitative, nonintrusive method for the measurement of fluid velocity in large areas. For the last 15 years, due to the strong improvement of laser, video camera and computer, PIV has undergone major developments and has become a powerful technique to investigate fluid mechanics [1–4]. The conventional implementation of PIV uses only one camera to record the motion of small tracer particles in a thin lightsheet. By using such a configuration, only two inplane components of the fluid velocity can be obtained in the plane of observation. The two components provide a wealth of information for many flows, however, it is sometimes rather difficult to understand the true physical significance of the observed flow phenomena without the third component. This is particularly true in turbulence. Therefore, it is necessary to measure all three components of the velocity in order to understand the organization of flow. Moreover, the out-of-plane component can introduce errors due to the optical projection [5]. Stereoscopic PIV have been developed to resolve these problems [6]. In SPIV, a stereoscopic camera system is used, in which the motion of the tracer particles is viewed from two different directions. Due to the out-of-plane motion, the two cameras see the tracer

A. Schroeder, C. E. Willert (Eds.): Particle Image Velocimetry,
Topics Appl. Physics **112**, 191–221 (2008)

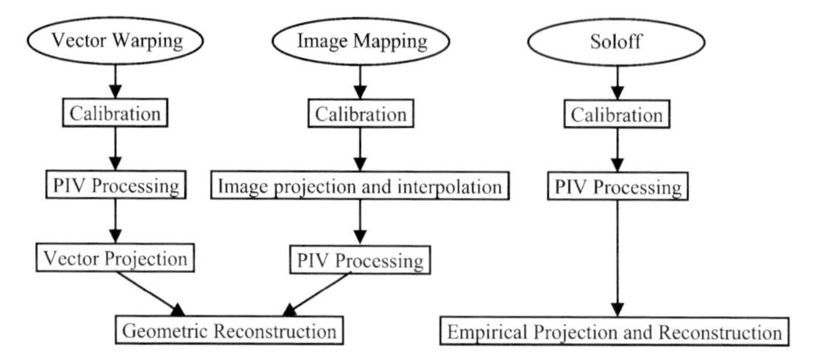

Fig. 1. Flowchart of vector warping, image mapping and Soloff method

particles travel over slightly different distances. From the differences in the apparent inplane motion it is possible to reconstruct all three components of the displacement. In the past decade, this method has been well developed and applied by a number of researchers [6–10]. It has been shown that the SPIV method can reach as good an accuracy as standard PIV [11]. However, the assessment of turbulence statistics with this technique has not been characterized in detail. In this chapter, the details of different methods for processing SPIV images are discussed and applied to near-wall turbulence.

The chapter proceeds as follows: Sect. 2 reviews and describes briefly the theory of stereoscopic-PIV algorithms. It is followed by a presentation of the experimental facility and setup in Sect. 3. In Sects. 4 and 5, SPIV processing is carried out in order to compare different algorithms and to select the most suitable one for the present database. Section 6 presents the turbulent statistics of the database using the selected algorithm and compares the results with those from hot-wire anemometry and numerical simulations. Section 7 finally concludes the present contribution.

2 Stereoscopic PIV Algorithms

Beside the standard pinhole model, recently discussed by *Wieneke* [12], which was not tested in the present study, three main algorithms are presently available to process SPIV images: vector warping [10], image mapping [10] and the Soloff technique [7]. These three methods are detailed in [11] and are summarized in the flowchart in Fig. 1.

2.1 Vector-Warping and Image-Mapping Methods

2.1.1 Empirical Backprojection

To process SPIV measurements, one needs to build an accurate relationship between the image plane of each camera and the object space. This is referred to as backprojection. The function for this relationship is usually generated empirically by using a calibration grid. It allows us to map each point of the image plane onto the corresponding point in the object space, which corresponds to the measurement point. The perspective backprojection function was proposed by *Raffel* et al. [3] as a ratio of second-order polynomials. In general, a least-squares fit between a large number of couples (object-image points) is used to determine the coefficients of this function. Except for the variations in the analytical form, this procedure is considered as standard. Recently, *Fei* and *Merzkirch* [13] proposed a third-order polynomial function in order to increase the accuracy of this projection. In the present study, a ratio of second-order polynomials was chosen [3]. This polynomial method allows us to take into account some optical distortion. In addition to its higher order (at least equivalent to fourth order), the ratio of polynomial function is based on an analytical projection function that takes into account the perspective effect of the Scheimpflug deformations. Both vector warping and image mapping use the empirical backprojection to project the vectors and images, respectively, into the object space. Both methods use the geometric reconstruction to obtain the 2D3C velocity fields [8]. However, the procedures of the two methods are different and described in detail below.

2.1.2 Vector Warping

A uniform mesh is firstly generated in the object plane and projected to obtain a deformed mesh in each camera image plane. This eliminates the need for any vector interpolation processing during the reconstruction process. On each point of the deformed mesh, the 2D2C-vector field is then calculated by using a standard PIV processing for each camera. After this processing the two vector fields are backprojected into the object space. The velocity vectors of each camera are referenced at the same point of the initial uniform mesh. Finally, a geometrical reconstruction method is applied to obtain a 2D3C-vector field from the 2D2C-vector fields from each camera (1–2):

$$\begin{aligned}
U &= \frac{U_1 \tan\alpha_2 - U_2 \tan\alpha_1}{\tan\alpha_2 - \tan\alpha_1} \\
V &= \frac{U_2 - U_1}{\tan\alpha_2 - \tan\alpha_1} \\
W &= \frac{1}{2}\left(W1 + W2 + (U_2 - U_1)\frac{\tan\beta_1 - \tan\beta_2}{\tan\alpha_1 - \tan\alpha_2}\right),
\end{aligned} \tag{1}$$

with

$$\begin{aligned}
\tan\alpha_1 &= \frac{X_1 - x}{y - Y_1} & \alpha_1 &= \frac{X_2 - x}{y - Y_2} \\
\tan\beta_1 &= \frac{Z_1 - x}{y - Y_1} & \tan\beta_1 &= \frac{Z_2 - x}{y - Y_2}.
\end{aligned} \tag{2}$$

The coordinate system for reconstruction is presented in Fig. 2. U, V and W are the resulting three velocity components after reconstruction along

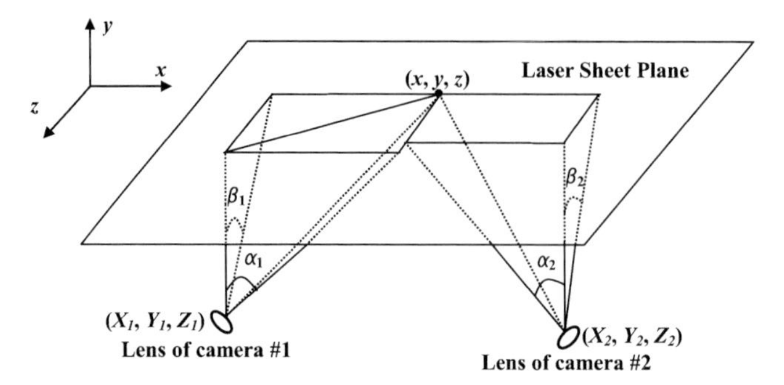

Fig. 2. Reconstruction in a stereoscopic PIV configuration

the X, Y and Z coordinates axis, respectively, while U_1 and W_1 refer to the two inplane components obtained by 2D2C analysis from camera #1, U_2 and W_2 are obtained from camera #2. (X_1, Y_1, Z_1) and (X_2, Y_2, Z_2) are the positions of the lenses of camera #1 and camera #2, respectively, in the object space, while (x, y, z) is the position of the measurement point.

2.1.3 Image Mapping

The recorded PIV images are firstly backprojected, or "mapped", to the object space pixel by pixel and interpolated on a new regular grid. The fields of 2D2C vectors are then calculated directly from these images by using standard PIV processing on a common regular grid. Subsequently, the geometric reconstruction method is used to obtain a 2D3C-vector field from two 2D2C-vector fields as with the vector warping method. The same reconstruction process is used for both methods. It should be mentioned that for both image mapping and vector warping, the geometrical reconstruction needs the value of some geometrical parameters (such as the position of the lens, see (2)), which are difficult to measure accurately on the experimental setup.

2.2 Soloff Method

Optical distortion due to inaccurate optical alignment, lens nonlinearity, refraction by optical windows, fluid interfaces and other optical elements of an experiment can generate inaccuracy by introducing spatial variations of magnification. It is important to compensate for these distortions because fractional changes in the magnification have a one-to-one effect on the accuracy of the measured velocity. *Soloff* et al. [7] introduced a general empirical calibration procedure, which allows us to obtain a specific matrix of the distorted imaging system, and an algorithm to accurately compute the velocity fields from measurements of distorted PIV images. From the calibration that is made by recording several images of a target, *Soloff* et al. [7] proposed to optimize a mathematical formalism that combines the projection and the reconstruction (3).

$$\begin{Bmatrix} \overline{\Delta X_1^1} \\ \overline{\Delta X_2^1} \\ \overline{\Delta X_1^2} \\ \overline{\Delta X_2^2} \end{Bmatrix} = \begin{Bmatrix} F_{1,1}^1 & F_{1,2}^1 & F_{1,3}^1 \\ F_{2,1}^1 & F_{2,2}^1 & F_{2,3}^1 \\ F_{1,1}^2 & F_{1,2}^2 & F_{1,3}^2 \\ F_{2,1}^2 & F_{2,2}^2 & F_{2,3}^2 \end{Bmatrix} \begin{Bmatrix} \overline{\Delta x_1} \\ \overline{\Delta x_2} \\ \overline{\Delta x_3} \end{Bmatrix} . \tag{3}$$

Here, the superscripts 1 and 2 indicate the camera #1 and #2, respectively. $\overline{\Delta X}$ is the displacement in the image plane (two dimensions) and $\overline{\Delta x}$ is the displacement in the object plane (three dimensions). F refers to the corresponding mapping function. The subscripts 1 and 2 represent the two inplane displacements, while the subscripts 3 stand for the out-of-plane displacement. The Soloff method is based on a third-order polynomial function for the inplane components and a second-order one for the out-of-plane component. At least two target images (with an accurately known spacing) are necessary to calibrate the Soloff method. In this study, three or five target images were used.

2.3 Comparison of the Three Methods

The main difference among the three methods is that the Soloff technique uses empirical optical projection and reconstruction, while vector warping and image mapping use empirical optical projection but geometrical reconstruction. Considering vector warping and image mapping only, vector warping projects the vectors and image mapping interpolates images. The drawback of the warping or Soloff methods is that the PIV analysis is conducted in the image plane, which leads to a local magnification that is not the same along the field. The interrogation window, which has a constant size in pixels in the image plane, then varies in size in the object space. On the contrary, the mapping method projects the images in the object space, making the magnification constant, but it distorts the particle images and introduces interpolation errors.

2.4 Calibration and Correction of Positions of the Image Planes

Calibration is a way to determine the relationship between the position in the object space and that in the image plane. For this purpose, it is necessary to acquire images of a calibration target whose location in the object space is known. By using these calibration targets, a generalized function to project the data from the image plane onto the object space can be found. According to the literature, a second-order polynomial [9] and a second-order ratio of polynomials [8] were used for a 2D calibration, while a cubic and a quadratic polynomial [7] and bicubic splines [14] were developed for a 3D calibration. In an experiment, however, it is difficult to make the position of the lightsheet and the calibration plane exactly the same. There are always small offsets and tilts between calibration and measurement planes. *Coudert* and *Schön* [10] proposed a method to correct the offset and tilt between them. The method works as follows. A set of single-exposure PIV images from each camera recorded at the same time is firstly backprojected as in the mapping method. A standard PIV processing is then used to calculate the displacement fields of the particle images illuminated at the same time by the same laser pulse. If the measurement and the calibration planes are perfectly superimposed, the mean displacement is zero (computed from about 50 vector fields). Normally, one can observe a small displacement from which an offset and tilt between the calibration and measurement planes can be deduced. Following this, a correction procedure is carried out to improve the projection function.

3 Experimental Setup

3.1 Wind Tunnel

The experiment was carried out in a boundary-layer wind tunnel (see [15] for details on the wind tunnel). This wind tunnel is $1 \times 2\,\mathrm{m}^2$ in cross section and 21.6 m in length. In order to use optical methods, the last 5 m of the working section is transparent on all sides. An air-water heat exchanger is located in the plenum chamber to keep the temperature within $\pm 0.2\,^{\circ}\mathrm{C}$. The turbulent boundary layer is studied on the bottom wall of the wind-tunnel test section. This flow presents a tiny longitudinal pressure gradient that is negligible and has no effect on the near-wall turbulence. The Reynolds number based on the momentum thickness Re_θ can reach 20 600 with a boundary layer thickness δ of about 0.3 m. The external velocity in the testing zone of the wind tunnel can vary from 0 to 10 m/s with a stability better than 0.5 %.

3.2 SPIV Setup

The purpose of the experiment was to obtain 3C velocity fields in planes parallel to the wall of a boundary layer, as close as possible to the wall. A Nd:YAG pulsed laser, with $2 \times 250\,\mathrm{mJ}$ of energy at 15 Hz, was used to generate the lightsheet. This lightsheet was shaped using a conventional optical setup (one sperical and one cylindrical lens) with a thickness of about 0.75 mm. The lightsheet passed through a lateral window located 1 m away from the measurement area. Two PCO SENSICAM cameras (1280×1024 pixel^2) were positioned under the wind tunnel as shown in Fig. 3. The cameras were arranged so that the Scheimpflug condition [8] was satisfied. The H and L parameters defined in Fig. 3 are: $H \cong 52\,\mathrm{cm}$ and $L \cong 50\,\mathrm{cm}$. These distances, which are necessary for the geometrical reconstruction, are measured with

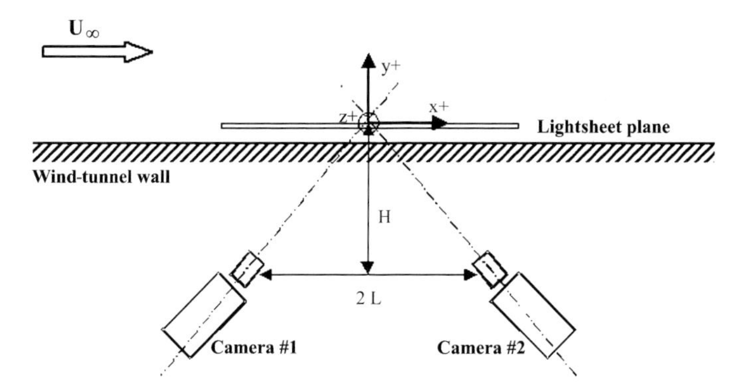

Fig. 3. Setup of the experiment and frame of reference for data analysis

respect to the center of the field of view. The line joining the two cameras is parallel to the main flow (camera #1 being upstream). The flow is from left to right in the images delivered by both cameras. The lightsheet propagates in the test section along z. Both cameras stand on the same ground under the wind tunnel upstream and downstream of the lightsheet in order to obtain symmetric light-scattering conditions. The focal length of the camera lenses was 105 mm. The field of view extends over $6.5 \times 4.0\,\text{cm}^2$ and $f_{\#} = 5.6$ was used for both cameras during the experiments. The average magnification is approximately 50 m/pixel in the object space. The depth of field was 3.5 mm. The focus was set at the middle value of the explored y domain and kept there for the remainder of the experiment (including the acquisition of calibration images). In this configuration, the Airy disk diameter is about 8.2 µm, which gives a size of the order of 1.3 pixels [1]. In order to measure this size more precisely, the autocorrelation of a single image was calculated. From the correlation result, the particle image size at 2σ is about 1.5 pixels and 1.3 pixels in streamwise and spanwise directions, respectively. This slight anisotropy is attributed to the stereoscopic distortion. The experiments were performed at $U = 3\,\text{m/s}$ (free-stream velocity). With this velocity, the Reynolds number Re_θ, based on the momentum thickness, is 7800. The friction velocity u_τ is of the order of 0.12 m/s. A wall unit ($\Delta y^+ = 1$) is 0.125 mm. Ten planes parallel to the wall were characterized. A total of 500 image pairs in each plane were recorded for each camera. The first plane was placed as near as possible to the wall, while avoiding too many reflections. The spacing between two neighboring planes was about 4 wall units.

4 Stereoscopic PIV Processing

As mentioned above, various methods including image mapping, vector warping and Soloff are available to obtain 2D3C results from SPIV. For each method, there are also several different choices of tools or parameters. There-

Table 1. Methods description

Short name	Method	Specialty	Shift method
MSI	image mapping	surfacial interpolation	integer shift
MSW	image mapping	surfacial interpolation	Whittaker shift
MWI	image mapping	Whittaker interpolation	integer shift
MWW	Image mapping	Whittaker interpolation	Whittaker shift
S3I	Soloff	3 calibration planes	integer shift
S3W	Soloff	3 calibration planes	Whittaker shift
S5I	Soloff	5 calibration planes	integer shift
S5W	Soloff	5 calibration planes	Whittaker shift
WI	vector warping		integer shift
WW	vector warping		Whittaker shift

fore, it is necessary to select one method with the best set of parameters for computation. In this section, the image number 10 of plane 5 was used in the first step to make a comparison of the instantaneous velocity fields provided by the different processing choices. Then, a statistical comparison was conducted using PDFs and spectra computed on the first 100 images of the plane 5.

The following methods were compared: image mapping with surfacial [16] or Whittaker [17] method to interpolate the image, vector warping and Soloff. For each method, the PIV analysis was performed with integer and subpixel Whittaker shift. In the case of Soloff, 3 and 5 calibration planes were taken. Table 1 shows the details of the methods and their abbreviations used in this chapter. For all the methods, a three-step multigrid approach was employed (window sizes: 64×64, 32×32 and 32×32 pixel^2).

As described earlier, calibration and its correction should be carried out before PIV processing. In the present study, 60 pairs of the first images of camera #1 and camera #2 are sufficient to calculate the average offset and tilt between calibration and measurement planes with a good convergence. By taking this value into account, the projection functions were corrected and the real position of the lightsheet was determined (see Table 2). Figure 4 shows the misalignment error between calibration plane 7 and laser plane 5 as an example. The vectors in Fig. 4 are quite constant, which means that the calibration and measurement planes are nearly parallel. The mean displacement in Fig. 4 is 0.58 mm, which implies a separation between calibration plane 7 and laser plane 5 of about 0.29 mm in depth (as the Scheimpflug conditions was used). This distance is taken into account in the correction process [10]. The gain in accuracy provided by the correction will be discussed further downstream.

In order to select suitable calibration plane(s) for each method, the location of each calibration and laser plane are listed in Table 2. Laser plane 5 is considered as an example. Based on Table 2, calibration plane 7 was used for image mapping and vector warping, calibration planes 6, 7, 8 were used

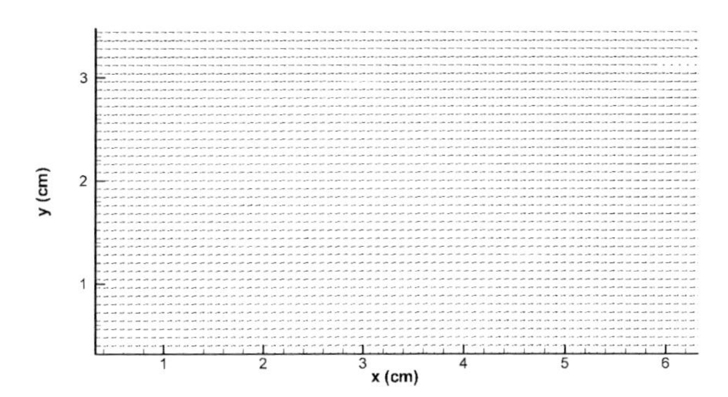

Fig. 4. Offset and tilt between calibration (No. 7) and measurement planes (plane 5)

Table 2. Absolute positions of the calibration and measurement levels

#	Calibration position (mm)	Laser position (mm)	Laser position (wall unit)	Dt (s)
1	0.68	1.81	14.5	600
2	1.15	2.32	18.5	600
3	1.61	2.78	22.2	400
4	2.07	3.29	26.3	400
5	**2.53**	**3.71**	29.7	350
6	**2.99**	4.16	33.3	350
7	**3.42**	4.63	37	350
8	**3.91**	5.07	40.6	350
9	**4.37**	5.5	44	300
10	4.83	5.99	48	300
11	5.29			

for the Soloff method with 3 calibration planes and calibration planes 5, 6, 7, 8, 9 were used for the Soloff method with 5 calibration planes. Table 2 also gives the PIV time delay for each plane, which was optimized to give a mean displacement of the order of 10 pixels in each field.

The comparison starts by comparing the accuracy of computation of the different methods. As the exact result is unknown, only relative comparisons between the different methods are possible. For this purpose, the following two error estimations were computed:

– Mean value of the modulus:

$$E_1 = \frac{\sum_{i=1}^{N} \sqrt{(u_1^i - u_2^i)^2 + (v_1^i - v_2^i)^2 + (w_1^i - w_2^i)^2}}{N} . \quad (4)$$

Table 3. Accuracy of the different methods

# line	Reference	Comparing with	E_1 (pixel)	E_2 (pixel)
1	WI	WW	0.12	0.35
2	S3I	S3W	0.11	0.22
3	MSI	MSW	0.09	0.58
4	MWI	MWW	0.11	0.59
5	MSI	MWI	0.15	0.61
6	MSW	MWW	0.18	0.6
7	MWI	MSW	0.17	0.59
8	S3I	S5I	0.04	0.26
9	MSI	WI	0.59	0.7
10	WI	S3I	0.13	0.63
11	MSI	S3I	0.75	0.67

– Standard deviation of the modulus:

$$E_2 = \sqrt{\frac{\sum_{i=1}^{N} \left(\sqrt{(u_1^i - u_2^i)^2 + (v_1^i - v_2^i)^2 + (w_1^i - w_2^i)^2} - E_1\right)^2}{N-1}} \,. \quad (5)$$

Here, (u_1, v_1, w_1) and (u_2, v_2, w_2) are the three instantaneous velocity components for, respectively, the reference and compared method. N is the total number of velocity vectors in the field. The values of E_1 and E_2 for the different methods are presented in Table 3.

Table 3 shows that the differences of the two parameters E_1 and E_2 between integer shift and Whittaker (subpixel shift) are quite small for both vector warping (line 1) and Soloff methods (line 2). This can be explained by the fact that the Whittaker shift, which is expected to reduce the peak locking, does not show any strong improvement from a statistical point of view [18]. The peak-locking effect appears mainly on the PDF of the velocity. This will be discussed in the next paragraph. As far as the image-mapping method is concerned, the comparison between two different shifts (line 3 and 4) gives a smaller value of E_1 than the comparison of two different interpolation methods (line 5 and 6). The values of E_2 of lines 3 and 4 keep the same order of magnitude but are generally two times higher than those with vector warping or Soloff methods (line 1 or 2). This suggests that both interpolation and shift methods have a strong influence on image mapping. Clearly, several successive interpolations (i.e., MWW or MSW) can damage the shape of the particle images and thus the correlation peak. However, it would be possible to couple both interpolations necessary for the projection and the subpixel shifting at the same step. The fact of using only one interpolation for both operations should improve the accuracy but will increase the computational time (interpolation is necessary at each pass of computation). As shown in line 8 of Table 3, the results of the Soloff method with 3 calibration planes and 5 calibration planes are similar to each other. This implies that, in the present configuration (lightsheet thickness, low distortion), it is not necessary to use 5 planes to calibrate the Soloff method. The comparison between MSI and WI in line 9 shows high values of both E_1 and E_2. These are introduced by the projection of images and strong differences are evident in the results. The comparison between vector warping and the Soloff methods (line 10) shows a small value of E_1 and a large value of E_2. The difference between Soloff and vector warping can only be attributed to the reconstruction that decreases the noise effect in the case of Soloff. This is confirmed by the comparison between MSI and S3I (line 11), which shows high values of E_2 but also of E_1. Besides the influence of the projection of images already evidenced in line 9, the reconstruction is the other source of this difference.

As shown in Table 3, the effect of subpixel shift is not evidenced by the comparison of lines 1 to 4. The reconstruction probably has a filtering effect, which decreases the peak locking. Therefore, the histogram of the decimal part of the velocity, in pixels, is calculated to analyze the effect of this peak locking. Figure 5 compares the histogram of the 2D2C PIV analysis (before reconstruction) for cameras #1 and #2 (C1 and C2) in the case of Soloff methods: S3I and S3W. *Foucaut* et al. [18] show a strong improvement when Whittaker interpolation is used for subpixel shifting. In the present case, this improvement is less visible, probably due to a particle image diameter smaller than two pixels [18]. Even if the Whittaker interpolation is used, a small peak-locking effect can still be shown for the u-component (u_{2C}). This effect is still larger for the w-component (w_{2C}). Figure 6 shows the histogram of the decimal part of the velocity after reconstruction by the Soloff method for the out-of-plane component v (which behaves like the u-component) and for the spanwise component w (less affected by the stretching of the SPIV). Due to the perspective effect of SPIV, a mean magnification was used to convert physical units to pixels. It is clear that the peak locking is filtered and that S3I and S3W give essentially the same histogram for both components presented in Fig. 6. Only some small oscillations remain due to a combination of peak locking, projection (variation of magnification along x) and reconstruction (between both 2D2C fields). To analyze the effect of these oscillations, the PDF of the velocity fluctuations can be studied. Furthermore, to characterize the measurement noise level, it is also interesting to look at the influence of the processing algorithm on the spectrum [4] of the velocity fluctuations. For this purpose, the S3I, S3W, MWI and WI methods were selected from the previous analysis to process 100 images pairs, which were used to obtain the spectrum and PDF of the three velocity components.

The components u and v are computed from U_1 and U_2 (1). As the angle of view is close to 45° in the present experiment, the behavior of these two components is similar. The w-component is perpendicular to the plane of the cameras and it is thus less affected by the stereoscopic reconstruction. Therefore, all the results of the spectrum and PDF are presented only for the components v and w. In Fig. 7, E_{22} and E_{33} are spectra of the v- and

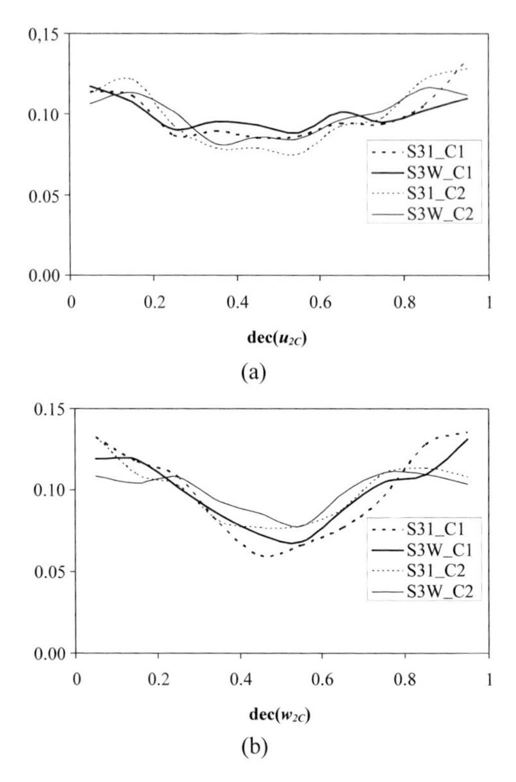

Fig. 5. Decimal part histogram of the u-(**a**) and w-(**b**) components of each camera before projection and reconstruction

w-components, respectively, and k is the wave number. According to the theory of *Foucaut* et al. [4], k_c is the PIV cutoff wave number ($k_\mathrm{c} = 2.8/S_\mathrm{IW}$, where S_IW is the interrogation window size). The PIV results are qualified only in the region $k \leq k_\mathrm{c}$. Figure 7 shows that the results of the four selected methods are almost the same in the valid region of PIV. The noise level, that is attained by the spectrum in the high-frequency part, is very close for each method. As in 2D2C PIV [18], a subpixel shift does not improve the spectrum as compared to the integer shift. This means that from the spectral point of view, all these methods can be used.

In Fig. 8, the PDFs of WI, S3I and S3W are very similar but MWI seems rather different from the others. The differences cannot be attributed to the reconstruction because image mapping and vector warping use the same reconstruction process. They may arise from the fact that the image-mapping method interpolates the deformed images before PIV processing, whereas the vector warping and Soloff methods are performed directly on the CCD images. As explained before, the perspective effect generates a difference in physical window size along the field and thus introduces a kind of smoothing

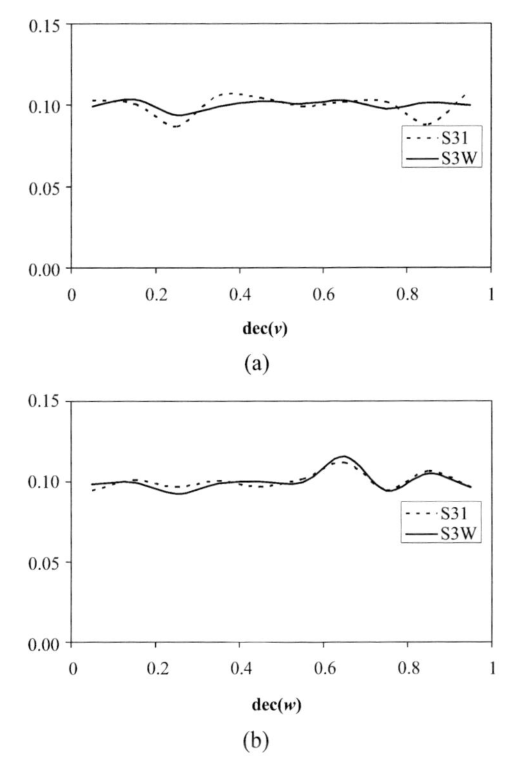

Fig. 6. Decimal part histogram of the v-(**a**) and w-(**b**) components after projection and reconstruction

in the statistics. This effect is clearly visible in Fig. 8a, as in Fig. 5. Moreover, the image-mapping method makes the magnification constant. As a consequence, particle images that are slightly affected by the interpolation are now deformed. This probably induces the modification of the peak locking that appears in Fig. 8b. The method S3W using the Whittaker subpixel shift does not efficiently remove the peak locking. Figures 9a–c show the PDF in pixels of the 3 components for three planes, plane 1 ($y^+ = 14.5$) is closest to the wall and plane 10 ($y^+ = 48.0$) is the furthest. Some fluctuations similar to peak locking are clearly visible in Figs. 9a and c. However, the fluctuations do not appear in Fig. 9b because the out-of-plane component presents a smaller dynamic range. They are probably smoothed out in this case. In Fig. 9a, the result of plane 1 shows different behavior from plane 5 and plane 10. This difference mainly arises from the fact that the dynamic range and the ratio of this range to the mean velocity of the u-component, is very large in this position, which causes the large range of the oscillations. Besides, the large velocity gradient at this position also has some influence. In Fig. 9c, small peaks can be found in the PDF of plane 1, which results mainly from

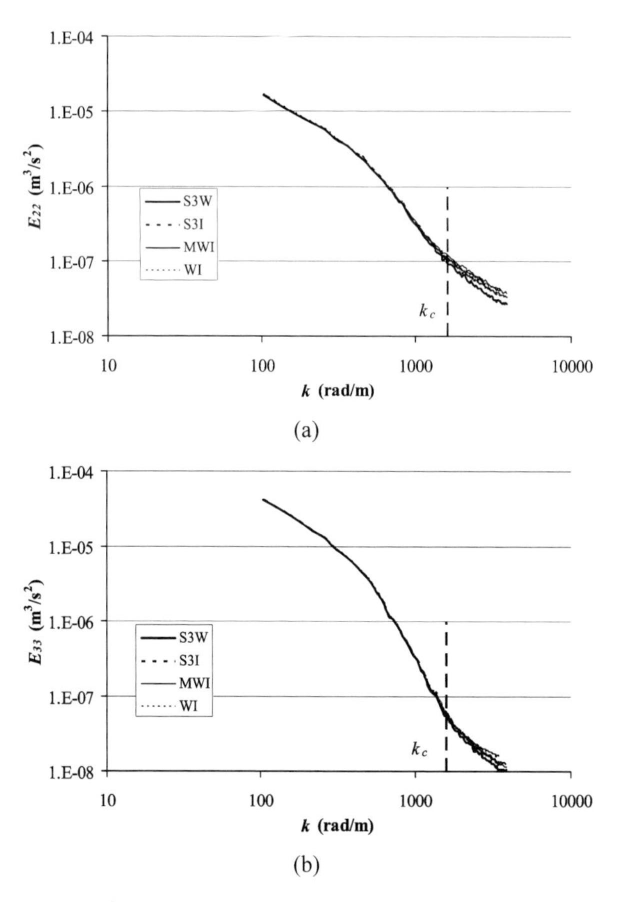

Fig. 7. Comparison of the velocity spectrum of the v-(**a**) and w-(**b**) components for different methods

the large velocity gradient at this position. The reduced amplitude of the peak locking benefits from the decrease of the velocity gradient away for the wall. Furthermore, the periodic distance of the main peaks is equal for all three planes, which implies a nearly constant dynamic range of the velocity fluctuation w'.

To study the magnitude of the velocity gradient and its variation with the wall distance, Fig. 10 shows the difference of particle displacement normalized by the particle image size between the top and the bottom of the lightsheet that has a thickness of about 0.75 mm [18, 19]. In 2D2C PIV, the criterion proposed by these authors to minimize the effect of gradient is $Du/di < 0.5$. In the present experiment, this parameter, computed from the mean gradient using the Van-Driest model [20], decreases as the distance to the wall increases and it seems acceptable when reaching plane 4. But there are still oscillations

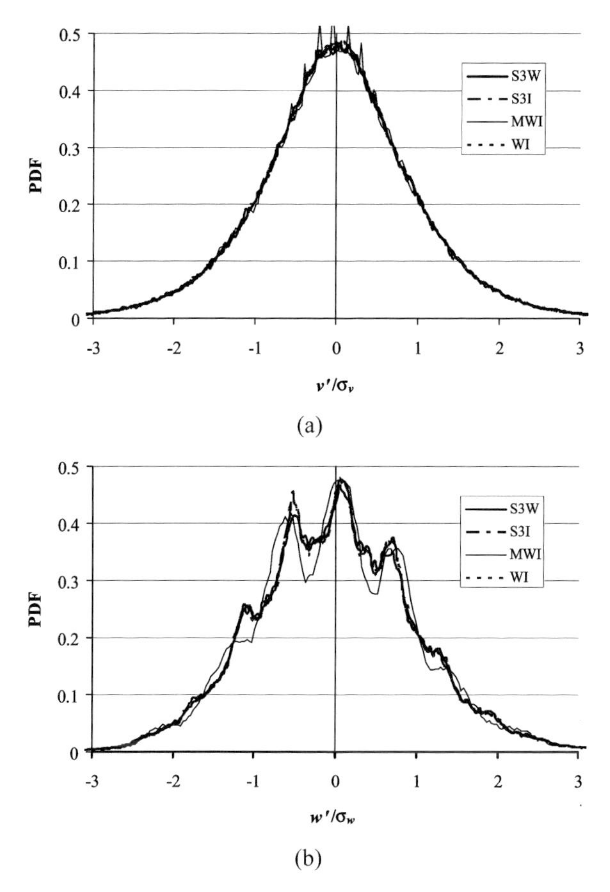

Fig. 8. Comparison of the PDF of the normalized fluctuations of the v-(**a**) and w-(**b**) components for different methods

of the PDF in this plane. This is probably due to the fluctuation of the instantaneous velocity gradient around this mean value and to a residual peak locking due to the particle image size (about 1.4 pixels).

The time needed by the different methods was computed and is listed in Table 4. It is estimated from a computation on a small sample of images (based on a computer with a PIII 800 processor and 256M RAM). The methods with Whittaker shift take the longest time for computation, about 4 times longer than the rest. For the image-mapping method, Whittaker interpolation needs much more time than surfacial interpolation when the computation is carried out with the same shift method.

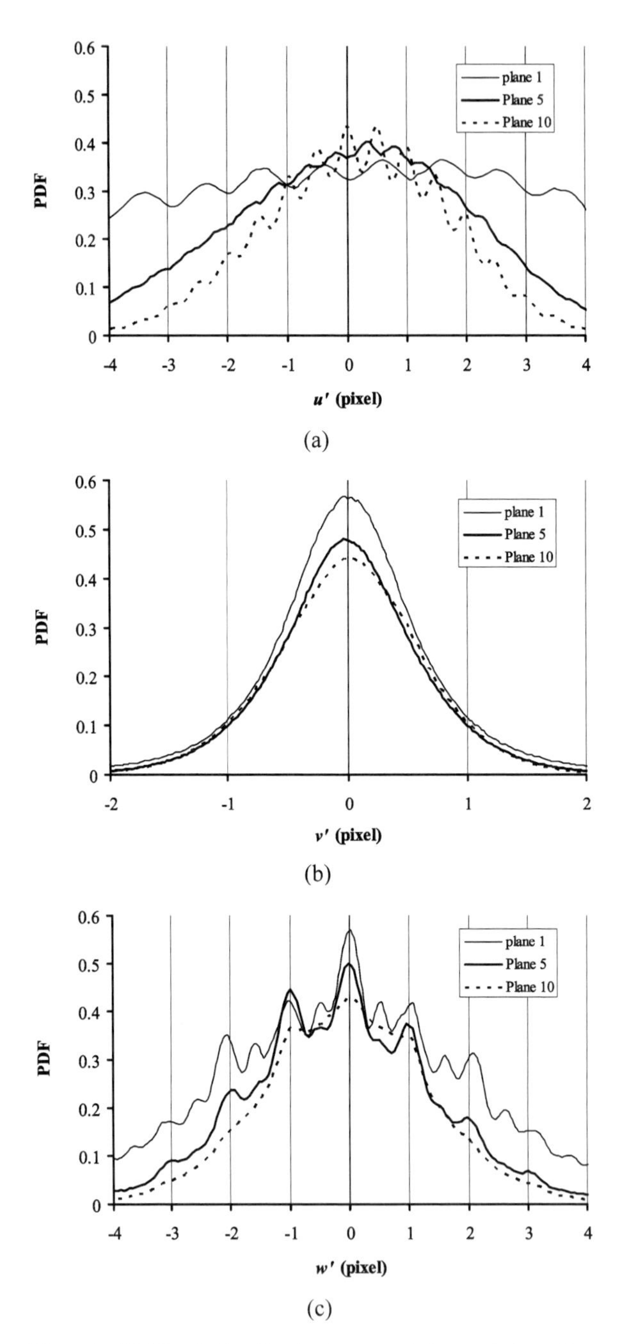

Fig. 9. PDF of the fluctuations of the u-(**a**), v-(**b**) and w-(**c**) components for different planes

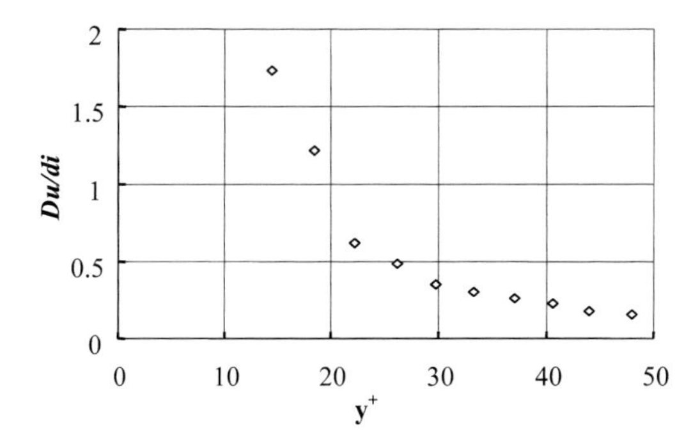

Fig. 10. Difference of velocity between the top and the bottom of the lightsheet for different wall distances

Table 4. Time consumption of the different methods for 500 image pairs

Method	MSI	MSW	MWI	MWW	S3I	S3W	S5I	S5 W	WI	WW
Time (h)	20	100	50	130	20	110	20	110	25	110

5 Method Selection

Based on the above results, the following arguments can be put forward:

Regarding the image-mapping method, the procedure of image interpolation can introduce errors that are impossible to avoid. A proper interpolation method such as Whittaker reduces the errors and thus improves the accuracy of the results. In addition, for window shifting, Whittaker is a subpixel shift that in principle is more accurate than the integer shift. As discussed before, the image-mapping method using Whittaker interpolation and Whittaker shift cannot be used together because two successive interpolations affect the correlation peak shape. Therefore, the most accurate method should retain Whittaker interpolation only once: for the image mapping or for the window shifting. However, it needs much more computational time in both cases compared to other interpolation techniques. Focusing on the PDFs in Fig. 8, it is clear that the mapping method causes more peak locking than the methods based on vector projection.

With respect to the vector warping method, when Whittaker shift is used, the computation is heavier and the accuracy of the results does not improve much. Therefore, to save computer time it is recommended to use integer shift when the vector warping method is selected.

As for the Soloff method, Table 3 shows that the Soloff with 5 calibration planes and with 3 calibration planes give very similar results. Therefore, it is not necessary to use 5 calibration planes. Additionally, the difference between the Whittaker and integer shifts is so small that the Whittaker shift is useless in view of the extra computation effort. Consequently, the Soloff method with

3 calibration planes and integer shift appears as the best compromise when the Soloff method is considered.

When using image-mapping and vector warping methods, it is required to measure geometric parameters such as the angle of the camera and the distances between the optical centers of camera lenses and calibration planes. The errors on the measurement of these parameters will affect the result of both methods. For the Soloff method, these parameters are not required. In this regard, the Soloff technique avoids these measurement errors and thus possibly provides more accurate results. *Fei* and *Merzkirch* [13] found a method for determining the viewing direction in the ángular displacement stereoscopic system by means of a digital imaging procedure. The method appears to improve the accuracy of results by avoiding the direct measurement of geometrical parameters of the setup. They found that their results are quite similar to that of the Soloff method, which supports the reliability of the Soloff approach.

It should be noted that Whittaker shift can improve significantly the accuracy of the result when the particle image size is sufficient large (normally > 2 pixels). In the case of small particles, this shift method can hardly perform well. This is supported partly by the present results: almost no difference between integer and Whittaker shift, because the present particle image size is only about 1.4 pixels.

As a conclusion of this synthesis, it appears that the Soloff method with 3 calibration planes and an integer shift (S3I) is the best choice in the present state-of-the-art. It was thus used for the present analysis of all 10 planes. Using this method, the correction technique proposed by [10] could be studied. Figures 11 and 12 show the efficiency of this correction. The SPIV algorithm was applied to plane 5 using two different sets of calibration planes: (4 to 6) called N°5 expected when the experiment was done and (6 to 8) called N°7 that was selected according the result of correction (Table 2). As shown in Table 2, the calibration plane N°5 is shifted by about 1.2 mm from the lightsheet location N°5, while the calibration plane No. 7 is much closer. The results are computed with correction (noted C) or without correction (noted NC). Figure 11 illustrates the PDF of the v- and w-components in each case. When the correction is applied, the PDF is comparable whatever the calibration plane is. Figure 12 leads to the same conclusion from the spectrum. Table 5 presents the comparison of the parameters E_1 and E_2 (see (2)) for the correction. Line 1 shows a small difference between the results obtained with correction from the two calibration planes N°5 and No°7. If calibration plane N°7 is used without correction, the differences from the corrected results increase a little but stay acceptable (line 2). The small increases of E_1 and E_2 come from the distance of 0.3 mm (about 6 pixels) between the calibration plane N°7 and the measurement plane No. 5. In lines 3 and 4, when the calibration plane N°5 is used without correction the results are remarkably different from other combinations. Considering the fact that the correction

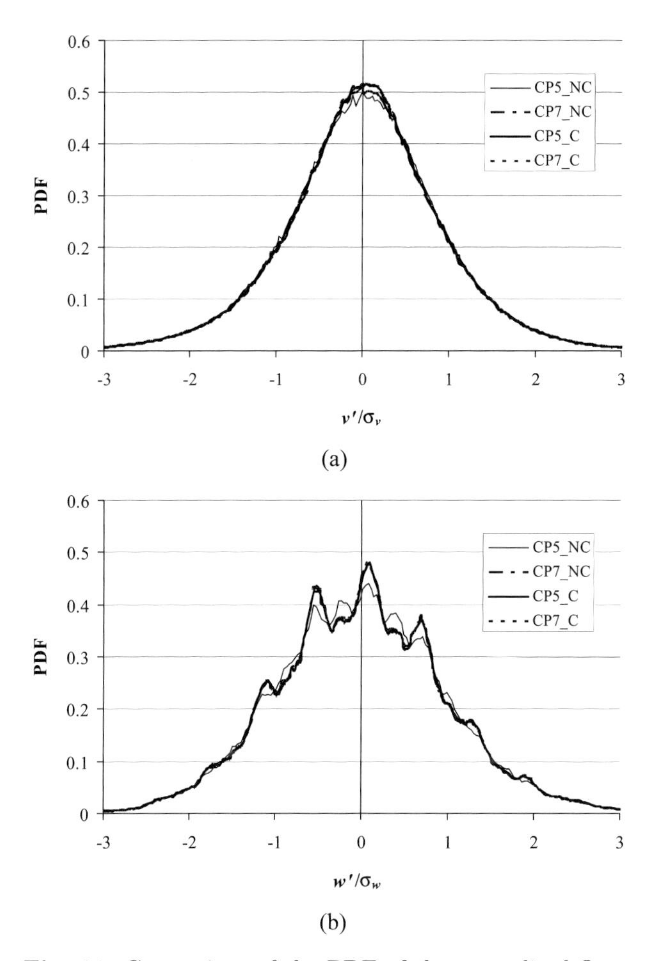

Fig. 11. Comparison of the PDF of the normalized fluctuations of the v-(**a**) and w-(**b**) components effect of correction

Table 5. Accuracy of the correction process

# line	Reference	Comparing with	E_1 (pixel)	E_2 (pixel)
1	CP7_C	CP5_C	0.11	0.4
2	CP7_NC	CP7_C	0.24	0.5
3	CP5_NC	CP5_C	0.66	0.63
4	CP5_NC	CP7_C	0.67	0.66

of calibration only takes a little time but can improve the accuracy of the results, correction is strongly recommended.

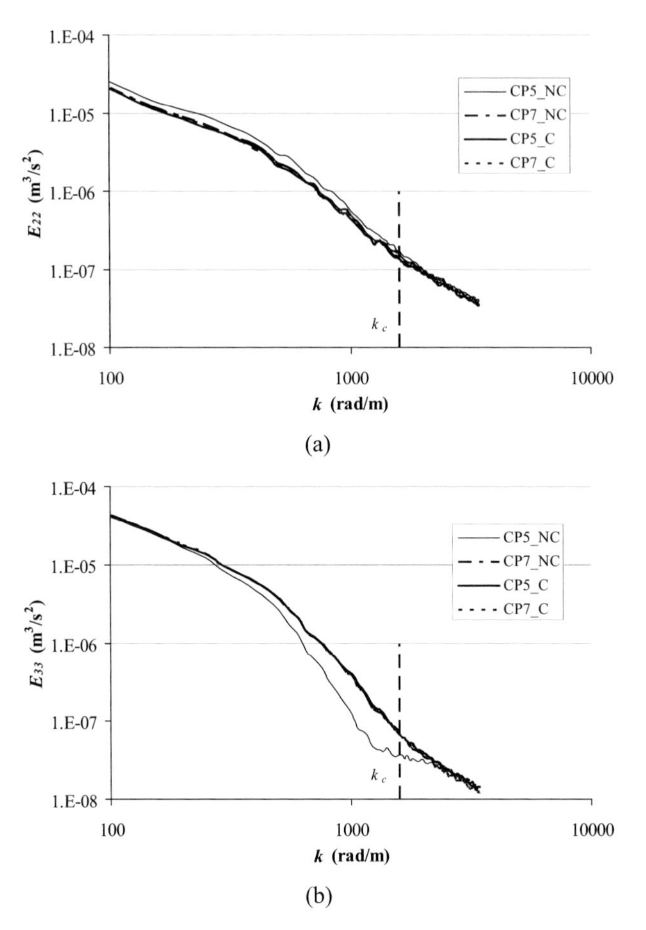

Fig. 12. Comparison of the spectrum of the v-(**a**) and w-(**b**) component effect of correction

6 Statistical Results for the 10 Planes

Using the selected S3I method with correction, the recorded SPIV images of the 10 planes were processed to obtain the instantaneous 2D3C velocity fields. In this process, three passes are used to calculate the standard 2D2C vector field. The window sizes are, respectively, 64×64, 32×32, and 32×32 (pixel^2) with a final spacing of 12 pixels (0.6 mm) corresponding to a mean overlapping of 67.5 %. Here, the 67.5 % overlapping is used to obtain better spatial resolution for detection of coherent structures in future work. The results were saved in a database built using the Pivnet 2 Netcdf format [21]. Then, a statistical analysis was performed to obtain the mean streamwise velocity and the Reynolds stresses, the velocity spectrum and PDF as well as skewness and flatness in the 10 planes. Here, these statistical results are

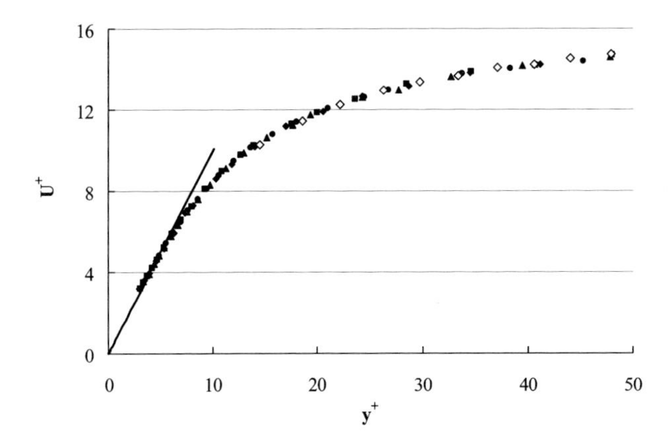

Fig. 13. Comparison of mean streamwise velocity distributions of SPIV (Re_θ = 11 400 (◊)) and hot-wire anemometry [15] (Re_θ = 11 400 (▲), 114 800 (■) and 20 600 (•))

compared with those of hot-wire anemometry [15]. Plane 4 was chosen to look at the spectra and PDF because it is in the middle of the range of wall distances studied ($y^+ = 26.3$) and it corresponds to the limit of validity for the velocity gradient inside the lightsheet (Fig. 10). In this section, the result that is normalized using the skin friction velocity u_τ $(= \sqrt{\tau_w/\rho}$, where ρ is the density of the fluid) and ν, is denoted with a superscript +.

Figure 13 shows the mean streamwise velocity profile in the near-wall region ($y^+ < 50$). In this figure the solid symbols correspond to hot-wire measurement [15] for different Reynolds numbers (Re_θ = 11 400 (▲), 14 800 (■) and 20 600 (•)). The hollow symbols correspond to the 10 planes measured with PIV. The straight line represents the viscous sublayer equation $u^+ = y^+$. This figure shows that the mean velocity obtained by SPIV is in perfect agreement with that of hot-wire anemometry.

Besides the mean streamwise velocity, the fluctuations of all three components are also basic characteristics of the turbulent boundary layer and thus need to be analyzed. Figure 14 shows a comparison of the profiles of the fluctuations obtained by the two methods (SPIV and HWA). These results are also compared with the results of the DNS by *Spalart* [22]. The Reynolds number of this simulation is $\mathrm{Re}_\theta = 1410$. The $\sqrt{v'^2}$ profile is very similar for all the methods down to $y^+ = 15$. Below this value, no PIV measurements are available and the hot wire starts to show a wall interference due to the probe size. For $\sqrt{u'^2}$ and $\sqrt{w'^2}$, the results of the hot-wire measurement are slightly higher than those of the DNS. The result obtained with SPIV ($\mathrm{Re}_\theta = 7800$) is between both, but closer to the HWA. The differences with DNS are attributed to the low Reynolds number influence. However, when it is very close to the wall, the difference of $\sqrt{v'^2}$ or $\sqrt{w'^2}$ between the results of SPIV

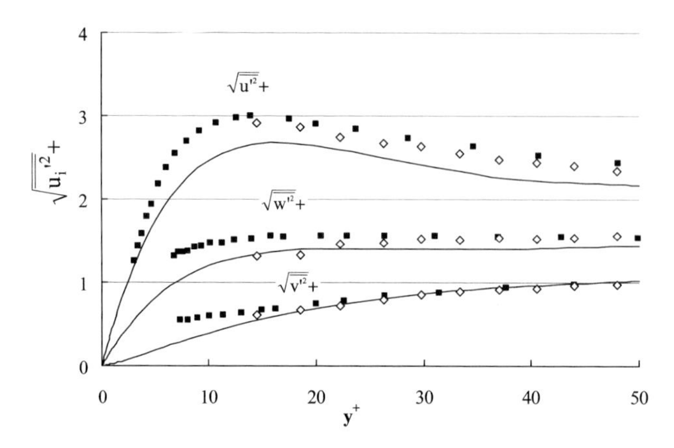

Fig. 14. Comparison of the profiles of fluctuations of SPIV (◊), hot-wire anemometry [15] (■) and DNS [22] (-)

and HWA increases. The main reason is that X-wire probes show an increasing bias when approaching the wall. This is due to wall interference and velocity gradients at the scale of the probe (0.5 mm). Considering this effect, the results reveal an excellent behavior of the SPIV measurement.

As is well known, Reynolds shear stress is a critical parameter of the turbulence. Figure 15 shows the data obtained by the two experimental methods (HWA and SPIV) compared with the results of DNS [22] and with the *Van Driest* model [20]. This model has been improved taking into account the weak pressure gradient of the test section $\partial p/\partial x = 0.057\,\mathrm{Pa/m}$ ($\partial p^+/\partial x^+ = 3.65 \times 10^{-4}$). The results of SPIV are similar to those of the Van Driest model and of the DNS. However, the results of HWA deviate considerably from the others, which once again shows the influences of the near-wall interference and gradients at the scale of the probe. This explains the low values of the turbulent shear stress of HWA. Small oscillations are visible in the PIV results due to the lack of convergence on this small term.

Figure 16 presents the comparison of spectra obtained from SPIV and from HWA using a local Taylor hypothesis [23]. In Fig. 16, k_{min} is the minimum wave number accessible with PIV ($k_{\mathrm{min}} = 2\pi/L_{\mathrm{f}}$, L_{f} being the field size) and k_{c} is the cutoff wave number of PIV due to the windowing effect (see [4]). According to *Foucaut* et al. [4], the PIV results are qualified to compare with the results of HWA only in the region between k_{min} and k_{c} (PIV cutoff wave number). In Fig. 16b, E_{33} shows a perfect fit with the result of the hot-wire anemometry. The E_{22} PIV spectrum (Fig. 16a) shows a slightly higher noise level than the E_{33} one at high wave number. This is probably due to the variation in magnification across the field and to the stretching effect in the x-direction linked to the stereoscopic setup. According to the experimental setup used, the w-component is always perpendicular to the axis of the lenses

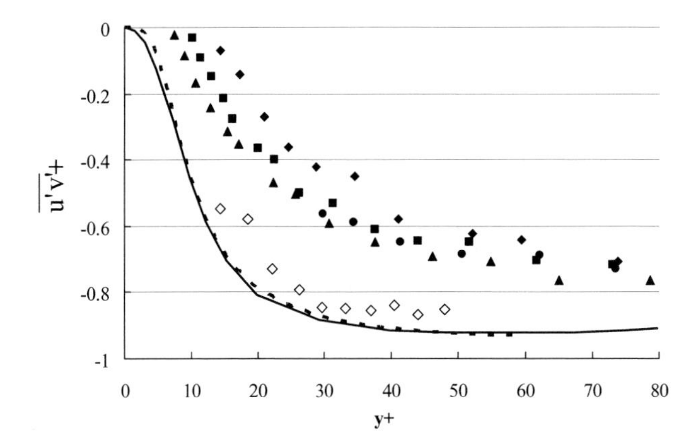

Fig. 15. Comparison of mean Reynolds shear stress of SPIV ($Re_\theta = 11\,400$ (◊)), hot-wire anemometry [15] ($Re_\theta = 11\,400$ (▲), 114 800 (■) and 20 600 (•)), Van Driest model [20] (...) and DNS [22] (-)

and is less affected than the v-component by the reconstruction process. The u-component (E_{11} spectrum not shown) is build from the same elements as v. It shows a similar behavior as E_{22} [24].

Figures 17a,b show the PDF of the v'- and w'-component, respectively. The PDF of w' shows much higher oscillation than that of v' (which is comparable to v'). As explained before, this is due to a peak-locking effect amplified by the gradient through the lightsheet. For v' (or u'), these oscillations are smoothed out by the stretching in the reconstruction procedure. Only a small difference in the height of the peak of v' is observable due to the noise caused by the strong velocity gradient near the wall. This difference disappears above the plane 4 (above this wall distance Fig. 10 shows that the gradient effect is negligible).

As is well known, the third-order moment of a random signal (e.g., signal A) S_A describes the asymmetry or skewness of the corresponding probability density function, while the fourth-order moment F_A (also referred to as flatness) reveals the frequency of occurrence of events far from the axis. These parameters are defined as:

$$S_\mathrm{A} = \frac{\overline{A^3}}{\overline{A^2}^{3/2}}\,,$$
$$F_\mathrm{A} = \frac{\overline{A^4}}{\overline{A^2}^{2}}\,. \tag{6}$$

For the skewness factor, $S_\mathrm{A} = 0$ is expected if the probability density function of A is symmetric. In turbulence, the three velocity fluctuations (u, v and w) often have a nearly Gaussian distribution. For such a distribution, $S_\mathrm{A} = 0$ and $F_A = 3$ are obtained.

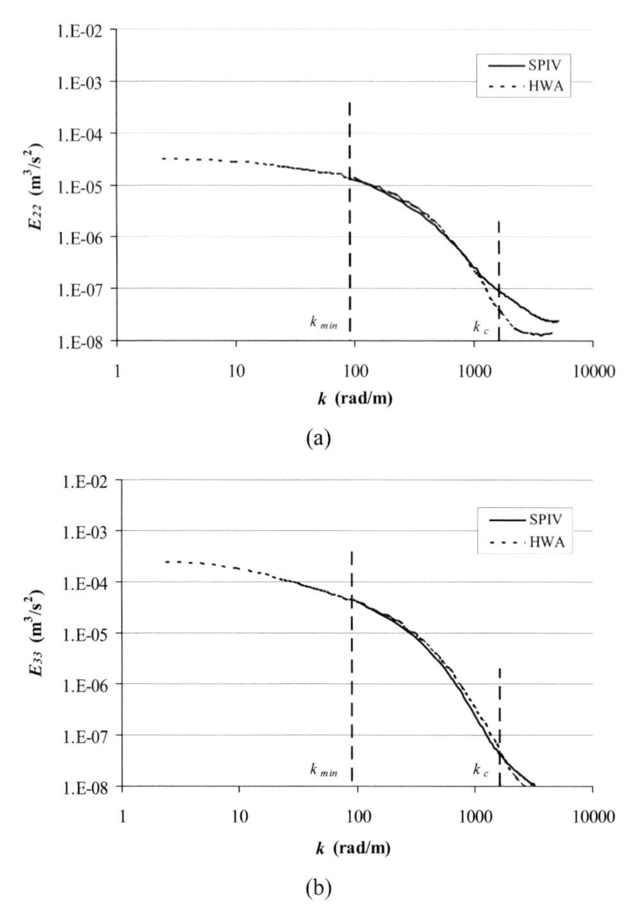

Fig. 16. Velocity spectrum of the v-(**a**) and w-(**b**) components of plane 4, comparison with HWA

Figures 18 to 20 show, respectively, the profiles of skewness factors $S_{u'}$, $S_{v'}$ and $S_{w'}$ for the three velocity fluctuations. In Fig. 18, the skewness factor for the streamwise fluctuations $S_{u'}$ is in very good agreement with the results of HWA [15]. For $S_{u'}$, both results indicate an increase toward the wall known to be due to the strong intermittency in the viscous sublayer. Above $y^+ \cong 15$, $S_{u'}$ is more or less constant and nearly zero, indicating a Gaussian behavior that is confirmed by the shape of the PDF. These results are in agreement with those of [25], who found that this location varies between $y^+ = 15$ and 20 for various Reynolds numbers. The positive value near the wall indicates that the frequency of occurrence of high positive streamwise fluctuations (high-speed streaks and sweeps) is higher than that of high negative fluctuations in this region.

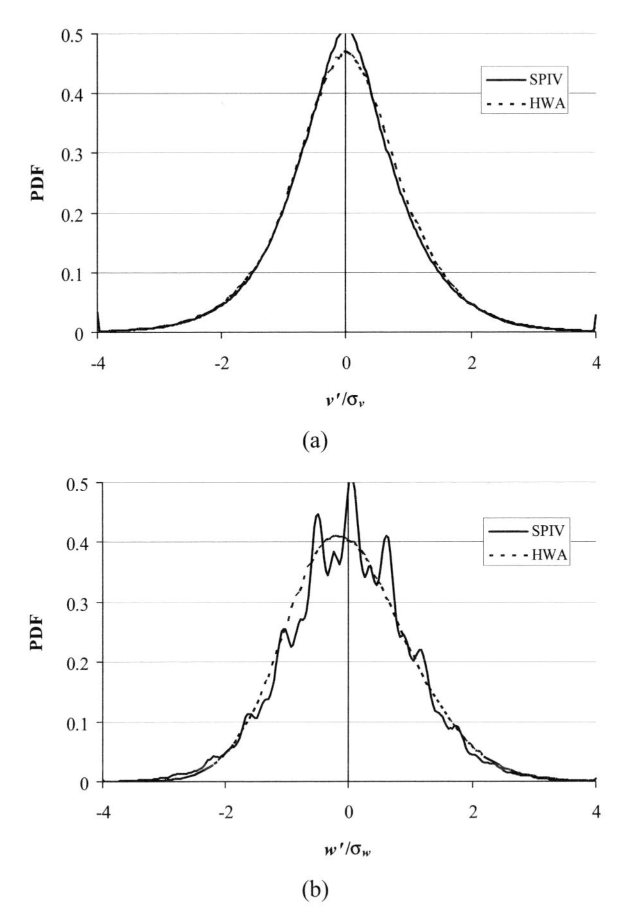

Fig. 17. PDF of the normalized fluctuations of the v-(**a**) and w-(**b**) components of plane 4, comparison with HWA

In Fig. 19, $S_{v'}$ agrees also very well with HWA. Again, it is nearly zero above $y^+ \cong 20$. The larger scatter compared to $S_{u'}$ is attributed to the small value of this component compared to u' (Fig. 14). The positive value of $S_{v'}$ evidences the asymmetry of the PDF close to the wall, indicating the predominance of ejections on the statistical behavior of this component.

The skewness factor $S_{w'}$ should be zero in a truly two-dimensional boundary layer due to the symmetry of the mean flow in the spanwise direction. This is confirmed by the present SPIV results in Fig. 20, where the factor $S_{w'}$ is nearly zero. The HWA results have a slightly positive value of about 0.25 that is comparable to that found by *Fernholz* and *Finley* [25]. It is attributed to a bias in the HWA measurements due either to a slight rotation

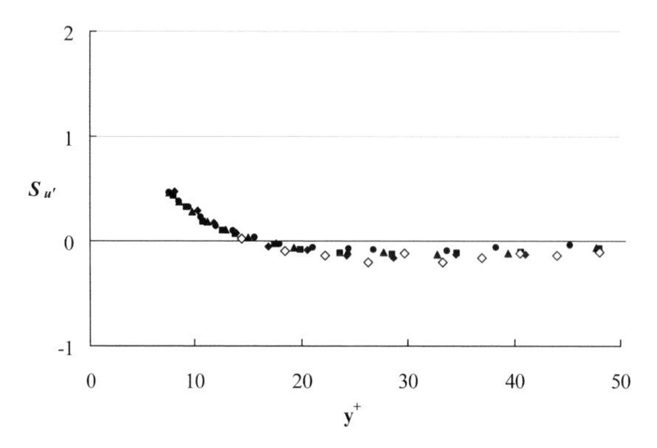

Fig. 18. Skewness factor $S_{u'}$. SPIV ($Re_\theta = 7800$ (◊)), hot-wire anemometry [15] ($Re_\theta = 11\,400$ (▲), $114\,800$ (■) and $20\,600$ (•))

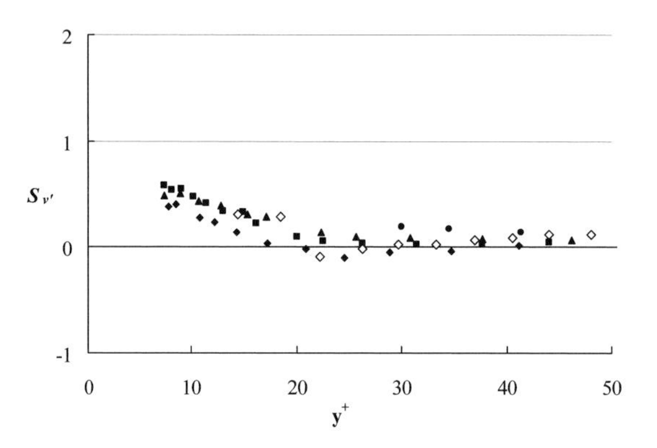

Fig. 19. Skewness factor $S_{v'}$. SPIV ($Re_\theta = 7800$ (◊)), hot-wire anemometry [15] ($Re_\theta = 11\,400$ (▲), $114\,800$ (■) and $20\,600$ (•))

of the probe around its axis or to the local velocity gradient at the scale of the probe.

Figures 21 to 23 show, respectively, the profiles of skewness factors $F_{u'}$, $F_{v'}$ and $F_{w'}$. Figure 21 compares the flatness factor $F_{u'}$ obtained in the present study with that obtained by HWA [15]. Again the SPIV results are in very good agreement with HWA. For $y^+ \geq 15$, the present SPIV results show that $F_{u'}$ increases slightly from 2.4 at $y^+ = 15$ to 2.8 at $y^+ = 39.7$ and then levels off afterwards. *Ueda and Hinze* [26] found a relationship between the position of the maximum of the streamwise normal Reynolds stress $(\overline{u'^2})$, the zero value of $S_{u'}$, and the minimum of $F_{u'}$. These characteristic points are at the same distance from the wall. In the present case, the maximum of $(\overline{u'^2})$ is at about $y^+ = 14$ (Fig. 14), the zero crossing of $S_{u'}$ is at around $y^+ = 16$

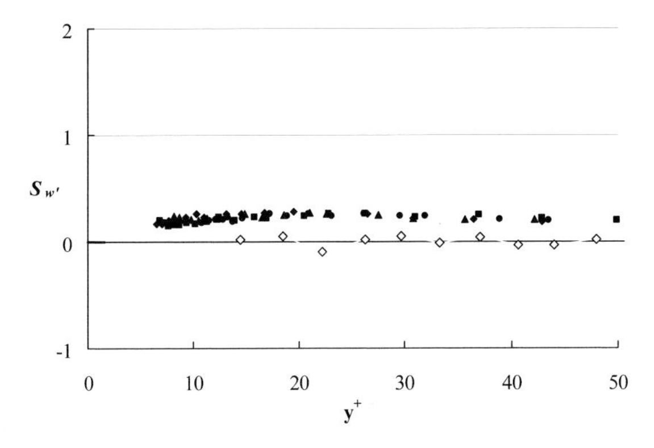

Fig. 20. Skewness factor $S_{w'}$. SPIV (Re$_\theta$ = 7800 (◊)), hot-wire anemometry [15] (Re$_\theta$ = 11 400 (▲), 114 800 (■) and 20 600 (•))

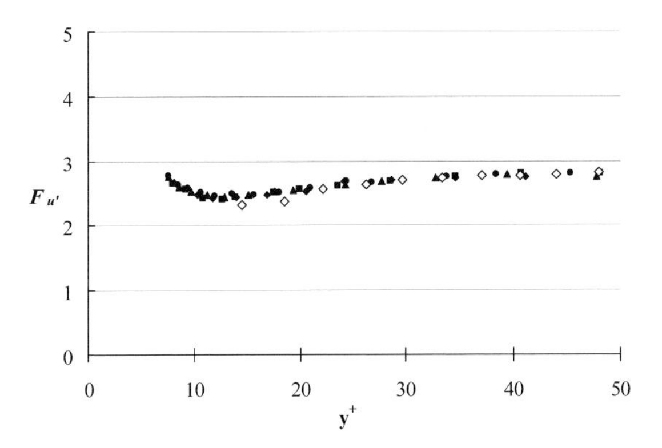

Fig. 21. Flatness factor $F_{u'}$. SPIV (Re$_\theta$ = 7800 (◊)), hot-wire anemometry [15] (Re$_\theta$ = 11 400 (▲), 114 800 (■) and 20 600 (•))

in Fig. 18, and the minimum value of $F_{u'}$ is near to $y^+ = 12$. Considering the experimental errors involved, the results confirm the relationship obtained by [26]. Similar to $S_{u'}$, $F_{u'}$ is known to increase toward large positive values in the viscous sublayer ($y^+ \leq 5$) due to intermittency.

Figure 22 presents the flatness factor $F_{v'}$ compared with HWA. In SPIV, this parameter decreases sharply between $y^+ = 14.5$ and 22.2. When $y^+ > 22.2$, $F_{v'}$ decreases slowly with increasing wall distance and is in good agreement with the results of HWA for various Reynolds numbers. This result was also obtained by other researchers [22, 25, 27]. The large values at $y^+ = 14.5$ and 18.5 can be associated with the intermittent character of near-wall flow in the buffer layer. The differences with the results of HWA at $y^+ = 14.5$ and

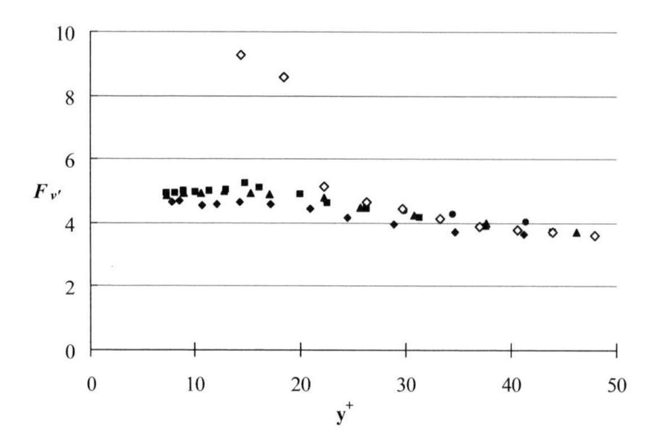

Fig. 22. Flatness factor $F_{v'}$. SPIV ($\mathrm{Re}_\theta = 7800$ (◊)), hot-wire anemometry [15] ($\mathrm{Re}_\theta = 11\,400$ (▲), 114 800 (■) and 20 600 (•))

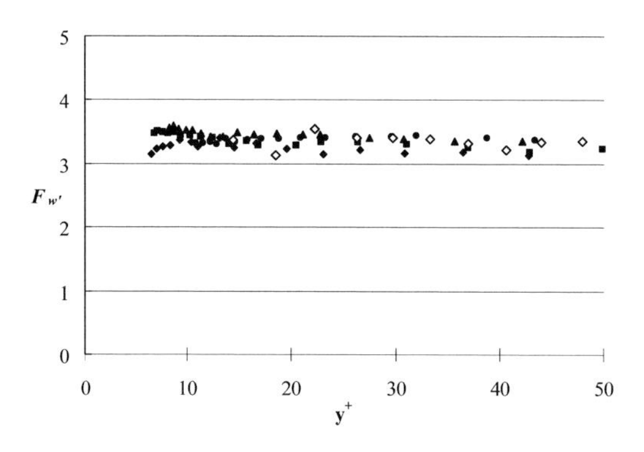

Fig. 23. Flatness factor $F_{w'}$. SPIV ($\mathrm{Re}_\theta = 7800$ (◊)), hot-wire anemometry [15] ($\mathrm{Re}_\theta = 11\,400$ (▲), 114 800 (■) and 20 600 (•))

18.5 are attributed to the velocity gradient at the size of the X-wire probe close to the wall.

Figure 23 compares $F_{w'}$ obtained by SPIV and HWA. There is a very good agreement. $F_{w'}$ is nearly constant. The value of 3.4 is the same as suggested by *Fernholz* and *Finley* [25].

7 Conclusions

In summary, an experiment of stereo-PIV was carried out to investigate near-wall turbulence. This experiment recorded 500 image pairs for each of 10 planes of a fully developed turbulent boundary layer flow along a flat plate. The first plane was placed at 14.5^+ from the wall. The spacing between neighboring planes was about 4^+. The Reynolds number based on the momentum thickness Re_θ was 7800. Three methods are available to analyze the database, namely image mapping, vector warping and the Soloff technique. They were compared in order to select the most suitable method of analysis for our database. The comparison took into consideration different interpolation and shift methods. The whole comparison was based on the estimation of computation time, the estimation of accuracy, the spatial spectra and velocity PDFs. The results favored the Soloff method over all the others. For the Soloff method, the difference between Whittaker and integer shift PIV processing and the difference between using 3 and 5 calibration planes was negligible. As a result, the Soloff method with 3 calibration planes for projection and reconstruction, using integer shift for PIV analysis was chosen as the most suitable method for the database. The improvement provided by the correction process [10] was presented. Recently, *Calluaud* and *David* [28] and *Wieneke* [12] proposed a method based on the pinhole model. This model is based on previous work in the field of computer vision. It can incorporate some limited optical distortion and has the advantage of using less parameters in the least square fit than the Soloff method (24 instead of about 80). *Scarano* et al. [29] compared this method with the image warping method with misalignment correction. They found that the two methods are practically equivalent for a correctly aligned system. In the present study, the Soloff method was found to be the best compromise to analyze turbulent PIV data, but the differences from the other two methods (mapping and warping) were fairly limited. The main advantage for the moment of the Soloff method is its generality and overall accuracy and it seems that any reconstruction method, properly applied leads to errors smaller than the PIV processing errors.

Using this processing method, the whole database was analyzed. The results were presented in terms of the mean streamwise velocity, velocity fluctuations, Reynolds shear stresses,, the spectra and PDFs, and skewness and flatness. They were compared with those of hot-wire anemometry, DNS and the Van Driest model. The comparisons showed that the results of SPIV are in good accordance with those of other methods. In general, the results of SPIV are closer to those of the Van Driest model and DNS than to HWA in the very near wall region. This chapter concludes that SPIV is a suitable method to study near-wall turbulence.

References

[1] R. Adrian: Particle-imaging techniques for experimental fluid mechanics, Ann. Rev. Fluid Mech. **23**, 261–304 (1991)

[2] J. Westerweel: Fundamentals of digital particle image velocimetry, Meas. Sci. Technol. **8**, 1379–1392 (1997)

[3] M. Raffel, M. Gharib, O. Ronneberger, J. Kompenhans: Feasibility study of three-dimensional PIV by correlating images of particles within parallel light sheets, Exp Fluids. **19**, 69–77 (1995)

[4] J. Foucaut, J. Carlier, M. Stanislas: PIV optimization for the study of turbulent flow using spectral analysis, Meas. Sci. Technol. **15**, 1046–1058 (2004)
[5] L. Lourenco: *Some Comments on Particle Image Displacement Velocimetry*, Von Karmann Institute for Fluid Dynamics, Lecture Series **1988-06** (Von Karmann Institute for Fluid Dynamics 1988)
[6] A. Prasad, R. Adrian: Stereoscopic particle image velocimetry applied to liquid flows, Exp. Fluids **15**, 49–60 (1993)
[7] S. Soloff, R. Adrian, Z. Liu: Distortion compensation for generalized stereoscopic particle image velocimetry, Meas. Sci. Technol. **8**, 1441–1454 (1997)
[8] C. Willert: Stereoscopic digital particle image velocimetry for applications in wind tunnel flows, Meas. Sci. Technol. **8**, 1465–1479 (1997)
[9] J. Westerweel, J. van Oord: Stereoscopic PIV measurements in a turbulent boundary layer, in *EUROPIV: Progress Towards Industrial Application* (Kluwer, Dordrecht 2000) pp. 459–478
[10] S. Coudert, J. Schön: Back projection algorithm with misalignment corrections for 2D3C stereoscopic PIV, Meas. Sci. Technol. **12**, 1371–1381 (2001)
[11] N. Pérenne, J. Foucaut, J. Savatier: Study of the accuracy of different stereoscopic reconstruction algorithms, in M. Stanislas, J. Westerweel, J. Kompenhans (Eds.): *Proceeding of the EUROPIV 2 Workshop on Particle Image Velocimetry: Recent Improvements* (Springer, Berlin, Heidelberg 2004) pp. 375–390
[12] B. Wieneke: Stereo-PIV using self-calibration on particle images, Exp. Fluids **39**, 267–280 (2005)
[13] R. Fei, W. Merzkirch: Investigations of the measurement accuracy of stereo particle image velocimetry, Exp. Fluids **37**, 559–565 (2004)
[14] N. Lawson, J. Wu: Three-dimensional particle image velocimety experimental error analysis of digital angular stereoscopic system, Meas. Sci. Technol. **8**, 1455–1464 (1997)
[15] J. Carlier, M. Stanislas: Experimental study of eddy structures in a turbulent boundary layer using particle image velocimetry, J. Fluid Mech. **535**, 143–188 (2005)
[16] T. Ursenbacher: *Traitement de vélocimétrie par images digitales de particules par une technique robuste de distortion d'images*, Ph.D. thesis, Ecole Polytechnique de Lausanne (2000)
[17] F. Scarano, R. Riethmuller: Advances in iterative multi-grid PIV image processing, Exp. Fluids [**Suppl.**], S51–S60 (2000)
[18] J. Foucaut, B. Miliat, N. Pérenne, M. Stanislas: *Characterization of Different PIV Algorithms Using the EUROPIV Synthetic Image Generator and Real Images from a Turbulent Boundary Layer* (Springer, Berlin, Heidelberg 2004)
[19] R. Keane, R. Adrian: Optimisation of particle image velocimeters-part I double pulsed systems, Meas. Sci. Technol. **1**, 1202–1215 (1990)
[20] E. R. van Driest: On turbulent flow near a wall, J. Aero. Sci. **23**, 1007–1011 (1956)
[21] C. Willert: Prososal for netCDF (re)implementation for use with planar velocimetry data, in M. Stanislas, J. Westerweel, J. Kompenhans (Eds.): *Proceeding of the EUROPIV 2 Workshop on Particle Image Velocimetry: Recent Improvements* (2004) pp. 251–262
[22] P. Spalart: Direct simulation of a turbulent boundary layer up to $\mathrm{Re}_\theta = 1410$, J. Fluid Mech. **1878**, 61–98 (1988)

[23] J. O. Hinze: *Turbulence*, Series in Mechanical Engineering (McGraw-Hill 1975)
[24] C. Willert, M. Gharib: Digital particle image velocimetry, Exp. Fluids **10**, 181–193 (1991)
[25] H. Fernholz, P. Finley: The incompressible zero-pressure-gradient turbulent boundary layer: An assessment of the data, Prog. Aerospace Sci. **32**, 245–311 (1996)
[26] H. Ueda, J. Hinze: Fine-structure turbulence in the wall region of a turbulent boundary layer, J. Fluid Mech. **61**, 125–143 (1975)
[27] P. Vukoslavcevic, J. Wallace, J. Balint: The velocity and vorticity vector fields of a turbulent boundary layer, J. Fluid Mech. **228**, 25–51 (1991)
[28] D. Calluaud, L. David: 3D PIV measurements of the flow around a surface-mounted block, Exp. Fluids **36**, 53–61 (2004)
[29] F. Scarano, L. David, M. Bsibsi, D. Calluaud: S-PIV comparative assessment image dewarping+misalignment correction and pinhole+geometric back projection, Exp. Fluids **39**, 257–266 (2005)

Index

Joint Numerical and Experimental Investigation of the Flow Around a Circular Cylinder at High Reynolds Number

Rodolphe Perrin[1,2], Charles Mockett[2], Marianna Braza[1], Emmanuel Cid[1], Sébastien Cazin[1], Alain Sevrain[1], Patrick Chassaing[1] and Frank Thiele[2]

[1] Institut de Mécanique des Fluides de Toulouse, CNRS/INPT UMR No. 5502, Toulouse, France
perrin@imft.fr

[2] Institute for Fluid Mechanics and Engineering Acoustics (ISTA), TU-Berlin, Berlin, Germany
mockett@cfd.tu-berlin.de

Abstract. A collaborative study is presented for the flow past a circular cylinder at high Reynolds number, which makes use of both experimental and numerical approaches. The case setup was designed specifically to maximize the level of comparison between experiment and simulation, incorporating a confined environment enabling simulation on a domain of moderate size and avoiding problems associated with "infinite conditions". The two principle goals of the investigation are the detailed validation of the simulation using the experiment and *vice versa* and the exploitation of the respective advantages of both approaches in order to provide a better understanding of the dynamics of the flow. The experiment was carried out using PIV, stereoscopic PIV and time-resolved PIV, and the simulation was performed using detached eddy simulation (DES), a hybrid method for turbulence treatment. A very good agreement between the simulation and the experiment is achieved for the steady mean motion, the coherent motion and the turbulence quantities. In the analysis, the experiment and simulation are shown to be complementary and to allow a more complete description of the flow studied.

1 Introduction

Considerable progress has been made during the last decades in both experimental and numerical techniques for the investigation of fluid flows. On the experimental side, there has been a general evolution from intrusive single-point measurements to nonintrusive planar measurements such as PIV (see, e.g., [1]) and more recently even time-resolved volume measurements.

Concerning computational fluid dynamics (CFD), increases in computer power have allowed the reduction of discretization errors through computation on increasingly fine grids, such that the dominant error source remaining stems from the assumptions met in the underlying physical models, particularly for the treatment of turbulence. In this field, a recent hybrid method known as detached eddy simulation (DES) has been shown to be particularly promising for bluff-body flows [2, 3].

A. Schroeder, C. E. Willert (Eds.): Particle Image Velocimetry,
Topics Appl. Physics **112**, 223–244 (2008)

Despite the impressive progress made in both experimental and numerical techniques, considerable limitations still remain. This makes a combination of approaches particularly attractive, as this can serve to alleviate the individual disadvantages, giving rise to a more complete understanding of the flow-physical problem.

This chapter presents an overview of a joint numerical and experimental study, arising from a collaboration between IMFT in Toulouse and ISTA, TU-Berlin, in the context of the European "DESider" program. As a generic bluff-body case, the flow past a circular cylinder is considered, which was chosen for several reasons. Strong flow separation characterized by coherent structures is present, and the nonlinear interaction of these with the turbulent motion represents a considerable challenge to standard turbulence models. Furthermore, the symmetries of the flow allow a better understanding of the dynamics and the use of postprocessing methods that are well adapted to the configuration. The objective of this study is therefore twofold. Firstly, a validation of the simulation is sought by a detailed comparison of the motions. Following this, an investigation of the dynamics of the flow is conducted, exploiting the respective advantages of both approaches. Sections 2 and 3 present the flow configuration and the methods used, Sect. 4 presents the results obtained and a conclusion and discussion of future work is to be found in Sect. 5.

2 Flow Configuration

The experimental configuration was designed at the outset to allow direct comparison with CFD. To this end, the cylinder was placed in a square channel with a high blockage and low aspect ratio. This confined environment was chosen to avoid the uncertainties typical in most experimental investigations of nominally two-dimensional configurations, such as endplates and the emulation of "infinite" boundary conditions. As a result, computations are possible on a domain of moderate size, corresponding precisely to the experimental geometry. A cylinder diameter of $D = 140\,\mathrm{mm}$ and a square channel cross section measuring $L = H = 670\,\mathrm{mm}$ were chosen, giving an aspect ratio of $L/D = 4.8$ and a blockage coefficient of $D/H = 0.208$ (Fig. 1a). The uniform inlet velocity U_0 of $15\,\mathrm{m \cdot s^{-1}}$ gives a Reynolds number based on the cylinder diameter of $\mathrm{Re} = 140\,000$. The free-stream turbulence intensity at the inlet, measured by hot wire, is $1.5\,\%$, normalized by the free-stream velocity. Although results presented here are for $\mathrm{Re} = 140\,000$, other Reynolds numbers between $70\,000$ and $190\,000$ have been investigated [4] and a decrease of the mean drag coefficient (Fig. 1b) and a degradation of the von Kármán vortices has been observed with increasing Re, indicating that the flow is at the beginning of the critical regime, where the transition location reaches the separation point. This regime is believed to occur at a lower Reynolds number than in other studies because of a combination of blockage

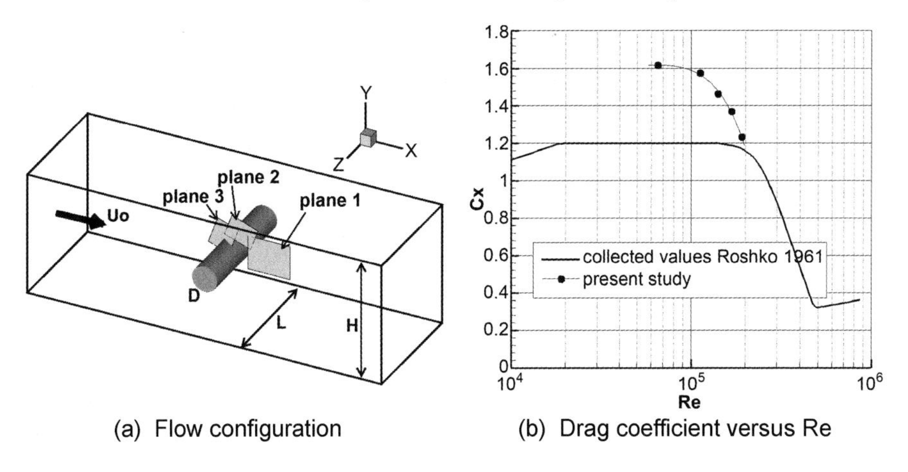

Fig. 1. Case setup and illustration of shedding regime

effects and the free-stream turbulence intensity. In the following, x, y and z denote the longitudinal, transverse and spanwise directions, respectively, and the origin is located at the center of the cylinder. U, V and W are the velocity components along these directions and all presented quantities have been nondimensionalized using U_0 and D.

3 Method

3.1 Experimental Measurement Techniques

The experiment was conducted in the S1 wind tunnel at IMFT. The majority of the measurements carried out were of wall pressure on the cylinder and velocity measurements in the near wake. The velocity measurements employed PIV, stereoscopic PIV and time-resolved PIV (TRPIV) techniques, which were chosen because of their suitability to flow cartography offered by the access to spatial information. The 2- and 3-component low data rate measurements were performed using a Nd:YAG (Quantel) laser, delivering an energy of $2 \times 200\,\text{mJ}$, and PCO-Sensicam cameras with a resolution of 1280×1024 pixels that were operated at a frequency of 4 Hz. Stereoscopic measurements were performed in the near wake both to check the influence of the velocity component normal to the plane, which was found to be negligible, and to evaluate all the components of the turbulent stress tensor. The Scheimpflug angular configuration was employed, with a camera placed on each side of the laser sheet. Details of the procedure employed for the reconstruction of the three velocity components can be found in [5]. The time-resolved measurements were carried out with a Pegasus (New Wave) laser delivering an energy of $2 \times 10\,\text{mJ}$, and a CMOS APX (Photron) camera with a resolution of 1024×1024 pixels. The system allows the acquisition of image pairs at a frequency of 1 kHz. In each case, the flow was seeded with DEHS (with a typical particle size of 1 µm).

The low data rate 2C measurements were carried out in the near wake of the cylinder and near the separation, as seen in Fig. 1a (planes 1, 2 and 3), whereas the stereoscopic measurements were carried out in plane 1 and the TRPIV measurements were limited to a smaller domain in the region ($0.76 < x/D < 1.24, 0 < y/D < 1.24$), because of the lower energy of the laser. The particle images were analyzed using an inhouse development of IMFT ("Services Signaux Images"), which uses an algorithm based on a 2D FFT cross-correlation function implemented in an iterative scheme with image deformation [6]. In the iterative procedure, the interogation windows are symetrically shifted on each image, and the displacements are corrected at each iteration. The chosen convergence criteria is the sum of the corrections on the whole field. Typically, 10 or 12 iterations are necessary. When the convergence criteria start to diverge at the Nth iteration, the validated result is taken at iteration $N - 1$. The flow was analyzed by crosscorrelating 50 % overlapping windows of 32×32 pixels, which contain roughly 40 particles, in the case of the low data rate acquisitions, yielding fields of 77×61 vectors with a spatial resolution of 3.13 mm ($0.002\,24D$), and of 64×64 pixels in the case of the TRPIV acquisitions, yielding a resolution similar to that of the low data rate measurements. Typical inter-image particle displacements are about 5 pixels. The maximum estimated velocity gradients are below 0.1 pixel/pixel and the estimated maximum error on the instantaneous velocity is of the order of 0.1 pixel. From the resulting vector fields, about 2 % of the vectors are detected as outliers and are removed using a sort based on the norm, the signal-to-noise ratio and the median test. These vectors are then replaced by a second-order least square interpolation scheme. More details on the experimental setup can be found in [4] and [5].

3.2 Numerical Simulation Setup

A particularly relevant technique for the simulation of such strongly separated flows is the hybrid approach known as detached eddy simulation (DES) proposed by Spalart and coworkers [7], and that is a topic of research at ISTA and in the DESider project in general. DES combines the advantages of RANS in attached boundary layers with LES in areas of massively separated flow to result in a technique that is more accurate than pure RANS approaches and less computationally expensive than pure LES. In order to achieve this, the length scale in the underlying turbulence model is replaced by the DES length scale:

$$L_{\mathrm{DES}} = \min(L_{\mathrm{RANS}}, C_{\mathrm{DES}}\Delta)\,,$$

where C_{DES} is a model constant analogous to that of the Smagorinsky constant in LES, and Δ is an appropriate grid scale, which takes the role of the

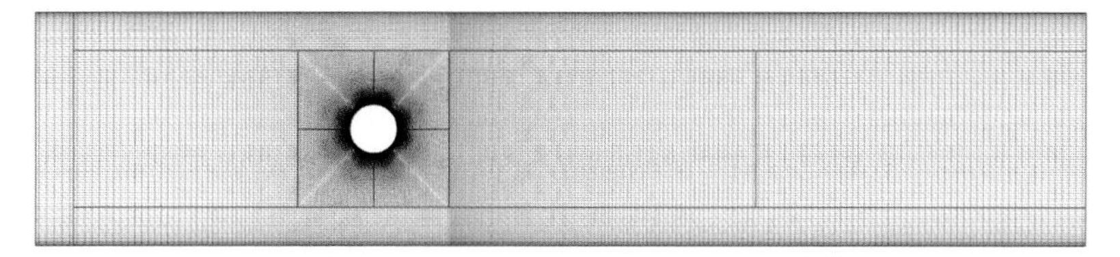

Fig. 2. 2D slice of the computational grid used for the simulation

filter width in LES-mode. The present computations have been carried out using the ISTA inhouse finite-volume Navier–Stokes solver ELAN [8]. The procedure is implicit and of second-order accuracy in space and time, with variables stored in the cell centers of curvilinear, block-structured grids. Diffusive terms are approximated with central schemes, whereas a hybrid blending of central and upwind-biased limited schemes is used for the treatment of convective terms for DES, as suggested by [9]. The pressure is obtained using a pressure-correction scheme of the SIMPLE type, which assures mass conservation [10] and a generalized Rhie and Chow interpolation is used [11].

The DES is implemented to a compact explicit algebraic Reynolds stress (CEASM) RANS model. The CEASM-DES formulation [12] incorporates a mechanism to ensure that the entire turbulent boundary layer is handled by the RANS mode of the DES. A modification has been made to the k-equation of the model, which suppresses the model's natural transition to ensure that the boundary layer on the cylinder remains laminar, and that transition occurs immediately following separation. The computational domain corresponds precisely to the experimental geometry, with the walls treated using no-slip boundary conditions. As the inlet to the channel is located approximately 7 diameters upstream of the cylinder, it is assumed that upstream unsteadiness effects are negligible. Accordingly, a steady inflow boundary condition is used. The full grid, a 2D slice of which is shown in Fig. 2, consists of some 5 million points. Two different time step sizes (3×10^{-4} s and 5×10^{-4} s, corresponding to approx. $0.05D/U_0$ and $0.03D/U_0$) have been computed, the effect of which will be discussed in the results section.

3.3 Postprocessing Techniques

For the PIV data, global averaging has been carried out using about 3000 instantaneous fields. These can be considered uncorrelated in time due to the low sampling rate, and uncertainties have been determined to the order of 2 % [5]. Concerning the simulation, global averaging has been performed on the entire domain during the calculation, giving access to information about the homogeneity of the flow in the spanwise direction and wall effects. Averaging has been performed over 3000 time steps for the large time step, and 14 000 time steps for the fine time step, which correspond to roughly 30 and 100 shedding periods, respectively. Although the coarse time step computation cannot be considered statistically converged, it will be seen that the deviation in the results is large enough to demonstrate the detrimental influence of the large time step.

To reduce uncertainty in the comparison of experiment and simulation, the same processing techniques were applied to both. To achieve this, unsteady flow data has been stored every time step on a 2D slice located at the midspan position $z/D = 0$, corresponding to the PIV measurements. The experimental drag coefficient was estimated by integration of the pressure on the cylinder at the midspan position, assuming homogeneity of the flow on a significant spanwise portion, and neglecting the contribution of the viscous forces. In the simulation, the drag has been calculated both as in the experiment as well as over the entire cylinder including pressure and friction components, allowing an evaluation of the experimental assumptions. Spectra of the experimental velocity component have been estimated at selected points in the TRPIV plane using Welch's averaged periodogram method [13]. Segments of length 1024, corresponding to 1.024 s are taken from the signal, to which a Hamming window is applied and the periodogram obtained from Fourier transforms of these segments are averaged, giving a spectral resolution of 0.98 Hz. Owing to the sampling frequency (1 kHz), the spectra are evaluated up to the frequency 500 Hz. The nearly periodic nature of the vortex shedding in the wake allows the use of phase averaging to characterize the unsteady mean motion and the turbulence quantities. This method allows the classical decomposition of the flow into a mean part, a quasiperiodic fluctuation, and a random fluctuation, which can be written as $U_i = \overline{U_i} + \tilde{u}_i + u'_i$ [14]. The phase-averaged motion is then written as $\langle U_i \rangle = \overline{U_i} + \tilde{u}_i$. The initial method used to obtain this decomposition employed the pressure on the cylinder at an angle of $\theta = 70°$ with the forward stagnation point as a trigger signal from which the phase angle was determined. This was achieved using the Hilbert transform [15], following which the instantaneous fields were classified into phase intervals of width $2\pi/128$ and then averaged at each angle. For the time-resolved data (TRPIV and simulation), the fields could also be resampled at each desired phase angle, which was shown to give very similar results. As will be presented in Sect. 4.1, it was seen that strong irregularities were present in the time signals from both simulation and experiment. These instants of altered vortex shedding behavior have been detected using a threshold sorting technique based on the amplitude and period of the trigger signal and excluded from the phase-averaging process for both the experimental and simulated data. Roughly 170 fields have been averaged for each phase angle in the case of the experiment, and 90 for the simulation. More details on the averaging procedure can be found in [5].

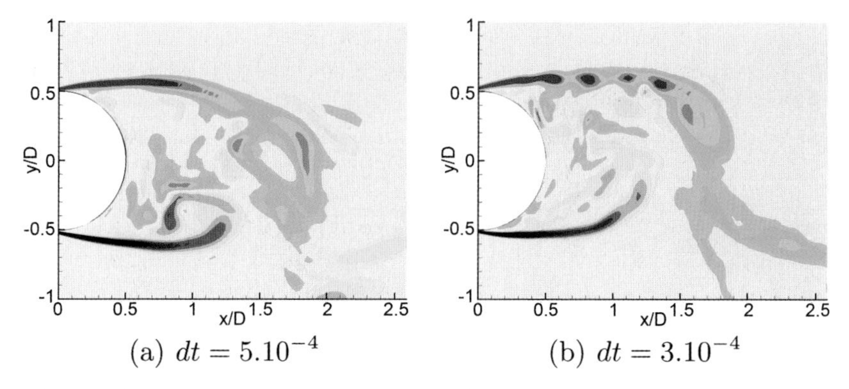

Fig. 3. Comparison of the shear-layer behavior obtained for each time step size, contours of Ω_{21} component of rotation rate tensor

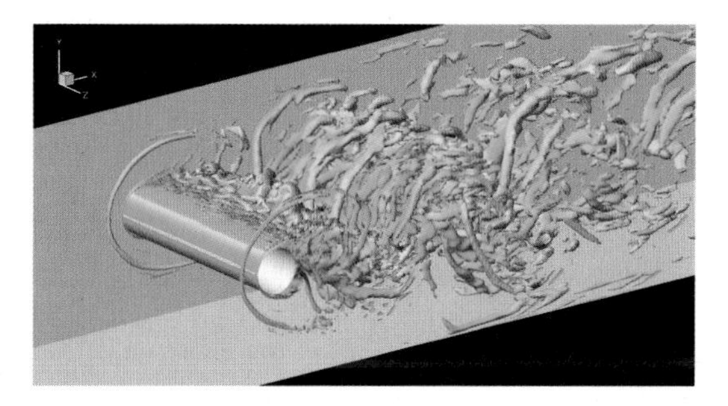

Fig. 4. Instantaneous visualization of the flow obtained from DES. Isocontours of λ_2 colored by the longitudinal velocity

4 Presentation and Discussion of Results

4.1 Analysis of the Instantaneous Motion

Before presenting the processed data, certain features can be observed by analysis of the instantaneous motion from both the experiment and the simulation. Figure 3 shows two snapshots of the Ω_{21} component of the rotation rate tensor in the near wake obtained with each of the simulation time steps, dt. It can be seen that the coarse time step leads to an additional filtering effect in the separated shear layer, while small-scale vortices develop for the fine time step. In Sects. 4.2 and 4.3, a comparison of time and phase-averaged motions with those from the experiment will show the effect this has on the global flow.

The complexity and three-dimensionality of the flow in the complete domain can be observed in Fig. 4, where the Von Kármán vortex shedding(can be identified together with longitudinal vortices connecting the primary vor-

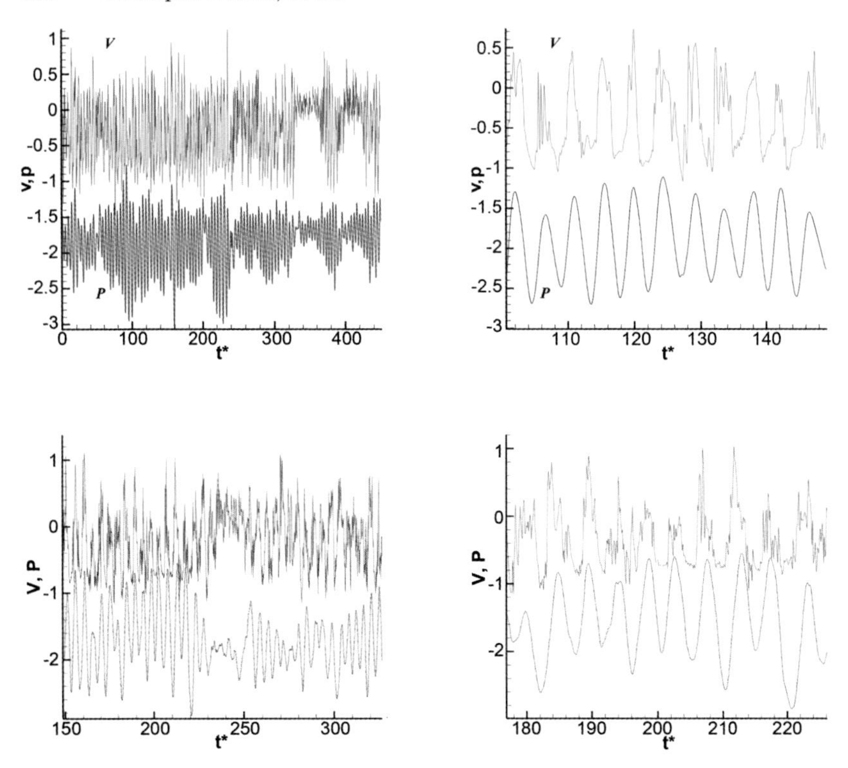

Fig. 5. Time signal of the V velocity at $x/D = 1$, $y/D = 0.5$ and of the pressure coefficient at $\theta = 70°$ on the cylinder, from the simulation (*above*) and TRPIV measurements (*below*)

tices and small-scale vortices in the separated shear layer. Horseshoe vortices are also seen at the junctions between the cylinder and the walls.

To compare the temporal evolution of the flow, time traces of the velocity at $x/D = 1$, $y/D = 0.5$ and of the pressure coefficient at $\theta = 70°$ on the cylinder are represented in Fig. 5. The signals exhibit similar qualitative character and levels of fluctuation, with the velocity presenting a strong quasiperiodic component corresponding to the vortex shedding and a superimposed random fluctuation corresponding to the turbulent motion. At certain instants (at about $t^* \simeq 350$ and $t^* \simeq 400$ for DES), irregularities occur in the signal; the "mean value" tends toward zero and the periodic component seems to disappear. The same kind of event occurs in the experiment, as can be seen at $t^* \simeq 240$. These events, furthermore, seem to coincide with an irregular loss of amplitude in the pressure signal. It has been shown [4] that these events correspond to a breakdown of the shedding in the near wake (Fig. 6a), a phenomena also observed in the DES of other bluff bodies such as an airfoil in deep stall [16]. The numerical simulation allows the visualization of the flow on a larger domain, and it appears that the formation of the vortices occurs

Table 1. Drag coefficient

	Pressure drag at $z/D = 0$	Total drag, whole cylinder
Experiment	1.45	–
DES $\mathrm{d}t = 3 \times 10^{-4}$	1.48	1.50
DES $\mathrm{d}t = 5 \times 10^{-4}$	1.32	1.33

further downstream in the wake (Fig. 6b), the velocity signal at a position further downstream still exhibiting a periodic component. Although a more detailed investigation of these events is necessary, this gives an example of how a joint approach can be used to analyze such flow-physical phenomena. From a more practical standpoint, these events also are of consequence for the postprocessing conducted. For the global averaging, all available data have been processed, including the irregularities. Although for the experiment, the actual time of acquisition is long enough to achieve convergence (due to the low data rate), these intermittent irregularities lead to a very long required computation time for statistical convergence, which constitutes a key difficulty for this flow case. Concerning the phase averaging, it is clear that these events differ significantly from the dominant, more periodic, shedding and as such were excluded from the phase averaging for the experiment and simulation, as described in Sect. 3.3.

4.2 Steady Mean Motion

4.2.1 Integral Forces and Pressure Distribution

To assess the simulation quality, usually the first quantities to be checked are the drag coefficient and the pressure distribution on the cylinder. As explained in Sect. 3.3, the experimental drag coefficient was evaluated from the pressure on the cylinder at the midspan. Table 1 summarizes the drag coefficient value found for the experiment and simulations. It is seen that the drag is underestimated by the coarse time step simulation, whereas a good agreement is found between the experimental value and the fine time step simulation. It should be noted that the blockage effect leads to an increased drag compared with a nonconfined environment. It is also seen that the pressure drag is very similar to the total drag in the simulations, therefore confirming the validity of experimental assumptions met (see Sect. 3.3).

Figure 7 shows the distribution of the mean and RMS pressure coefficient on the cylinder. As with the drag values, the pressure profile exhibits much better agreement for the fine time step, the base pressure coefficient (i.e., the pressure coefficient at $\theta = 180°$) being overestimated for the coarse time step. The RMS values are also in better agreement for the fine time step, although slightly underestimated.

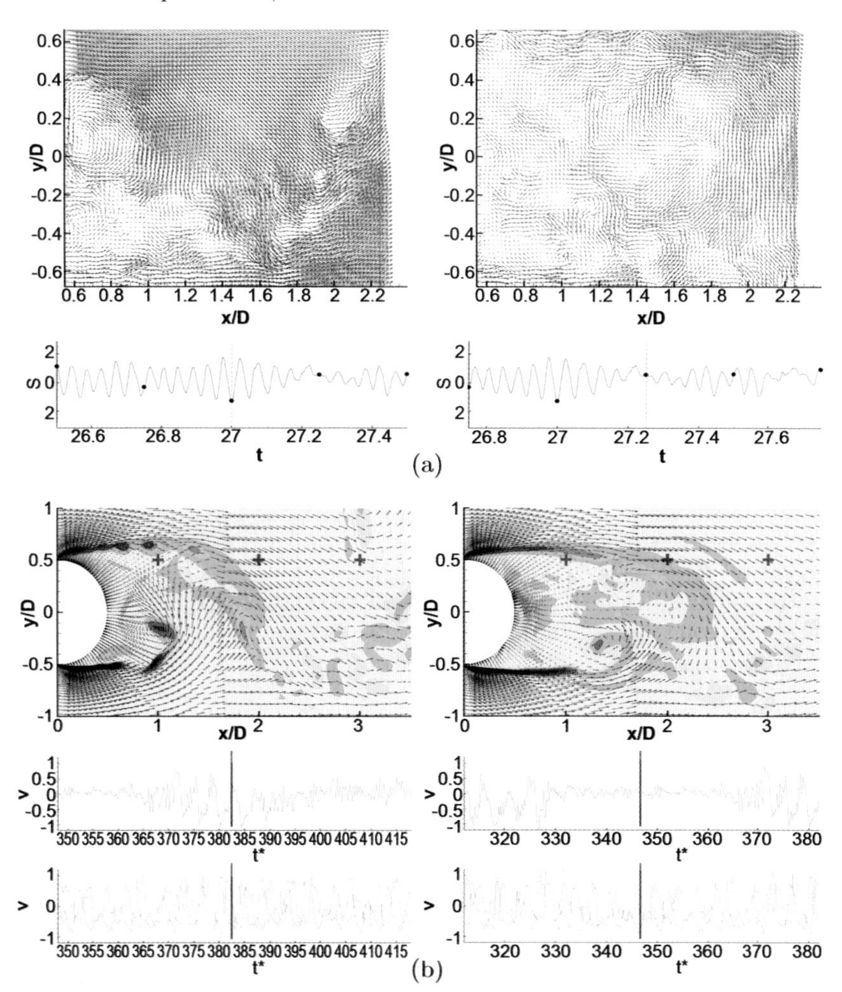

Fig. 6. Near-wake behavior at instants where the shedding is regular (*left*) and irregular (*right*). The time trace depicted from the experiment (**a**) is of pressure, whereas the velocity at $y/D = 0.5$ and $x/D = 1, 2, 3$ (locations denoted by crosses) is shown for the simulation (**b**)

4.2.2 Spanwise Homogeneity and Wall Effects

As the simulation gives access to the mean field over the whole domain, some quantities on the (x, z) slice at $y/D = 0$ are presented (Fig. 8) to examine the effect of the side walls on the flow, and to assess the level of spanwise homogeneity. It is seen that the flow is almost two-dimensional over a large portion of the span in the central region, up to about $1D$ from the wall. The regions of visible near-wall effect then extend slowly with increasing downstream distance. It is also apparent that the averaged quantities exhibit some

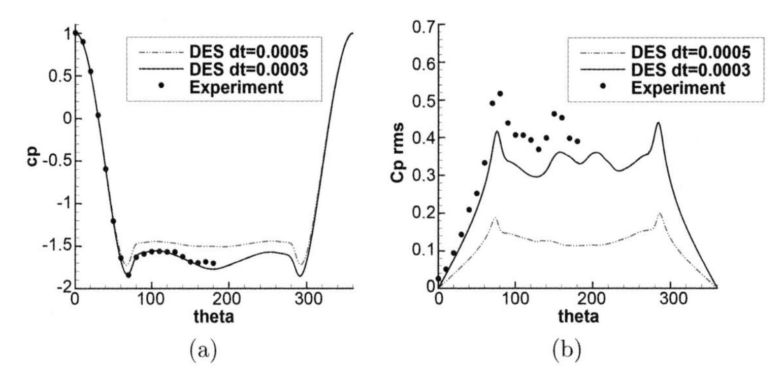

(a) (b)

Fig. 7. Pressure-coefficient profile on the cylinder and its *rms* value

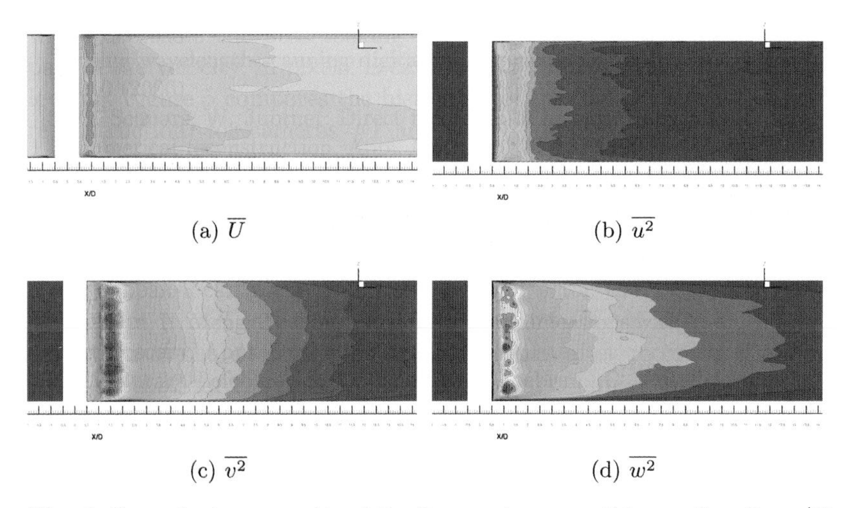

(a) $\overline{U}$ (b) $\overline{u^2}$

(c) $\overline{v^2}$ (d) $\overline{w^2}$

Fig. 8. Spanwise homogeneity of the flow: various quantities on the slice $y/D = 0$

statistical scatter, particularly the second-moment quantities. In order to enhance the statistical convergence, additional averaging has been performed over the observed homogeneous central region (up to a distance of $1D$ from each side wall).

4.2.3 Near-Wake Region

A direct comparison is possible between the near-wake regions of the experiment and simulation. Figure 9 shows some profiles of the $\overline{U}$ velocity in the near-wake and on the rear axis for the experiment and the two simulated time step sizes. It is shown that, in agreement with the visualizations of Fig. 3, the filtering effect of the coarse time step leads to excessive shear in the separated shear layer due to the suppression of resolved turbulent mixing (the

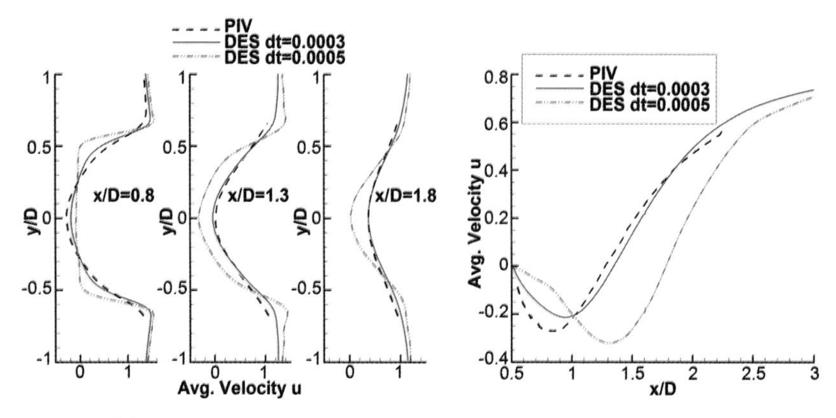

Fig. 9. $\overline{U}$ velocity profiles in the near-wake and on the rear axis. Comparison between experiment and simulations

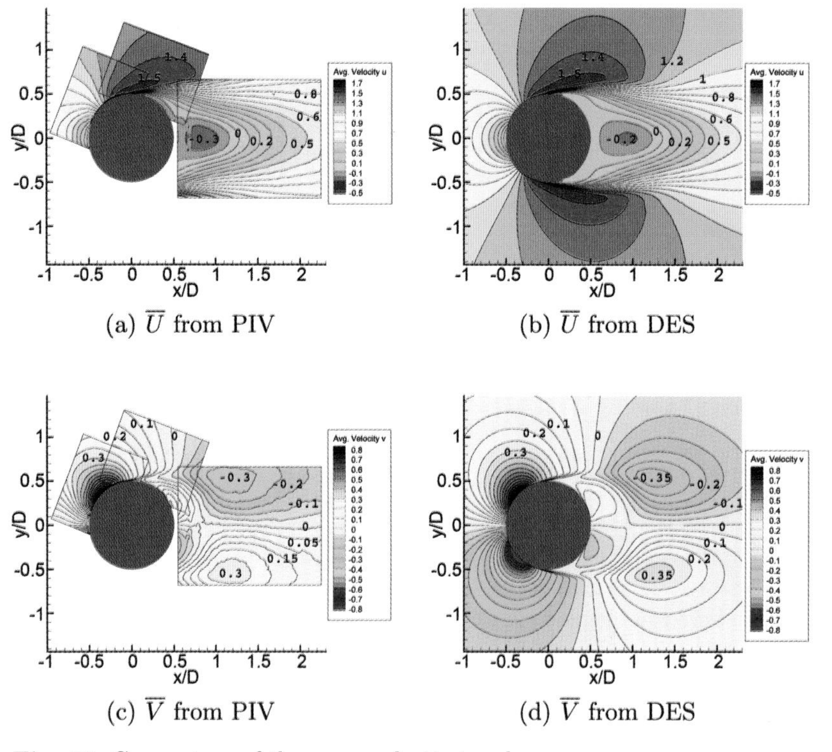

(a) $\overline{U}$ from PIV (b) $\overline{U}$ from DES

(c) $\overline{V}$ from PIV (d) $\overline{V}$ from DES

Fig. 10. Comparison of the mean-velocity topology

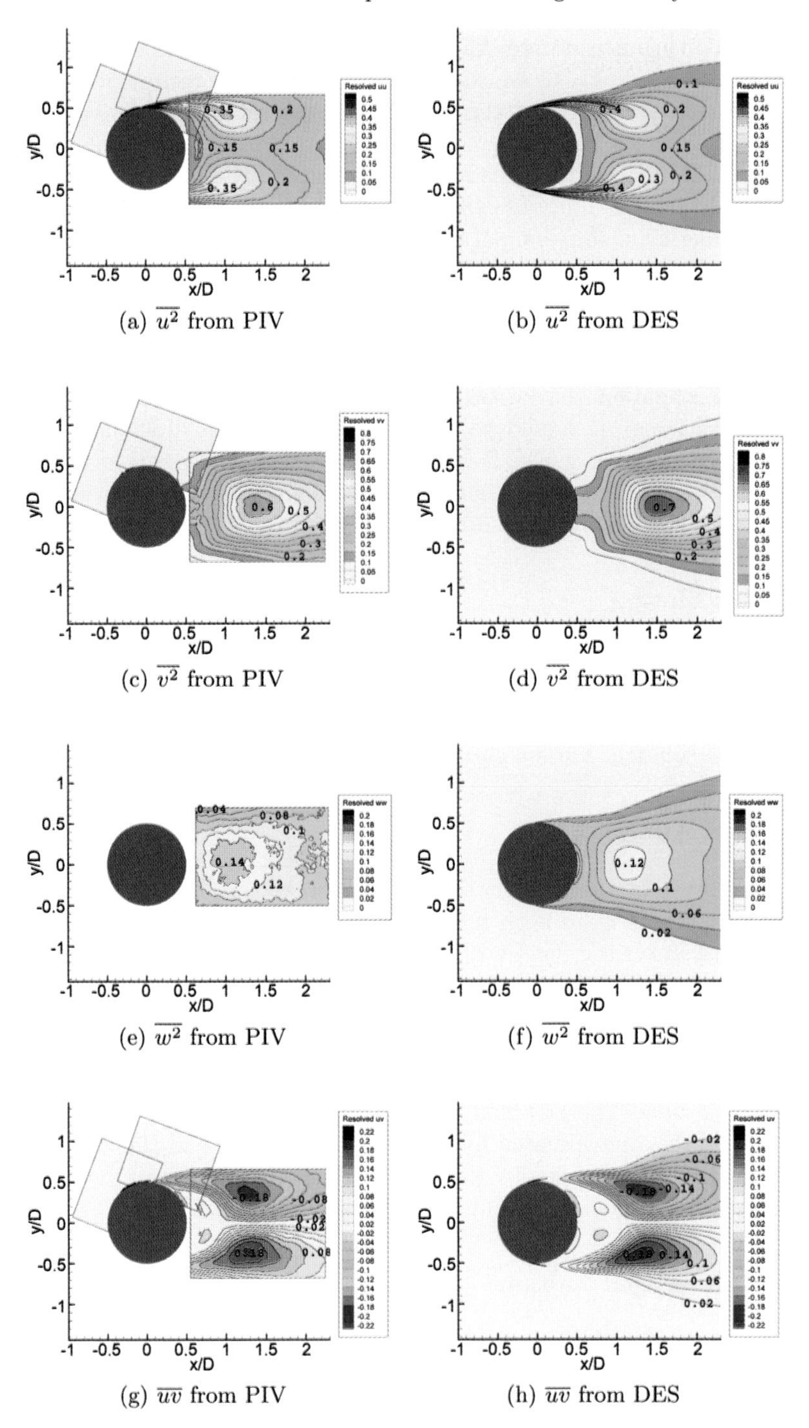

(a) $\overline{u^2}$ from PIV (b) $\overline{u^2}$ from DES

(c) $\overline{v^2}$ from PIV (d) $\overline{v^2}$ from DES

(e) $\overline{w^2}$ from PIV (f) $\overline{w^2}$ from DES

(g) $\overline{uv}$ from PIV (h) $\overline{uv}$ from DES

Fig. 11. Comparison of the mean stress topology

levels of modeled turbulence were examined and no significant difference was found), and correspondingly to an exaggerated length of the recirculation region. The recirculation length, determined by the position where $\overline{U} = 0$ on the rear axis, is $l_c \simeq 1.8$ for the coarse time step, $l_c \simeq 1.35$ for the fine time step and $l_c \simeq 1.28$ for the experiment. The topology of the mean quantities shown in Fig. 10 further illustrate the good agreement between the experiment and the fine time step simulation, with only a slight extension of the wake seen in the simulation. Turning to the turbulent stresses, the levels of modeled stress in the simulation were found to be almost negligible compared to the resolved stress. Therefore, the stresses presented for the simulation are from the resolved motion only. The mean turbulent stresses are compared in Fig. 11. It is apparent that stress magnitudes are slightly overpredicted for $\overline{u^2}$ and $\overline{v^2}$, although it should be acknowledged that this difference may arise from statistical convergence. Despite this, however, the agreement is encouraging, the general topology being displaced slightly downstream in the simulation, as seen for the mean motion.

4.3 Coherent Motion

Having validated the mean flow field, it is now of interest to examine whether the different components of the motion are well represented by the simulation, and then to use the joint experimental and numerical approach to study the dynamics of the coherent motion subjected to turbulence.

4.3.1 Spectral Analysis

A spectral analysis of the velocity in the near-wake has been performed at selected points present in both the simulation and the TRPIV measurements. Figure 12 shows the spectra of the u- and v-components at the points $x/D = 1, y/D = 0.5$ and $x/D = 1, y/D = 0$. The spectra demonstrate the well-documented strong peak corresponding to the vortex shedding, and a continuous part corresponding to the turbulent motion. As a classical result, the peak is also no longer visible in the u-spectrum on the rear axis due the symmetry of the flow. The Strouhal number, which is $\mathrm{St} = 0.21$ in the experiment, is slightly overestimated by the DES ($\mathrm{St} = 0.23$), and the effect of the LES spatial filtering is seen by the damped high frequencies in the DES spectra. Otherwise, a good spectral agreement is observed.

4.3.2 Phase-Averaged Motion

As has been discussed, the quasiperiodicity of the vortex shedding enables analysis using phase averaging, which facilitates the separation of the shedding from the turbulent motions. According to the goals of this study, this should allow an assessment of whether the shedding is well represented by the

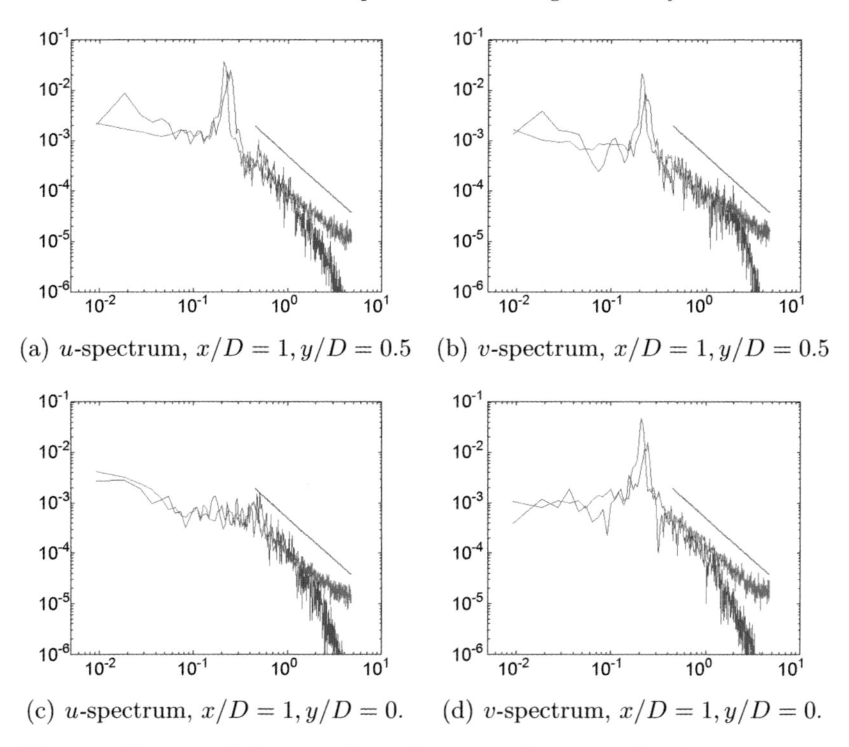

(a) u-spectrum, $x/D = 1, y/D = 0.5$ (b) v-spectrum, $x/D = 1, y/D = 0.5$

(c) u-spectrum, $x/D = 1, y/D = 0.$ (d) v-spectrum, $x/D = 1, y/D = 0.$

Fig. 12. Spectra of the u and v velocity in the near-wake (*green:* experiment, *blue:* simulation, straight line depicts $-5/3$ gradient)

simulation, as well as offering a more detailed description of the flow using both approaches. Figure 13 shows isocontours of the $\langle \Omega_{21} \rangle$ component of the rotation rate tensor at a phase angle of $\varphi = 45°$ for the experiment and for both investigated simulation time step sizes. While a very good agreement is achieved between the fine time step simulation and the experiment, it is seen that the effect of the additional filtering introduced by the coarse time step leads to a delayed formation of the vortices downstream. This also explains the exaggerated recirculation length obtained in the steady mean motion in this case. Figure 14 depicts the phase-averaged stresses issued from the experiment and the fine time step simulation. The topology and magnitudes agree rather well, and as such the simulation can be considered validated. It can be seen that the simulation allows a visualization on a larger domain and therefore an enhanced description of the topology of the stresses compared to the experiment. While in the very near wake region, high values of $\langle u^2 \rangle$ and $\langle uv \rangle$ are classicaly located in the separated shear layer (in agreement with the result of [17] in the case of a vertical flat plate), in the region of vortex convection, the normal stress peaks $\langle u^2 \rangle$ and $\langle v^2 \rangle$ are located near the vortex centers, while the shear stress maxima $\langle uv \rangle$ are found near the saddle points, in agreement with [18, 19] and other authors.

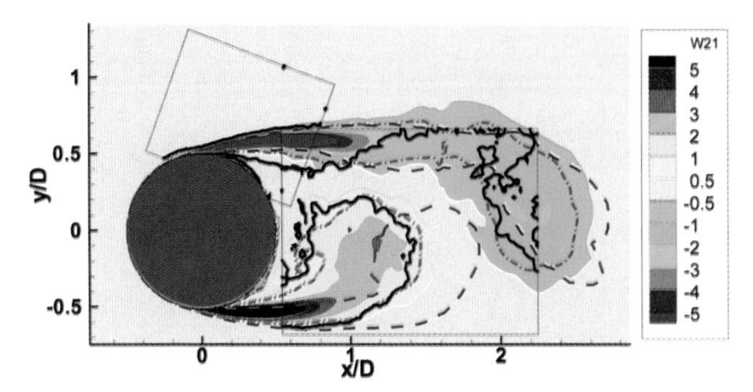

Fig. 13. Comparison of the phase-averaged $\langle \Omega_{21} \rangle$ component of the rotation rate tensor at $\varphi = 45°$. Contour shading: simulation with fine $\mathrm{d}t$, lines depict values of $\langle \Omega_{21} \rangle = \pm 1$ for the PIV (*black solid*), DES with fine $\mathrm{d}t$ (*blue dashed-dotted*) and coarse $\mathrm{d}t$ (*red dashed*)

4.3.3 POD

A brief presentation of preliminary results concerning the extraction of the coherent motion using POD will now be presented. From a dataset $U(X)$, the POD consists of searching for the function $\phi(X)$ that is most similar to the members of $U(X)$ [20]. For the experiment, POD has been performed on the fluctuations using the snapshot method with the 3000 instantaneous fields available for PIV plane 1. For a first comparative step, POD has been performed in the same way for the simulation, and on the same reduced domain corresponding to plane 1. Figures 15a,b show the first two modes of the experiment and simulation, which are seen to agree very well. Furthermore, although not presented here, the modes exhibit strong similarity up to $N = 8$, indicating that the most energetic part of the flow is well represented by the simulation. As a classical result for bluff-body flows, the first two modes are linked with the Von Kármán vortex convection. Figure 15c shows a comparison of the eigenvalues obtained, showing that the energy of each mode is also well represented, the energy of the two first modes being slightly higher in the case of the experiment. The time evolution of the first two coefficients represented in Fig. 15d confirms the quasiperiodicity assumed by [21–23] from low data rate PIV measurements around bluff bodies, also demonstrated at low Reynolds number by [24, 25] among others. This, furthermore, illustrates the possibility to define a phase angle representative of the vortex shedding from the velocity field itself, as has been done in the first cited studies.

5 Conclusions and Future Work

The flow past a circular cylinder at high Reynolds number has been investigated by both experimental and numerical means. The comparability

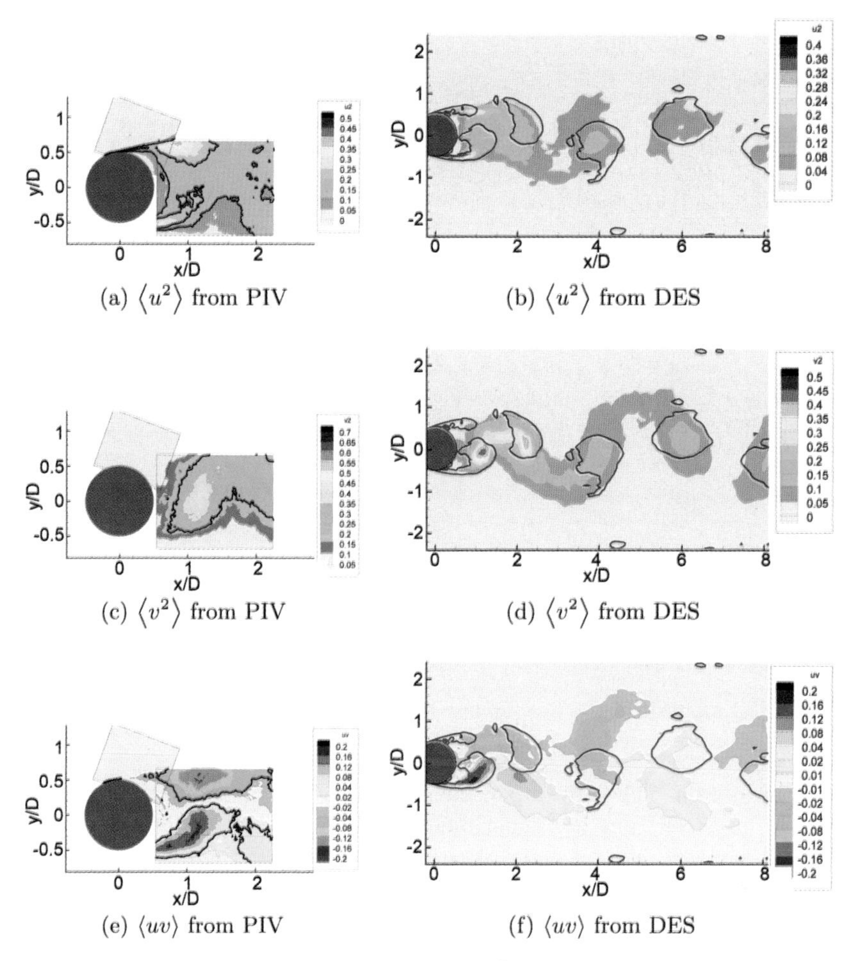

(a) $\langle u^2 \rangle$ from PIV (b) $\langle u^2 \rangle$ from DES

(c) $\langle v^2 \rangle$ from PIV (d) $\langle v^2 \rangle$ from DES

(e) $\langle uv \rangle$ from PIV (f) $\langle uv \rangle$ from DES

Fig. 14. Phase-averaged stresses at $\varphi = 45°$. *Red lines* on the simulation plots (*right*) represent the value 0.5 of the Q-criterion

between these has been greatly facilitated by the use of the same clearly defined boundary conditions, therefore allowing statements on the methods themselves. A very good agreement has generally been achieved for the mean motion, as well as for the coherent motion and the turbulence quantities. However, it has been clearly demonstrated that such simulation techniques are highly sensitive to user-specified parameters, and a strong degradation in solution quality has been witnessed for relatively small changes in these (the time step size in this case). The availability of high-quality experimental data, against which a direct comparison is possible, is therefore of high value to the validation of numerical methods, and for the generation of best-practice guidelines for their application. By the same token, information available only from simulations can be of use in analyzing any assumptions made in the ex-

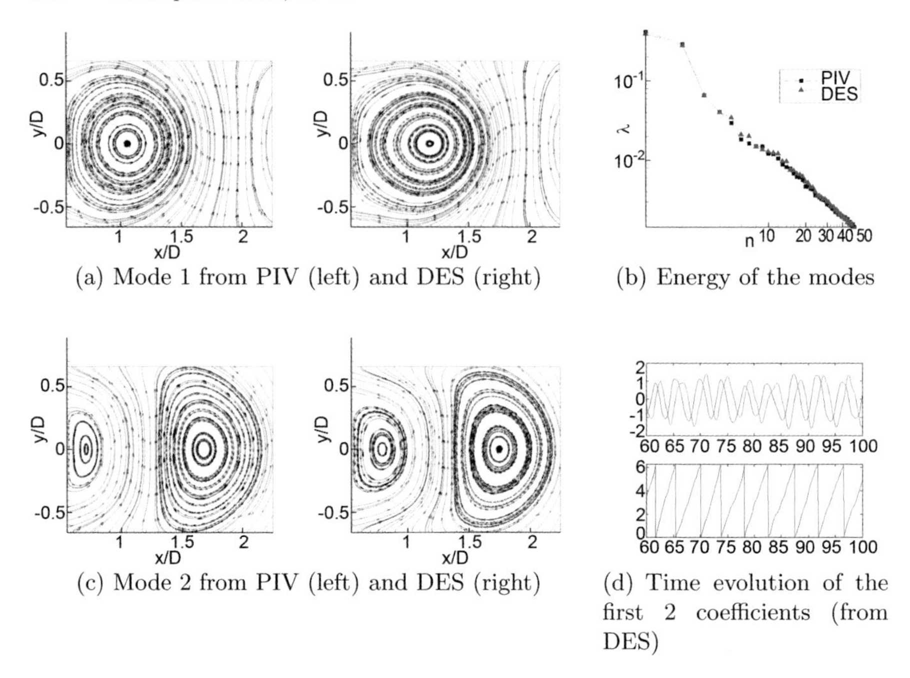

(a) Mode 1 from PIV (left) and DES (right)

(b) Energy of the modes

(c) Mode 2 from PIV (left) and DES (right)

(d) Time evolution of the first 2 coefficients (from DES)

Fig. 15. POD analysis of the flow on the PIV measurement domain

perimental approach. For example, having validated the simulation results, it could be asserted that the experimental assumption of negligible friction forces and spanwise homogeneity in the computation of the drag coefficient was valid. Such combined approaches can therefore provide a crossflow of information inaccessible to each method in isolation.

In addition to crossvalidation of the methods, the respective advantages and limitations of each have been experienced. The simulation gives access to time-resolved 3D information of the entire flow field, whereas the PIV remains restricted to measurements in some limited region of the flow. Conversely, very long computation times result from the need to capture enough samples for statistical convergence (particularly exacerbated by the intermittency of the vortex shedding in this case), which is not a problem for an experiment with a low data sampling rate. The combination of approaches has been shown to reduce the effect of these inherent disadvantages, and a more complete picture of the flow physics has been achieved as a result.

The encouraging level of comparability observed opens the door to a number of future areas for investigation. Concerning the simulation method, the investigation of a still finer time step would be desirable to confirm the positive trend seen in this study, to ascertain whether this could be responsible for the remaining deviations seen, as well as to establish the presence of a monotonic convergence. Furthermore, the reliable and detailed experimental data offers the opportunity to conduct a study of the sensitivity to grid refinement, although such a study would involve very high numerical cost. Finally, the forced transition treatment used here reflects a serious limitation common to any existing turbulence closure, namely the issue of transition prediction. Computation of this case at varying Reynolds numbers would present a challenging test for current and future transition models. From the point of view of the flow analysis, the availability of the complete time-resolved flow field given by the simulation allows more detailed investigations, which have as yet not been completely exploited. Particularly interesting such studies would be the analysis of three-dimensional effects, in particular the longitudinal vortices, and of the shedding intermittency observed by both approaches. As such, the simulation could serve as a preliminary study for further experimental investigations. Furthermore, an assessment of the post-processing techniques performed is possible; for example, it has been shown that phase averaging using the pressure signal leads to an overestimation of the turbulent motion, because of phase lags occuring at some instants between the pressure and the velocity in the wake [23]. This effect was alleviated by the use of a phase angle obtained directly from the velocity fields using the POD coefficients, and the simulation data should allow the confirmation of this tendency. Additionally, the influence of the domain size on the extracted POD modes and the evolution of their coefficients could be ascertained using the simulation data.

Finally, some comments are necessary concerning the case dependency of the findings. This study was deliberately conducted on a simple geometry, the comparisons and postprocessing being facilitated by the symmetries of the flow. It is acknowledged that some of the methods used to enhance the comparability are not universally applicable (e.g., the use of porous windtunnel walls is necessary for transonic investigations). Having established the feasibility and advantages of such joint investigations on this simplified case, a future task will be to conduct combined studies on more complex geometries.

Acknowledgements

The authors gratefully acknowledge the partial funding by the European Community represented by the CEC, Research Directorate-General, in the 6th Framework Program, under Contract No. AST3-CT-2003-502842 ("DESider" project), and by the German Research Foundation (DFG) within the scope of the Collaborative Research Center SFB 557.

The DES computations were conducted on the IBM pSeries 690 at the Zuse-Institute Berlin (ZIB) part of the Norddeutschen Verbund für Hoch- und Höchstleistungsrechnen (HLRN), the support of which is acknowledged with gratitude.

Particular thanks are also due to Thorsten Reimann, for his valuable support in the computational work and data manipulation.

References

[1] R. J. Adrian: Twenty years of particle image velocimetry, Exp. Fluids **39**, 159–169 (2005)
[2] W. Haase, B. Aupoix, U. Bunge, D. Schwamborn: Flomania: Flow-physics modelling - an integrated approach, Notes on Numerical Fluid Mechanics and Multidisciplinary Design, Springer Verlag **94** (2006)
[3] A. Travin, M. Shur, M. Strelets, P. R. Spalart: Detached-eddy simulations past a circular cylinder, Int. J. Flow, Turb. Combus. **63**, 293–313 (2000)
[4] R. Perrin: *Analyse physique et modélidation d'ecoulements incompressibles instaionnaire turbulents autour d'un cylindre circulaire à grand nombre de Reynolds*, Thèse de doctorat, Institut National Polytechnique de Toulouse (5 Juillet 2005)
[5] R. Perrin, E. Cid, S. Cazin, A. Sevrain, M. Braza, F. Moradei, G. Harran: Phase averaged measurements of the turbulence properties in the near wake of a circular cylinder at high Reynolds number by 2C-PIV and 3-C PIV, Exp. Fluids **42(1)**, 93–109 (2007)
[6] B. Lecordier, M. Trinite: Advanced PIV algorithms with image distortion – validation and comparison from synthetic images of turbulent flows, in *PIV03 Symposium* (Busan, Korea 2003)
[7] P. R. Spalart, W. H. Jou, M. Strelets, S. R. Allmaras: Comment on the feasability of LES for wings, and on a hybrid RANS/LES approach, in *C. Lieu and Z.Liu, editors,* 1st *AFOSR Int. Conf. on DNS/LES, in Advances in DNS/LES, Columbus, OH, Aug 4–8,* Grynden Press (1997)
[8] L. Xue: *Entwicklung eines effizienten parallel Lösungsalgorithmus zur dreidimensionalen Simulation komplexer turbulenter Strömungen*, Ph.D. thesis, Technical University of Berlin (1998)
[9] A. Travin, M. Shur, M. Strelets, P. R. Spalart: Physical and numerical upgrades in the detached-eddy simulation of complex turbulent flows, in R. Friederich, W. Rodi (Eds.): *Advances in LES of Complex Flows*, vol. 65 (2002) pp. 239–254
[10] K. Karki, S. Patankar: Pressure based calculation procedure for viscous flows at all speeds, AIAA J. **27**, 1167–1174 (1989)
[11] S. Obi, M. Perić, M. Scheurer: Second moment calculation procedure for turbulent flows with collocated variable arrangement, AIAA J. **29**, 585–590 (1991)
[12] U. Bunge, C. Mockett, F. Thiele: New background for Detached Eddy Simulation, in *in Proceedings of the Deutscher Luft- un Raumfahrtkongress, DGLR-JT2005-212, p.9, Friedrichshafen* (2005)
[13] P. D. Welch: The use of fast Fourier transform for the estimation of power spectra: A method based on time averaging over short, modified periodograms, IEEE Trans. Audio Electroacous. **15**, 70–73 (1967)
[14] W. C. Reynolds, A. K. M. F. Hussain: The mechanics of an organized wave in turbulent shear flow. Part 3. Theoretical models and comparisons with experiments, J. Fluid Mech. **54**, 263–288 (1972)
[15] R. W. Wlezien, J. L. Way: Techniques for the experimental investigation of the near wake of a circular cylinder, AIAA J. **17**, 563–570 (1979)
[16] C. Mockett, U. Bunge, F. Thiele: Turbulence modelling in application to the vortex shedding of stalled airfoils, in 6th *ERCOFTAC International Symposium on Engineering Turbulence Modelling and Measurements, ETMM6* (Sardinia, Italy 2005)

[17] A. Leder: Dynamics of fluid mixing in separated flows, Phys. Fluids A **3**, 1741–1748 (1991)
[18] B. Cantwell, D. Coles: An experimental study of entrainment and transport in the turbulent near wake of a circular cylinder, J. Fluid Mech. **136**, 321–374 (1983)
[19] A. K. M. F. Hussain, M. Hakayawa: Eduction of large-scale organized structures in a turbulent plane wake, J. Fluid Mech. **180**, 193–229 (1987)
[20] G. Berkooz, P. Holmes, J. L. Lumley: The proper orthogonal decomposition in the analysis of turbulent flows, Ann. Rev. Fluid. Mech. **25**, 539–575 (1993)
[21] M. B. Chiekh, M. Michard, N. Grosjean, J. C. Bera: Reconstruction temporelle d'un champ aérodynamique instationnaire partir de mesures PIV non résolues dans le temps, in 9e *Congrès Francophone de Vélocimétrie Laser* (2004)
[22] B. W. van Oudheusden, F. Scarano, N. P. van Hinsberg, D. W. Watt: Phase-resolved characterization of vortex shedding in the near wake of a square-section cylinder at incidence, Exp. Fluids **39**, 86–98 (2005)
[23] R. Perrin, M. Braza, E. Cid, S. Cazin, A. Barthet, A. Sevrain, C. Mockett, F. Thiele: Phase averaged turbulence properties in the near wake of a circular cylinder at high Reynolds number using POD, in 13th *Int. Symp. on Applications of Laser Techniaues to Fluid Mechanics* (Lisbon, Portugal 2006)
[24] A. E. Deane, I. G. Kevrekidis, G. E. Karniadakis, S. A. Orzag: Low-dimensional models for complex geometry flows: Application to grooved channels and circular cylinders, Phys. Fluids A **3**, 2337–2354 (1991)
[25] B. Noack, K. Afanasiev, M. Morzynski, G. Tadmor, F. Thiele: A hierarchy of low-dimensional models for the transient and post-transient cylinder wake, J. Fluid Mech. **497**, 335–363 (2003)

Index

Natural Gas Burners for Domestic and Industrial Appliances.

Application of the Particle Image Velocimetry (PIV) Technique

Lucio Araneo, Aldo Coghe, Fabio Cozzi, Andrea Olivani, and Giulio Solero

Dipartimento di Energetica, Politecnico di Milano, Via La Masa 34, 20156 Milano, Italy
fabio.cozzi@polimi.it

Abstract. This contribution presents some examples of the application of the particle image velocimetry (PIV) technique to domestic appliances and small-scale burners, with the aim of discussing relevant results together with problems encountered. Combustion efficiency and pollutant emissions of gas burners are strongly influenced by the fluid dynamics of the mixture in both premixed and nonpremixed flames. For these reasons the Combustion Laboratory of Politecnico di Milano started using laser diagnostic techniques (LDV and PIV) many years ago.

The first PIV application was on a premixed V-flame attached over a burner plate with rectangular twin slots, developed for domestic appliances. The autocorrelation of a double-exposed photograph was used to define the 2-D velocity flow field and high spatial resolution was obtained with a 1 : 1 magnification and a zoom Nikkor objective.

More recently, a crosscorrelation CCD camera with a double-pulse Nd:YAG laser was used to characterize high-swirl flows under both nonreacting and reacting conditions.

In the following, special emphasis will be given to the discussion of the most relevant results and the main problems encountered with PIV applications to the investigated cases.

1 Introduction

The performances and efficiency of gas burners depend on the fluid dynamics of the mixture in both premixed and nonpremixed cases. In premixed conditions flame stability and gas-temperature distribution are strongly dependent on the mixture flow pattern and the mixing with secondary air. In the nonpremixed case the combustion is strongly influenced by the fluid dynamics that controls mixture formation and chemical reactions. Today, the reduction of pollutant emissions from fossil-fuel combustion processes is one of the most important aims of scientific research and this problem is again strictly related to the fluid dynamics of the combustible mixture. Therefore, direct evaluation of the flow pattern in reacting flows is of great significance, both from the design point of view and the modeling of the combustion devices with CFD

A. Schroeder, C. E. Willert (Eds.): Particle Image Velocimetry,
Topics Appl. Physics **112**, 245–257 (2008)

simulations. In this respect, particle image velocimetry (PIV) has the potentialities for fast and accurate evaluation of the instantaneous two-dimensional flow pattern even in reacting regimes, and the results are in a format easily comparable with numerical data from stationary CFD codes. This type of information cannot be obtained by using point-measurement techniques such as laser Doppler velocimetry (LDV). Moreover, the two-dimensional nature of the measured velocity information can be used to estimate fluid-mechanics-relevant quantities by means of differentiation (i.e., vorticity) or integration (i.e., stream function, circulation, mass flow rates, etc.).

However, as also evidenced by many authors (see for example [1]), problems may arise from the seeding requirements of tracer particles and their ability to follow the gas flow field and survive in the high-temperature environment. Further problems are posed by the spatial resolution, generally inadequate to resolve the smallest turbulent (mixing) scales, and the difficulties originated by spatial velocity and density gradients due to the flame front and the temporal fluctuations induced by turbulent structures.

Proper selection of the PIV operating parameters is needed for each application, with the constraints imposed by the experimental setup and the availability of the optical tools.

The examples reported in the chapter refer to different conditions: laminar premixed and turbulent nonpremixed flames, both fuelled with methane or natural gas and air.

2 PIV Measurements on a V-Flame: The First Attempt

We started applying the PIV technique in early 1997 to a lean premixed methane-air laboratory flame similar to those used in a conventional domestic burner. In this atmospheric premixed burner, a lean methane-air mixture is fed through an arrangement of rectangular twin slots and gives rise to an inverted flame (V-shaped), attached on the burner plate. This flame typology, being particularly stable and easy to handle, was selected as a test case for different nonintrusive diagnostic techniques, including LDV and PIV. The main results of this study were published in a previous paper [2].

Figure 1 shows the double-exposed photograph of the V-flame illuminated by a pulsed Nd:YAG laser sheet. Photographic recording was used to achieve a higher spatial resolution near the flame front. Micrometric oil particles were introduced as tracers in the reactants mixture, allowing the measurements only in the preflame region. This study was intended to quantify the modification induced by the flame on the incoming flow and to identify the flame front as the region where the particles rapidly disappear due to the high temperature.

All the information concerning the PIV setup is summarized in Table 1.

With the present photographic PIV configuration the focusing process is quite difficult, due to the reduced depth of field. For best focusing, the 35 mm

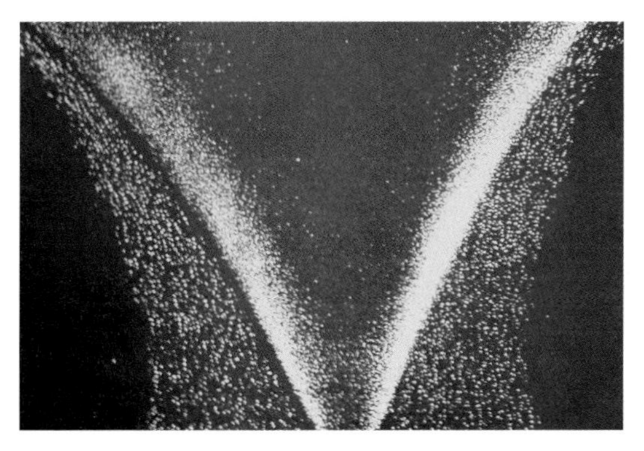

Fig. 1. Double-exposed photograph of a V-flame illuminated by a pulsed Nd:YAG laser sheet

Table 1. PIV recording parameters for the V-flame configuration

Maximum inplane velocity	$U = 2\,\mathrm{m/s}$
Field of view	$20 \times 13\,\mathrm{mm}^2$
Interrogation volume	$0.24 \times 0.24 \times 0.5\,\mathrm{mm}^3$
Recording method	single frame/double exposure
Recording medium	Kodak 400 ASA photographic film, 35 mm
Frame digitalization	Polaroid scanner (2700 dpi)
Recording lens	Zoom Nikkor 80–200 mm, $M = 1$, $f_{\#}$ 5.6
Illumination	Nd:YAG laser, $\lambda = 532\,\mathrm{nm}$, 100 mJ/pulse
Pulse delay	$96\,\mu\mathrm{s}$
Seeding particles	silicone-oil droplets ($d_\mathrm{p} \approx 1\,\mu\mathrm{m}$)
Image processing	digital autocorrelation
Depth of field	0.3 mm

photo-camera was mounted on a traversing slit with 0.1 mm resolution and different recordings were taken at different observation distances.

No attempt was made to remove the directional ambiguity arising from the double-exposed/single-frame images, since the flow direction was known and previous LDV analysis of the turbulent field and timescales revealed that the flame typology falls in the laminar flamelet regime, that is a laminar flame slightly distorted by turbulence effects. Some uncertainty in flow direction was only visible at the exit plane, in correspondence of the small area between two consecutive slots, where a wake-like flow is generated.

Measurement uncertainty due to background noise (flame luminosity) could produce a sensible effect due to the relatively long exposure time (1/8 s) needed to record the two laser pulses on the same frame, in the absence of any device for laser–camera synchronization. In fact, the image quality was quite good for the isothermal flow and less satisfactory in the reacting case, due to the blue-flame luminosity. No interference filter in front of the cam-

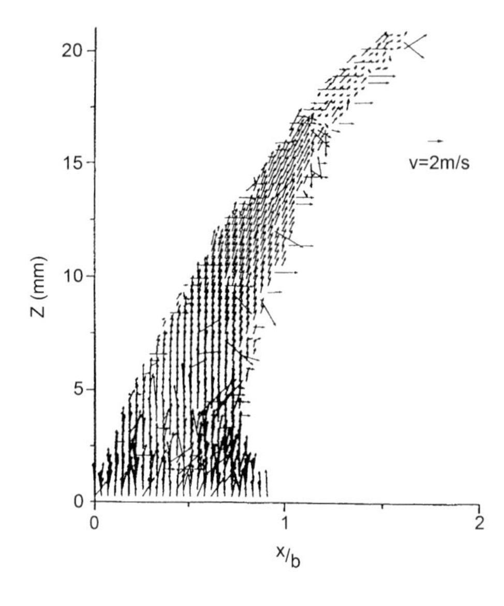

Fig. 2. Single-frame PIV reconstruction of the instantaneous velocity field

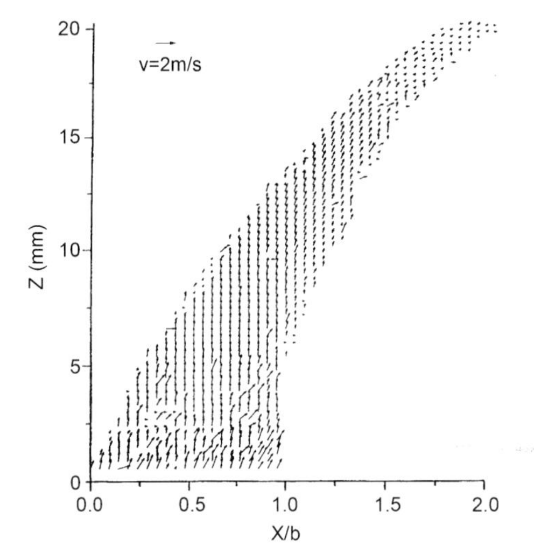

Fig. 3. Velocity field inside reactants obtained by averaging 5 frames

era, centered on the laser wavelength, was used at the time, although it could have reduced the problem.

The gas flow was densely seeded by oil particles, allowing us to achieve a high valid detection rate and a low measurement uncertainty even with small interrogation windows. The results shown in Fig. 2 indicate that single-frame PIV reconstruction of the instantaneous velocity field was very satisfactory for the isothermal flow, with a limited number of outliers, and in the reacting case the average over 5 frames was enough to achieve a reasonable flow pattern (Fig. 3), without using specific postprocessing of PIV data.

The high spatial resolution allowed the evaluation of the mixture velocity close to the flame front, as already deduced from the photographic image of Fig. 1. However, approaching the flame front a noticeable acceleration of the flow (up to 2 m/s) produced less accurate results due to thermal expansion and to the loss of particles pairs due to progressive evaporation of the seeding droplets. A similar effect is noticed on the external boundary of the flow, caused by the reduced particle density due to dilution with the ambient air.

This experiment was our first attempt to apply PIV in the investigation of a premixed combustion process and it was found of great help in revealing the potentialities of the technique and in understanding the main problems associated with its use in flames.

3 PIV Measurements on a Swirl Burner

In combustion systems, the strong favorable effects of applying swirl to injected air and fuel are extensively used as an aid to stabilization of the high-intensity combustion processes and efficient clean combustion in a variety of practical situations: internal combustion engines, gas turbines, industrial furnaces, utility boilers and many other practical heating devices [3]. In particular, nonpremixed swirling flows are widely used in industrial combustion systems, particularly gas turbines, boilers and furnaces, for safety and stability reasons. Since swirling motion is regarded as an efficient way to improve and control the mixing rate between fuel and oxidant streams and to improve flame stabilization through the swirl-induced recirculation of hot products [4], the PIV technique was used in order to characterize high-swirl flows under both nonreacting and reacting conditions.

In the present study, swirl motion was generated through axial plus tangential air entries and variation of the relative amounts of axial and tangential airflows controlled the swirl strength. The swirl burner was characterized by the presence of a coaxial swirling air stream, around a central fuel injector. Two different fuel-injection typologies, coaxial and radial (i.e., transverse with respect to the swirl air stream), were tested in order to study the effects of different mixing mechanisms. Under nonreacting conditions, this injector was moved further upstream the throat of the burner of about 5 injector diameters from its original position, and was used only for flow seeding. More details on the experimental apparatus can be found in [5] for the reacting configuration and in [6] for the nonreacting one.

4 Swirl Flow – Nonreacting Conditions

Under nonreacting conditions, PIV measurements were performed using a double-pulsed Nd:YAG laser, operating at 532 nm, with a pulse energy of 200 mJ/pulse. Images were acquired with a Dantec HiSense PIV Camera. All

Table 2. PIV recording parameters for the free swirling jet

Maximum inplane velocity	$W = 20\,\mathrm{m/s}$
Field of view	$75 \times 60\,\mathrm{mm}^2$
Interrogation volume	$1 \times 1 \times 0.5\,\mathrm{mm}^3$
Recording method	double frame/single exposure
Recording camera	Dantec HiSense PIV/PLIF camera full frame CCD (1280×1024 pixel) $6.7\,\mu\mathrm{m}$ pixel pitch
Interrogation area	16×16 pixel, overlap 50 %
Recording lens	Micro-Nikkor 60 mm, $M = 0.11$, $f_{\#}$ 5.6
Depth of field	6.8 mm
Illumination	Nd:YAG laser, $\lambda = 532\,\mathrm{nm}$, 200 mJ/pulse, 8 Hz
Pulse delay	$10\,\mu\mathrm{s}$
Seeding particles	silicone-oil droplets ($d_\mathrm{p} \approx 1\,\mu\mathrm{m}$)
Image processing	crosscorrelation FFT algorithm

measurements were performed on a plane normal to the burner axis and close to the burner head. The spatial resolution of PIV measurements in the plane of the laser sheet was estimated to be about 1 mm, and the spatial resolution in the direction normal to this plane is estimated from the average thickness of the laser sheet, which is about 0.5 mm. Experimental data were acquired at the frequency of 1 Hz, with a time interval of $10\,\mu\mathrm{s}$ between the two laser pulses. The interrogation area and the laser pulse delay were selected in order to avoid out-of-boundary particle motion. All parameters concerning the PIV setup are summarized in Table 2.

The highly three-dimensional nature of the flow in a swirl burner makes the measurements difficult using any diagnostic technique and even more so when using the PIV technique due to the reduced spatial resolution of CCD-camera-based instruments. The spatial resolution of any measurement technique must be able to properly resolve the velocity gradients, but the PIV has also a limitation related to a measurement bias when velocity and seeding density gradient are present in the interrogation area. A measure of the spatial resolution in a velocity gradient is the shear parameter α [7],

$$\alpha = \frac{D_\mathrm{p}}{R_\mathrm{c}} \frac{\partial(W/W_\mathrm{max})}{\partial(r/R_\mathrm{c})} ,$$

where W is the vortex tangential velocity, R_c is the core radius of the vortex, D_p the effective probe dimension, and r the radial coordinate. For a vortex flow the parameter reduces to

$$\alpha \approx \frac{D_\mathrm{p}}{R_\mathrm{c}}$$

and general guidelines for the maximum value of α are documented in the literature. The more stringent limitation on α for PIV is dictated by a limiting mean flow velocity gradient across D_p, which should be interpreted as the

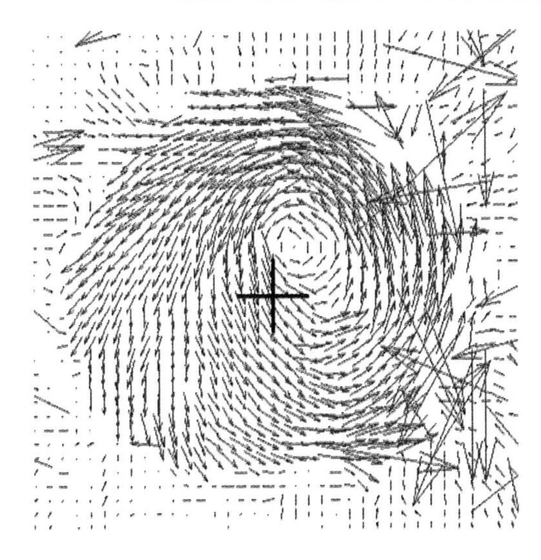

Fig. 4. Isothermal swirling jet: instantaneous velocity field in a cross section close to the nozzle exit

dimension of the interrogation area. Since in a solid-body rotating vortex the velocity changes by 100 % of the peak swirl velocity within the length R_c, the usually recommended condition $\Delta W/W < 0.2$ for flows without streamline curvature would give $\alpha < 20\,\%$. The streamline curvature in the vortex core actually makes this requirement more stringent [7]. In the present case the value of the shear parameter was estimated to be less than 10 %, based on $D_p = 1\,\text{mm}$ and $R_c = 12\,\text{mm}$.

Under nonreacting conditions, since seeding particles were uniformly distributed in the flow field, it was possible to obtain high-quality instantaneous velocity maps that confirmed the precession motion of the vortex center [8]. Indeed, the precessing vortex core (PVC) instability occurs in high Reynolds and swirl number flows, and is characterized by the regular precession of the vortex core around the geometrical axis of symmetry. An example of an instantaneous velocity field at the burner exhaust is reported in Fig. 4, while in Fig. 5 the ensemble-averaged velocity field is reported, based on 200 couples of images, under the same experimental conditions. The validation rate (according to the standard validation criterion of the PIV system, i.e., peak-ratio validation) is quite high in the region of interest. It is evident that the instantaneous velocity field does not show the symmetry of the average one; moreover, it is possible to easily localize the vortex core, since the core structure is evident.

5 Swirl Flow – Reacting Conditions

Under reacting conditions, PIV measurements were obtained with the same experimental apparatus described in the previous section. A Micro-Nikkor

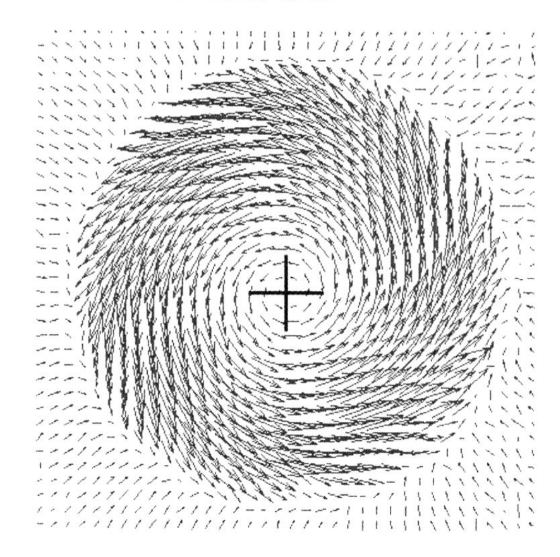

Fig. 5. Isothermal swirling jet: average velocity field, based on 200 couples of images

($f = 60\,\mathrm{mm}$) objective was used to collect the scattered light, and the resulting field of view was approximately 100 (horizontal) by 80 (vertical) mm^2. Considering an interrogation area of 32×32 pixels, with a 50 % overlap, the spatial resolution in velocity measurements was about $2.5 \times 2.5\,\mathrm{mm}^2$. At least 200 couples of images were taken to ensure accurate data analysis.

Background light and flame luminescence were removed using a 10-nm bandpass filter centered at 532 nm. However, the time-delayed readout of the second PIV image (in the present case $> 100\,\mathrm{ms}$), an artifact of the double-shuttered interline transfer CCD, resulted in the accumulation of the flame luminosity on the sensor, and this fact strongly limited the acquisition of good-quality PIV recordings.

In the reacting case, one more difficulty arises from inconsistent, inhomogeneous seeding levels. In general, seeding in gas flows is very complex, especially when the observed flow region is characterized by mixing layers between incoming and recirculating streams, or products and reactants streams. Thus, the seeding density in different areas of the measurement plane may not necessarily be the same. In the present study, to completely characterize the flow field inside the nonpremixed flame a double seeding was used: silicone-oil droplets (mean diameter = $1\,\mu\mathrm{m}$) were dispersed in the fuel flow; while alumina particles (mean diameter = $5\,\mu\mathrm{m}$) were injected in the swirl air flow using a fluidized-bed generator. The high melting temperature of Al_2O_3 (melting point 2300 K) allows us to measure the velocity field not only inside reactants, but also inside the flame region and inside combustion products. However, it should be noted that the relatively high density of alumina particles ($\rho_{Al_2O_3} = 3990\,\mathrm{kg/m}^3$ and their size ($5\,\mu\mathrm{m}$) limits their capability to follow high-frequency flow fluctuations, but anyway allowed correct definition of the mean flow field.

Table 3. PIV recording parameters for the swirl burner under reacting conditions

Maximum inplane velocity	$U = 16\,\mathrm{m/s}$
Field of view	$100 \times 80\,\mathrm{mm}^2$
Interrogation volume	$2.5 \times 2.5 \times 0.5\,\mathrm{mm}^3$
Recording method	double frame/single exposure
Recording camera	Dantec HiSense PIV/PLIF camera full frame CCD (1280 × 1024 pixel) 6.7 µm pixel pitch
Interrogation area	32 × 32 pixel, overlap 50 %
Recording lens	Micro-Nikkor 60 mm, $M = 0.09$, $f_{\#}$ 5.6
Depth of field	9.8 mm
Illumination	Nd:YAG laser, $\lambda = 532\,\mathrm{nm}$, 200 mJ/pulse, 8 Hz
Pulse delay	30 µs
Seeding particles	silicone-oil droplets ($d_\mathrm{p} \approx 1\,\mu\mathrm{m}$) in the fuel jet Al_2O_3 particles ($d_\mathrm{p} \approx 5\,\mu\mathrm{m}$) in the air flow
Image processing	crosscorrelation FFT algorithm

All parameters concerning the PIV measurements performed on the swirl burner under reacting condition are summarized in Table 3.

Considering the tendency of alumina particles to agglomerate into larger clumps and the intermittent operation of the seeding generator, the flow field was not uniformly seeded and thus it was difficult to obtain significant instantaneous measurements of the entire flow field from each couple of images. However, the average of at least 200 couples of images shows good results as will be presented in the following.

Streamlines, obtained by two-dimensional (2D) PIV measurements, are superimposed on the flame picture and are reported in Fig. 6. The reactants outlet is located in the bottom center of the velocity field, between radial coordinates of −18 mm and +18 mm. The combustion chamber wall is outside the field of view. Figure 6 clearly indicates the existence of the fuel jet (going upward), of the central toroidal recirculation zone (CTRZ) and the corner recirculation zone (CRZ). The CTRZ is the broad region of reverse flow confined by the expanding air stream and extending back toward the burner head. It is attributed to a nonsteady phenomenon occurring at high swirl numbers ($S > 0.6$) and called vortex breakdown [3]. The CRZ, which is partially outside the PIV field of view, is produced by the confinement due to the chamber wall and the bottom plane and it extends externally to the air stream.

As clearly shown by the streamlines in Fig. 6, the fuel jet penetrates into the backflow in the middle of the combustion chamber and then is forced outwards and redirected into the air flow. Moreover, a toroidal structure, generated by the entrainment of the fuel jet on the surrounding stream, is clearly visible in the bottom center of the velocity field in Fig. 6 and controls the mixing between fuel and air at the entrance of the combustion chamber.

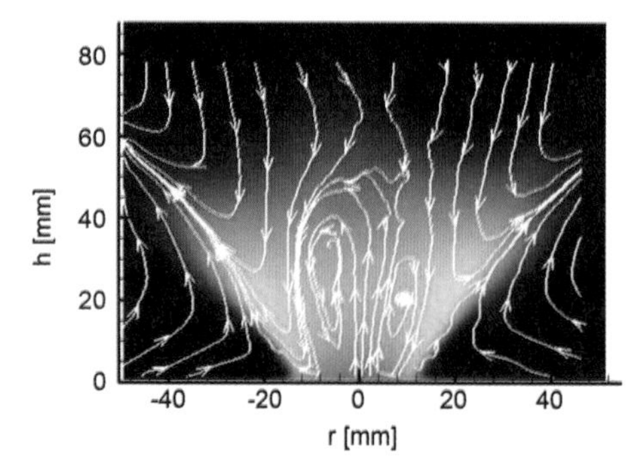

Fig. 6. Superimposition of streamlines and picture for a nonpremixed swirl flame with the axial fuel-injection configuration

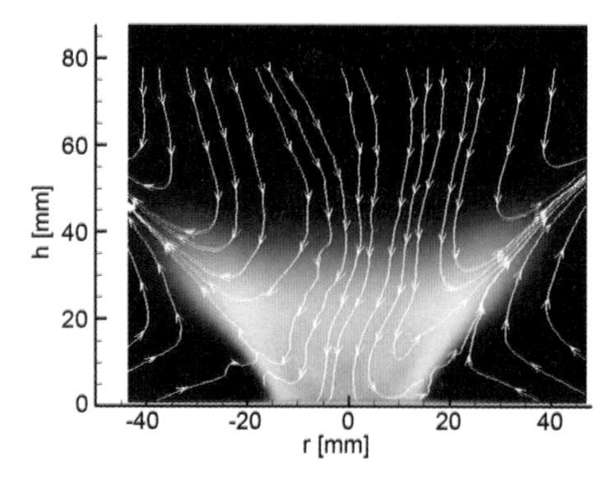

Fig. 7. Superimposition of streamlines and picture for a nonpremixed swirl flame with the transverse fuel-injection configuration

In order to improve the mixing between reactants right at the entrance of the combustion chamber, a different fuel-injection configuration was considered, and the corresponding flow field is illustrated in Fig. 7. In this case the fuel mixture is injected transversally with respect to the swirl air flow, and thus the fuel jet does not interact with the formation of the CTRZ. As a consequence, the reverse-flow region, carrying hot combustion products, moves upstream closer to the reactants outlet, improving the mixing between reactants and products. As reported in many previous works [5, 9, 10], the fuel-injection procedure has a strong impact on pollutants formation and the PIV technique was very useful in order to gain information on mixing processes between reactants and products and to evidence the interaction between the fuel jet and the surrounding stream.

Considering that U and V are, respectively, the axial (z) and radial (r) velocity components, the out-of-plane vorticity component can be calculated from the average 2D velocity maps as,

$$\omega = \frac{\partial U}{\partial r} - \frac{\partial V}{\partial z} \, .$$

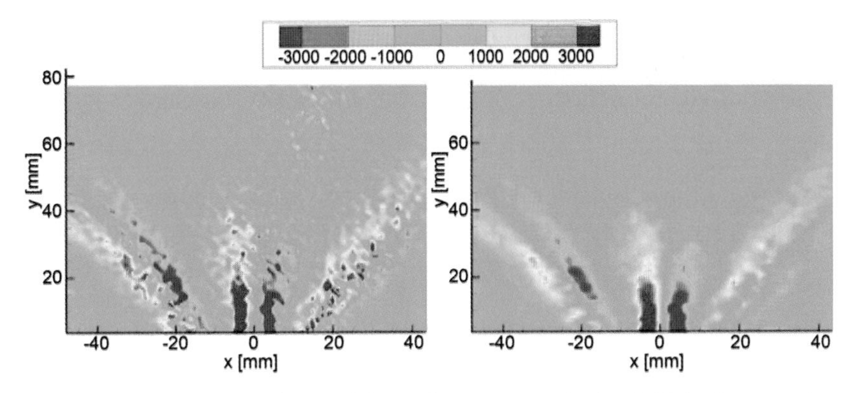

Fig. 8. Vorticity maps obtained from the average velocity field in the experimental conditions of Fig. 6: center-difference approach (*left*); least-squares approach (*right*)

Two finite-difference schemes, based on the center-difference and the least-squares approaches, were used to obtain estimates for the first derivate and to calculate the vorticity field [11]. Vorticity maps estimated from average velocity fields are summarized in Fig. 8 for the case with the axial fuel-injection configuration (center difference approach on the left; least-squares approach on the right). As can be observed, the least-squares approach reduces the noise but it also reduces the spatial resolution since the first derivate is estimated over a larger number of adjacent velocity data. Anyway, in the present study the resolution of the least-squares approach is adequate to evidence the large-scale vortical structure around the fuel jet.

Regions of positive vorticity are associated with counterclockwise rotating flows, while regions of negative vorticity are associated with clockwise-rotating flows. For all cases, peak-vorticity magnitudes of about $3000\,s^{-1}$ are observed in the shear layers between the CTRZ and the CRZ. This result underlines that in this region an intense mixing process takes place between reactants coming from the burner head and hot products from recirculation zones. So, it is mainly in this region of the flow field that hot combustion products and radicals provide energy for the ignition of the incoming fuel-air stream. Finally, close to the burner head, an area of intense fuel recirculation and high vorticity is clearly visible and this fact evidences the appearance of a toroidal vortex around the axial fuel jet.

6 Conclusions

Sophisticated laser-based diagnostic techniques, such as PIV, allow detailed investigation of the combustion process, but their use in large-scale combustion devices may be problematic. If limited to small-scale combustor models, even in the presence of bias errors, the large quantity of experimental data available in a "short time" by PIV can help in providing a better understanding of the most important combustion phenomena and a useful indication for improved design of industrial burners. Moreover, PIV measurements can also be used in order to develop and validate specific combustion submodels and CFD codes.

However, improvements of the PIV spatial resolution are probably necessary, together with more advanced software and more efficient seeding generators, in order to implement the use of the PIV technique in full-scale combustion devices.

Acknowledgements

The authors wish to thank Mr. Fulvio Martinelli for performing PIV measurements on the isothermal swirling jet.

References

[1] C. Willert, M. Jarius: Planar flow field measurements in atmospheric and pressurized combustion chambers, Exp. Fluids **33**, 931–939 (2002)

[2] A. Coghe, G. Solero, N. Riva, V. Santoro: Misure di velocità tramite PIV in una fiamma premiscelata magra, in *52° Congresso Nazionale ATI, Cernobbio (CO)* (1997)

[3] A. K. Gupta, D. G. Lilley, N. Syred: *Swirl Flows* (Abacus, Tunbridge Wells 1984)

[4] D. Shen, J. M. Most, P. Joulain, J. S. Bachman: The effect of initial conditions for swirl turbulent diffusion flame with straight-exit burner, Combus. Sci. Technol. **100**, 203–224 (1994)

[5] A. Olivani: *Thermo-fluid-dynamic analysis of methane/hydrogen/air mixtures under reacting conditions by laser diagnostics*, Joint Ph.D. Thesis, Politecnico di Milano & Université d'Orléans (2006)

[6] F. Martinelli: *Analysis of the precessing vortex core instability in a free swirling jet*, Master thesis in aerospace engineering, Politecnico di Milano (2005)

[7] P. B. Martin, J. G. Leishman, G. J. Pugliese, S. L. Anderson: Stereoscopic piv measurements in the wake of a hovering rotor, in *56th Annual Forum of the American Helicopter Society* (2000)

[8] F. Martinelli, A. Olivani, A. Coghe: Experimental analysis of the precessing vortex core in a free swirling jet, Exp. Fluids **42**, 827–839 (2007)

[9] G. Solero, A. Olivani, F. Cozzi, A. Coghe: Experimental analysis of fuel injection procedure in a natural gas swirling flame, in *European Combustion Meeting* (2005)

[10] A. Olivani, G. Solero, F. Cozzi, A. Coghe: Experimental analysis of a swirl burner, in *8th Int. Conf. on Energy for a Clean Environment* (2005)

[11] M. Raffel, C. Willert, J. Kompenhans: *Particle Image Velocimetry – A Practical Guide* (Springer, Berlin, Heidelberg 1998)

Index

PIV Application to Fluid Dynamics of Bass Reflex Ports

Massimiliano Rossi, Enrico Esposito, and Enrico Primo Tomasini

Department of Mechanics, Università Politecnica delle Marche
{massimiliano.rossi,enrico.esposito,ep.tomasini}@univpm.it

Abstract. A bass reflex (or vented or ported) loudspeaker system (BRS) is a particular type of loudspeaker enclosure that makes use of the combination of two second-order mechanic/acoustic devices, i.e., the driver and a Helmotz resonator, in order to create a new system with reinforced emission in the low frequency region. The resonator is composed by the box itself in which one or more ports are present with suitable shapes and dimensions.

This category of loudspeaker presents several advantages compared to closed-box systems such as higher efficiency and power, smaller dimensions and reduced distortion at lower frequencies. Notwithstanding these advantages, they present some drawbacks like more complexity and unloading of the cone below the tuning frequency. Moreover, at high power levels the airflow in the port(s) may generate unwanted noises due to turbulence as well as distortion and acoustic compression.

In this work we will present and compare a series of experiments conducted on two different bass reflex ports designs to assess their performance in terms of flow turbulence and sound-level compression at high input power levels. These issues are quite important in professional sound systems, where increasing power levels and sound clarity require exponentially growing cost and weight. For these reasons it is vital to optimize port design.

To the knowledge of the authors there does not exist an accurate, nonintrusive experimental full-field study of air flows emitting from reflex ports in operating conditions. In this work, the experimental fluid dynamic investigation has been conducted by means of PIV and LDA techniques.

1 Introduction

Bass reflex loudspeaker systems (BRS) are analogous to a 24 dB per octave cutoff highpass filter and are generally well regarded when compared to closed-box systems for a number of reasons [1–3]:

1. Greater maximum acoustic output (up to 5 dB) ,
2. Higher efficiency (max +3 dB) ,
3. Lower cutoff frequency ,
4. Reduced distortion at lower frequencies .

Low-frequency added extension comes from exploiting the Helmholtz resonator created by the compliance of the air inside the enclosure and the inertance of the air in the port. An acoustic transformer created in this way

A. Schroeder, C. E. Willert (Eds.): Particle Image Velocimetry,
Topics Appl. Physics **112**, 259–270 (2008)

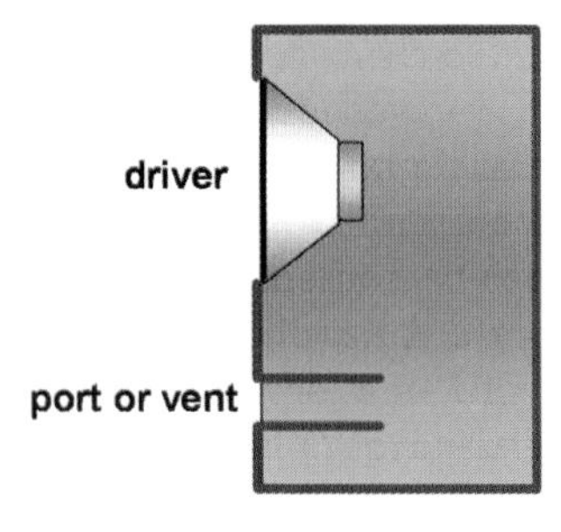

Fig. 1. Schematic of a bass reflex loudspeaker system

has its own resonant frequency, F_b, at which most (or all, if there were no losses) of the system acoustic output comes from the port. In fact, at this frequency, the driver is blocked, while the air volume of the port is oscillating and acts like the driver itself producing the sound. A fairly obvious implication of this is the significant velocity associated with the air flow through the port. A schematic of a BRS is depicted in Fig. 1.

The first patent describing BRS was granted to Thuras in 1932, followed by the contributions of *Locanthi* [4], *Beranek* [5], *De Boer* [6], *Novak* [7] in the 1950s. The turnkey arrived with Thiele's paper of 1961 (republished in 1971 [8]), and twelve years later contributions by *Small* [9]. We must also mention the contribution of *Benson* [10], who was the examiner of the Ph. D. thesis of Small. Thanks to the work of these researchers, modern design practice of BRS was born and brought to a more practical implementation including design tables by *Bullock* in 1981 [11]. Vented enclosures, however, also present two major drawbacks when compared to other types of loudspeaker systems:

1. Design procedure is more complicated than usual, and design errors, "misalignments", have serious consequences on sound quality.
2. Airflow in the port(s) may cause all sorts of unwanted noises to be generated by the port and also distortion and acoustic compression.

Thanks to the work of the above-mentioned researchers, point 1 has been resolved and nowadays there are many different software packages to help loudspeaker-systems designers and also WEB calculators [12] or ready-to-print online guides [13]. Preceding studies [2, 3, 14–17], have tackled point 2 but to the knowledge of the authors there does not exist an accurate, non-intrusive experimental full-field study of air flows coming out of reflex ports under operating conditions. This work will present a set of PIV and LDA measurements made on a sample professional BRS driven at different power levels. This system has also been compared with a custom-built BRS equipped with standard cylindrical tubes instead of the almost rectangular ones used in the professional system. Part of this work has already been presented in SMAC-03 [18].

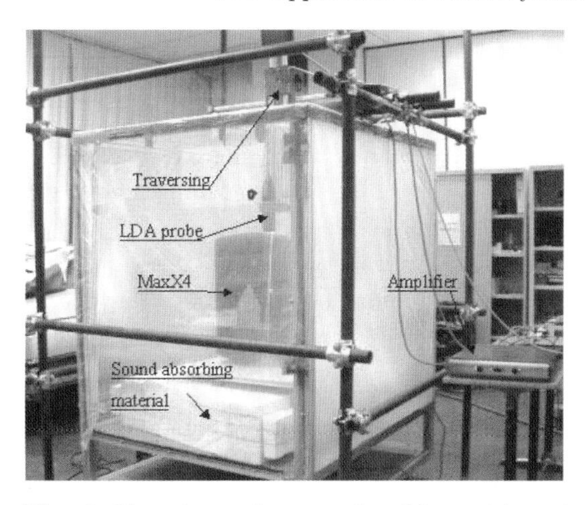

Fig. 2. Experimental setup: closed box with optical access for the PIV and LDA measurements

2 Experimental Setup

To perform our PIV and LDA measurement sessions, a particular ambient was prepared in order to fulfill the following, mandatory specific requirements:

1. seeding had to remain inside the ambient for a time long enough to allow measurements to be done;
2. no external influence should disturb the flow;
3. flow should be optically accessible;
4. due to the employed high-power lasers, safety measures had to be implemented, especially to preserve operator eyes.

Following these indications, the structure of Fig. 2 has been realized, consisting of an aluminum frame closed by plexiglas and light plastic panels. The realized closed box was checked to verify if the presence of flow reflections at box walls could lead to modifications of the external flow generated by the port. Preliminary PIV measurements have been carried out, by means of which we observed that the airflow totally disperses 40–50 cm downstream of the exit of the port. Thus, as the distance between the port and the wall is more than 1 m, the presence of flow reflections, in the opinion of the authors, can be neglected. Moreover, we checked for the presence of vibrations of the walls induced by loudspeakers, which could generate flow inside the box as well: accelerometers placed at the wall measured negligible vibrations. LDA instrumentation was mounted on top of it, while the PIV system was positioned on tripods in front of it. The seeding particles must faithfully follow the fluid flow. In the case of turbulent or high-speed gas flows, seeding-particle diameters of about 1 μm or smaller are necessary [19]. For this reason a fog generator has been used with a nominal particle size of 1 μm.

Fig. 3. FBT MaxX4 (*left*) and custom built bass reflex system (*right*)

In Fig. 3 we have two photos of studied samples, namely a FBT MaxX4 and our custom-built reflex box, employing the same loudspeaker, and with the same volume (about 40 l) and tuning frequency of the former system. The MaxX4 port has a rectangular cross section with rounded corners, 182 mm in height and 41 mm in width. The custom box has an equivalent cylindrical port with 77 mm diameter.

Both boxes have been tested at ten power levels, starting from about 1.5 W up to power-amplifier clipping, in the vicinity of 185 W. The excitation signal has been fixed to a 63.25 Hz sinusoid, coinciding with the BRS tuning frequency.

2.1 The LDA System

The LDA system used for the measurements was a Dantec two-component LDA system working in a backscatter configuration. The laser source was a Coeherent Innova 70 argon-ion laser operating in the fundamental TEM_{00} mode. The overall output of the laser source was adjusted to 1 W to ensure sufficient light intensity.

The transmitting optics provided four beams to measure two components: two blue beams at 488.8 nm wavelength and two green beams at 514.5 nm. A 40-MHZ frequency shift was applied to two of them for way detection.

The probe was equipped with a 160-mm focal lens that means a working distance of 160 mm from the edge of the probe to the measurement point. The probe was mounted in a triaxial traversing system with a submillimetre resolution for an accurate positioning of the measurement point in the measurement grid. Each point has been measured with the probe in two different configurations in order to detect the three velocity components.

The receiving optics was connected to a preamplified photomultiplier connected in turn to the BSA (burst spectrum analyzer) that detected, processed and validated the signals.

At the onset of each cycle (63.25 Hz sinusoid, period equal to 15.81 ms) a TTL trigger signal was sent from the signal generator to the synchronization input of the BSA in order to allow a "phase-resolved" data acquisition [20]. The measurements have been performed with an average data rate

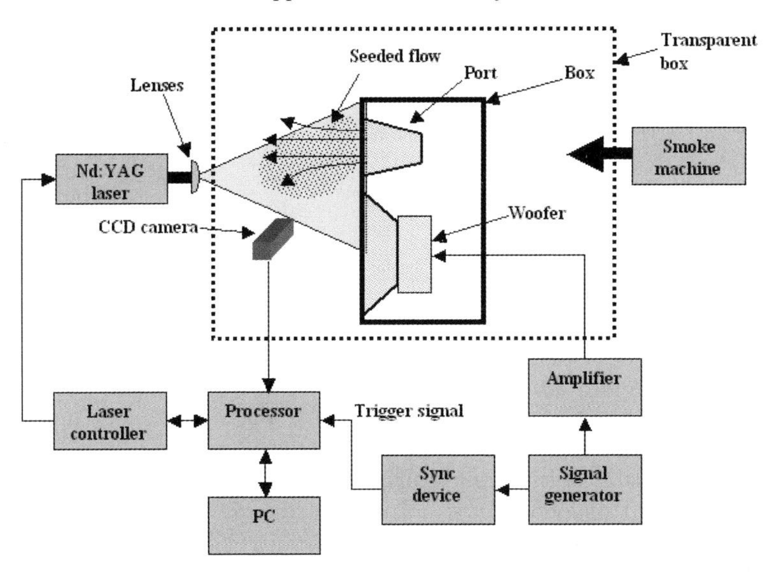

Fig. 4. Schematic of the PIV set-up

of 2000 bursts per second, thus about 125 bursts per cycle. In each acquisition 20 000 bursts were collected altogether corresponding to an average of 75 burst per phase angle.

2.2 The PIV System

The PIV measurements were performed by a 2D Digital PIV system. The illuminating source was provided by a 50-mJ, double-cavity Nd:YAG laser (NewWave Research). A 1280×1024 pixel, 15 Hz repetition rate, CCD camera has been used for the image recordings with a Nikon AF Micro Nikkor objective with 60 mm focal length.

Commercial software, DANTEC Flow Map, has been used for the acquisition and the processing of the particle images. The PIV evaluation has been performed using an adaptive multipass algorithm with an interrogation window size of 128 × 128 pixels in the first pass and of 32 × 32 pixels in the final pass with 50 % overlap applied to all passes, corresponding to a grid of 80 × 64 vectors. A standard three-point Gaussian fit has been applied for correlation-peak detection. Moving-average validation has been applied for outlier detection.

The particle image pair acquisition has been synchronized with the input signal of the loudspeaker in order to allow a “phase-averaged” acquisition. Since the Δt between the particle images pair is of the order of 100 µs and the cycle period is about 16 ms the PIV measurements can be considered representative of the instantaneous velocity field.

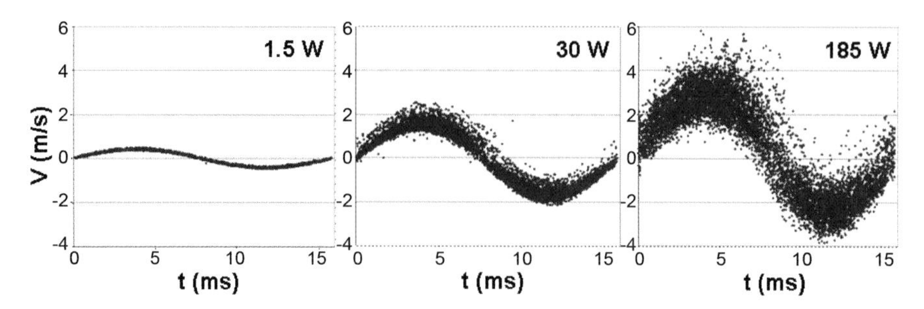

Fig. 5. Axial velocity in one cycle in a single point downstream of the port of MaxX4 box at three different power levels

The cycle has been scanned at 16 different time instants equally spaced by 0.99 ms. The flow has been assumed to be periodic and for each time instant an average of 100 PIV measurement has been taken. A schematic of the PIV setup is shown in Fig. 4.

3 Results

3.1 Measurement Results: LDA

In Fig. 5 is shown the time development of axial velocity in one cycle in a single point in the center of the port of MaxX4 box at three different power levels.

These measurements were acquired in "phase-averaged" mode, i.e., samples are not part of the same cycle but they are put "in phase" using the reference trigger input. The velocity oscillates following a sinusoidal trend, as expected from theory. On increasing the power level the velocity intensity increases as well as the dispersion of the samples, i.e., the turbulence (we measured a turbulence intensity of 23.6 % at maximum power level).

The same plot is presented in Fig. 6 relative to the custom box. In this case, the sinusoidal trend is present only at very low power. As the power is increased a distortion occurs and the dispersion of the flow increases as well (we measured a turbulence intensity of 32.0 % at maximum power level, larger than in the MaxX4 box). Moreover, the net flow in the cycle is positive, meaning that at this point the port is globally blowing air. Since the integration of the overall flow rate at the end of one cycle has to be zero there will be points where the port is globally sucking air. These distortions and three-dimensional effects contribute to markedly degrade the loudspeaker performance.

The behavior of the phase-averaged velocity as a function of the power level is presented in Fig. 7, where a cumulative graph of air velocity vs. input power for the MaxX4 at four different instants in the cycle is shown. The

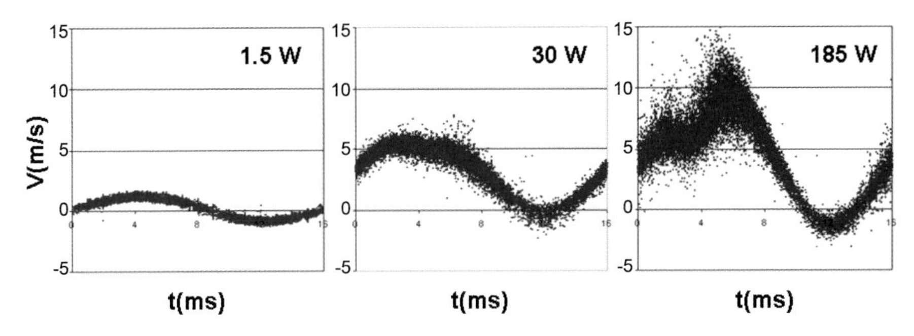

Fig. 6. Axial velocity in one cycle in a single point downstream of the port of the custom box at three different power levels

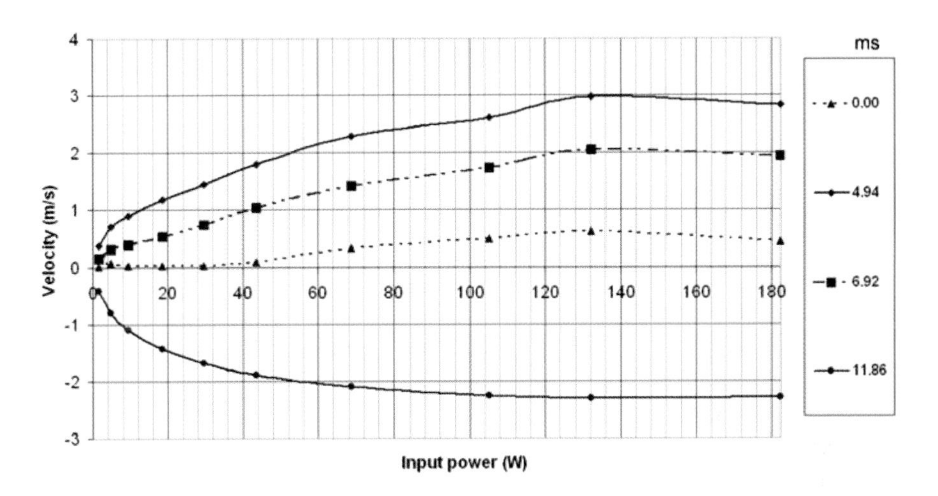

Fig. 7. Air velocity vs. input power for the MaxX4 box at the center of the vent at three different time instants

measurement point has been taken in the central position, 5 cm downstream of the port. The maximum value of flow velocity is 3 m/s, which is reached at 130 W: after this value, fluid flow saturates and we cannot observe any increase.

In Fig. 8 the same graph for the custom box is reported. In this case the fluid-flow saturation happens much earlier and beyond 30 W the flow velocity tends to be constant. The flow velocity is larger than in MaxX4 with values around 5 m/s.

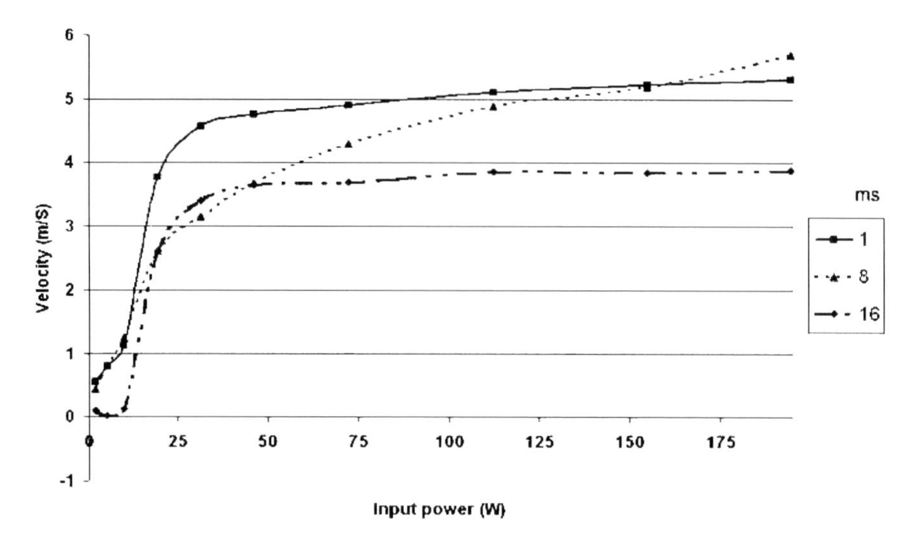

Fig. 8. Air velocity vs. input power for the custom box at the center of the vent at three different time instants

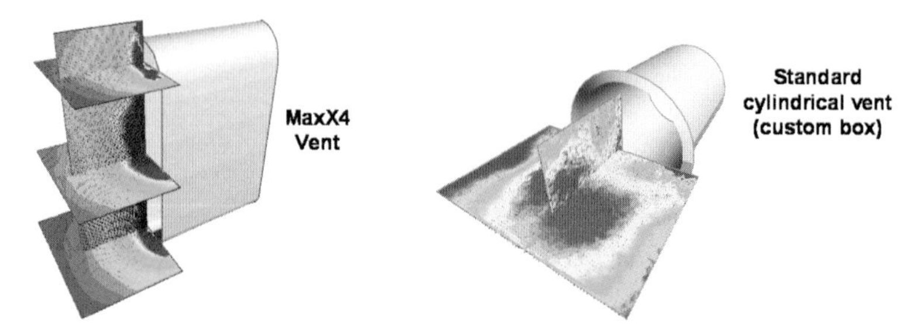

Fig. 9. Global view of the PIV measurement planes taken into account for the MaxX4 vent and the custom box vent

3.2 Measurement Results: PIV

A global view of the PIV measurement planes taken into account for the MaxX4 vent and the custom box vent are shown in Fig. 9. We recall that all the PIV measurements correspond to phase-averaged velocities, i.e., each velocity vector field is an average of instantaneous vector fields taken at the same point in a cycle, along different cycles. In Fig. 9 the velocity fields are taken at maximum power, at the point in the cycle corresponding to the maximum outgoing flow.

The boxes are both equipped with two ports and as already mentioned in the previous paragraph they are equivalent in terms of volume and tuning

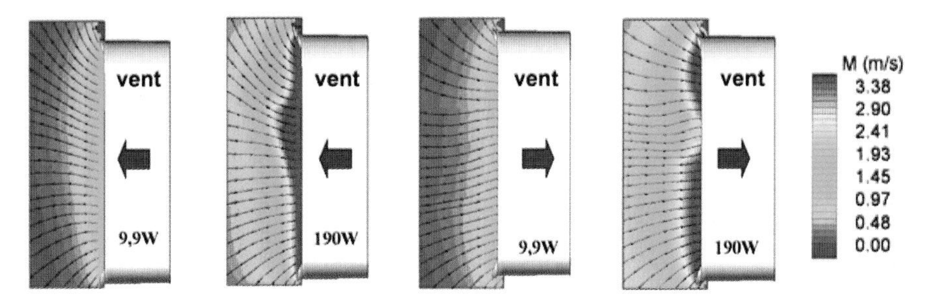

Fig. 10. Streamlines and colormap of velocity magnitude of the outgoing (on the left, low and high power) and incoming (on the right, low and high power) flow through the MaxX4 vent

frequency. The PIV measurements have been conducted on the right port of each box.

Figure 10 shows the streamlines and velocity magnitude contour map of the flow field in the vertical median plane of the MaxX4 port at two power levels in two points of the cycle, corresponding, respectively, to the maximum outgoing and incoming flow that are the most critical points due to the largest velocity involved.

At high power levels an asymmetry in the velocity magnitude can be observed above the center of the port. Nevertheless, the flow continues to stay organized and the streamlines present the same trend at low and high power. Also, no vortices are observed on the tip of the port. The maximum velocity value is about 3.5 m/s. In conclusion, the PIV measurements do not point to any strong difference between the flow at low and high power, testifying to the good performance of the port.

The outgoing and incoming flows through the custom-box port in the horizontal plane at two different power levels are depicted in Fig. 11. Velocity streamlines are shown together with a color map of the velocity magnitude as in the previous image.

In this case, the flow presents a significantly different behavior between high power and low power. In particular, at high power the outgoing flow becomes a jet that continues to accelerate downstream of the port. When the port is sucking air the jet continues to remain at a certain distance from the port. The incoming air is sucked from the sides and two big vortices are visible. Noticeably, the maximum velocity is larger than in MaxX4, with values around 8 m/s.

Finally, we investigated the presence of jet flapping in the MaxX4. Figure 12 displays the map of the velocity fluctuations at low and high power levels, both for the incoming and outgoing flows. It can be clearly noticed that: low and quite uniform fluctuations characterize the incoming flow at low power level; high power levels induce high, but uniform, fluctuations imme-

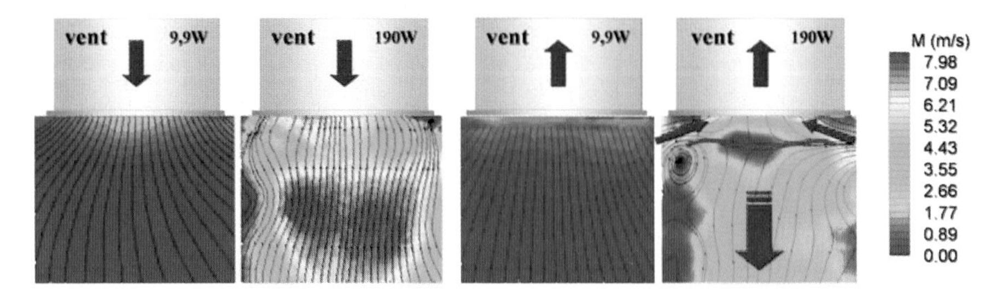

Fig. 11. Streamlines and colormap of velocity magnitude of the outgoing (on the left, low and high power) and incoming (on the right, low and high power) flow through the custom box vent

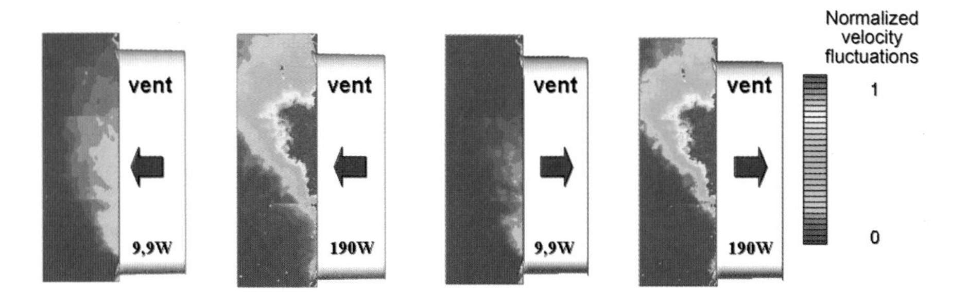

Fig. 12. Normalized velocity fluctuations of the outgoing (on the left, low and high power) and incoming (on the right, low and high power) flow through the MaxX4 vent

diately downstream of the port, slightly skewed up. This information could be used to optimize the design of the port, in order to obtain a more uniform distribution of the velocity fluctuations at the exit, thus avoiding the onset of undesired jet-flapping phenomena.

4 Discussion

The airflow through the ports of bass reflex systems is really complex because of turbulence, instabilities, and three-dimensional effects. In this work LDA and PIV have been used to measure such flows in bass reflex systems at operating conditions. In particular, the application of whole-field techniques such as PIV can locate the presence of vortices or flow instabilities due to bad port design, which can lead to noise or decreased performance.

We compared an industrial product with the cabinet produced by gas-injection molding, with a more traditional approach based on a wood cabinet and glued circular plastic tubes.

Although the latter system showed greater maximum velocities, its flux appears quite distorted. In both cases we observed flow saturation after a certain input power level: also in this respect the MaxX4 demonstrated a superior performance reaching 130 W before flow saturation occurred instead of the 20 W of the other box.

We see that modern technologies offer superior performance but also require a careful design, because each mistake is costly due to the necessity of modifying the cabinet mold. From this point of view numerical simulations of the flow can significantly improve the efficiency of the bass reflex port design if used to predict flow behavior prior to any prototype realization. In these cases experimental measurements are also needed for the validation of the numerical code.

References

[1] M. Colloms: *High Performance Loudspeakers* (Pentech, London 1991)

[2] V. Dickanson: *The Loudspeaker Design Cookbook* (Audio Amateur Press, Peterborough 1997)

[3] B. Raczynski: How good is your port?, audioXpress **Sept.** (2001)

[4] B. N. Locanthi: Application of electronic circuit analogies to loudspeaker design problems, J. Acoust. Eng. Soc. **19**, 778–785 (1971) originally published as Caltech paper

[5] L. L. Beranek: *Acoustics* (McGraw-Hill, New York 1954)

[6] E. De Boer: Acoustic interaction in vented loudspeaker enclosures, J. Acoust. Soc. Am. **31**, 246 (1959)

[7] J. F. Novak: Performance of enclosures for low-resonance high-compliance loudspeaker, IRE Trans. Audio **AU-7**, 5–13 (1959)

[8] N. Thiele: Loudspeakers in vented boxes, J. Acoust. Eng. Soc. **May–June** (1971) also available in Loudspeakers Vol. 1, (Ed.): R.E. Cooke, AES book store

[9] R. Small: Vented-box loudspeakers systems, J. Acoust. Eng. Soc. **June–October** (1973) also available in Loudspeakers Vol. 1, (Ed.): R.E. Cooke, AES book store

[10] J. E. Benson: *The Theory and Design of Loudspeakers Enclosures* (Prompt Publications, Indianapolis 1996) originally published by Amalgamated Wireless Australia Technical Review in three parts in 1968, 1971, and 1972

[11] R. Bullock, III, Thiele: Small and vented loudspeaker design, Speaker Builder **3** (1981)

[12] URL: `www.kbapps.com/audiodesign/calculators/`

[13] URL: `www.ciare.it`

[14] J. Backman: The nonlinear behaviour of reflex ports, in *Proc. 98th AES Convention* (1995) preprint #3999

[15] J. A. Pedersen, J. Vanderkooy: Near-field acoustic measurements at high amplitudes, in (1998) preprint #4683

[16] N. B. Roozen, J. E. M. Veal, J. A. M. Nieuwendijk: Reduction of bass-reflex port nonlinearities by optimizing the port geometry, in *Proc. 98th AES Convention* (1998) preprint #4661

[17] J. Vanderkooy: Nonlinearities in loudspeaker ports, in *Proc. 98th AES Convention* (1998) preprint #4748

[18] E. Esposito, M. Marassi: Quantitative assessment of air flow from professional bass reflex systems ports by particle image velocimetry and laser Doppler anemometry, in *Stockholm Music Acoustics Conf.* (2003)

[19] A. Tonddast-Navaei, D. B. Sharp: The use of particle image velocimetry in the measurement of sound fields, in *Proc. Int. Symp. on Musical Acoustics* (2001) pp. 379–382

[20] DANTEC: *Dantec Burstware Software Manual*

Index

Overview on PIV Application to Appliances

Enrico Primo Tomasini, Nicola Paone, Massimiliano Rossi, and Paolo Castellini

Department of Mechanics, Università Politecnica delle Marche
{ep.tomasini,n.paone,m.rossi,p.castellini}@mm.univpm.it

Abstract. The term appliances describes a wide range of products that are used to perform a wide variety of tasks. In the home environment, household appliances like refrigerators, ovens, washing machines, dishwashers, vacuum cleaners, hair driers are sold in million pieces per year; they also have commercial and industrial application. Typical products of the appliance industry have a relatively low industrial cost and the appliance market shows a strong competition, in which appliance technical performance plays a role together with aesthetics and costs. Therefore, efforts in applied research for product technical improvement can be done only if the ratio cost to benefits is advantageous. Many such appliances have complex fluid-dynamic problems; it is important that any experimental technique bears inherent characteristics of simplicity, provides a rapid means to collect experimental data, provides information that engineers can readily exploit for product enhancement, and mostly for computational fluid-dynamic (CFD) code validation. Particle image velocimetry (PIV) satisfies most of the above-mentioned requirements and therefore appears as a good candidate experimental technique to be proposed in the field of appliances. Nevertheless, PIV applications are rare. The PIVNet2 workshop on "PIV application to appliances" that took place in Ancona on June 2003 intended to illustrate the great potential for application of the PIV technique in this sector. This chapter presents an overview of PIV application to appliances according to the outcome from this workshop.

1 Introduction

Fluid dynamics plays a crucial role for a large fraction of household appliances, you have just to think about how important air recirculation is in an oven for the perfect cooking of a meal or the complex multiphase flow that takes place inside a washing machine.

Heat transfer also plays a crucial role in most household applications that is strictly connected with the type of flow involved.

In order to make the product competitive in the current appliances market it is important to improve the efficiency of the system but it is also very important to take care of the aesthetics of the object and sometimes these two things are in contradiction with each other. For this reason, nowadays it is getting more and more important to get detailed knowledge of the phenomena behind each single product, according to its particular shape and working principles.

A. Schroeder, C. E. Willert (Eds.): Particle Image Velocimetry,
Topics Appl. Physics **112**, 271–281 (2008)

Although PIV is widely used by researchers in sectors such as aeronautics, automotive, etc., the appliances field can be considered as a new sector, where at the state-of-the-art applications are almost nonexistent. The PIVNet2 Workshop on "PIV application to appliances" that took place in Ancona on June 2003 intended to illustrate the great possibilities of application of the PIV technique in this sector. This chapter presents an overview of PIV application to the appliance industry according to the outcome from this workshop.

In the following pages the most relevant fluid-dynamic aspects connected with appliances will be indicated together with the problems of applying PIV to this specific sector. In this way, it intends also to be a sort of starting point for researchers and designers that want to use PIV to investigate fluid dynamic of appliances. Moreover, the state-of-the-art of PIV application on this kind of products is shown with some examples of activities already carried out.

2 Fluid Dynamics of Appliances

The large amount of different types of appliances with different types of fluids and flows involved produces a great variety of different fluid-dynamic problems. The temperature is also critical and optimizing heat transfer is important for most applications. We can find a wide range of temperatures from the subzero temperature of the refrigerating applications to the hundreds of degrees Celsius in ovens. The flow patterns are often complicated by walls and structures of the devices and the modulus of velocity can vary from a few mm/s of natural convection flow in a refrigerator to several m/s of the jet of a hair dryer.

In order of give an overview of these problems we present a list of the most common fluid-dynamic characteristics that can be found when investigating flows in appliances.

Type of fluid involved:

- air (almost all types of appliances),
- water (washing machines, dishwashers, boilers),
- gas: (gas ovens, gas cookers),
- other fluids: (refrigerating fluids).

Type of flows:

- free convection (refrigerators, ovens),
- forced convection (ovens, range hoods),
- multiphase flows (washing machines, dishwashers),
- jets (hair driers, dishwashers),
- flames (gas ovens, gas cookers).

Environment:

- open air (air-conditioning systems, range hoods),
- closed environment (refrigerators, ovens).

Temperature:

- subzero (refrigerators),
- below 100 °C (air-conditioning systems, washing machines),
- above 100 °C (ovens).

We can also classify the appliances into several groups depending on their main functions.

Cooking. This group contains appliances used for cooking like electrical ovens, gas ovens and gas cookers. The main problem for this category is to exchange heat in the most efficient way. For this reason air circulation (that can be both free or forced convection) has to be investigated. For gas ovens and gas cookers, a study of flames has to be also taken into account.

Refrigeration. This category contains appliances used to refrigerate and stock food, like home refrigerators, freezers, ice-cream cabinets [1]. In this case we have on the one hand the fluid dynamics of the air circulating inside the device that has to keep the food at the desired temperature, on the other hand the fluid dynamics of the cooling circuit and the refrigerating fluid, with the compressor, the expansion valve, the serpentine and so on

Washing. In this group there are the appliances used for washing, basically washing machines and dishwashers. The fluid dynamics in this case is very complex, a multiphase flow composed of water, soap, air that interact with movable surfaces.

Air conditioning. All the appliances used for air conditioning in houses, offices, shops. In this group we can also include fans and range hoods. In these appliances it is important to investigate the behavior of air both in the environment outside the device to control the efficiency of air recirculation both inside the device to control the efficiency of the air conditioning.

Others. There are a lot of other household appliances with different uses in which fluid dynamics plays a crucial role, like hair driers, vacuum cleaners, but also air recirculation around lamps and hi-fi systems [2].

In Table 1 the fluid dynamics problems in several appliances are summarized.

3 PIV Applied to Appliances

Applying PIV on appliances is not always straightforward. Of course, appliances are not designed for PIV and in many cases they have to be modified to allow the setup of the experiment. On the other hand, on modifying the

Table 1. Fluid dynamics problems in appliances

	Oven	Washing machine	Refrigerator	Air conditioner	Hair drier	Range hood
Fluid	air, gas	water, air, soap	air, refrigerator fluid	air, refrigerator fluid	air	air, steam, smoke
Typical temperature range	60–220 °C	40–90 °C	–30–4 °C	10–20 °C	20–50 °C	room temperature
Type of flow	natural convection, forced convection	multiphase flow jets	natural convection, forced convection	forced convection	jet	natural convection, forced convection
Environment	closed	closed	closed	open/ closed	open/ closed	open/ closed

devices, their normal operating conditions change as well. For this reason it is really crucial when setting up the PIV experiment to try to maintain the product functionality as close as possible to the operating condition in real life.

As seen in the previous paragraph the working fluid in most appliance applications is air, but also water and other fluids are involved, especially in washing machines and in the cooling system of air conditioners and refrigerators. In this chapter only applications with air or gas are considered that are the most suitable for PIV applications.

In order to setup a reliable and safe experiment in this field it is necessary to apply the same rules valid for every PIV experiment [3]. The most relevant aspects are:

3.1 Seeding

In many cases the standard seeding materials used for gas flows are acceptable. Smoke generators and atomizers operating with oil or water can be successfully applied in most applications. In appliances operating at high temperature, like ovens or air driers, oil should be avoided. In this case smoke or atomized water solutions (e.g., water–sugar) are safer and equally effective. In flames, solid tracers like aluminum or titanium oxides can be used.

In some applications, like air conditioners, a very large measurement area is required. Using standard seeding, if tracer particles are small with respect to pixels and are not evenly distributed across the area we can not have clear single-particle images. In this case we obtain gray-level images that can be used to obtain velocity vectors. Anyway, it is advisable to improve the spatial resolution dividing the measurement area in a set of windows.

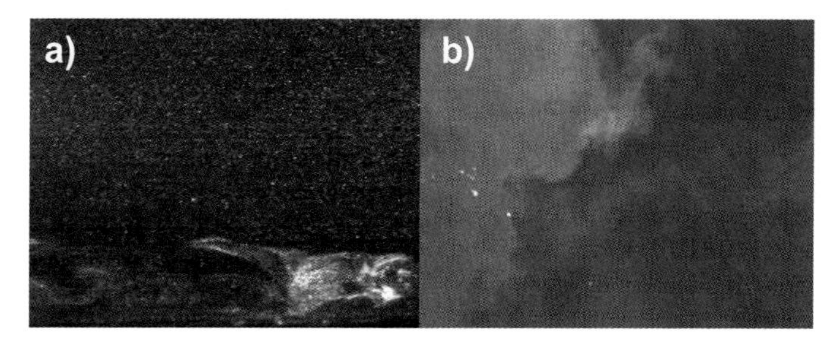

Fig. 1. Difference between a particle image (**a**) and a gray-level image (**b**). In a particle image single particles can be detected, while that is not possible in a gray-level image

The difference between a particle image and a gray-level image is shown in Fig. 1.

3.2 Optical Accesses

When it is necessary to investigate internal fluid dynamic of appliances optical accesses are needed for the laser and the camera. Creating a suitable optical access can be a very critical issue and can compromise the final reliability of the experiment. For instance, in appliances where heat transfer is involved, making transparent windows changes the boundary condition of the problem, introducing a different behavior of the device with respect to its usual functioning.

3.3 Wall Reflections

The most interesting flows in appliances are very often located close to surfaces, like inlets, outlets, grids, fans. Reflections from surfaces introduce a relevant source of optical noise. Mat paint on the surfaces is required in this case to reduce reflections. Fluorescent paint has also been successfully applied. In this case, the wavelength of the reflections is shifted with respect to the illumination source and can be filtered out. Using a Nd:YAG laser ($\lambda = 532\,\text{nm}$) Rhodamine B can be used as fluorescent paint together with a highpass filter.

3.4 Working Conditions

In order to have a reliable experiment it is important to try to reproduce the operating conditions of the appliance in real life. For instance, evaluating the performance of a range hood, the environment around it, like furniture, cookers, boiling pots, should be taken into account as well. In this case, building up an environment simulating a small kitchen can be useful.

3.5 Limitations of PIV on Appliances

The application of PIV on appliances offers many advantages, nevertheless, some limitations have to be accounted for. For instance, the use of high-power laser beams may introduce the need for specific safety regulation. But the main concern is that PIV is a complex technique, which requires qualified staff to be successfully implemented; therefore companies interested in using it should provide training programs for the personnel. PIV is a complex technique because it measures in 2 or 3-dimensional space the multiple components of the velocity (2 or 3). The simplest and common 2D-2C PIV might not be suitable for apparently simple applications, which are actually complex owing to turbulence or the three-dimensionality of the flow.

4 Examples

In this section, some real cases of PIV application to appliances are reported with the double intent of illustrating some aspects of what was explained in the previous sections and showing the state-of-the-art in this field. All the following cases have been presented in the PIVNet2 workshop on “PIV application to appliances”.

4.1 Ovens [4, 5]

In the design of gas ovens the uniformity of fume flow patterns in the cooking chamber is very important to obtain the optimum cooking conditions. In Fig. 2a is shown the setup built at Ikerlan to investigate the flow inside a gas oven. To simulate the oven a prototype has been manufactured. This prototype is supplied with orthogonal optical accesses for the laser and camera as shown in Fig. 2b.

The PIV measurements have been acquired by means of a 1248×1024 CCD Camera, a Nd:YAG laser ($\lambda = 532\,\mathrm{nm}$, energy: $90\,\mathrm{mJ/pulse}$) and Aerosil (SiO_2, $0.6\,\mu\mathrm{m}$) as tracer. The oven walls were painted with black and fluorescent paint (Rhodamin B) to avoid reflections.

Figure 3 shows the velocity vectors of the fumes in the tray area of the center of the gas oven obtained by PIV (on the left) and by numerical simulation in fluent (on the right). The comparison between the two vector maps exhibits a good agreement between numerical and experimental data.

4.2 Air Conditioning Systems [6, 7]

Air conditioners are typical appliances where we have problems of external and internal fluid dynamics. Internal fluid dynamics concerns all the flows inside the device, i.e., problems connected with fluid dynamics of the heat

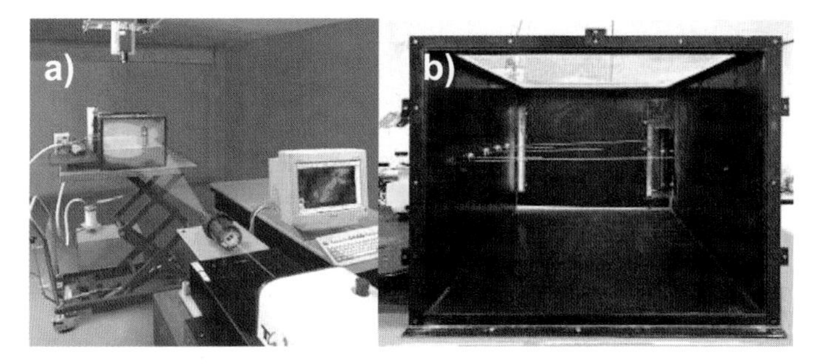

Fig. 2. (**a**) Experimental setup for the PIV measurements in a gas oven at Ikerlan. (**b**) Oven prototype with optical access for PIV measurements

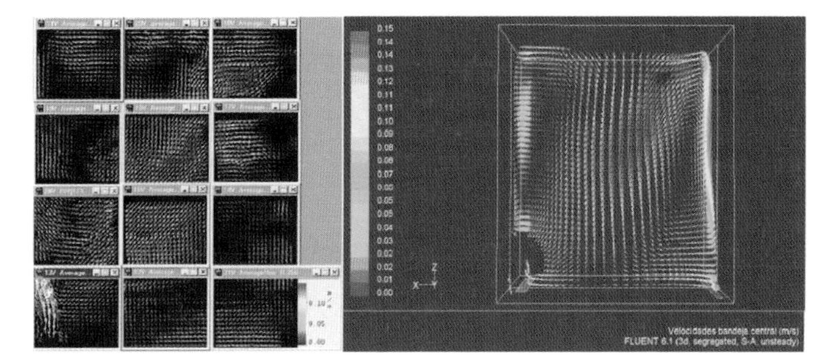

Fig. 3. Comparison between experimental data (*left*) and numerical data (*right*)

exchanger, the fan and the other internal components. External fluid dynamics mostly concerns the mixing of the air coming out from the air conditioner with the air of the external environment.

An example of PIV applied to the internal fluid dynamics of air-conditioning systems is shown in Fig. 4. In this case the region under investigation is the area between the heat exchanger and the fan of a large-size Libert–Hiross air-conditioning system. The PIV images are obtained using a smoke generator for seeding.

Regarding PIV applied to external fluid dynamics of air-conditioning systems the mixing region near the outlet is reported. The PIV measurements have been performed at Sunmoon University Fluid Machinery Lab and a velocity and vorticity map is shown in Fig. 5. In this case, vegetable-oil droplets provided by a six-jet atomizer have been used as tracer particles.

4.3 Range Hoods [8]

Range hoods are very common low-cost household appliances with a rather simple technology. Nevertheless, the design of modern range hoods has to satisfy specific criteria of functionality, ergonomics and aesthetics that can

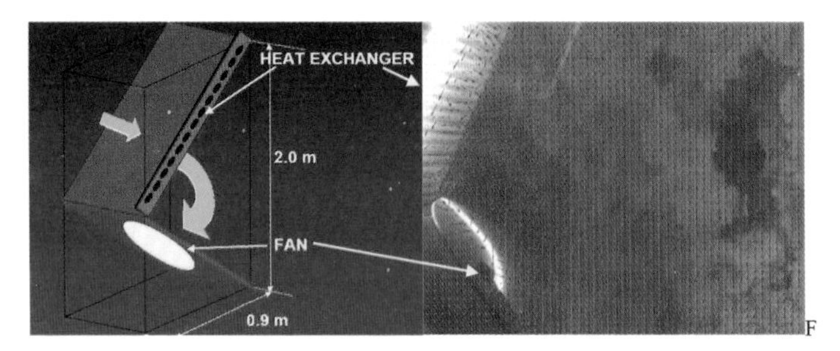

Fig. 4. Velocity vector field and vorticity contour map of the flow in the mixing region downstream of the outlet of an air-conditioning system

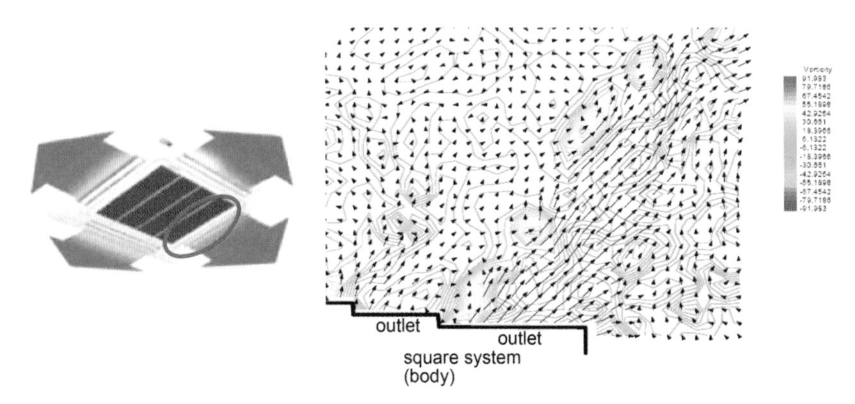

Fig. 5. PIV measurement inside an air-conditioning system. The measurement zone is between the heat exchanger and the fan, as shown in the schematic on the left

not ignore consideration of fluid dynamics performance. A feasibility study of PIV application to the intake flow of kitchen evacuation hoods in operating conditions has been carried out by the Università Politecnica delle Marche. A small environment simulating a small kitchen has been built up (Fig. 6).

An electrical heater and a pot with boiling water has been used to simulate the standard operating conditions of a range hood in a kitchen. The seeding was provided by the steam produced by the boiling pot with the addition of smoke provided by a fog generator.

Figure 7 shows the scalar map of velocity magnitude with streamlines of a small region close to the intake grid. This is part of a study of the interference of the hood's structures with the incoming flow.

4.4 Lamps [9]

The last example is a study carried out by iGuzzini (manufacturer of lighting fitting) on air recirculation around a halogen lamp (Fig. 8). The purpose of this study was to characterize the cooling of the lamp by convective flow in

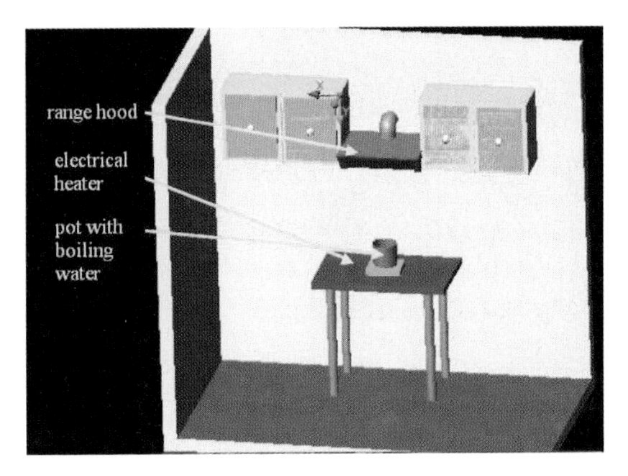

Fig. 6. Environment simulating a small kitchen. The range hood is positioned between wall cupboards and an electrical heater, while a boiling pot simulates a working cooking stove

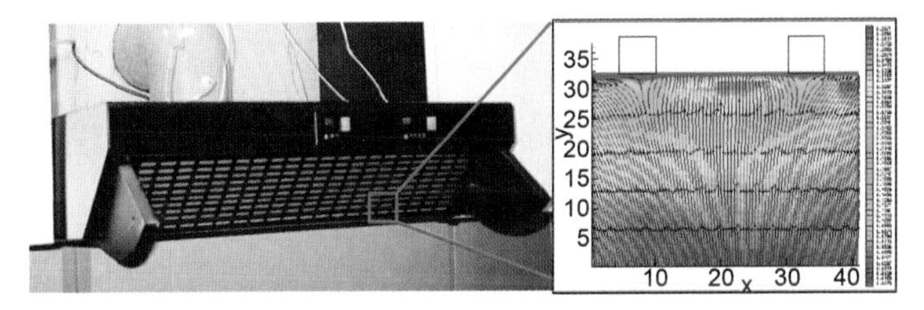

Fig. 7. Scalar map of velocity magnitude with streamlines of a small region close to intake grid

Fig. 8. Halogen lamp produced by iGuzzini object of the PIV investigation

order to design and optimize the temperature distribution in the lamp. In this case, 3DPIV measurements are performed together with CFD simulation.

In Fig. 9 the flow field of two different regions is shown. The top image shows the flow in an axial plane on the top of the lamp that is placed in a vertical position; the bottom image shows the flow in a radial plane with

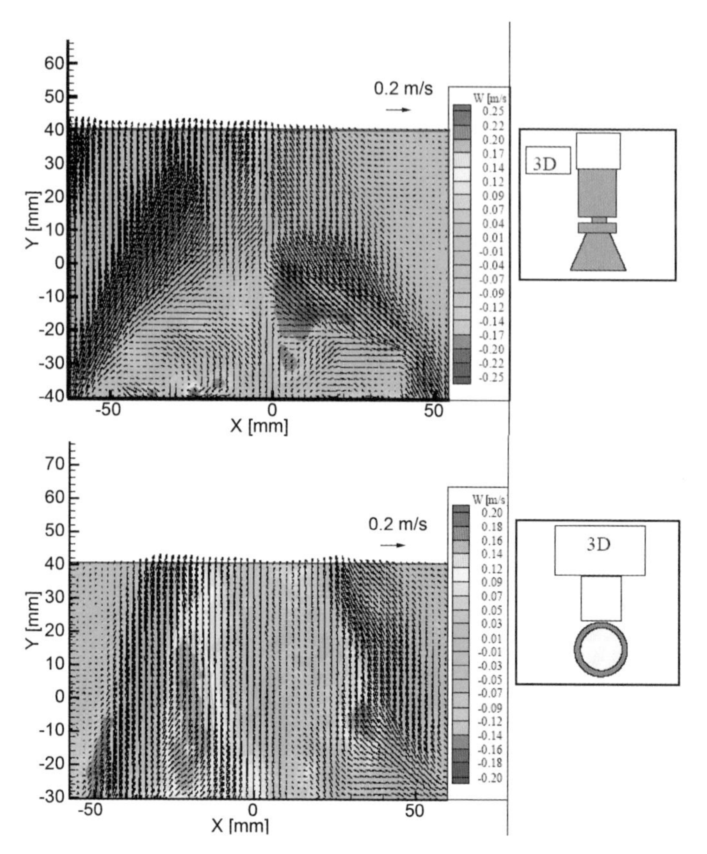

Fig. 9. Velocity vector fields measured by means of 3DPIV in two different regions of the lamp. The contour map shows the out-of-plane component. The velocity field of the two regions shows the strong three-dimensionality of the flow

the lamp placed in a horizontal position. The color map is the out-of-plane component. The flow is a convective flow driven by the temperature gradient between the surface of the lamp that can reach 70 °C and the air. The geometry of the lamp makes the flow strongly three-dimensional, as can be seen in Fig. 9, so that the use of 3DPIV is advisable.

5 Conclusions

Particle image velocimetry has extremely high potential to be used as a tool in the development and design of household appliances and appliances in general. The interest in using PIV in this sector is growing in research institutes and companies, as testified by the contributions presented at the PIVNet2 workshop on "PIV application to appliances".

The success reached by the event is testified also by the large number of participants.

The event, in fact, was attended by 55 delegates from Universities, enterprises and Research Institutes from all around Europe (Belgium, Denmark, Germany, Italy, Portugal, Spain, Turkey, United Kingdom), from Mexico and from South Korea. Of all the participants only 18 % were partners in the PIVNet2 European thematic network. This testifies to the great interest in this topic also in the world of appliance producers.

References

[1] G. Piersigilli: Optical techniques applied to the study of internal fluid dynamics of ice-cream refrigerators, in *PIVNet2 Workshop on PIV application to appliances* (2003)
[2] E. Esposito, M. Rossi, E. P. Tomasini: PIV application to fluid dynamics of bass reflex ports, in *PIVNet2 Workshop on PIV application to appliances* (2003)
[3] M. Raffel, C. Willert, J. Kompenhans: *Particle Image Velocimetry: A Practical Guide* (Springer, Berlin, Heidelberg 1998)
[4] I. Alava, U. Onederra, R. Marin, A. Laresgoiti: Comparison between PIV measurements and CFD modelling on a domestic gas oven, in *PIVNet2 Workshop on PIV application to appliances* (2003)
[5] L. Araneo, A. Coghe, G. Solero: Natural gas burners for domestic appliances: Application of the particle image velocimetry technique (PIV), in *PIVNet2 Workshop on PIV application to appliances* (2003)
[6] M. Lazzarato, G. Meneghin, M. Scattolin: PIV applied to the internal fluid dynamics of close control air-conditioning systems, in *PIVNet2 Workshop on PIV application to appliances* (2003)
[7] J. W. Kim: Mixed flows near inlet air-conditioner, in *PIVNet2 Workshop on PIV application to appliances* (2003)
[8] N. Paone, M. Marassi, P. Santonicola: PIV applied to kitchen evacuation hoods, in *PIVNet2 Workshop on PIV application to appliances* (2003)
[9] S. Rita, A. Guzzini: Study of lamp cooling by 3D PIV, in *PIVNet2 Workshop on PIV application to appliances* (2003)

Index

Selected Applications of Planar Imaging Velocimetry in Combustion Test Facilities

Christian Willert[1], Guido Stockhausen[1], Melanie Voges[1], Joachim Klinner[1], Richard Schodl[1], Christoph Hassa[1], Bruno Schürmans[2], and Felix Güthe[2]

[1] Institute of Propulsion Technology, German Aerospace Center (DLR), 51170 Köln, Germany
chris.willert@dlr.de
[2] ALSTOM, 5401 Baden, Switzerland
bruno.schuermans@power.alstom.com

Abstract. This chapter provides an overview on the application of particle image velocimetry (PIV) and Doppler global velocimetry (DGV) in combustion test facilities that are operated at pressures of up to 10 bar. Emphasis is placed on the experimental aspects of each application rather than the interpretation of the acquired flow-field data because many of the encountered problems and chosen solution strategies are unique to this area of velocimetry application. In particular, imaging configurations, seeding techniques, data-acquisition strategies as well as pre- and postprocessing methodologies are outlined.

1 Introduction

Aeroengines as well as stationary power generation will continue to rely heavily on fossil and substitute fuels in the coming decades. Increasing fuel costs demand higher efficiencies, while new governmental regulations require a further reduction of emissions. The development of modern aeropropulsion and gas-turbine technology increasingly relies on combustion-related research projects that are driven by similar motivations, namely to provide experimental data to validate advanced simulation methods that model the extremely complex flows found in modern combustors. To offer a reliable alternative to expensive full-scale rig tests, confidence in numerical models has to be gained on the basis of validation experiments that capture certain aspects of the flow where a corresponding physical model is tested in isolation. In order to judge the performance of computational fluid dynamics (CFD) tools in realistic applications, data must also be obtained from facilities that are capable of capturing the rather comprehensive range of effects found inside the combustion chamber in relevant geometries and at higher pressures. The complexity and operational cost of full-scale combustor sectors prevents the application of advanced optical measurement techniques to fully capture the range of phenomena, such that compromises must be made by using simplified and/or downsized combustors.

On the other side, combustion CFD codes that are currently used in the design of combustors are based on Reynolds-averaged Navier–Stokes (RANS)

A. Schroeder, C. E. Willert (Eds.): Particle Image Velocimetry,
Topics Appl. Physics **112**, 283–309 (2008)

equations and rely on a steady-state solution of time-averaged flow equations. Although these codes have proven themselves valuable in the design process, the RANS approach has well-known deficiencies in accurately predicting features of turbulent flows like the spreading and mixing of turbulent jets, highly turbulent and genuinely unsteady flows. A more accurate prediction of the detailed flow and temperature field and distributions of chemical species will result in higher-accuracy predictions of emissions (NO_x, CO, unburned hydrocarbons, etc.). Some of the limitations of RANS codes can be overcome by the large eddy simulation (LES) method, where time-dependent turbulent fluctuations of a size larger than the size of the numerical grid are resolved accurately in time.

Confidence in improved CFD techniques cannot be solely built on validation experiments that represent one aspect of the flow where the corresponding physical model is tested in isolation. To qualify the numerical codes for practical calculations, highly resolved experimental data must be obtained from test facilities capable of presenting a rather comprehensive range of effects found inside the combustion chamber in relevant geometries and at higher pressure. In practice, these test facilities should represent the conventional combustor architecture with a diffusion burning primary zone, a mixing zone with colliding jets and film-cooled liner walls.

While spectroscopic methods are generally used for acquisition of temperature, mixture fraction and species data, velocity data obtained by laser Doppler velocimetry, PIV and DGV provides insights into the kinetic effects influencing the combustion processes, especially under the influence of increased pressure or combustion oscillations.

This chapter intends to summarize experiences gained in bringing both PIV and DGV to (routine) application on combustion test facilities. While the experimental data obtained in these applications are described elsewhere with respect to their physical significance, the emphasis here is to give some background on the general solution strategies that led to the successful acquisition of this data.

2 Challenges on Diagnostics

A successful application of both PIV as well as DGV in combustion facilities depends on a careful optimization of three equally important issues: 1. optical access, 2. imaging aspects and 3. seeding techniques. In the following, each of these three aspects are briefly described with particular emphasis on PIV.

2.1 Optical Access

Clearly, any optical diagnostic method, be it of pointwise (e.g., LDA, PDA, Raman scattering) or planar (e.g., LIF, PIV, DGV) nature, requires adequate optical access to the generally confined flow of the combustion test facility.

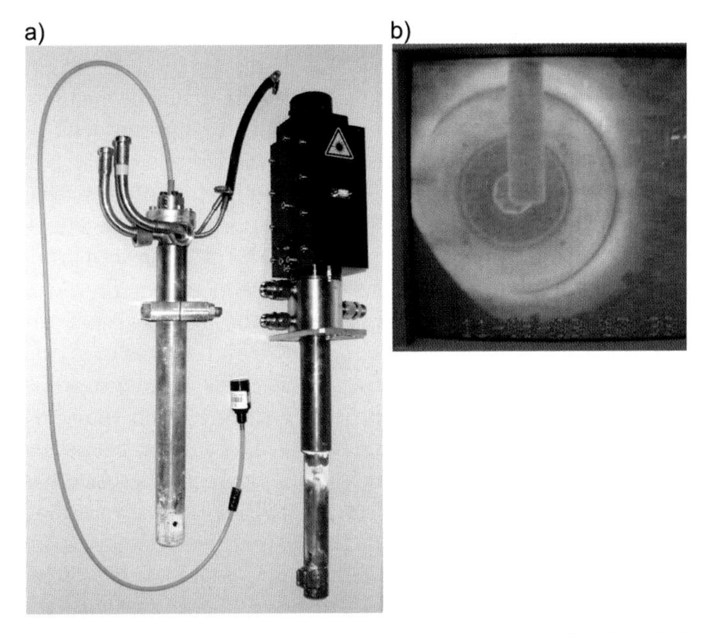

Fig. 1. (**a**) Water-cooled imaging fiber periscope (*left*) and lightsheet generator probe (*right*) used for DGV in combustors; (**b**) video image of an imaging periscope mounted about 0.7 m downstream of a gas-turbine combustor

In the case of PIV a laser lightsheet has to be introduced that is observed by the imaging optics of a recording camera. While periscope probes as shown in Fig. 1 are feasible for the introduction of the lightsheet into the facility, the rather low resolution of imaging periscopes may allow intensity-based DGV measurements, as this technique does not require that particles are discretely resolved. However, the low optical resolving power of these imaging periscopes is unsuited for PIV imaging. So, in practice at least, one viewing window is required to properly resolve particles suspended in the flow.

However, these windows must also sustain high temperatures as well as elevated pressures that often results in rather complex designs consisting of multiple windows (thin liner window combined with thick pressure window). Active film cooling is frequently used to protect the window from the high temperatures near the reaction zone with the drawback that the cooling film may influence the reaction processes. A close collaboration between the facility designers, combustion experts and optical measurement specialists is of utmost importance during the design phase of the facility to ensure a reasonable balance between adequate optical access and its possible influence on the flow physics.

2.2 Imaging Aspects

Figure 2 shows a PIV recording obtained from a kerosene flame inside a single-sector combustor and illustrates a number of issues that are representative of the application of PIV to combustion facilities. Here, the lightsheet plane is parallel to the centerline of the double-swirl fuel nozzle that itself is hidden from view at the lower edge of the image. The lower portion of the image shows a cross section through the kerosene spray cone that is ejected from the nozzle and contains areas of significant overexposure of the CCD sensor as exhibited by saturated pixels. The vertical stripes in these overexposed areas result from pixel blooming on the CCD sensor – the high-intensity signal spreads the overflowing charge into neighboring pixels. This overexposure is caused by the rather large droplets (typically $\geq 50\,\mu$m) for which the scattering cross section is significantly larger than for the micrometer-sized flow seeding for which the PIV recording system is optimized. The problem is limited to the lower half of the image since the droplets evaporate as they propagate upward into the flame. In order to distinguish droplet velocity from gas velocity of this two-phase flow various strategies are possible. In the present case, highpass filtering combined with binarization of the acquired images matched the intensity levels of the seeding with that of the kerosene droplets such that areas normally biased to the velocity of the kerosene droplets could be corrected in favor of the gas velocity [1]. The image-enhancement procedures and their effect on the recovered displacement data are illustrated in Fig. 3. An alternative approach using the fluorescence of both particles and fuel was proposed and applied by *Kosiwczuk* et al. [2, 3]. However, the applicability of the technique to pressurized combustion has yet to be tested with regard to stability of the fluorescent dyes at higher temperatures.

The sensor saturation in the upper half of the image of Fig. 2 has a different cause: this is the region of highest flame luminosity that cannot always be removed solely through the use of a laser line filter. As illustrated in Fig. 2c the signal levels in this area differ significantly between the frames of the image pair (1500 vs. 4000 counts) that results from the fact that the second frame of current PIV cameras stays sensitive for the time required for the readout of the previously recorded first-image frame. This readout time is on the order of 50–100 ms compared to less than 1 ms for the first frame. Flames with high fuel-to-air ratios at elevated pressures are characterized by strong luminosity (e.g., glowing soot) such that narrow-banded laser line filters are not always sufficient for its suppression. A solution to this problem could be the use of a pair of CCD cameras – each individually shuttered electronically – with a common viewing axis [4]. A more practical solution is the use of mechanical or electro-optic shutters. While mechanical shutters provide complete extinction, their lifetime is limited when operated at high frame rates and steep rise/fall times. Electro-optic shutters have rise times in the submicrosecond range but either have low transmission ($< 40\,\%$) as for ferroelectric devices [5] or have low extinction ratios as for liquid crystal scattering shutters [1].

The recording shown in Fig. 2 contains a third problem that is not immediately obvious: Even in the absence of both flame luminosity and particulate seeding there is significant background illumination that not only lowers the overall contrast but may even result in a complete loss of signal due to sensor saturation. This background lighting is primarily caused by laser light scattered from window surfaces and illuminates features within the field of view such as the circular port in the background of Fig. 2a. The undesired signal generally increases during facility operation due to the increased deposition of seeding on the windows. The use of light traps and recessed windows reduces the problems [6] but is not always possible on a given facility. Aside from the saturation of the sensor, measurements close to windows are difficult due to the stationary speckle signal resulting from light scattered by the deposited seeding.

Image quality in terms of resolution ultimately depends on the integral distortion effects between the individual light-scattering particle and the imaging sensor. The use of high-quality recording lenses and plain quartz glass windows is generally sufficient in isothermal PIV applications. Temperature gradients inside and outside of combustion facilities introduce light refraction along the imaging path that results in both the displacement of the particle image on the sensor (similar to background-oriented Schlieren, BOS) and, even worse, a blurring of the particle image. This beam-steering effect also strongly affects other point-resolving techniques such as laser/phase Doppler anemometry (LDA, PDA) as well as spectroscopic techniques such as Raman scattering. The blurring can be roughly considered as a sum or product of temperature, pressure and penetration depth, such that the application of PIV may fail with increasing facility size and operating pressures. Blurring can be controlled to some degree by two simple techniques. First, an increased distance between the recording lens and the optically disturbed media (by using zoom lenses) reduces beam steering, which is similar to reducing the sensitivity in a Schlieren setup. Secondly, a reduced lens aperture confines the light-collecting cone such that all light rays pass similar refracting media resulting in a sharper image. This is particularly useful for highly turbulent flows with locally strong modulations of density. The disadvantage of this approach is the increased depth of field that may bring (grainy) areas into focus that disturb the PIV correlation signal (e.g., seeding deposits on windows or walls).

2.3 Seeding of High-Temperature, Reacting Gas Flows

Aerosol seeding using atomized liquids, which is common in many aerodynamic applications, is not possible in high-temperature reacting flows due to evaporation and/or combustion of the droplets. In these cases, seeding based on solids must be used. Metal oxide powders are especially well suited for

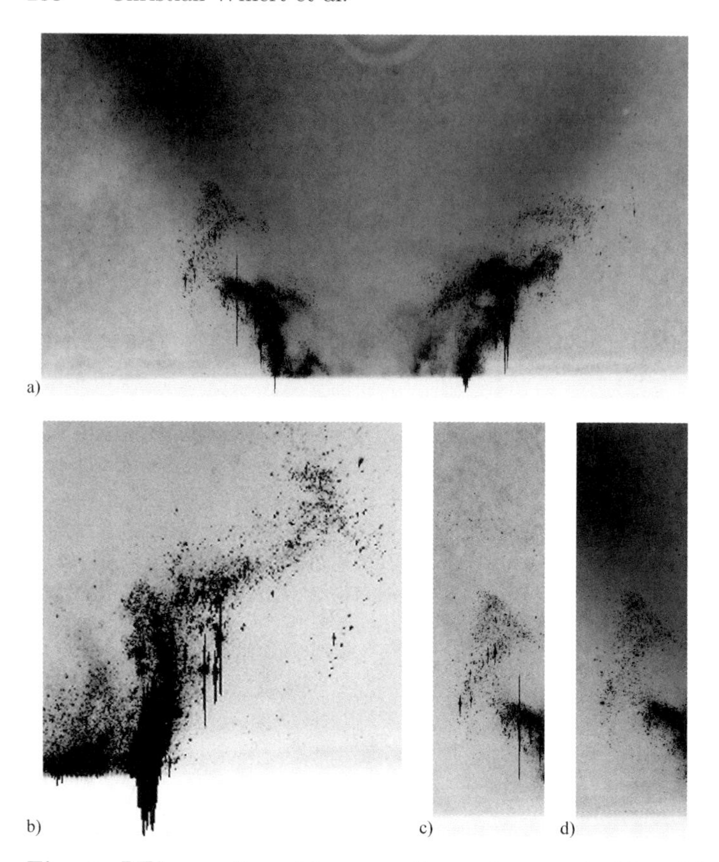

Fig. 2. PIV recording of kerosene flame above a double-swirled fuel nozzle at 3 bar (to enhance contrast, the negative image is shown: *black areas* are brightest), (**b**) detail of the kerosene fuel spray exhibiting strong saturation of sensor, (**c, d**) corresponding regions from each frame showing effects of flame luminosity

this purpose due to their inertness, high melting point and rather low cost. Titanium dioxide, alumina and silica powders are some of the most commonly used materials. However, the controlled dispersion of these powders is more challenging than for liquid materials since the powders have a strong tendency to form agglomerates, especially for small grain sizes in the submicrometer range. Therefore, the seeding device has to either break up the agglomerates or remove them from the powdered aerosol prior to delivery into the facility such as through the use of a cyclone separator [7, 8].

Another approach to deagglomerate the bulk seed material was proposed by *Wernet* and *Wernet* [9]: the interparticle forces responsible for the agglomeration can be directly influenced by controlling the acidity of liquid suspensions of the seed material. They suggest the use of acidic (pH $\sim$ 1) dispersions of alumina/water or alumina/ethanol that can be dispersed us-

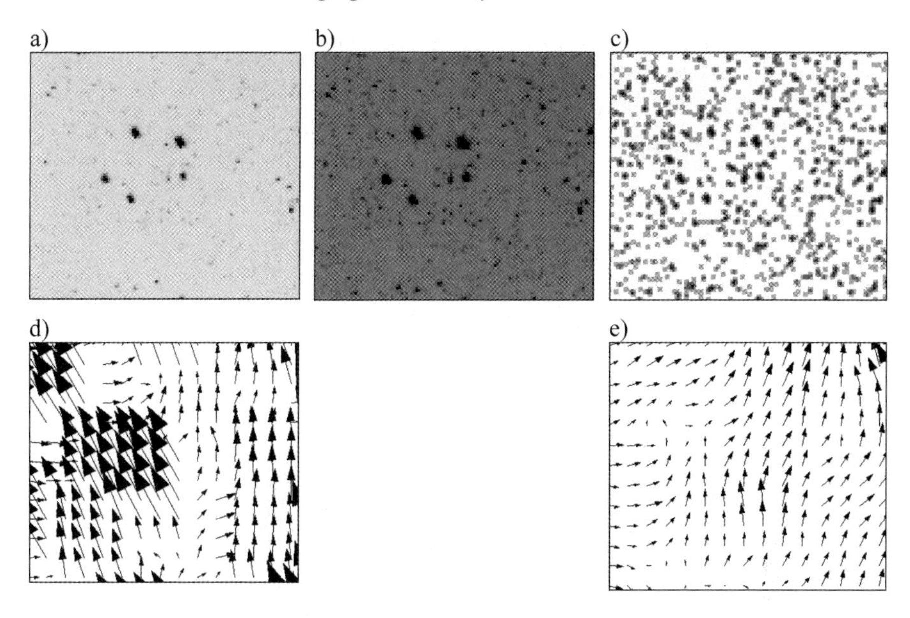

Fig. 3. Portion of PIV recordings – inverted for clarity – and processed PIV data: (**a**) fuel droplets, (**b**) brightened version of (a) with seeding visible, (**c**) preprocessed image, (**d**) flow field after standard PIV analysis, (**e**) flow field obtained after enhancement of PIV images, (from [1])

ing liquid atomization. The solid seed material remains after evaporation of the carrier liquid.

Depending on the relative mass-flow rates between seeded and unseeded flows the use of liquid-particle suspensions is not always feasible, especially when the flow may be disturbed due to evaporation cooling and/or changes in the reactive chemistry. Therefore, the aerosol should be created through dispersion of the dry powder. A common approach is to aerate the powder inside a vertical tube from below, resulting in a so-called fluidized bed. The flow rate through the seeder is chosen just large enough such that the bed of particles is fluidized, carrying smaller particles into the region above the bed (known as the freeboard) toward the exit orifice and from there into the facility under investigation. Figure 4 shows a simple fluidized-bed seeding device for use in elevated-pressure applications [1, 10]. This generator has two noteworthy features: the strong shear flow present in a sonic orifice at the exit breaks up larger agglomerates. The size of this orifice is chosen to ensure sufficient flow rate to aerate the powder. The second feature is a bypass line that can be used to maintain constant mass flow rates into a test facility even when no seeding is required.

The following gives some recommendations for the successful operation of fluidized-bed seeders:

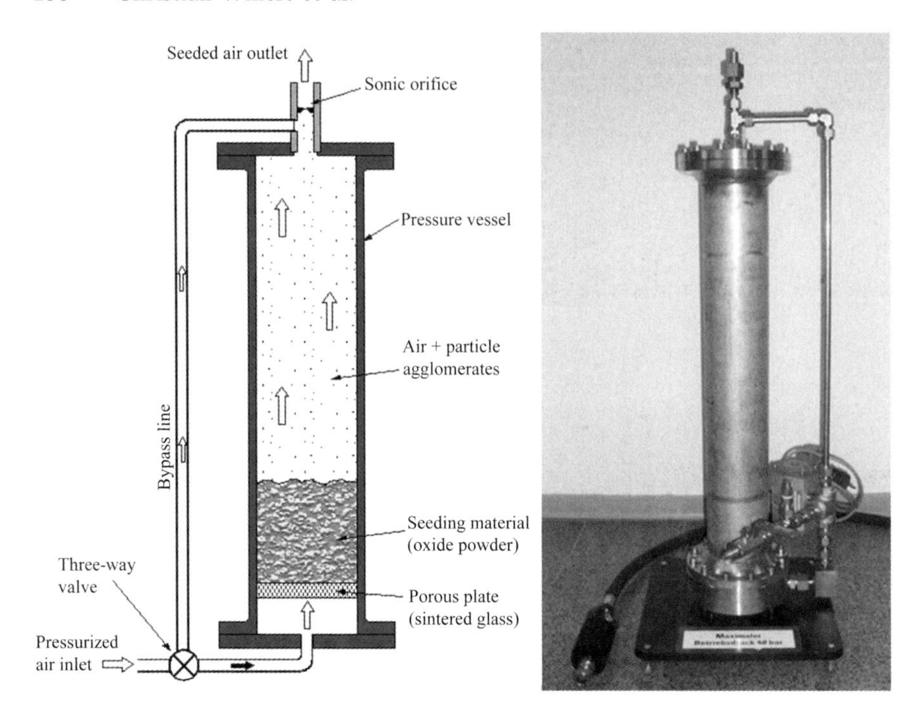

Fig. 4. Fluidized-bed seeding device for high-pressure applications

- The seeding powder should be kept dry, preferably by heating the material to remove excess moisture before filling the seeding device. Dry air or nitrogen should be used to operate the seeder.
- Short supply lines between seeder and facility should be used to prevent the formation of agglomerates. If possible, additional carrier air should be used to reduce the relative seeding concentrations.
- Frequent agitation of the seeding system reduces the chance of channel formation within the fluidized bed.
- The mechanical interaction of the seed material with small brass spheres (100–500 μm) added to the fluidized bed also helps to break up agglomerates. This configuration is referred to as a two-phase fluidized bed.

In the various applications described herein the fluidized-bed seeding generator was operated using mainly alumina particles (Al_2O_3). Operation of the seeder was not always reproducible, which in part could be attributed to moisture in the carrier air stream. Although the manufacturer had specified a mean particle size in the 0.8 μm range, scanning electron microscopy (SEM) of the powder revealed the presence of a significant number of very small particles in the 100–300 nm range (Fig. 5). Particles of this size have very low light scattering (Rayleigh scattering) and are considered to be unsuitable

Fig. 5. Micrographs of alumina powder (Martinswerk, Martoxid MR70) used in the presented PIV applications (Images courtesy of R. Borath, DLR Institute of Materials Research, Köln)

Fig. 6. Micrographs of titanium dioxide (Kemira, UV Titan L830). *Right:* porous silica spheres for PIV seeding (Images courtesy of R. Borath, DLR Institute of Materials Research, Köln)

for PIV. It is believed that only a fraction of the particles are resolved by the PIV camera, while the bulk of the material adds to a mean background intensity and faster window contamination. These observations prompted a search for better seeding materials that offer a narrow size distribution in the 1 µm range. A few examples obtained during this search are shown in the SEM images of Fig. 6.

Titanium dioxide (TiO_2) particles are used rather frequently in velocimetry although their availability is limited to primary sizes in the 50–300 nm range (for color-enhancement demands). In supersonic-flow applications of PIV this material has been used successfully in conjunction with cyclone separators [7]. However, in combustion applications titanium dioxide was observed to drastically reduce in scattering signal within the reaction zone, which most likely can be attributed to a breakup of larger agglomerates into the primary size particles during the rapid thermal expansion in the reaction zone. This could not be observed with the alumina material used in the present applications.

Alternatively, silicon dioxide (SiO_2) particles are available as ground material or as spheres. Although more expensive than polydisperse particles, spherical monodisperse particles hold the biggest promise with regard to optimizing the tradeoff between facility contamination and signal yield. This was confirmed by preliminary atmospheric tests with this material. Other materials yet to be tested in a combustion environment are made of porous silica and have bulk specific weights as low as 0.1, which should produce very good aerodynamic properties (Fig. 6, right).

3 Sample Applications

With the exception of the previously described seeding device, rather standardized, commonly available PIV hardware was used for all applications that are described in the following sections. Illumination was provided by frequency-doubled, double-cavity Nd:YAG lasers (wavelength $\lambda = 532\,\mathrm{nm}$) with an energy of 100–150 mJ per light pulse. An articulated light-guiding arm was generally used to improve overall handling and laser safety. PIV recordings were obtained using thermoelectrically cooled CCD cameras with resolutions of either 1280×1024 or 1600×1200 pixel. A laser line filter placed in front of the aperture of the photographic imaging lens was generally sufficient to suppress the flame luminosity. Table 1 summarizes the applications with regard to experimental aspects.

3.1 PIV in a Pressurized Single-Sector Combustor

Initial efforts of applying PIV in a pressurized combustor at the DLR Institute of Propulsion Technology date back to 2000 and were undertaken on a single-sector facility (SSC), shown in Fig. 7. With a rather small square cross section of $100 \times 100\,\mathrm{mm}^2$ the facility is primarily used for the investigation of new burner designs using optical techniques, especially spectroscopy and velocimetry. Aside from good optical access from at least three sides, one advantage is that the facility can be traversed in all three directions, which greatly simplifies the use of optical diagnostics, even those with high setup complexity. The facility can be operated at up to 20 bar with mass flows up to 1.5 kg/s at 850 K preheating. The pressure and mass flow in the chamber is controlled through a sonic orifice at the exit. Jets in crossflow arrangement at a roughly midlength position of the 250-mm long combustor provide preheated mixing air and confine the primary zone to a roughly cubic volume.

Optical access is granted from three sides through windows consisting of a thick pressure window and a thin liner window. The gap between the windows is purged with cooling air, while the inside of the liner window is film cooled using a portion of the plenum air.

For PIV the lightsheet was aligned with the burner axis with the camera arranged in a typically normal viewing arrangement. By allowing the laser

Table 1. PIV parameters for the described combustion applications

		SSC1	GenRig	SSC2	Alstom
facility type		double annular swirl nozzle			swirl-stabilized premix burner
fuel		kerosene	gas	gas	gas
operating pressure	(bar)	3	atmospheric	2 and 10 bar	atmospheric
imaging setup		2C PIV	2C / 3C PIV	DGV-PIV	2C PIV
field of view	(mm^2)	100×50	85×110	100×60	200×250
recording lens	(mm)	$55, f_{\#}8$	$55, f_{\#}4$ $105, f_{\#}4$	$55, f_{\#}2.8$	$55, f_{\#}4$
camera resolution	(pixel)	1280×1024	1200×1600	1280×1024	1600×1200
acquisition rate	(Hz)	≈ 2	15	≈ 2	15
laser pulse delay	(µs)	3	10	2	7
exposure time (1st frame)	(µs)	≈ 10	10	≈ 10	10
exposure time (2nd frame)	(ms)	≈ 120	≈ 33	≈ 120	≈ 33
acquisition mode		continuous	phase-resolved	continuous	phase-resolved
flame oscillation frequency	Hz	–	≈ 430	–	≈ 100–200
maximum velocity	(m/s)	≈ 70	≈ 50	≈ 120	n.a.
interrogation window size	(pixel)	32×32	48×48	32×32	64×64
spatial resolution	(mm^3)	$1 \times 1 \times 1$	$1 \times 1 \times 1$	$1 \times 1 \times 1$	$2 \times 2 \times 1$
images per sequence		13	188	100 (1000 for selected cases)	188
total acquired image pairs		≈ 500	≈ 29000	≈ 10000	≈ 25000
seeding material		SiO_2, 0.8 µm	Al_2O_3	Al_2O_3	Al_2O_3
reference		[1]	[11]	[10]	–

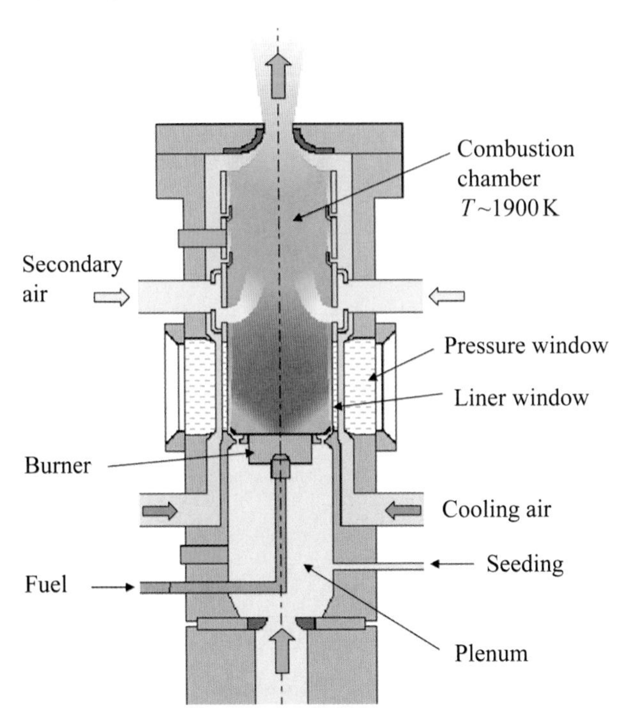

Fig. 7. Longitudinal cross section of the pressurized single-sector combustor (SSC)

lightsheet to pass straight through the combustor the amount of laser flare on imaging windows and walls could be kept at an acceptable level.

Seeding, consisting of amorphous SiO_2 particles, was introduced to the plenum upstream of the burner through a porous annular tube. As the window film cooling is supplied directly with seeded air from the plenum, the windows are unfortunately subjected to an accelerated buildup of seeding deposits. Preferably the film-cooling air should have been separated from the main burner air. Fired with kerosene the combustor provided PIV images of the type shown in Fig. 2, exhibiting strong Mie scattering off the kerosene spray as well as strong flame luminosity. A corresponding PIV result obtained for a slightly leaner operating condition with less kerosene spray is provided in Fig. 8. Image enhancement, as shown in Fig. 3, was applied prior to PIV processing and reduced the influence of droplet velocities on the air flow velocity estimates by equalizing the droplet image intensities with the much weaker intensity of the seeding particles [1].

3.2 Stereoscopic PIV in a Generic Gas Combustor

In order to gain further insights into the aerothermodynamical processes that govern the limit cycle of combustion driven oscillations frequently found in

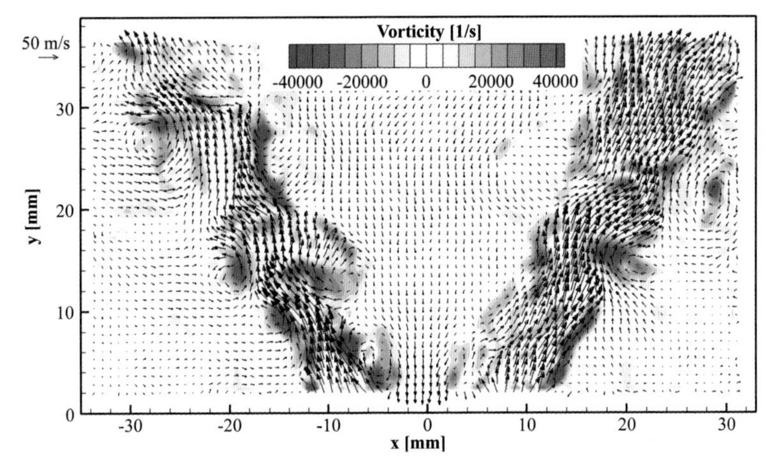

Fig. 8. Single PIV data set for the flow above a coswirled, kerosene air blast fuel nozzle operated at 3 bar obtained from the facility shown in Fig. 7

lean combustion systems a generic combustor was designed. The dynamics of the combustion process were influenced through a variation of preheat temperature, air mass flow/pressure drop over the burner and stoichiometry. The goal of the parameter variation was to find operation points with combustion-driven oscillation at realistic preheat temperatures and pressure drops over the burner. In the course of the project, phase-resolved measurements of the three-component velocity field were to be obtained by stereoscopic PIV (SPIV) measurements at specific points of operation.

The generic gas combustor (Fig. 9, left) consisted of a rectangular combustion chamber $85 \times 85 \times 114\,\mathrm{mm}^3$ in size with the double swirl nozzle centrally located at the bottom. The combustor offered optical access from three sides through thin (2 mm) quartz glass windows. The fourth, metal side wall contained thermocouples, a pressure sensor and an ignition unit. Two highly sensitive sound-pressure sensors were located in one corner pillar of the combustion chamber and provided acoustic signals associated with the combustion-driven oscillations (dominant frequency $\approx 430\,\mathrm{Hz}$). The filtered chamber-pressure signal obtained from these sensors was used to phaselock the PIV image acquisition.

The SPIV setup was optimized during a feasibility study prior to the measurement campaign in order to reduce direct laser-light reflections and scattered light to an acceptable minimum. In the final setup one camera was arranged in a classical (2C) normal viewing position combined with a second camera inclined to a 45° offaxis view, as shown in Fig. 10. One important advantage of this arrangement was that the entire chamber could be imaged in a classical 2C PIV setup that would not have been possible for symmetrically arranged stereo-PIV configurations. Figure 11 shows two PIV results

Fig. 9. Generic gas combustor (*left*) operating under atmospheric conditions (*right*)

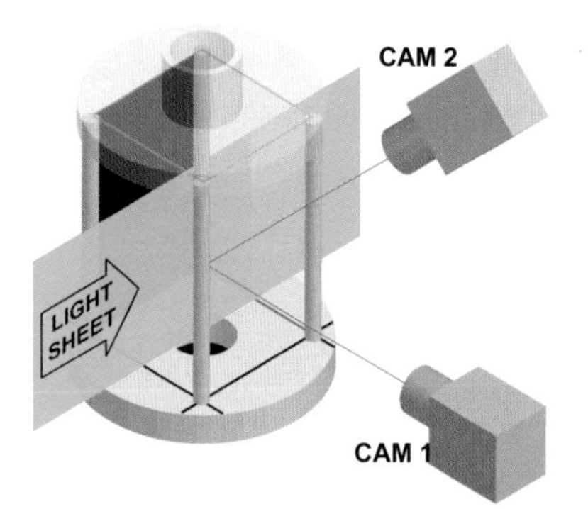

Fig. 10. Imaging configuration for stereoscopic PIV on small-scale generic model combustor

of phase-averaged 2C velocity fields acquired in the symmetry plane of the burner.

For the stereoscopic configuration the combined viewing area was restricted to the lower half of the combustor due to viewing obstruction caused by the lid of the chamber. Therefore, the 3C flow field could only be reconstructed for the lower half of the combustion chamber in the vicinity of the burner exit (see Fig. 12). As the entire PIV setup was installed on a three-axis traversing bench, different coplanar measurement planes could be acquired efficiently without requiring additional recalibration.

The PIV measurements were initially hampered due to strong temperature gradients outside the combustor, which were associated with density fluctuations and a natural convection in the surrounding air: This in turn led to significant optical blurring and displacement of the particle images, especially for the obliquely viewing camera – even though the imaging optics were already optimized for the minimization of these effects (e.g., long focal lengths, reduced apertures). The problem could be solved effectively by forcing cooler ambient air into the imaging path outside of the burner by means of a blower. The combustion, its oscillation in particular, was not affected by this measure.

As the air-fuel ratio was the most sensitive parameter for the adjustment of the combustion-driven oscillation, the additional mass flow injected into the plenum upstream of the burner by the fluidized-bed seeding device had to be taken into consideration. To maintain a stable operating condition regardless of whether seeding was injected or not, the seeding device was operated continuously with a bypass line, allowing a supply of particles only when required (seeding duration $\approx$ 20 s). Nonetheless, the lack of purging air on the windows resulted in unavoidable deposition of particles on the windows, which in turn limited the PIV acquisition to 3 to 5 sequences per burner test run. Here, the higher acquisition rate of the current camera systems (15 Hz) was a clear advantage, resulting in 200 images per sequence within 13 s.

The most challenging part of the stereo-PIV processing was the treatment of the stray light scattered from the chamber base plate, which illuminated the image background of the oblique view camera. Laser flare on the metal surface of the nozzle could not be avoided nor suppressed with laser line filters and/or background image subtraction during postprocessing. This resulted in a loss of signal in these areas. Consequentially, the affected image region had to be masked, as is visible in Fig. 12. This problem is representative of the application of stereo-PIV in confined facilities in general where the necessary oblique viewing arrangement causes additional flare regions on the imaged area either from windows or other surfaces that normally can be avoided in classical 2C imaging arrangements. As will be described in a different application later, the combination of PIV with DGV can partially alleviate this problem by relying on a single (lightsheet-normal) viewing direction only.

3.3 Phase-Resolved Measurements of a Gas-Turbine Combustor

In this application, the heat release and flow field of a gas-turbine burner was studied using phaselocked PIV and OH-chemiluminescence. Here as well, the aim of the experiments was to obtain a better understanding of thermoacoustic interactions in the combustion process. Therefore, the interaction between heat-release fluctuations, acoustic fluctuations and velocity/vorticity fluctuations in the combustion process was investigated experimentally.

The atmospheric test facility consisted of a single swirl-stabilized burner that issued into a combustion chamber of $380 \times 280\,\mathrm{mm}^2$ cross section. The experimental configuration, shown in Fig. 13, employed a standard, two-component PIV imaging arrangement with the laser lightsheet introduced to the facility through a narrow window at the bottom. The camera observed the flow immediately downstream of the nozzle in the axial symmetry plane of the burner. A rather large ($\sim 220 \times 250\,\mathrm{mm}^2$), air-film-cooled, viewing window provided optical access to most of the cross section. Seeding was

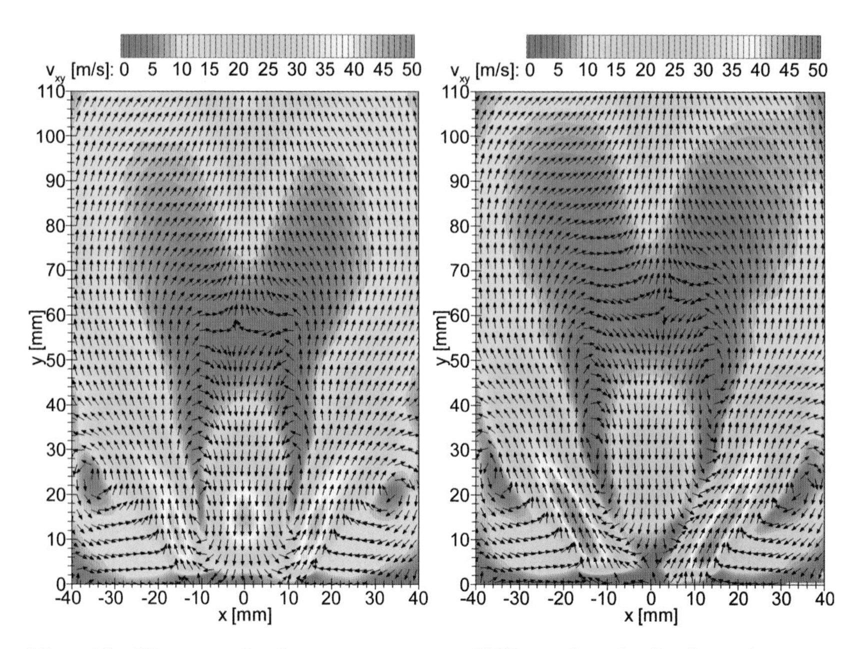

Fig. 11. Phase-resolved, two-component PIV results obtained at the symmetry plane (*color coding* represents inplane velocity magnitude, *vectors* indicate flow direction only)

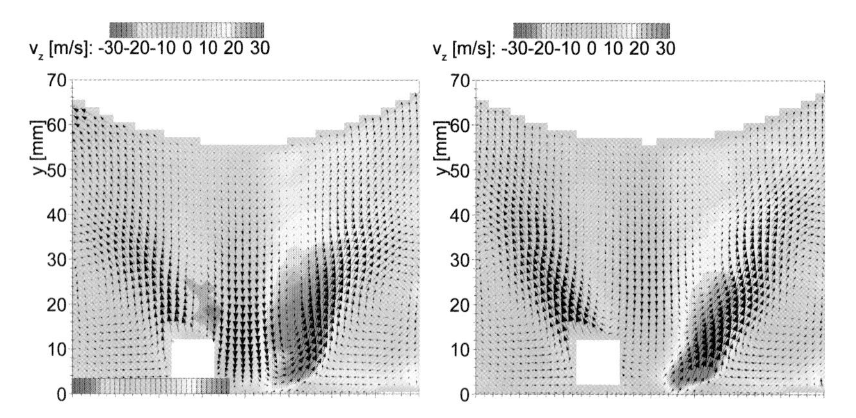

Fig. 12. Phase-resolved, stereoscopic PIV results at two different phase angles (color coding represents out-of-plane velocity)

provided by a pair of fluidized-bed particle seeders (Fig. 4) and was injected in the main flow upstream of the burner using a perforated annular tube. Due to the high mass flow rates in the facility two particle generators were used in parallel and supplied with dry, compressed nitrogen (1 % of total mass flow) to prevent possible clumping of the seeding powder.

As in the previously described application, the PIV camera and laser could be operated at 15 Hz repetition rate acquiring sequences of 190 images each ($\approx$ 12.6 s). Although the seeders were operated in a bypass mode for most of the time except for PIV acquisition, accumulation of seeding on the imaging window limited the data yield to about 10 useful PIV sequences before the system had to be shut down for window cleaning. This problem was compounded by the fact that the lightsheet directly hit the opposite wall, thereby illuminating the entire facility and especially that background area viewed by the camera. Clearly, an additional window for the passage of the lightsheet would have been advantageous here but was avoided in order to maximize the hot liner surface area that is known to have a direct influence on the combustion.

Because of the rather short operational times as well as the unsteadiness of the combustion oscillations (dominant frequency $\approx$ 100–200 Hz) at certain critical points of operation, the idea of acquiring phase-resolved data directly was abandoned in favor of a novel phaselocking method that allowed phase sorting of the PIV (and chemiluminescence) data in a postprocessing step (Fig. 14, 15). This was made possible by simultaneously recording a sound-pressure signal from the combustion chamber alongside with the PIV image acquisition instances on synchronized data tracks. The so-called Hilbert-Huang transform method (HHT) allows a decomposition of the sound pressure signal into instantaneous phase and frequency signals such that each individual PIV recording can be associated with a certain phase angle (and frequency).

The offline phaselocking method makes use of simultaneously recorded time traces of both the pressure signal and the camera (or laser) trigger signals as described in [12]. First, a bandpassed Hilbert transform of the microphone signal is calculated. The phase of this complex-valued time trace represents the instantaneous phase of the acoustic signal within the frequency band of interest. Because the filtering procedure is done offline, zero phase distortion can be achieved for the bandpass filtering operation. As the time instance of acquisition for each PIV image is available from the time trace, each image can then be assigned with an instantaneous phase for one (or more) frequency bands. Phase sorting and subsequent averaging of the instantaneous PIV data is performed across a number of equally spaced phase intervals (typically 8). Finally, the averaged images for each phase interval yield the phaselocked sequence. Because the phaselocking is done offline, the method is not restricted to the calculation of mean values at each phase angle. An additional advantage of this method is that the dynamic behavior

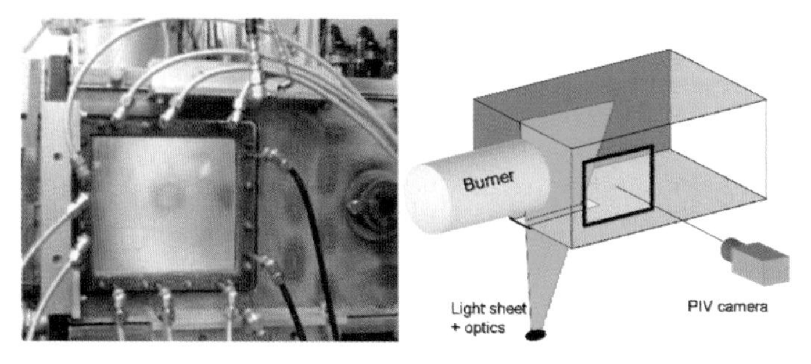

Fig. 13. Imaging window used for PIV measurements immediately downstream of a full-scale gas-turbine combustor

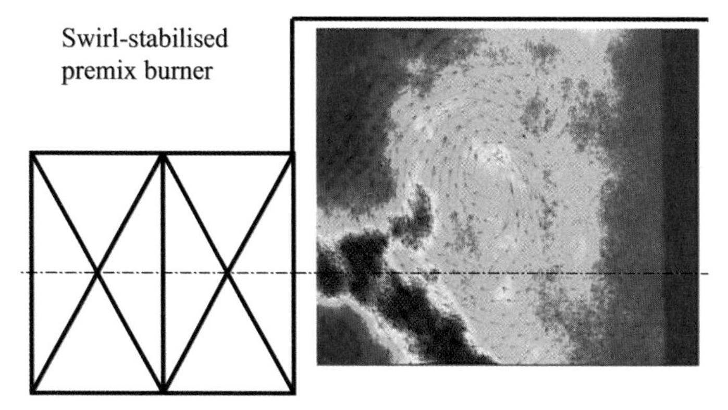

Fig. 14. Single-shot PIV data overlayed on a simultaneously acquired chemiluminescence signal (pseudocolored)

at several frequencies can be obtained from one experiment, which is not possible with standard phase-averaging methods.

In the present case the processed instantaneous PIV datasets were conditionally averaged for phase angle ranges of $\pm 22.5°$. Histograms of the phase distribution showed 15–30 samples per bin. The averaged velocity fields were then decomposed into their solenoid and irrotational parts. The unsteady solenoid part represents the fluctuating vorticity field, the irrotational part represents the acoustic motion. This, together with the chemiluminescence data – which correlates with the heat-release rate – allowed the analysis of thermoacoustic interactions at various operating conditions.

The postprocessed phaselocked results show the velocity field's fluctuation with respect to phase angle for the dominant frequency. Quantities such as the velocity components, the root mean square of the velocity vector, the magnitude of velocity and the curl of the velocity vector field (i.e., out-of-plane vorticity component) can then be derived. The unsteady part of all the aforementioned quantities is also found by subtracting the total mean

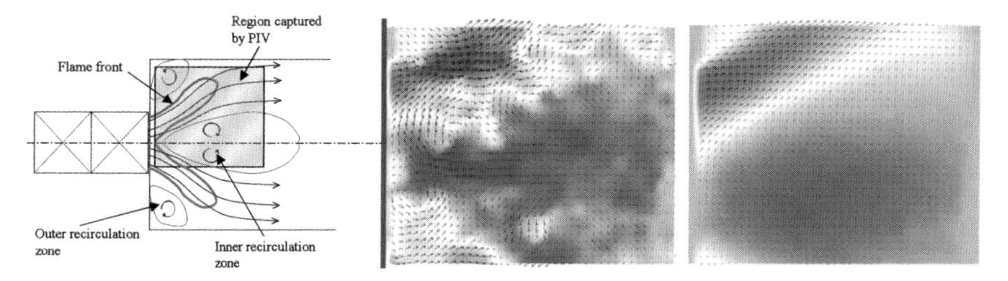

Fig. 15. Instantaneous and ensemble-averaged velocity maps obtained with PIV (*color coding* represents the horizontal velocity component)

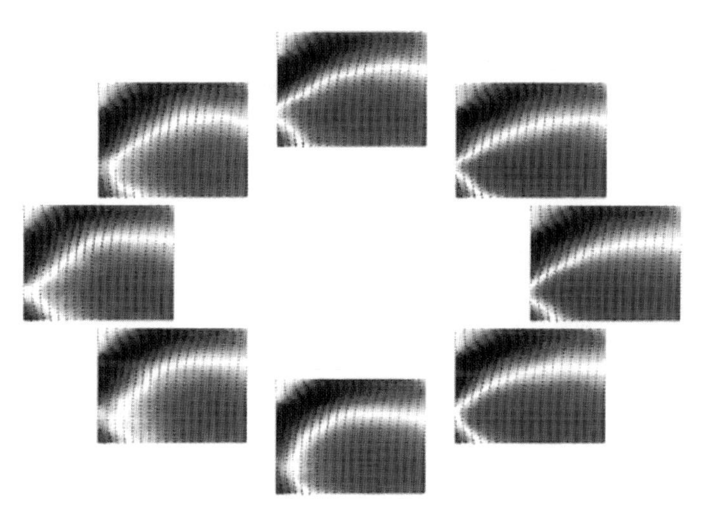

Fig. 16. Phase-resolved mean velocity maps obtained with PIV (*color coding* represents the axial velocity component). Note the movement of the vortex stagnation point

from each phaselocked average. The axial velocity and the unsteady axial velocity are shown, respectively, in Figs. 16 and 17. The arrows represent the velocity fields, the colors indicate the axial component of the velocity. The white-colored region indicates an axial velocity of zero, the blue a negative axial velocity and the red a positive axial velocity.

For a given burner operating condition each PIV data series was decomposed into its rotational and irrotational velocity components prior to performing the phase averaging. The phase-resolved total component of axial velocity is shown in Fig. 16, and the unsteady picture is shown in Fig. 17. The unsteady part is representative of the acoustic motion of the fluid. So, it is not surprising to see a predominantly axial motion in Fig. 16, since the frequency of oscillation was well below the acoustic cuton frequency of the combustion chamber. This is also visualized by a phase-dependent movement

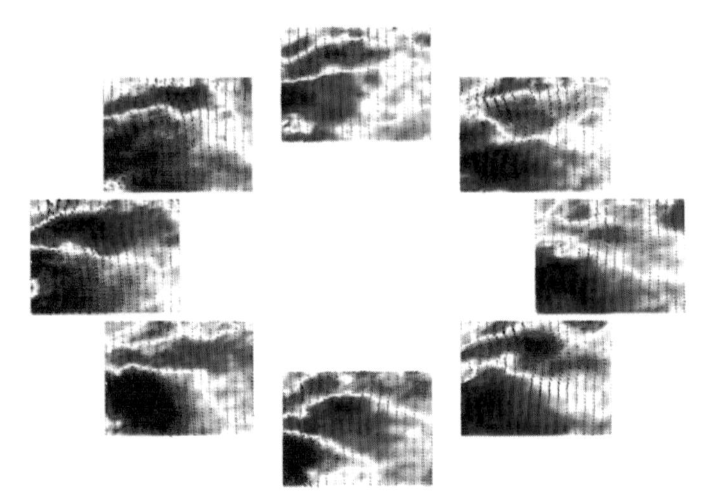

Fig. 17. Unsteady velocity data corresponding to the phase-resolved data shown in Fig. 16. (*color coding* represents the axial, horizontal, velocity component)

of the stagnation point. However, the gradient of the acoustic velocity is much larger than what would be expected based on the acoustic wavelength for this frequency of oscillation. This can be explained by analyzing the chemiluminesence data: it is found that the unsteady heat release is responsible for the gradients of velocity.

Further examples of the application of the Hilbert-Huang transform method to combustion research are described in [12]. In general the HHT method is applicable to essentially any type of randomly sampled experimental data for which a continuous time trace is available [13, 14].

3.4 Combined DGV and PIV in a Pressurized Gas-Turbine Combustor

The following planar velocimetry measurements were performed on the pressurized single-sector combustor facility (SSC) described in Sect. 3.1. The overall aim of the underlying project was to provide a comprehensive data set for validation of RANS and LES codes at isothermal as well as combusting conditions at 2 and 10 bar with 650 K preheat using natural gas as fuel [15]. While part of the investigation was focused on the primary zone, additional data was also to be provided of the secondary – or mixing – zone. This required a new combustor with optical access to the secondary zone as outlined in Figure 18. Due to the presence of mixing-air supply ports, optical access could only be facilitated by two windows facing each other. To, nonetheless, introduce a lightsheet parallel to the imaging windows, the new rig was additionally fitted with two narrow windows next to the burner's exit nozzle. In a first series of measurements the isothermal flow in planes parallel to the wall with the mixing-air jet orifices could be measured using a stereoscopic PIV setup. However, with increased distance from the wall the coincident region as observed by both cameras was limited due to optical obstruction. In order to obtain a nearly complete picture of the flow field in the mixing zone, a combination of PIV with DGV was chosen.

Following Fig. 19 it is possible to reconstruct all three velocity components from a single viewing direction and single lightsheet: While PIV essentially provides the inplane velocity components, DGV measures the component projected on the bisector between observation and lightsheet vector. A straightforward vector transformation reconstructs the Cartesian velocity components from the three measured velocity components. The combined technique is very attractive as it considerably reduces the complexity regarding optical access (one view and one lightsheet). This concept was first successfully demonstrated by *Wernet* on a free jet using pulsed DGV together with PIV [15].

In the present application inside the combustor, a combination of DGV and PIV was chosen that is slightly different from the one presented by *Wernet* [16] since a suitable, frequency-stabilized laser for pulsed DGV was not available. Here, it was necessary to apply PIV and DGV in succession.

The current limitation of the combined PIV-DGV technique presented here is that the DGV laser system and camera was designed for acquisition of time- or phase-averaged velocity data [6,17]. While the PIV measurements inherently provide unsteady velocity data, DGV images are obtained through on-camera integration of many pulses (at least 1000) of scattered light from a laser with rather low pulse energy ($\approx 2\,\mathrm{mJ}$, [6]). This means that information regarding temporal fluctuations is lost during acquisition of the DGV images. Nevertheless the corresponding PIV data sets can at least provide two components of the fluctuating data but should also be treated with caution since 100 images acquired for each plane as in the present case are insufficient to properly estimate the fluctuating components.

The utilized PIV hardware and seeding approach was equivalent to that of the previously described applications. The lightsheet was introduced through one of the narrow windows of the combustor's exit flange and aligned orthogonally to the viewing direction of the PIV camera. The unique capability of traversing the combustor rig with respect to a stationary optical diagnostic setup allowed the efficient acquisition of the multiple coplanar datasets by PIV and DGV, thereby mapping a complete volume of the flow field with three velocity components.

Data acquisition took place in two phases, first acquiring a volumetric PIV dataset, followed by exchange of hardware and subsequent acquisition of the corresponding DGV frequency shift images under equivalent operation conditions and corresponding spatial positions. In total, 41 coplanar slices of combined DGV-PIV data were acquired, resulting in a volumetric dataset of time-averaged, three-component velocity measurements as shown in Fig. 20. The spatial resolution of the velocity data is roughly $1 \times 1 \times 2\,\mathrm{mm}^3$. The

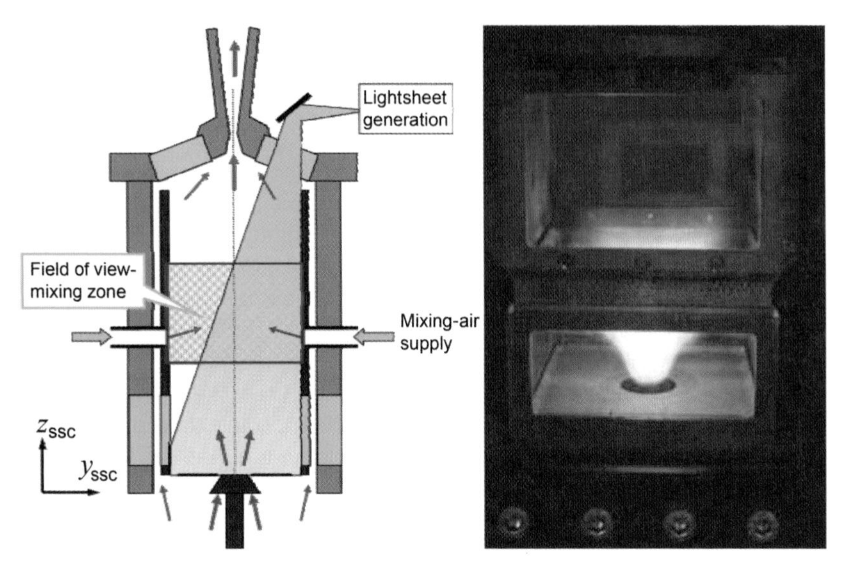

Fig. 18. Imaging arrangement used for mapping of the flow in the mixing zone by means of combined DGV and PIV. *Right:* photograph of the flame at a pressure of 4 bar and air preheating temperature of 700 K

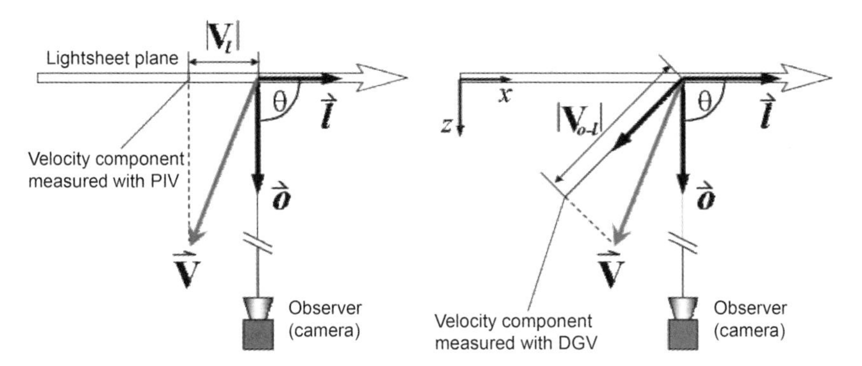

Fig. 19. Combination of PIV with DGV can be used to recover the out-of-plane velocity component from a single viewing direction (from [10])

influence of the mixing jets up until the middle of the combustor is clearly visible in Fig. 20b. The mixing jets also limit the recirculation region, as indicated by the abrupt change of the vertical velocity component displayed in Fig. 20c. Further details about this application and its results can be found in [10] as well as [15].

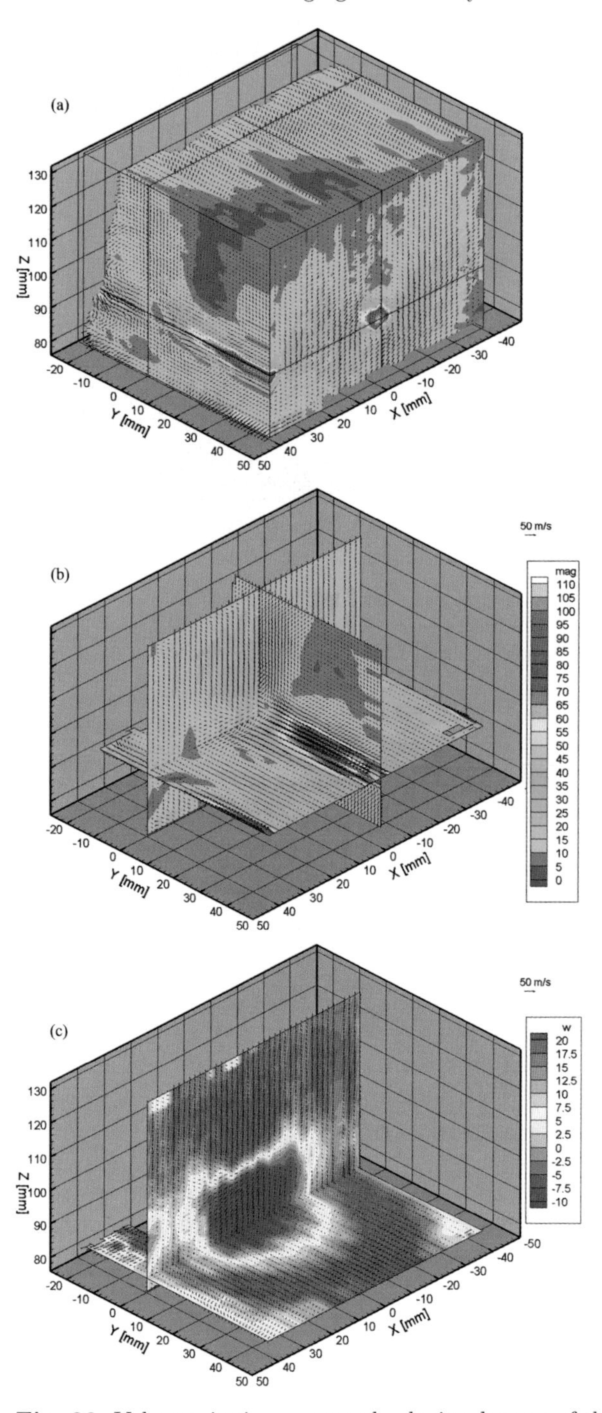

Fig. 20. Volumetric time-averaged velocity dataset of the mixing zone obtained at 2 bar (from [10])

4 Conclusions

As exemplified by the four previously described applications, the development and qualification of planar velocimetry (PIV and DGV) for use in combustion test facilities has undergone significant progress in the past 6 years, taking profit not only from technological advances in PIV as a whole, but also through a better understanding of the problems specific to this particular range of applications. In particular, seeding techniques, relying on the dispersion of powdered, noncombusting solid particles, have been improved sufficiently to allow seeding of significant mass flows ($\geq$ 1 kg/s) at elevated pressures (up to 20 bar). Further investigations in this area are still in progress to improve the continuity of the particle dispersion and to develop strategies that limit the undesired accumulation of seeding deposits in the facility, especially on the windows. Here, a close collaboration between the facility designers and optical metrology personnel is mandatory during the design phase of new test facilities to allow, for instance, the introduction of seeding only to relevant areas in the facility that leave window-cooling films unseeded.

Nonetheless, a gradual accumulation of seeding deposits in the facilities can not be avoided entirely. These deposits increase light scattering from normally matte surfaces (e.g., combustor liner walls) and may result in a loss of signal for PIV due to saturation of the sensor. This stray light may also constitute a significant error source in intensity-based imaging techniques such as DGV since it cannot be separated from the desired signal by *a-priori* measurements in the absence of seeding. One recently proposed solution to this problem essentially relies on estimating the background luminosity from additional DGV images of a striped (spatially intensity modulated) light-sheet [18]. With this correction method it was then possible to rectify a velocity bias exceeding 30 m/s of a flow field measured inside a large-scale combustor facility.

The described problems associated with increased Mie scattering off large fuel droplets may be less of an issue in future measurement campaigns as new burner designs increasingly utilize premixing and pre-evaporation of liquid fuels for improved combustion. Nonetheless, approaches based on fluorescence imaging can allow a nearly complete separation of the two-phase flow in gas velocity and droplet velocity fields. One such approach has been recently suggested by *Kosiwczuk* et al. [2,3] and relies on the different fluorescence signals of the fuel and particles that are captured by two cameras with different color sensitivity. The applicability of the technique to pressurized facilities depends on the stability of the fluorescence molecules at elevated temperatures and pressures.

While new approaches are being developed to improve the applicability of both DGV and PIV to combustion flows, the rapid technological development of lasers, camera hardware and imaging tools is of immediate benefit for the application of these techniques. Faster cameras and lasers enable higher image acquisition rates and are less influenced by flame luminosity due to the reduced digitalization time of the images. Increased frame rates (can) reduce facility operating costs and may eventually make time-resolved measurements of the velocity field possible, thus allowing unsteady phenomena to be captured, for example, flame-out and reignition processes during lean-combustor operation.

In parallel with the continuous hardware advance, there is also an immediate need to improve data processing, data handling and subsequent postprocessing techniques and archival. The vast amount of PIV image data acquired in rather short times, often with marginal signal-to-noise ratios, has to be processed efficiently and preferably without too much user intervention. While the recent literature reports considerable advances in improving the precision of PIV image processing, there is still significant work to be done toward faster, self-optimizing PIV algorithms capable of delivering data with reliable quality estimates. Of course this requirement by no means is restricted to the previously described field of application, but is of significant importance for widespread future utilization of PIV in industry and research alike.

Acknowledgements

Portions of the presented material were produced in the framework of the following European Community research projects:

MOLECULES "Modelling of Low-Emission Combustors Using Large Eddy Simulations" funded under the "Competitive and Sustainable Growth" Programme, Contract N: G4RD/CT 2000/00 402.

FUELCHIEF "Demonstration of a Low NOx Fuel-staged Combustor in a High Efficiency Gas Turbine Action F: Gas Power Generation", funded under the "Energy, Environment and Sustainable Development" Programme, Contract N: NNE5/2001/382.

PRECCINSTA "Prediction and Control of Combustion Instabilities in Turbular und Annular GT Combustion Systems" funded under the "Energy, Environment and Sustainable Development" Programme, Contract N: ENK5/CT/2000/00 060.

The support through each of these projects is gratefully acknowledged.

Also, we would like to acknowledge the efforts by R. Borath of DLR's Institute of Materials Research in providing the micrographs of the seeding materials.

References

[1] C. Willert, M. Jarius: Planar flow field measurements in atmospheric and pressurized combustion chambers, Exp. Fluids **33**, 931–939 (2002)

[2] W. Kosiwczuk, A. Cessou, M. Trinité, B. Lecordier: Simultaneous velocity field measurements in two-phase flows for turbulent mixing of sprays by means of two-phase PIV, Exp. Fluids **39**, 895–908 (2005)

[3] W. Kosiwczuk, A. Cessou, M. Trinité, B. Lecordier: Simultaneous measurements of gas and droplet velocity fields for turbulent mixing of GDI sprays by mean of two-phase PIV, in *13th Int. Symp. on Applic Laser Techniques to Fluid Mechanics* (2006)

[4] C. Willert, B. Stasicki, M. Raffel, J. Kompenhans: A digital video camera for application of particle image velocimetry in high-speed flows, in *5th Int. Symposium on Optical Diagnostics in Fluid and Thermal Flows*, Proc. SPIE **2546** (1995) pp. 124–134

[5] D. Honore, S. Maurel, A. Quinqueneau: Particle image velocimetry in a semi-industrial 1 MW boiler, in *4th Int. Symp. on Particle Image Velocimetry* (2001)

[6] M. Fischer, G. Stockhausen, J. Heinze, M. Seifert, M. Müller, R. Schodl: Development of Doppler global velocimetry (DGV) measurement devices and combined application of DGV and OH*-chemiluminescence imaging to gas turbine combustor, in *12th Int. Symp. Applic Laser Techniques to Fluid Mechanics* (2004)

[7] F. F. J. Schrijer, F. Scarano, B. W. van Oudheusden: Application of PIV in a Mach 7 double-ramp flow, Exp. Fluids **41**, 353–363 (2006)

[8] W. D. Urban, M. G. Mungal: Planar velocity measurements in compressible mixing layers, in *35th Aerospace Sciences Meeting*, AIAA Paper 97-0757 (1997)

[9] J. H. Wernet, M. P. Wernet: Stabilized alumina/ethanol colloidal dispersion for seeding high temperature air flows, in *ASME Symp. on Laser Anemometry: Advances and Applications* (19–23 June 1994)

[10] C. Willert, C. Hassa, G. Stockhausen, M. Jarius, M. Voges, J. Klinner: Combined PIV and DGV applied to a pressurized gas turbine combustion facility, Meas. Sci. Technol. **17**, 1670–1679 (2006)

[11] O. Diers, D. Schneider, M. Voges, P. Weigand, C. Hassa: Investigation of combustion oscillations in a lean gas turbine model combustor, in *ASME Turbo Expo 2007 - Power for Land, Sea and Air,* (2007) pp. GT2007–27360

[12] F. Güthe, B. Schuermans: Phase-locking in post-processing for pulsating flames, Meas. Sci. Technol. **18**, 3036–3042 (2007)

[13] N. E. Huang, Z. Shen, S. R. Long, M. C. Wu, H. H. Shih, Q. Zheng, N.-C. Yen, C. C. Tung, H. H. Liu: The empirical mode decomposition and the Hilbert spectrum for nonlinear and non-stationary time series analysis, Proc. Royal Society A **454**, 903–995 (1998)

[14] N. E. Huang, M.-L. C. Wu, S. R. Long, S. S. P. Shen, W. Qu, P. Gloersen, K. L. Fan: A confidence limit for the empirical mode decomposition and Hilbert spectral analysis, Proc. Royal Society A **459**, 2317–2345 (2003)

[15] C. Hassa, C. Willert, M. Fischer, G. Stockhausen, I. Roehle, W. Meier, L. Wehr, P. Kutne: Nonintrusive flowfield, temperature and species measurements on a generic aeroengine combustor at elevated pressures, in *ASME Turbo Expo – Power for Land, Sea, and Air* (2006) pp. GT2006–90213

[16] M. P. Wernet: Planar particle imaging Doppler velocimetry: A hybrid PIV/DGV technique for three-component velocity measurements, Meas. Sci. Technol. **15**, 2011–2028 (2004)
[17] R. Schodl, I. Röhle, C. Willert, M. Fischer, J. Heinze, C. Laible, T. Schilling: Doppler global velocimetry for the analysis of combustor flow, Aerosp. Sci. Technol. **6**, 481–493 (2002)
[18] R. Schodl, G. Stockhausen, C. Willert, J. Klinner: Komplementär-Streifen-Verfahren für die Doppler Global Velocimetry (DGV) zur Korrektur des Einflusses von Hintergrundstrahlung, in *14. Fachtagung d. GALA (German Association for Laser Anemometry)* (2006)

Index

Recent Applications of Particle Image Velocimetry to Flow Research in Thermal Turbomachinery

Jakob Woisetschläger and Emil Göttlich

Institute for Thermal Turbomachinery and Machine Dynamics, Graz University of Technology, Inffeldgasse 25, A – 8010 Graz, Austria
jakob.woisetschlaeger@tugraz.at

Abstract. During the past decade particle image velocimetry (PIV) has become a versatile tool in the investigation of flow fields in turbomachinery. In this overview a short summary on recent applications of PIV in these machines is given, with a focus on rotating turbine and compressor test rigs and the developments within the PivNet network funded by the European Union. Several topics discussed during the PivNet workshops are addressed. To summarize the capabilities of PIV in thermal turbomachinery, the application of PIV to flow investigations in two test rigs is presented. The first one is a transonic turbine operating at 10 600 rpm with 24 stator and 36 rotor blades at Graz University of Technology, Austria, and the second is a centrifugal compressor with a vaned diffusor and an impeller with 13 main and 13 splitter blades rotating at speeds up to 50 000 rpm at the German Aerospace Center DLR, Cologne, Germany. At both facilities, workshops were organized during the PivNet program.

1 Introduction

Subsonic and transonic regions with a high level of turbulence and a significant level of unsteadiness characterize the flow through modern thermal turbomachinery. The unsteadiness in the flow is related to the relative motion between rotor and stator, with the rotor blade passing frequencies up to 20 kHz and above. In multistage axial turbomachinery the unsteady mixing of wakes from stator and rotor blades during rotor–stator motion results in a complex three-dimensional (3D) flow field, especially when shock systems are present, being reflected by passing blades.

The trend for turbines is towards higher efficiency at constant and possibly decreasing costs per kW shaft power. Advanced 3D aerodynamic design and higher cycle temperatures and pressures tackle the objective of higher efficiency. In order to optimize costs and weight it is advantageous to reduce the number of stages, thus resulting in high-pressure (HP) ratios and transonic conditions for the remaining stages.

The modern compressor has to be compact, of high efficiency and often has to meet operation at variable speed with variable geometry. Surge and stall conditions are of special interest in these machines, with very complex internal flow structures often at extremely high rotational speeds.

A. Schroeder, C. E. Willert (Eds.): Particle Image Velocimetry,
Topics Appl. Physics **112**, 311–331 (2008)

To tackle these objectives a number of experiments have been performed worldwide applying various kinds of pressure and temperature sensors (e.g., [1]) as well as pointwise laser-optical velocimetry (e.g., [2]). On the other hand, the year 2004 marked the 20th anniversary of "particle image velocimetry" [3], with a number of early applications of PIV to flows related to turbomachines or turbomachinery components (e.g., [4–6]). The breakthrough for PIV in turbomachinery applications came with the development of digital PIV (DPIV, [7]) and stereoscopic PIV (SPIV, [8]), with commercial systems and a detailed discussion of the basics of this technique soon available [9]. Fast recording of three-component velocity providing ensemble-averaged as well as instantaneous data is an advantage of PIV, especially in turbomachinery research with its highly unsteady flows and test rigs that are expensive in operation. Thus, PIV offers a major advance for the experimenter, but is not without disadvantages. Due to the high flow velocities very small tracer particles have to be used and imaged well focused through windows of special design. To overcome these disadvantages a number of planar multiple-component techniques are proposed (e.g., [10–12]). The discussion and improvement of PIV as well as its comparison to other measurement techniques used in turbomachinery were objectives to the PivNet thematic network funded by the European Union.

2 Recent Flow Research in Thermal Turbomachinery

Most recently, considerable research has been done applying DPIV and SPIV to the investigation of turbomachinery components, especially in a nonrotating environment focusing on the one hand on cooling-flow investigations in models and turbine and compressor blade cascade flows on the other.

Chanteloup and *Bölcs* [13] and *Chanteloup* et al. [14] combined SPIV with heat-transfer measurements and applied this technique to the internal coolant passage of a turbine blade including film-cooling ejection modeled in acrylic glass. Using a similar experimental approach *Casarsa* and *Arts* [15] studied the influence of a high blockage rib-roughened cooling channel, *Elfert* et al. [16] investigated the flow through a transparent model of the leading-edge duct, *Servouze* and *Sturgis* [17] used DPIV in a rotating model of a two-pass duct and *Uzol* and *Camci* [18] investigated the wake flow field behind pin fins using DPIV. More recently, *Panigrahi* et al. [19] did a more basic study on the heat transfer behind ribs using SPIV. Films ejected from a model settling chamber without and with pulsating cooling flows were studied by *Bernsdorf* et al. [20] and [21] using SPIV, while *Yoon* and *Martinez-Botas* [22] focused on cooling flows ejected into the blade tip clearance using DPIV in a model.

The list of researchers using PIV for flow investigations in turbine blade cascades is much longer and only recent work can be cited within the scope of this chapter. Recently, *Palafox* et al. [23] performed flow measurements using DPIV in the tip gap of a low-speed cascade with moving endwall. In this research the camera observed the light-sheet through a transparent blade tip. *Rehder* and *Dannhauer* [24] used oil-film visualization, heat transfer measurements and velocity fields recorded by DPIV to investigate the turbine leakage flow and the flow in the wall and tip region. *Raffel* and *Kost* [25] discussed the interaction between cooling air ejected from the trailing edge and the shock system forming behind the blade. The authors gave detailed information on imaging particles through a strong density gradient. *Vicharelli* and *Eaton* [26] compared a detailed DPIV study of turbulence in a transonic turbine cascade to numerical results. DPIV and laser vibrometer were combined by *Woisetschläger* et al. [27] to discuss the influence of boundary layer state on vortex shedding from turbine blades. For ensemble averaging the PIV images were sorted by vortex shedding phase obtained from the vorticity calculated from the instantaneous recordings. Instantaneous DPIV recordings were also used by *Stieger* et al. [28] to quantitatively visualize the wake-induced transition in a turbine blade cascade. *Langford* et al. [29] used a triggered DPIV system to study the effect of moving shocks on a compressor stator flow field. *Estevadeordal* et al. [30] and *Zheng* et al. [31] focused their research on boundary-layer-based flow-control systems in compressor cascades supported by DPIV data. Due to its strength to record instantaneous as well as ensemble-averaged data, DPIV is often used in experiments controlling boundary layer separation, recently e.g., [32] and [33].

During the last decade PIV was developed and used for rotating flows at subsonic or transonic speeds in several unique testing facilities for turbomachines worldwide. Since these testing facilities are of special interest to the turbomachinery community a more detailed summary of worldwide activities is given in Table 1, focusing on research in axial and centrifugal compressors as well as on turbines done within the last decade.

At the von Karman Institute PIV and DPIV were used to measure the flow field in a compressor rotor demonstrating the use of periscope-type light-sheet probes and recording velocities in different radial heights [34,35]. In an international cooperation [36–38] a detailed DPIV study was presented in high-pressure single-stage transonic turbines in blowdown facilities located at the Massachusetts Institute of Technology, MA, USA and Defense Evaluation and Research Agency, Pyestock, UK. *Chana* et al. [37] inserted lightsheet probes into the nozzle guide vanes fitted with optical windows and *Ceyhan* et al. [36] used two probes and transparent nozzle guide vanes to illuminate the whole inter-blade passage. Finally, *Bryanston-Cross* et al. [39] compared the application of DPIV in this type of machines with other optical flow diagnostics tools.

At Purdue University, *Day* et al. [40] successfully applied DPIV to a low-speed two-stage turbine with and without film cooling in five span-wise locations. Later, *Treml* and *Lawless* [41] investigated the stator–rotor interaction in the same test rig. *Gallier* et al. [42] used DPIV to record the influence of the seal air flow on the secondary flows in a low-speed two-stage turbine. *Sanders* et al. [43, 44] focused on the blade-row interaction in a transonic axial compressor and discussed shock reflections at 20 000 rpm using DPIV results, with *Papalia* et al. [45] extending this research to offdesign conditions. To address offdesign unsteady aerodynamics including dynamic stall *Key* et al. [46] used DPIV to record data in an annular cascade with a motor-driven axial-flow rotor.

At NASA Glenn Research Center, demonstrations of DPIV in a single-stage transonic axial compressor and in a high-speed centrifugal compressor were performed [47,48]. The results are summarized and discussed in [49–51]. Recently, *Wernet* et al. [52] applied SPIV to record the tip clearance flow in a low-speed four-stage axial compressor.

Gogineni et al. [53] and *Estevadeordal* et al. [54] developed a two-color DPIV system using a 3 k × 3 k sensor to record the flow field in a low-speed axial fan at Wright–Patterson Air Force Base. Focusing on the interaction between vortex shedding from wake generators (stators) and bypassing blades in a transonic axial compressor, *Estevadeordal* et al. [55] observed phaselocking of vortex shedding to the bypassing rotor blades due to the strong pressure fluctuations caused in the flow field. A detailed discussion of the results can be found in [30]. This phaselocking allowed the studies of *Copenhaver* et al. [56] at near-stall conditions.

Locking of vortex shedding to the rotor-blade movement was first predicted by *Sondak* and *Dorney* [57] for transonic turbine stages. At Graz University of Technology this effect was found in a transonic single-stage turbine by SPIV [58]. In a more detailed analysis SPIV and a direct recording of density fluctuations by laser vibrometers were combined and *Woisetschläger* et al. [59] showed that shock reflections from the bypassing rotor blades enforce vortex shedding in these transonic machines. Recently, *Göttlich* et al. [60] presented a detailed study of vortex shedding and wake–wake interaction in this turbine stage under fully transonic conditions.

At Kyushu University, *Hayami* et al. [61,62] investigated shock waves and their strong fluctuations in a transonic centrifugal compressor by DPIV in cooperation with industry. The recordings were done in the inducer of the impeller and the low-solidity diffusor.

At the University of Karlsruhe, data were recorded from the preswirl cavity located between stator and rotor in the cooling-flow supply to a turbine rotor [63,64]. For these experiments a SPIV system and an endoscopic DPIV were used. Recently, *Childs* et al. [65] gave a conclusion for all internal air system test rigs used in the ICAS-GT2 European research program.

At Beijing University SPIV was developed to record the unsteady flow field in the tip region of a single-stage large-scale axial compressor. In this research the two cameras were mounted on each side of the lightsheet, which illuminated the flow through the observation window in the tip-to-hub direction. A detailed discussion of the optical setup is given in [71], a discussion on the flow phenomena can be found in [72] and [73]. Most recently, a DPIV investigation of the rotor–stator interaction was presented in [74].

Table 1. Recent flow research in thermal turbomachinery using PIV in rotating test rigs (worldwide activities in chronological order)

Authors	Institution	Experiments	PIV technique
Voges et al. [66]	German Aerospace Center DLR, Cologne, Germany	transonic centrifugal compressor, 35 000–50 000 rpm	DPIV SPIV
Ibaraki et al. [67]	Mitsubishi Heavy Industries MHI, Japan	transonic centrifugal compressor, 28 700 rpm	DPIV
Porreca et al. [68] *Yun* et al. [69]	ETH Zurich, Switzerland	turbine, 2625 rpm	SPIV
Wheeler et al. [70]	Whittle Laboratory, Cambridge Univ., UK	large-scale compressor, 500 rpm	endoscopic DPIV
Liu et al. [71–73] *Gong* et al. [74]	Beijing University of Aeronautics and Astronautics, China	axial compressor, 1200 rpm; axial compressor, 3000 rpm	SPIV DPIV
Childs et al. [65] *Bricaud* et al. [64] *Geis* et al. [63]	Univ. Karlsruhe, Germany	preswirl air flow, up to 7000 rpm,	SPIV, endoscopic DPIV
Hayami et al. [61, 62]	Kyushu Univ., Japan	transonic centrifugal compressor, up to 18 450 rpm	DPIV
Göttlich et al. [60] *Woisetschläger* et al. [59] *Lang* et al. [58]	Graz Univ. of Technology, Austria	transonic turbine, 9600–10 600 rpm	SPIV
Estevadeordal et al. [30, 54, 55] *Copenhaver* et al. [56] *Gogineni* et al. [53]	Wright-Patterson AFB, OH, USA	transonic axial compressor, 14 000 rpm; axial fan, 2200 rpm	DPIV
Wernet et al. [52] *Wernet* [47–51]	NASA Glenn Research Center, OH, USA	centrifugal compressor, $\approx$ 22 000 rpm; axial compressor, $\approx$ 17 000 rpm; 4-stage axial compressor, 980 rpm	DPIV, SPIV

Authors	Institution	Experiments	PIV technique
Papalia et al. [45] *Gallier* et al. [42] *Key* et al. [46] *Sander* et al. [43, 44] *Treml*, *Lawless* [41] *Day* et al. [40]	Purdue Univ., IN, USA	transonic axial compressor, 20 000 rpm; turbine, 2500 rpm	DPIV
Bryanston-Cross et al. [38, 39] *Ceyhan* et al. [36] *Chana* et al. [37]	Warwick Univ., UK	transonic turbine, 8200 rpm; transonic turbine, 6000–7800 rpm	DPIV
Balzani et al. [35] *Tisserant*, *Breugelmans* [34]	von Karman Inst., Belgium	axial compressor, 3000–6000 rpm	DPIV, PIV

At the Whittle Laboratory, Cambridge University, *Wheeler* et al. [70] presented the first results on the interaction between wakes and boundary layers in a large-scale axial compressor obtained by DPIV and hot-wire probe measurements.

At the Swiss Federal Institute of Technology ETH Zurich, SPIV in combination with fast response aerodynamical probes was applied in [69] and [68] to investigate leakage flows across shrouded turbine blades and their influence on the flow field downstream of the rotors in a two-stage axial turbine.

A detailed study in the vaned diffusor of a high-speed transonic centrifugal compressor was published in [67] at the Nagasaki R&D Center, Mitsubishi Heavy Industries. With the help of the DPIV and numerical results the authors discussed the unsteady flow field between shroud and hub.

Most recently, *Voges* et al. [66] presented results from the diffusor section in a transonic centrifugal compressor rotating at 50 000 rpm at the German Aerospace Center DLR. In this publication results by SPIV are also shown.

Although not in compressible medium, the work of *Uzol* et al. [75] at Johns Hopkins University has to be mentioned. This refractive-index-matched facility combines rotor and stator blades made of acrylic glass with a working fluid of the same index of refraction, so that an unobstructed view is possible for DPIV investigations. Results especially interesting for turbomachinery research are discussed in [76–79].

3 Optical Configuration

3.1 General Configuration of the PIV System for Use in Turbomachinery

During the PivNet workshops organized at the German Aerospace Center DLR, Cologne, Germany and Graz University of Technology, Austria, live

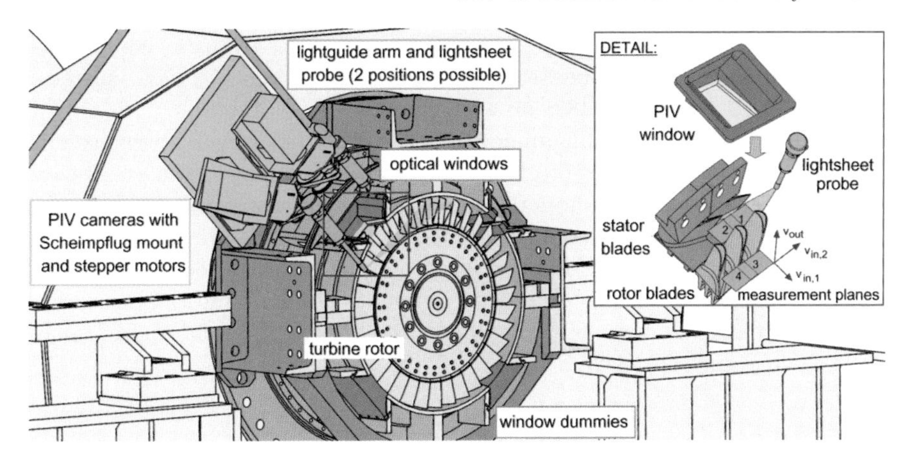

Fig. 1. Experimental setup for SPIV at the transonic test turbine facility (TTTF) at Graz University of Technology (one-stage configuration). The exhaust part was cut off for better visibility. The *detailed view* shows the two lightsheets used and the inplane and out-of-plane components recorded by SPIV, v_{in} and v_{out}. The optical window is a plane-concave quartz glass for the PIV (*upper left window*) and a plane quartz glass for laser Doppler velocimetry (*upper right window*)

demonstrations of the application of PIV to turbomachinery flows were performed. In Fig. 1, a cut through the transonic test turbine facility (TTTF) at Graz University of Technology is shown, presenting the setup commonly used in these types of machines consisting of a periscope-type lightsheet probe, a sufficiently large optical window and a platform for the SPIV camera system. A second window grants optical access for a laser Doppler velocimeter or a laser vibrometer. In this facility rotating the nozzle guide vane ring rather than traversing the probes adjusts the circumferential position for all measurement systems. Thus, the SPIV arrangement is fixed in the laboratory system.

3.2 Stereoscopic PIV

In rotating axial machinery, velocity data are usually provided in terms of axial, circumferential and radial velocity, so the yaw and pitch angles of the flow can be easily obtained. Only SPIV allows the calculation of these components from the in- and out-of-plane velocities recorded within the light-sheet plane. On the other hand, optical access into high-speed turbomachines is limited due to structural reasons. Depending on the maximum angle between the two cameras the sensitivity of the out-of-plane component is usually less than that of the inplane components. This sensitivity can be described in terms of sensitivity vectors (e.g., [80]) or as a first-order approximation by the tangent of the half-angle between the two camera axes (e.g., [12]). Since in most turbomachinery applications curved windows are used, careful calibration of the system is needed using calibration targets in the lightsheet position (e.g., [81]) with [82] or without [83] image dewarping. Based on the findings of [82] a misalignment of the calibration target might lead to an artificially increased out-of-plane component, especially in the presence of strong velocity gradients (shocks). *Lang* et al. [58] discussed this effect for the transonic flow through a turbine. *Wieneke* [84] presented a self-calibration correction scheme for SPIV application. To overcome this problem of limited angle of view, some authors propose a combination of PIV and Doppler global velocimetry (DGV), e.g., [85] and [86], or to combine PIV and digital image plane holography DIPH [87].

3.3 Seeding

A uniform seeding of a sufficiently high concentration is essential for PIV. Depending on the mass flow seeding is provided globally or locally by an injection tube or multiple jets. Since the tracer particles act as a lowpass filter to the turbomachinery flow velocity data, extremely small particles must be used when transonic flows or high-frequency flow phenomena are to be observed [27, 88]. Various types of nozzles and atomizers, often in combination with cyclone separators, are commercially available and often also used in high-speed flows [89]. In the test rig shown in Fig. 1 a PALAS AGF 5.0D aerosol generator (PALAS GmbH, Karlsruhe) injected DEHS oil droplets 500 mm upstream of the turbine stage. The seeding pipe with an inner diameter of 7 mm was mounted in the rotateable part of the casing and its end extended perpendicular to the flow direction in the tangential direction to the annular channel. To spread the seeding uniformly this end of the pipe was equipped with 7 rows of drilled holes distributed uniformly over the diameter along a length of 100 mm (hole diameter 1.5–1.8 mm, 140 holes, closed pipe end).

While diffraction effects associated with the chosen aperture dictate the particle image size, it has to be kept in mind that small particles scatter less light. At the limit of detectability insufficient particle image size will lead to "peak-locking" effects during subpixel interpolation in the evaluation [9]. The potential of nanospheres of sufficient size but smaller density was also discussed during the PivNet workshops, although this technology is not ready yet.

4 Lightsheet Delivery

In most turbomachinery applications a periscope-type lightsheet probe is used to deliver the lightsheet. The two designs used at Graz University of Technology and German Aerospace Center DLR, Cologne are shown in Fig. 2. The first one (Fig. 2a) is of basic design combining a spherical lens with a cylindrical lens, with a cover glass protecting the optics. All elements are clamped and glued using a three-component high-temperature resin (R&G Faserverbundwerkstoffe GmbH, Waldenbuch, Germany). Due to the curvature of the casing the probe towered only slightly into the flow and was inserted in the downstream section of the flow (see Fig. 1). To avoid any intrusion effect *Liu* et al. [71] proposed direct illumination through the window section with the two cameras arranged on opposite sides of the light-sheet, alternatively *Estevadeordal* et al. [90] developed a fiber-optic system for turbomachinery applications. The probe shown in Fig. 2b is a DLR in-house development and also allows an angular alignment of the periscope. The periscope is purged with compressed air to cool and protect the delicate optics.

4.1 Data Recording

In turbomachinery applications, the rotating shaft provides the trigger for the PIV system. Different stator–rotor positions are realized with different trigger delays in the PIV acquisition software. Using this trigger signal, ensemble averaging of data for each of the different stator–rotor positions investigated can be performed. *Woisetschlager* et al. [27] used the vorticity in the instantaneous recordings of a stator wake to sort and ensemble average the recordings without a trigger signal in order to investigate the shedding process for different boundary-layer states.

Due to the small dimensions of turbomachinery blading, the high laser power used and the high sensitivity of the camera light reflections might cause problems. In the setup shown in Fig. 1, the inner and outer walls and the blades were covered with a matt black paint to minimize reflections. To cover single reflections fluorescent dye was used when the machine operated in the moderate temperature range (e.g., red Edding® markers). Since the flow through a turbine is highly directional with only the secondary flow effects, vortex shedding and flow interactions being of special interest, an image-shifting technique was applied in the main flow direction (up to 11 pixels). In both applications presented here background images (without seeding) were recorded for all stator–rotor positions and subtracted from the recordings (with seeding). *Westerweel* [91] gives a theoretical analysis of the measurement precision in PIV.

During the PivNet meetings the application of high-speed camera systems was discussed (e.g., [92]). There was a general agreement that in the periodic turbomachinery flows with a trigger signal available high-speed PIV is not necessarily needed. On the other hand, these systems might significantly decrease the measurement time needed, which is of special interest in short-duration test facilities.

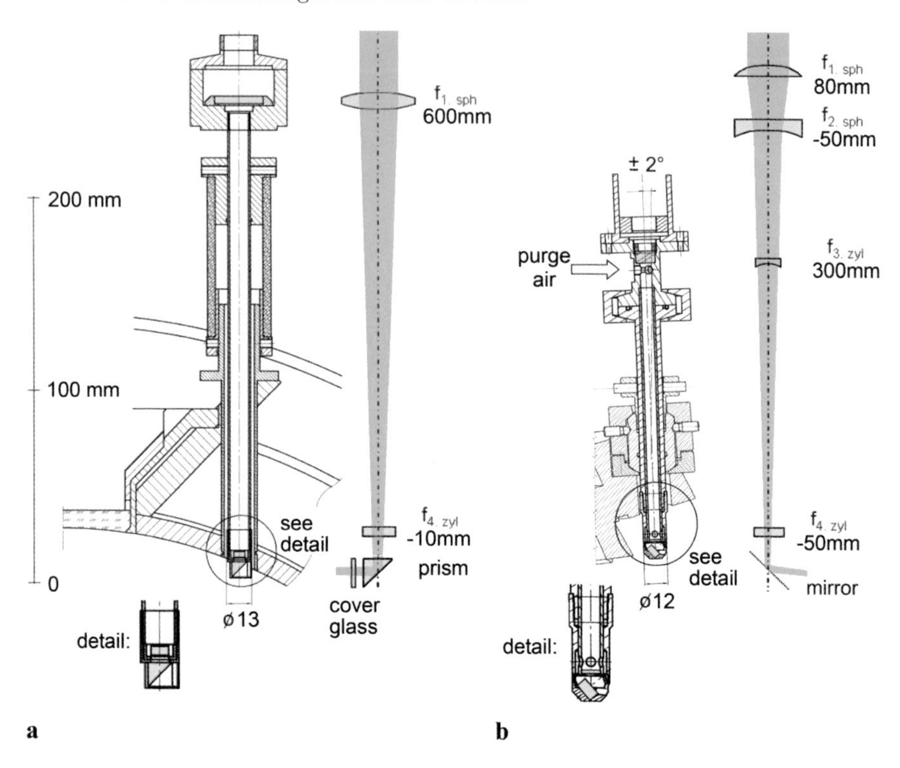

Fig. 2. Two different lightsheet probes used for PIV. (**a**) shows a periscope probe used at Graz University of Technology, (**b**) was developed at the German Aerospace Center DLR, Cologne, Germany (image courtesy of Voges, DLR, Cologne, Germany)

5 Results

To summarize the capabilities of PIV in thermal turbomachinery, the results from two demonstrations of PIV to flow investigations in thermal turbomachinery are presented. The first one is a transonic turbine operating at 10 600 rpm with 24 stator and 36 rotor blades at Graz University of Technology, Austria. The results of SPIV recordings at midspan are given in Figs. 3–5. For each of the six stator–rotor positions investigated, approximately 180 recordings (dual frame) were acquired (1280×1024 DANTEC 80C60 HiSense and DANTEC FlowMap 1500). Here, the trigger delays were chosen to realize six stator–rotor positions within one blade-passing period, in order to record all data within the same rotor-blade pitch. Thus, manufacturing precision does not influence the measured shock positions. A crosscorrelation technique with 64×64 interrogation area size and 50 % overlap was applied to the recordings resulting in the single vector fields. A 2D-Gaussfit in a 3×3 pixel matrix was used for subpixel resolution. The single vector fields

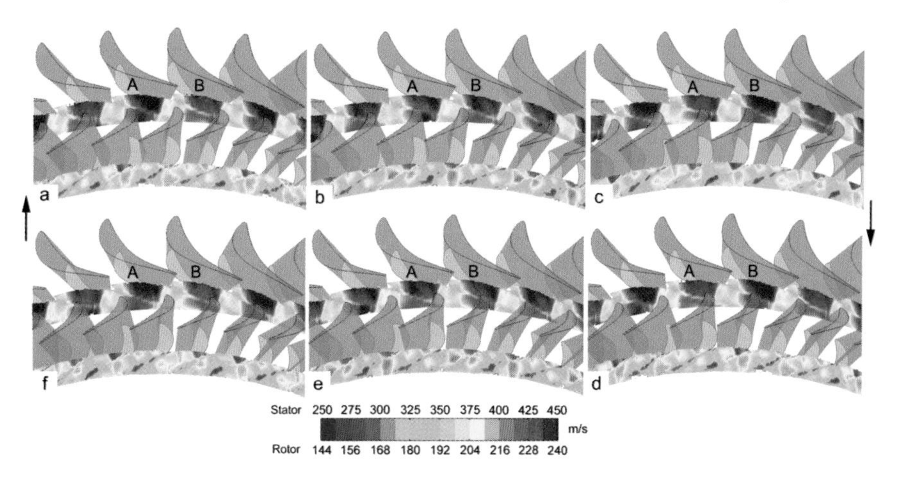

Fig. 3. (**a–f**) Ensemble-averaged velocity recorded by SPIV for six stator–rotor positions at 10 600 rpm, recorded in the transonic turbine shown in Fig. 1. A and B indicate two stator blades

were postprocessed using a peak-height ratio validation, a range validation and a moving-average filter applied to a 5×5 vector matrix. Due to the fact that the vortex-shedding from the trailing edge of the turbine blades contains boundary-layer fluid with little seeding the uncertainty in the estimation of the mean value of velocity might increase in these areas.

When looking at the ensemble-averaged velocity in Fig. 3 one can identify a pronounced shock system behind the stator blades (e.g., A and B in Fig. 3). The bypassing rotor blades modulate this shock system and the wakes behind the stator. Shock reflections by the rotor blades alter the yaw angle. In Fig. 4 this interaction can be seen when the yaw angle is plotted, calculated from the axial and circumferential velocities. For the same stator–rotor positions Fig. 5 gives the vorticity calculated from the axial and circumferential velocities. The shock reflection impinging at the stator-blade boundary layer triggers vortex shedding in Fig. 5d. Seven phases of vortex-shedding during one period of blade passing can be observed. This means the vortex-shedding frequency is about 40 kHz. A detailed comparison with interferometric measurements indicated that the tracer particles used started to act as a lowpass filter at about 80 kHz. Therefore, only the first harmonic of the vortex movement can be found in the PIV results. On the other hand, PIV provides the unique possibility to investigate the interaction between shocks, shock reflections, vortex shedding and wake–wake interaction in these turbulent and transonic flows. Further discussion of the results can be found in [60].

The second demonstration from which results are presented is a centrifugal compressor with a vaned diffusor and an impeller with 13 main and 13 splitter blades rotating at speeds up to 50 000 rpm at the German

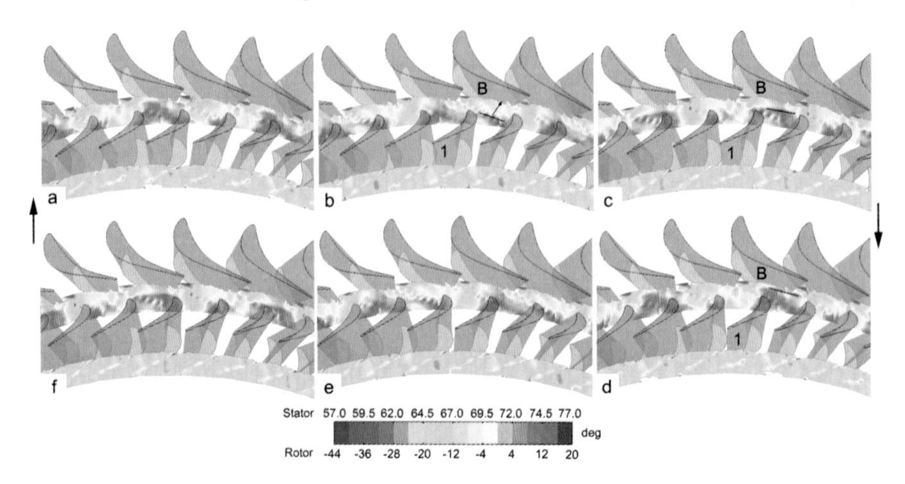

Fig. 4. (**a**–**f**) Ensemble-averaged yaw angle recorded by SPIV (see Fig. 3 for velocity). B indicates a stator blade, 1 a rotor blade. The *single line* in (**b**–**d**) marks the shock reflection

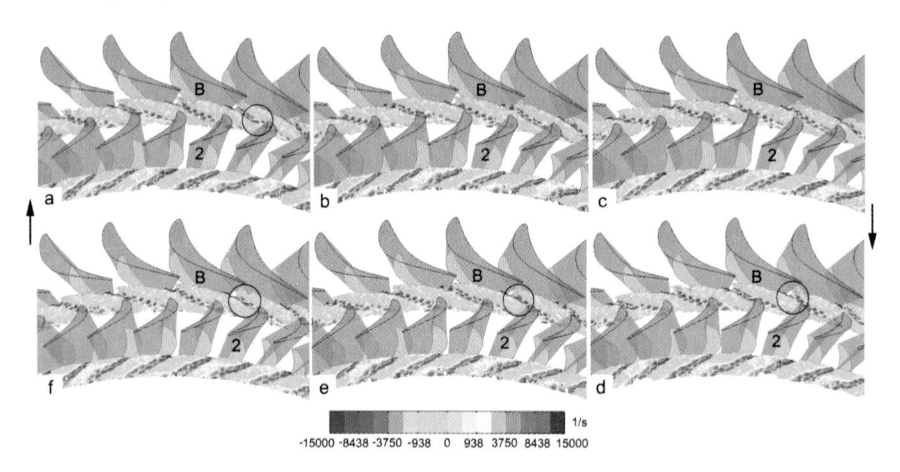

Fig. 5. (**a**–**f**) Ensemble-averaged vorticity recorded by SPIV (see Fig. 3 for velocity). B indicates a stator blade, 2 a rotor blade. The *circles* in (**d**,**e**,**f**,**a**) mark vortex shedding enforced by the shock reflection in Figs. 4b–d. Behind blade 2 vortex–vortex interference is visible in (a–f)

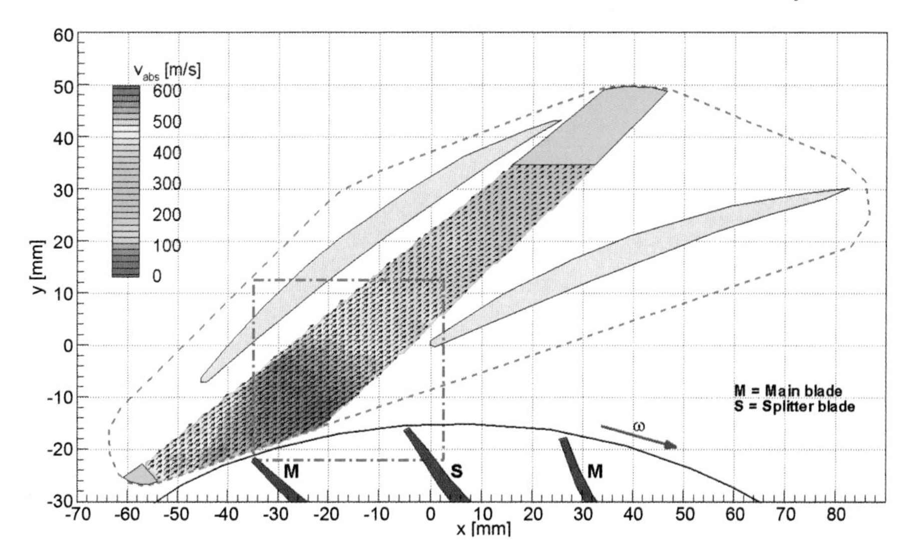

Fig. 6. Ensemble-averaged velocity recorded by PIV in the diffuser passage of a transonic centrifugal compressor at 50 000 rpm for the rotor position shown. The *dashed blue line* indicates the window section. The recording was done at the German Aerospace Center DLR, Cologne, Germany (image courtesy of Voges, DLR, Cologne, Germany)

Aerospace Center DLR, Cologne, Germany. Evaluation of the PIV image data was performed after preprocessing with a highpass filter, subtraction of background image and masking image areas without velocity information (e.g., diffuser casing or window support). A correlation algorithm with multigrid option resulting in a final 32×32 pixel interrogation window with 50 % overlap was used. The size of single interrogation areas in the lightsheet plane achieved during processing was 0.5×0.5 mm, potentially corresponding to flow structures passing at frequencies up to 1.4 MHz at a velocity up to 700 m/s. Here, the size of the particles has an important influence on the obtained velocity data. The particles used (oil droplets, below 1 μm) behave like a lowpass filter with a cut off frequency of 100 kHz. Given a blade-passing frequency around 20 kHz, this implies that only large-scale structures are faithfully captured, while smaller scales are damped out.

In the PIV test sequence phase-resolved measurements were carried out using eight phase angles per main-splitter passage. As the impeller exit flow was not expected to be symmetric between main-splitter and splitter-main blade passages, the number of phase angles was doubled. The resulting 16 phase angles allow for detailed flow investigations related to one complete main-splitter-main passage. For each angle 180 PIV recordings were averaged (ensemble averaging).

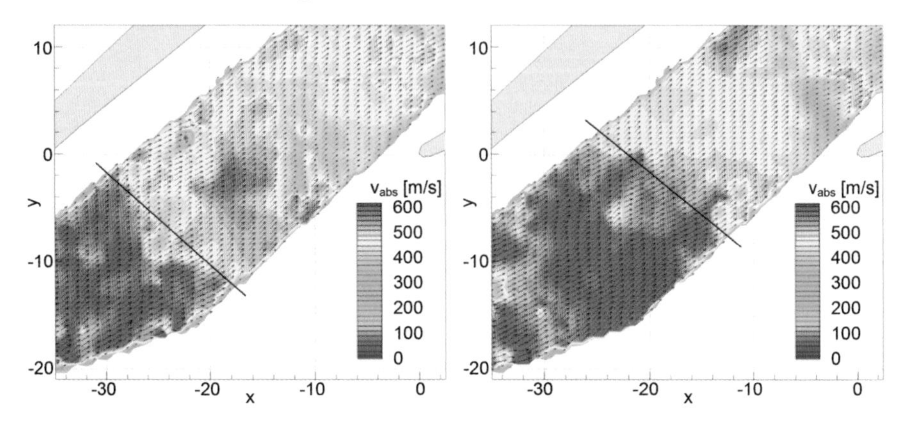

Fig. 7. Instantaneous velocity recordings by PIV for the same rotor–vane position as shown in Fig. 6. The *line* indicates the shock position. A strong shock oscillation is observed between the *left* and the *right* image. The recording was done at the German Aerospace Center DLR, Cologne, Germany (image courtesy of Voges, DLR, Cologne, Germany)

Following the transient diffuser passage flow while the impeller passes by, the flow structures emerging from impeller blades become visible. At a rotational speed of 50 000 rpm the impeller blades were passing at a frequency of 21.74 kHz. Based on the exit speed of 580 m/s such frequencies become visible as structures of 26.7 mm size. Such structures can be identified in the velocity recording in Figs. 6 and 7. While Fig. 6 gives the ensemble-averaged velocity, Fig. 7 plots two instantaneous recordings, showing a significant oscillation of the shock front for the same impeller position. Further discussion of the results can be found in [66].

6 Conclusions

Today, particle image velocimetry is established as a powerful tool in recording unsteady as well as ensemble-averaged velocity data from fast-rotating turbomachinery. However, some details of the basics of this technique have to be taken into account to obtain quantitatively correct data. These details and future developments were discussed during the PivNet workshops with mutual visits between the turbomachinery labs involved in this program. Additionally, the presentations and the reports had been made available to the European turbomachinery community through the European strategic research projects all of us participated in.

Throughout the workshops there was an ongoing discussion as to which structures can be observed within the turbomachinery flow field by the PIV technique. In most turbomachinery applications a size of 0.6×0.6–1×1 mm for the single interrogation areas was achieved. Depending on the flow speed this correlates to frequencies up to 1.2 MHz, compared to fast response aerodynamic probes a remarkably good result. On the other hand, the fact that PIV spatial resolution is reduced to corresponding frequencies below 100 kHz by the inertia even of small particles, was generally accepted and has to be mentioned in scientific work where fast-changing flow phenomena are observed. The development of novel tracer particles with lower density has to be watched carefully by the turbomachinery community.

Most turbomachinery applications are characterized by difficult optical access to the areas of interest. Especially when the windows are highly curved or perspective viewing cannot be neglected ray tracing and unwarping of images previous to the correlation procedure was recommended, including an estimation of the directional sensitivity within the lightsheet. The difficult optical access also limits the use of SPIV and recommends the combination with other techniques such as DGV or the development of miniaturized systems. Additionally, any combination of techniques improves the physical understanding of flow phenomena (i.e., DPIV in combination with pressure-sensitive paint, interferometry, surface temperature by infrared imaging or thermoliquids).

High-speed capabilities are welcomed, but mainly in order to speed up the measurements. Recordings sorted by the phase of the rotor are certainly sufficient to investigated stator–rotor interaction and correlated unsteady effects in turbomachines.

Summarizing, the above-mentioned new developments in the PIV technology will also break new ground in turbomachinery flow research. The PivNet triggered these developments.

Acknowledgements

The authors wish to acknowledge the support of Mrs. Melanie Voges and Dr. Chris Willert, both German Aerospace Center DLR, Cologne, Germany in the preparation of the manuscript with respect to the experiments in the high-speed compressor performed at DLR.

References

[1] C. H. Sieverding, T. Arts, R. Dénos, J.-F. Brouckaert: Measurement techniques for unsteady flows in turbomachines, Exp. Fluids **28**, 285–321 (2000)

[2] R. Schodl: Capabilities of optical point measurement techniques with respect to aero engine applications, in *Planar Optical Measurement Methods for Gas Turbine Components*, vol. 217, RT0 Lecture Series (1999) 16–17 Sept. 1999, Cranfield, UK and 21–22 Sept. 1999, Cleveland, Ohio RTO-EN-6, paper 1

[3] R. J. Adrian: Twenty years of particle image velocimetry, Exp. Fluids **39**, 159–169 (2005)

[4] N. Paone, M. L. Riethmuller, R. A. Van den Braembussche: Experimental investigation of the flow in the vaneless diffuser of a centrifugal pump by particle image displacement velocimetry, Exp. Fluids **7**, 371–378 (1989)
[5] L. P. Goss, M. E. Post, D. D. Trump, B. Sarka, C. D. MacArthur, G. E. Dunning: A novel technique for blade-to-blade velocity measurements in a turbine cascade, in *25th AIAA/ASME/ SAE/ASEE Joint Propulsion Conference and Exhibit* (1989) paper AIAA-1989-2691
[6] P. J. Bryanston-Cross, C. E. Towers, T. R. Judge, D. P. Towers, S. P. Harasgama, S. T. Hopwood: The application of particle image velocimetry (PIV) in a short-duration transonic annular turbine cascade, ASME J. Turbomach. **114**, 504–509 (1992)
[7] C. E. Willert, M. Gharib: Digital particle image velocimetry, Exp. Fluids **10**, 181–193 (1991)
[8] M. P. Arroyo, C. A. Greated: Stereoscopic particle image velocimetry, Meas. Sci. Technol. **2**, 1181–1186 (1991)
[9] M. Raffel, C. Willert, J. Kompenhans: *Particle Image Velocimetry, A Practical Guide* (Springer, New York 1998)
[10] R. W. Ainsworth, S. J. Thorpe, R. J. Manners: A new approach to flow-field measurements – a view of Doppler global velocimetry techniques, Int. J. Heat Fluid Flow **18**, 116–130 (1997)
[11] I. Roehle, R. Schodl, P. Voigt, C. Willert: Recent developments and applications of quantitative laser light sheet measuring techniques in turbomachinery components, Meas. Sci. Technol. **11**, 1023–1035 (2000)
[12] M. Samimy, M. P. Wernet: Review of planar multiple-component velocimetry in high-speed flows, AIAA Journal **38**, 553–574 (2000)
[13] D. Chanteloup, A. Bölcs: Flow characteristics in two-leg internal coolant passages of gas turbine airfoils with film-cooling hole ejection, ASME J. Turbomach. **124**, 499–507 (2002)
[14] D. Chanteloup, Y. Juaneda, A. Bölcs: Combined 3-D flow and heat transfer measurements in a 2-pass internal coolant passage of gas turbine airfoils, ASME J. Turbomach. **124**, 710–718 (2002)
[15] L. Casarsa, T. Arts: Experimental investigation of the aerothermal performance of a high blockage rib-roughened cooling channel, ASME J. Turbomach. **127**, 580–588 (2005)
[16] M. Elfert, M. P. Jarius, B. Weigand: Detailed flow investigation using PIV in a typical turbine cooling geometry with ribbed walls, in *Proc. ASME Turbo Expo 2004* (2004) GT2004-53566
[17] Y. Servouze, J. C. Sturgis: Heat transfer and flow field measurements in a rib-roughened branch of a rotating two-pass duct, in *Proc. ASME Turbo Expo* (2003) GT2003-38048
[18] O. Uzol, C. Camci: Elliptical pin fins as an alternative to circular pin fins for gas turbine blade colling applications – part 2: Wake flow field measurements and visualization using PIV, in *Proc. ASME Turbo Expo* (2001) pp. 2001–GT–0181
[19] P. K. Panigrahi, A. Schröder, J. Kompenhans: PIV investigation of flow behind surface mounted permeable ribs, Exp. Fluids **40**, 277–300 (2006)
[20] S. Bernsdorf, M. G. Rose, R. S. Abhari: Modeling of film cooling – part I: Experimental study of flow structure, ASME J. Turbomach. **128**, 141–149 (2006)

[21] S. Bernsdorf, M. G. Rose, R. S. Abhari: Experimental validation of quasi-steady assumption in modeling of unsteady film cooling, in *Proc. ASME Turbo Expo* (2006) GT2006-90166
[22] J. H. Yoon, R. F. Martinez-Botas: Film cooling performance in a simulated blade tip geometry, in *Proc. ASME Turbo Expo* (2005) GT2005-68863
[23] P. Palafox, M. L. G. Oldfield, J. E. LaGraff, T. V. Jones: PIV maps of tip leakage and secondary flow fields on a low speed turbin blade cascade with moving endwall, in *Proc. ASME Turbo Expo* (2005) GT2005-68189
[24] H.-J. Rehder, A. Dannhauer: Experimental investigation of turbine leakage flows on the 3D flow field and end wall heat transfer, in *Proc. ASME Turbo Expo* (2006) GT2006-90173
[25] M. Raffel, F. Kost: Investigation of aerodynamic effects of coolant ejection at the trailing edge of a turbine blade model by PIV and pressure measurements, Exp. Fluids **24**, 447–461 (1998)
[26] A. Vicharelli, J. K. Eaton: Turbulence measurements in a transonic two-passage turbine cascade, Exp. Fluids **41**, 897–917 (2006)
[27] J. Woisetschläger, N. Mayrhofer, B. Hampel, H. Lang, W. Sanz: Laser-optical investigation of turbine wake flow, Exp. Fluids **34**, 371–378 (2003)
[28] R. D. Stieger, D. Hollis, H. P. Hodson: Unsteady surface pressures due to wake-induced transition in a laminar separation bubble on a low-pressure cascade, ASME J. Turbomach. **126**, 544–550 (2004)
[29] M. D. Langford, A. Breeze-Stringfellow, S. A. Guillot, W. Solomon, W. F. Ng, J. Estevadeordal: Experimental investigation of the effects of a moving shock wave on compressor stator flow, in *Proc. ASME Turbo Expo* (2005) GT2005-68722
[30] J. Estevadeordal, S. Gogineni, L. Goss, W. Copenhaver, S. Gorrell: Study of wake-blade interactions in a transonic compressor using flow visualization and DPIV, J. Fluid. Eng.-T ASME **124**, 166–175 (2002)
[31] X. Q. Zheng, A. P. Hou, Q. S. Li, S. Zhou, Y. J. Lu: Flow control of annular compressor cascade by synthetic jets, in *Proc. ASME Turbo Expo* (2006) GT2006-90211
[32] E. Canepa, D. Lengani, F. Satta, E. Spano, M. Ubaldi, P. Zunino: Boundary layer separation control on a flat plate with adverse pressure gradients using vortex generators, in *Proc. ASME Turbo Expo* (2006) GT2006-90809
[33] C. Cerretelli, K. Kirtley: Boundary layer separation control with fluidic oscillators, in *Proc. ASME Turbo Expo* (2006) GT2006-90738
[34] D. Tisserant, F. A. E. Breugelmans: Rotor blade-to-blade measurements using particle image velocimetry, ASME J. Turbomach. **119**, 176–181 (1997)
[35] N. Balzani, F. Scarano, M. L. Riethmuller, F. A. E. Breugelmans: Experimental investigation of the blade-to-blade flow in a compressor rotor by digital particle image velocimetry, ASME J. Turbomach. **122**, 743–750 (2000)
[36] I. Ceyhan, E. M. d'Hoop, G. R. Guenette, A. H. Eppstein, P. J. Bryanston-Cross: Optical instrumentation for temperature and velocity measurements in rig turbines., in *Advanced Non-Intrusive Instrumentation for Propulsion Engines* (AGARD PEP 90th Symposium AGARD-CP-598 1997) paper 27
[37] K. S. Chana, N. Healey, P. J. Bryanston-Cross: Particle image velocimetry measurements from the stator-rotor interaction region of a high pressure transonic turbine stage at the DERA isentropic light piston facility, in *Advanced*

Non-Intrusive Instrumentation for Propulsion Engines (AGARD PEP 90th Symposium AGARD-CP-598 1997) paper 46

[38] P. J. Bryanston-Cross, D. D. Udrea, G. Guenette, A. Eppstein, E. M. d'Hoop: Whole-field visualisation and velocity measurement of an instantaneous transonic turbine flow, in *Int. Cong. Instrumentation in Aerospace Simulation Facilities,* (ICIASF 1997) pp. 278–286

[39] P. Bryanston-Cross, M. Burnett, B. Timmerman, W. K. Lee, P. Dunkley: Intelligent diagnostic optics for flow visualization, Opt. Laser Technol. **32**, 641–654 (2000)

[40] K. M. Day, P. B. Lawless, S. Fleeter: Particle image velocimetry measurements in a low speed two stage research turbine, in *32nd AIAA/ASME/SAE/ASEE Joint Propulsion Conf. and Exhibit* (1996) paper AIAA-1996-2569

[41] K. M. Treml, P. B. Lawless: Particle image velocimetry of vane-rotor interaction in a turbine stage, in *34th AIAA/ASME/SAE/ASEE Joint Propulsion Conference and Exhibit* (1998) paper AIAA-1998-3599

[42] K. D. Gallier, P. B. Lawless, S. Fleeter: Development of the unsteady flow on a turbine rotor platform downstream of a rim seal, in *Proc. ASME Turbo Expo 2004* (2004) GT2004-53899

[43] A. J. Sanders, J. Papalia, S. Fleeter: A PIV investigation of rotor-IGV interactions in a transonic compressor, J. Propul. Power **18**, 969–977 (2002)

[44] A. J. Sanders, J. Papalia, S. Fleeter: Multi-blade row interactions in a transonic axial compressor: Part 1 - stator particle image velocimetry (PIV) investigation, ASME J. Turbomach. **124**, 10–18 (2002)

[45] J. Papalia, P. B. Lawless, S. Fleeter: Off-design transonic rotor-inlet guide vane unsteady aerodynamic interactions, J. Propul. Power **21**, 715–727 (2005)

[46] N. L. Key, P. B. Lawless, S. Fleeter: Rotor-generated vane row off-design unsteady aerodynamics including dynamic stall, part I, J. Propul. Power **20**, 835–841 (2004)

[47] M. P. Wernet: Demonstration of PIV in a transonic compressor, in *Advanced Non-Intrusive Instrumentation for Propulsion Engines, AGARD PEP 90th Symp.* (1997) pp. AGARD–CP–598, paper 51

[48] M. P. Wernet: Application of digital particle image velocimetry to turbomachinery, in *Planar Optical Measurement Methods for Gas Turbine Components. RT0 Lecture Series 217* (1999) pp. RTO–EN–6, paper 2

[49] M. P. Wernet: Application of DPIV to study both steady state and transient turbomachinery flows, Opt. Laser Technol. **32**, 497–525 (2000)

[50] M. P. Wernet: Development of digital particle imaging velocimetry for use in turbomachinery, Exp. Fluids **28**, 97–115 (2000)

[51] M. P. Wernet: A flow field investigation in the diffuser of a high-speed centrifugal compressor using digital particle imaging velocimetry, Meas. Sci. Technol. **11**, 1007–1022 (2000)

[52] M. P. Wernet, D. V. Zante, T. J. Strazisar, W. T. John, P. S. Prahst: Characterization of the tip clearance flow in an axial compressor using 3-D digital PIV, Exp. Fluids **39**, 743–753 (2005)

[53] S. Gogineni, J. Estevadeordal, B. Sarka, L. Gose, W. Copenhaver: Application of two-color digital PIV for turbomachinery flows, in *Advanced Non-Intrusive Instrumentation for Propulsion Engines* (AGARD PEP 90th Symposium 1997) pp. AGARD–CP–598, paper 49

[54] J. Estevadeordal, S. Gogineni, W. Copenhaver, G. Bloch, M. Brendel: Flow field in a low-speed axial fan: A DPIV investigation, Exp. Therm. Fluid Sci. **23**, 11–21 (2000)

[55] J. Estevadeordal, S. Gogineni, L. Goss, W. Copenhaver, S. Gorrell: DPIV study of wake-rotor syncronization in a transonic compressor, in *31th AIAA Fluid Dynamics Conf. and Exhibit* (2001) paper AIAA-2001-3095

[56] W. Copenhaver, J. Estevadeordal, S. Gogineni, S. Gorrell, L. Goss: DPIV study of near-stall wake-rotor interactions in a transonic compressor, Exp. Fluids **33**, 899–908 (2002)

[57] D. L. Sondak, D. J. Dorney: Simulation of vortex shedding in a turbine stage, ASME J. Turbomach. **121**, 428–435 (1999)

[58] H. Lang, T. Mörck, J. Woisetschläger: Stereoscopic particle image velocimetry in a transonic turbine stage, Exp. Fluids **32**, 700–709 (2002)

[59] J. Woisetschläger, H. Lang, B. Hampel, E. Göttlich, F. Heitmeir: Influence of blade passing on the stator wake in a transonic turbine stage investigated by particle image velocimetry and laser vibrometry, J. Power Energy, ImechE **217**, 385–391 (2003)

[60] E. Göttlich, J. Woisetschläger, P. Pieringer, B. Hampel, F. Heitmeir: Investigation of vortex shedding and wake-wake interaction in a transonic turbine stage using laser-velocimetry and particle-image-velocimetry, ASME J. Turbomach. **128**, 178–187 (2006)

[61] H. Hayami, M. Hojo, S. Aramaki: Flow measurement in a transonic centrifugal impeller using a PIV, J. Visual. **5**, 255–261 (2002)

[62] H. Hayami, M. Hojo, N. Hirata, S. Aramaki: Flow with shock waves in a transonic centrifugal compressor with low-solidity cascade diffusor using PIV, in *Proc. ASME Turbo Expo* (2004) GT2004-53268

[63] T. Geis, G. Rottenkolber, B. Dittmann, K. Richter, K. Dullenkopf, S. Wittig: Endoscopic PIV-measurments in an enclosed rotor-stator system with pre-swirled cooling air, in *11th Int. Symp. on Applications of Laser Techniques to Fluid Mechanics,* (2002) paper 26-4

[64] C. Bricaud, B. Richter, K. Dullenkopf, H.-J. Bauer: Stereo PIV measurements in an enclosed rotor–stator system with pre-swirled cooling air, Exp. Fluids **39**, 202–212 (2005)

[65] P. Childs, K. Dullenkopf, D. Bohn: Internal air systems experimental rig best practice, in *Proc. ASME Turbo Expo 2006* (2006) GT2006-90215

[66] M. Voges, M. Beversdorff, C. Willert, H. Krain: Application of particle image velocimetry to a transonic centrifugal compressor, Exp. Fluids **43**, 371–384 (2007)

[67] S. Ibaraki, T. Matsuo, T. Yokoyama: Investigation of unsteady flow field in a vaned diffusor of a transonic centrifugal compressor, ASME Turbomach **129**, 686–693 (2006)

[68] L. Porreca, Y. I. Yun, A. I. Kalfas, S. J. Song, R. S. Abhari: Investigation of 3D unsteady flows in a two-stage shrouded axial turbine using stereoscopic PIV and FRAP – part I: Interstage flow interactions, in *Proc. ASME Turbo Expo* (2006) GT2006-90752

[69] Y. I. Yun, L. Porreca, A. I. Kalfas, S. J. Song, R. S. Abhari: Investigation of 3D unsteady flows in a two-stage shrouded axial turbine using stereoscopic PIV and FRAP – part II : Kinematics of shroud cavity flow, in *Proc. ASME Turbo Expo* (2006) GT2006-91020

[70] A. P. S. Wheeler, R. J. Miller, H. P. Hodson: The effect of wake induced structures on compressor boundary layers, ASME J. Turbomach **129**, 705–712 (2006)
[71] B. J. Liu, X. J. Yu, H. X. Liu, H. K. Jiang, H. J. Yuan, Y. T. Xu: Application of SPIV in turbomachinery, Exp. Fluids **40**, 621–642 (2006)
[72] B. J. Liu, H. W. Wang, H. X. Liu, H. J. Yu, H. K. Jiang, M. Z. Chen: Experimental investigation of unsteady flow field in the tip region of an axial compressor rotor passage at near stall condition with stereoscopic particle image velocimetry, ASME J. Turbomach. **126**, 360–374 (2004)
[73] B. J. Liu, X. J. Yu, H. W. Wang, H. X. Liu, H. K. Jiang, M. Z. Chen: Evolution of the tip leakage vortex in an axial compressor rotor, in *Proc. ASME Turbo Expo* (2004) GT2004-53703
[74] Z. Q. Gong, Z. P. Li, M. Y. Li, Y. J. Lu: An investigation of the IGV/rotor interaction in a low speed axial compressor using DPIV, in *Proc. ASME Turbo Expo* (2006) GT2006-90976
[75] O. Uzol, Y. C. Chow, J. Katz, C. Meneveau: Unobstructed particle image velocimetry measurements within an axial turbo-pump using liquid and blades with matched refractive indices, Exp. Fluids **33**, 909–919 (2002)
[76] O. Uzol, Y. C. Chow, J. Katz, C. Meneveau: Experimental investigation of unsteady flow field within a two-stage axial turbomachine using particle image velocimetry, ASME J. Turbomach. **124**, 542–552 (2002)
[77] O. Uzol, Y. C. Chow, J. Katz, C. Meneveau: Average passage flow field and deterministic stresses in the tip and hub regions of a multistage turbomachinery, ASME J. Turbomach. **125**, 714–725 (2003)
[78] Y. C. Chow, O. Uzol, J. Katz: Flow nonuniformities and turbulent "hot spots" due to wake–blade and wake–wake interactions in a multi-stage turbomachine, ASME J. Turbomach. **124**, 553–563 (2002)
[79] F. Soranna, Y. C. Chow, O. Uzol, J. Katz: The effect of inlet guide vanes wake impingement on the flow structure and turbulence around a rotor blade, ASME J. Turbomach. **128**, 82–95 (2006)
[80] A. Naqwi: Distortion compensation for PIV systems, in *10th Int. Symp. on Applications of Laser Techniques to Fluid Mechanics* (2000) paper 6.2
[81] F. Scarano, L. David, M. Bsibsi, D. Calluaud: S-PIV comparative assessment: Image dewarping + misalignment correction and pinhole + geometric back projection, Exp. Fluids **39**, 257–266 (2005)
[82] C. Willert: Stereoscopic digital particle image velocimetry for application in wind tunnel flows, Meas. Sci. Technol. **8**, 1465–1479 (1997)
[83] S. M. Soloff, R. J. Adrian, Z.-C. Liu: Distortion compensation for generalized stereoscopic particle image velocimetry, Meas. Sci. Technol. **8**, 1441–1454 (1997)
[84] B. Wieneke: Stereo-PIV using self calibration on particle images, Exp. Fluids **39**, 267–280 (2005)
[85] M. P. Wernet: Planar particle imaging doppler velocimetry: A hybrid PIV/DGV technique for three-component velocity measurements, Meas. Sci. Technol. **15**, 2011–2028 (2004)
[86] C. Willert, C. Hassa, G. Stockhausen, M. Jarius, M. Voges, J. Klinner: Combined PIV and DGV applied to a pressurized gas turbine combustion facility, Meas. Sci. Technol. **17**, 1670–1679 (2006)

[87] P. Arroyo, J. Lobera, S. Recuero, J. Woisetschläger: Digital image plane holography for three-component velocity measurements in turbomachinery flows, in *13th Int Symp on Applications of Laser Techniques to Fluid Mechanics* (2006) paper 34.1
[88] R. Mei: Velocity fidelity of flow tracer particles, Exp. Fluids **22**, 1–13 (1996)
[89] F. F. J. Schrijer, F. Scarano, B. W. van Oudheusden: Application of PIV in a mach 7 double-ramp flow, Exp. Fluids **41**, 352–363 (2006)
[90] J. Estevadeordal, T. R. Meyer, S. P. Gogineni, M. D. Polanka, J. R. Gord: Development of a fiber-optic PIV system for turbomachinery applications, in *43rd AIAA Aerospace Science Meeting and Exhibit* (2005) paper AIAA-2005-38
[91] J. Westerweel: Theoretical analysis of the measurement precision in particle image velocimetry, Exp. Fluids **29, Suppl**, S3–S12 (2000)
[92] A. Schröder, M. Herr, T. Lauke, U. Dierksheide: Measurements of trailing-edge-noise sources by means of time-resolved PIVeedings, in *6th International Symposium on Particle Image Velocimetry* (2005) paper S 15-2

Index

Two-Phase PIV: Fuel-Spray Interaction with Surrounding Air

Stefan Dankers, Mark Gotthardt, Thomas Stengler, Gerhard Ohmstede, and Werner Hentschel

Volkswagen AG, Postbox 1785, D-38436 Wolfsburg, Germany
Stefan.dankers@volkswagen.de

Abstract. The demand for improvement of combustion-engine processes leads to a great need for detailed experimental information about the complicated processes of injection, breakup and propagation of the fuel jet and the vaporization of the fuel.

Thus, in the presented work it was the aim to visualize the air flow that is induced by the injected fuel jet and also to explore the air entrainment into the fuel spray. This was done using two-phase PIV (particle image velocimetry).

The fuel jet was illuminated with a Nd:YAG PIV-laser and PIV analysis was done by simply measuring the elastically scattered light of the fuel droplets. For the investigation of the surrounding air propylene-carbonate doped with DCM-dye was dispersed and the droplets were added to the continuous gas flow upstream of the chamber. Scattered and fluorescence signals, respectively, were detected perpendicular to the laser sheet with a CCD camera. To detect the gas flow, the scattered light from the liquid fuel was suppressed by an OG 590 longpass filter glass that transmits the fluorescence signal of the DCM-dye.

It was possible to measure both the fuel jet and the gas flow in the presence of the fuel spray and a clear separation of the two phases could be achieved.

In both (fuel and air) vector pictures corresponding vortices could be identified near the air/fuel boundary layer. Maximum velocities in the jet are depending on the operation conditions up to $150\,\mathrm{m\,s^{-1}}$ and the gas flow has typically a velocity of 1 to $10\,\mathrm{m\,s^{-1}}$.

In the region next to the injector the air was pressed away during the injection. After the end of the injection a strong fast air entrainment flow into this region can be observed that compensates the pressure difference.

1 Introduction

The demand of people for flexible mobility is further increasing and thus the car remains an important factor to meet this demand. In spite of extensive research for alternative engines, the combustion engine will stay the most important solution for the next few decades. At the same time one is facing declining resources and increasing environmental stresses. This means that the requirement to improve combustion-engine processes to reduce both fuel consumption and emissions is greater than ever. This led, amongst others, to the development of modern direct-injection concepts for spark-ignited engines for which it is important to ensure the provision of a burnable air–fuel

A. Schroeder, C. E. Willert (Eds.): Particle Image Velocimetry,
Topics Appl. Physics **112**, 333–343 (2008)

mixture at the spark plug at the moment of ignition [1]. Further improvement necessitates comprehensive modeling of the fuel-mixture generation and there is a great need for detailed experimental information about the complicated processes of injection, breakup and propagation of the fuel jet and the vaporization of the fuel as input data for calculations and computational fluid dynamics (CFD) [2].

In the presented work it was the aim to analyze the air–fuel interaction. The injected fuel jet induces an air flow that was to be visualized and also the air entrainment into the fuel spray was to be explored. Furthermore, the motion of the fuel droplets and the development of vortices are of great interest regarding fuel vaporization and provision of a burnable mixture. To gather information about these aspects it is necessary to measure the flow of both phases, liquid fuel and gaseous air, for which the application of two-phase PIV is ideal [3]. Similar investigations were made by *Kubach* et al. [4] who combined PIV with shadowgraphy. They did not, however, examine the interaction between air and fuel.

2 Experimental

The investigations were performed in an optically accessible pressure chamber since the experimental effort is reduced in comparison with engine measurements and allows an easy variation of operating conditions. The relevant phenomena of the mixture generation in this part of the combustion process are only weakly affected by the geometry of the engine and can be observed in detail in a chamber.

Thermodynamic conditions similar to those in the engine can be realized and the optical accessibility is very good. Special air flows, e.g., a tumble changing with the crank angle that can be found in the engine, however, cannot be simulated in the used chamber. Nevertheless, for general investigations on mixture formation this does not represent a major drawback.

The injector was placed in the top of the chamber and the direction of the fuel spray is downwards. There is an air flow continuously streaming through the chamber to scavenge the remains of the injections. The resulting maximum frequency of alternating injections is about 3 Hz. The air can be heated to 400 °C and the pressure in the chamber can be varied from 0.2 to 42 bar.

The light of a frequency-doubled Nd:YAG PIV laser was formed into a lightsheet with a height of approximately 100 mm and a pair of pictures for PIV analysis of the fuel jet was taken by simply measuring the elastically scattered light of the fuel droplets. There are enough separated droplets and this approach works for most of the investigated operating points after 50 to 100 μs after the first fuel leaves the injector because of the high injection pressure results in a fast spray breakup.

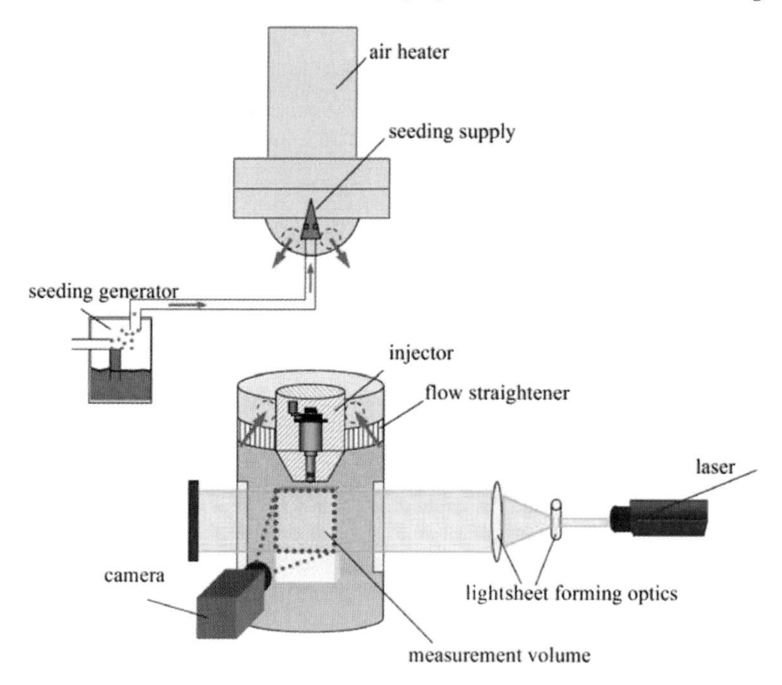

Fig. 1. Experimental setup: top of the chamber with seeding supply and optical setup

For the investigation of the surrounding air propylene carbonate doped with DCM-dye was dispersed by a commercial seeding device and the droplets were added to the continuous gas flow upstream of the chamber. The dye can be excited at 532 nm and emits in the region of 620 to 690 nm with a maximum emission around 660 nm.

The seeder was driven with pressurized air (less than 0.6 MPa) and worked without problems up to a counterpressure of 0.4 MPa.

Scattered and fluorescence signals, respectively, were detected perpendicular to the laser sheet with a double-shutter CCD camera. To detect the gas flow the scattered light from the liquid fuel was suppressed by an OG 590 longpass filter glass that transmits the fluorescence signal of the DCM-dye.

The PIV evaluation was done with LaVision Davis 7.1. In the preprocessing a constant background was subtracted. A crosscorrelation analysis was performed. The size of the interrogation areas was varied in a multipass approach starting with 128×128 decreasing to 32×32 pixel or 16×16 pixel for the fuel jet with an overlap of 25 or 50 % or to 64×64 pixel and 32×32 pixel for the gas flow.

In the postprocessing only a median filter was used to suppress noise vectors. The optimal setting of the calculation parameters had to be adjusted depending on the individual operating conditions.

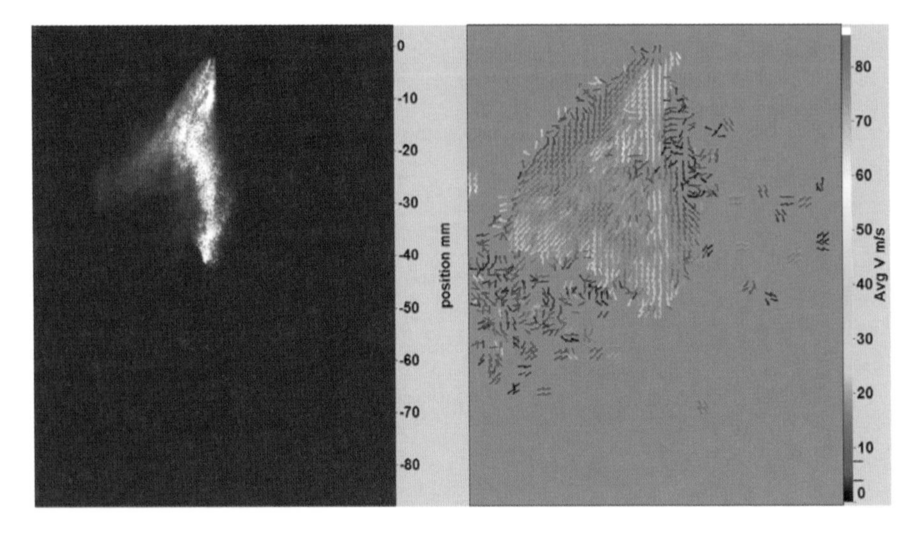

Fig. 2. Raw scattering signal of the fuel spray and corresponding vector picture

Fig. 3. Fluorescence signal of dye in the gas seeding and corresponding vector picture

The repetition rate of the PIV system is about a few Hz. Thus, only one pair of pictures can be detected for one injection and the temporal evolution of the investigated flows can only be observed in successive injections.

In Fig. 2 a typical detected scattering signal is shown. The laser lightsheet enters the chamber from the right side. This causes higher scattering intensity in the raw pictures on this side of the spray due to laser absorption when traveling through the dense spray. This did not affect the quality of the PIV analysis.

A fluorescence picture is depicted in Fig. 3. A dense distribution of seeding particles can be seen. This results in the fact that the outline of the fuel jet is clearly visible. This strong signal is not elastically scattered light that passes the longpass filter glass but the intense scattering in the fuel spray induces fluorescence between the spray and the camera outside the laser lightsheet. Furthermore, shadowing effects of the spray can be observed on the left side. The surrounding gas flow can be reconstructed by the PIV analysis. No vectors are found in the dense spray.

The temporal evolution of the flow was evaluated by shifting the recording time by 50 µs from 0.3 to 2 ms after the start trigger of the injection. The injection ends, depending on the operating conditions, around 1.3 ms.

The time delay between the two PIV pictures was chosen individually for each operating point, depending on the conditions, especially the average flow velocity. For the visualization of the fuel jet the optimal time delay was between 1.5 and 8 µs. For the analysis of the slower gas flow, delays between 30 and 100 µs were chosen. One possible option would be the use of an image splitter with the longpass glass filter in front of one side of the picture to detect simultaneously both flows within one pair of pictures. Another approach could be the use of one color camera, which was suggested by *Towers* et al. [5], but the limited dynamic range of the camera causes problems in this case. Besides these considerations, the difference in optimal delay time made the simultaneous detection unfortunately impossible.

Thus, a sequential method was employed, i.e., first the gas flow was detected with a filter in front of the camera and an optimized time delay, and in a second experimental set the fuel jet was observed with the same operating conditions of the pressure chamber. The reproducibility of the injection is quite good; nevertheless, there are cyclic fluctuations when using high-pressure injectors, making a statistical analysis of the measured data necessary. For each point, 20 vector pictures were generated and averaged for the fuel and the air flow, respectively.

3 Results

It was possible to measure both the fuel jet and the gas flow in the presence of the fuel spray and a clear separation of the two phases could be achieved. In both (fuel and air) vector pictures corresponding to vortices could be identified near the air/fuel boundary layer as shown as an example in Fig. 3. These are essential for the preparation of a burnable mixture since fuel droplets leave the spray and they slow down. Thus, more time for vaporization is available. Simultaneously, air enters the fuel jet.

Maximum velocities in the jet depend on the operation conditions up to $150\,\mathrm{m\,s^{-1}}$ and the gas flow has typically a velocity of 1 to $4\,\mathrm{m\,s^{-1}}$. In the region next to the injector the air was pressed away during the injection. After the end of the injection a strong fast air entrainment flow into this

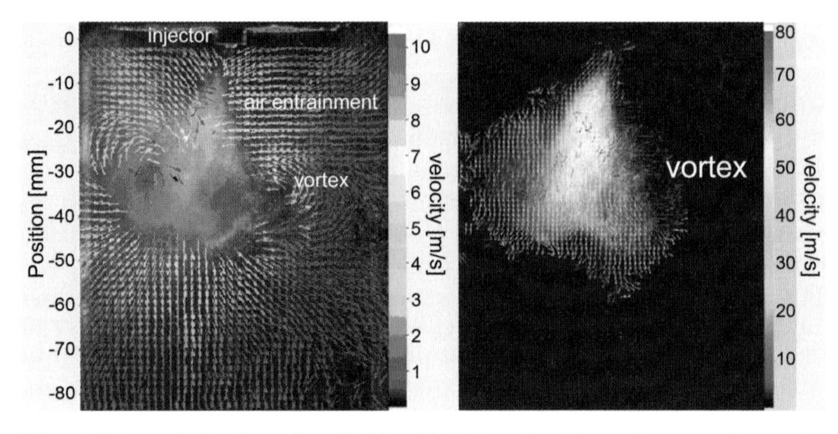

Fig. 4. Vector field of gas flow (*left*) with induced swirls and strong air entrainment (the fuel spray is displayed in the *background*) and corresponding flow field of fuel droplets (*right*)

region can be observed that compensates the pressure difference. The gas is accelerated up to $15\,\mathrm{m\,s^{-1}}$.

In Fig. 5 the influence of the chamber pressure on the fuel spray is shown. The penetration length decreases with increasing counterpressure. Additionally, the fuel jet is slowed down and becomes more compact. In the bottom part of the picture one can see that the velocity of the induced air flow corresponds to the velocity of the fuel droplets in the spray: where the fuel enters the chamber with high speed the air is stronger accelerated. The intensity of the air entrainment decreases with increasing chamber pressure. Also, the vortex in the region of the air/fuel boundary layer is less pronounced, which may lead to a worse mixture preparation. The development of the vortices in the fuel jet is similar to the results of calculations, e.g., from *Gavaises* [6].

The complete development of the air flow is shown in Fig. 6. In the first picture the undisturbed continuous scavenging gas flow is observable. Then the fuel jet begins to penetrate the chamber and the air is pushed away. The air is accelerated up to $7\,\mathrm{m/s}$. Later, the strong air-entrainment flow into this region with reduced pressure starts.

Furthermore, it was investigated whether the scavenging gas flow has an influence on the jet behavior and the fuel–air mixture generation.

A variation of the gas mass flow shows higher flow velocities in the continuous gas flow but does not affect the fuel jet flow, nor the induced gas flow, as shown in Fig. 7.

The mass flow is increased by a factor of 4 and, as one would expect, the maximum velocities of the gas flow before the start of injection increase from 0.25 to $1\,\mathrm{m\,s^{-1}}$, i.e., by almost the same factor.

The pictures of the fuel-droplet motion as well as the induced gas flow, however, are quite similar. In the bottom row differences appear only in the

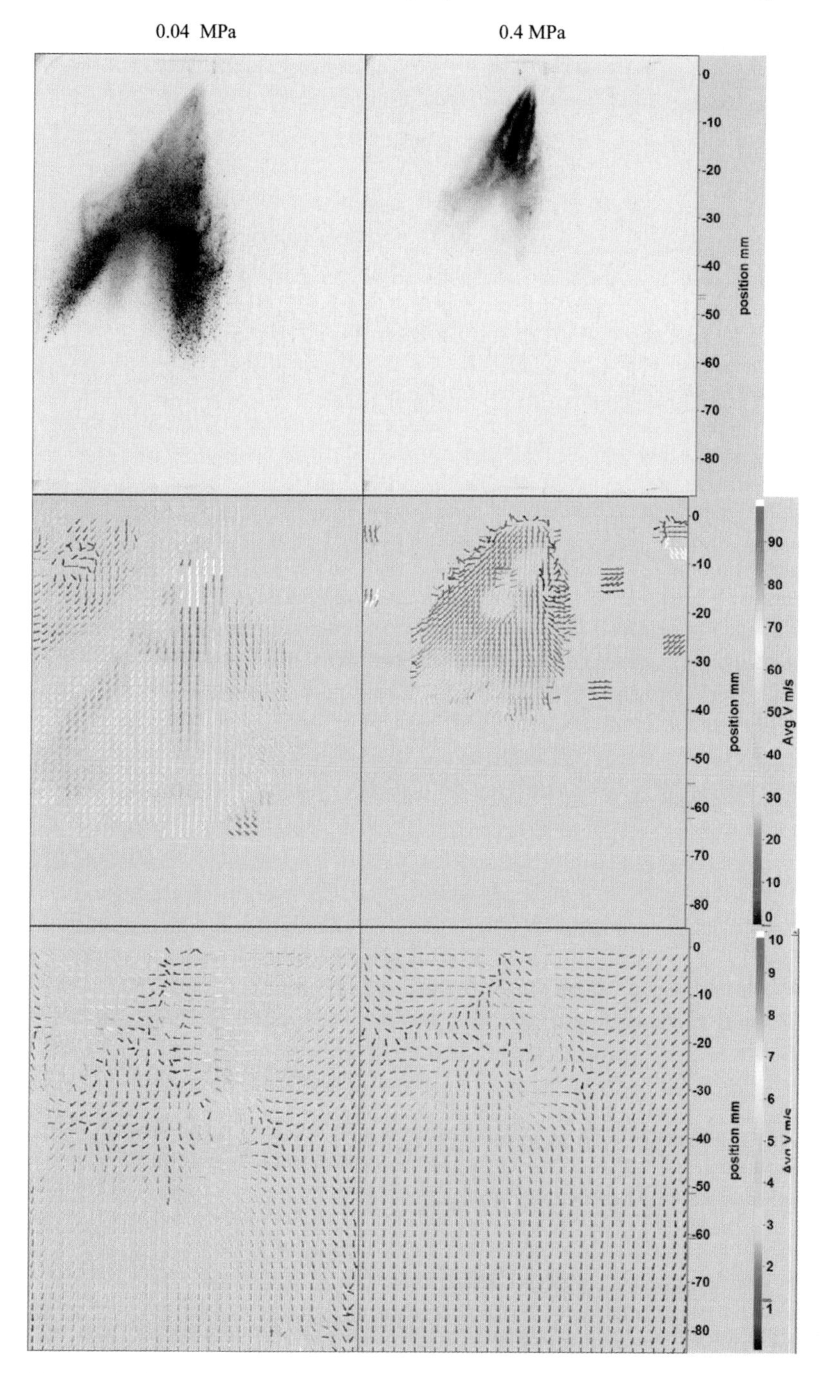

Fig. 5. Influence of the chamber pressure on fuel propagation: *top:* raw scattering images of the fuel jet, *center:* calculated fuel jet flow, *bottom:* calculated surrounding gas flow field

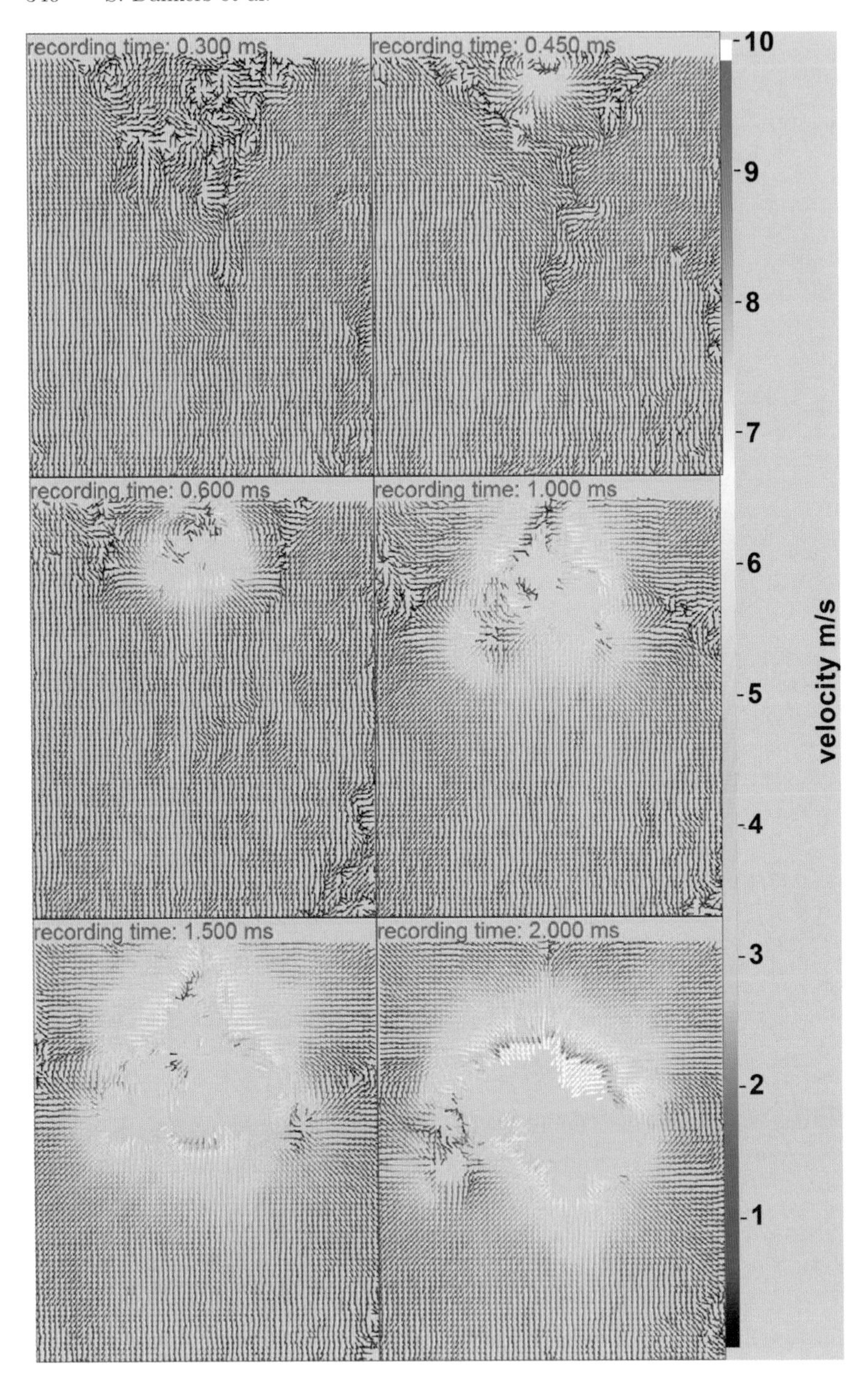

Fig. 6. Temporal evolution of the surrounding gas flow during the injection

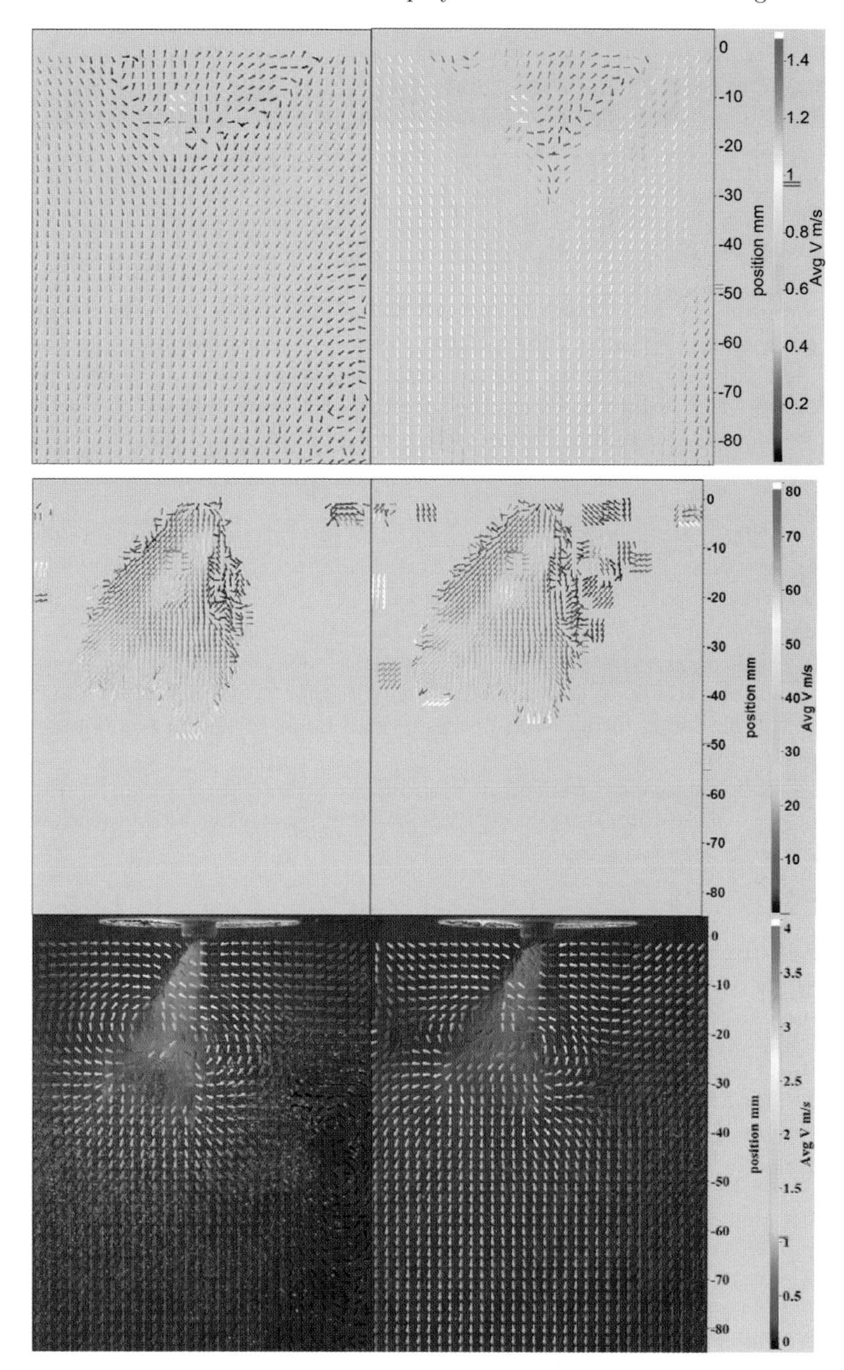

Fig. 7. Influence of the chamber air mass flow (*left side* $20\,\mathrm{m}^3/\mathrm{h}$ and right side $85\,\mathrm{m}^3/\mathrm{h}$): *top:* accelerated gas flow before start of injection, *center:* flow of the fuel jet, *bottom:* induced air flow

region below the spray that is not influenced by the injection. There, the velocities are higher for the greater mass flow.

This insight proves the applicability of the pressure-chamber investigations.

4 Conclusions

The two-phase PIV technique was successfully applied to the characterization of the air–fuel interaction and the building mechanisms of a burnable mixture using high-pressure injectors. The investigations were made in a pressure chamber and the results can be transferred to the improvement of direct-injection concepts for spark-ignited engines where mixture preparation is an essential step. Both fuel jet flow and the induced flow of the surrounding gas could be detected and analyzed.

The vortices that form at the boundary layer and the air-entrainment flow could be visualized. Many different operating conditions regarding temperature and pressure according real engine conditions and various injectors were investigated.

It could be shown that there is no influence of the scavenging air flow on the air–fuel interaction. The provided knowledge represents important input data for the modeling of injection processes and helps to design combustion processes.

For simultaneous measurement of both phases at least three pictures have to be taken using an image splitter, since there is no overlap in the possible time delays for the two phases.

References

[1] T. Honda, M. Kawamoto, H. Katashiba, M. Sumida, M. Fukutomi, K. A. Kawajiri: Study of mixture formation and combustion for spray guided DISI, in SAE 2004-01-0046 (2004)
[2] D. Probst, J. Ghandhi: An experimental study of spray mixing in a direct injection engine, Int. J. Engine Res. **4**, 27–45 (2003)
[3] J. Raposo, G. Rottenkolber, W. Dullenkopf, K. Hentschel, W. Merzkirch, S. Wittig, J. Gindele, U. Spicher: Phase separation technique for simultaneous two-phase flow measurements using PIV, in *8th Int. Conf. on Laser Anemometry – Advances and Applications* (1999)
[4] H. Kubach, J. Gindele, U. Spicher: Investigation of mixture formation and combustion in gasoline direct injection engines, in *Int. Fall Fuels and Lubricants Meeting and Exposition*, SAE-PAPER-Nr.: 2001-01-3647 (2001)
[5] D. P. Towers, C. E. Towers, C. H. Buckberry, M. Reeves: Directionally resolved two-phase flow measurements with fluorescent particles and colour recording, in *9th International Symposium on Applications of Laser Techniques to Fluid Mechanics* (1998)

[6] M. Gavaises, C. Arcoumanis: Modelling of sprays from high-pressure swirl atomizers, Int. J. Engine Res. **2**, 95–117 (2001)

Index

High-Speed PIV: Applications in Engines and Future Prospects

David Towers and Catherine Towers

School of Mechanical Engineering, University of Leeds, Leeds, LS2 9JT, UK
d.p.towers@leeds.ac.uk

Abstract. In this chapter we present the development of a high-speed imaging system for the measurement of automotive in-cylinder flows and sprays. High-speed imaging devices with framing rates of 12 000 images per second or more produce engine data at a resolution of at least one crank angle degree crankshaft rotation with engine speeds of up to 2000 rpm. In combination with high repetition rate pulsed lasers synchronized with the camera it is shown that data on liquid-fuel distribution can be obtained and analyzed. With the introduction of artificial tracer particles and appropriate data-processing techniques planar vector fields are obtained at the same temporal resolution. For both the spray and flow image sequences the data is considered as 4-dimensional: two spatial axes in the measurement volume, one temporal axis within an engine cycle and a further temporal axis from one engine cycle to another. A statistical analysis of the data produces information on the cyclic variability of the spray and flow fields that are critical to modern combustion system design. Whilst a high volume of information can readily be created, the industrial requirements are for validation of computational models or to provide insight into the mixture-preparation mechanisms as a means of evolving the design process. Therefore, an emphasis of this chapter is to show how the large quantities of detailed optical measurement data may be reduced in such a way as to pass on the most significant performance data to powertrain specialists.

1 Introduction

Automotive combustion remains a challenging area of research owing to increasingly strict emissions legislation coupled with ever more demanding consumer performance requirements. To make progress requires an increased understanding of combustible-mixture preparation, the combustion process itself and the formation of emissions. The ensemble-average development of the in-cylinder flow field and the breakdown of bulk flow features into turbulence during the compression stroke are known to affect burn rates. In addition to considering the temporal development of the flow field averaged over many engine cycles, the presence of cyclic variations in the bulk flow can critically affect combustion performance [1,2]. The affect of cyclic variability on combustion is amplified in direct-injection combustion, e.g., gasoline direct injection (GDI), as the flow governs fuel transport as well as the conditions for flame propagation.

A. Schroeder, C. E. Willert (Eds.): Particle Image Velocimetry,
Topics Appl. Physics **112**, 345–361 (2008)

There are three primary sources of information that form the basis for combustion system development: thermodynamic engines, computational fluid dynamics (CFD) and optical diagnostics. Each source provides unique information. Thermodynamic engines provide overall system-performance data and critically enable engine emissions to be quantified accurately. However, no information is obtained on the detailed mechanism of mixture preparation or combustion. Reynolds-averaged CFD codes potentially provide full 3-dimensional data on all fluid properties and are resolved through the engine cycle. Furthermore, the development of large eddy and direct numerical simulations offer the potential to examine cyclic variability effects. However, appropriate validation remains a concern particularly when trying to incorporate spray and combustion models. Optical diagnostics based on particle imaging (Mie scattering) provide both qualitative and quantitative information on spray development and in-cylinder flow. Spectroscopic techniques offer the potential for postignition and species-specific measurements, e.g., of trace-gas production. It has been demonstrated that appropriate optical engines can be designed that reproduce the geometry and compression ratio of thermodynamic engines. The major differences between thermodynamic and optical engines are in the heat-transfer rates and in the piston rings used, depending on whether the rings run on the glass barrel, as is the case for full-stroke optical access, or whether the rings run on a metal barrel below a small length of optical barrel giving access only to the top end of the stroke. In either case, the crevice volume is normally slightly increased. With modern lasers and detectors, information at high temporal and spatial resolution is available for validation of computational models, to correlate with the performance of equivalent geometry thermodynamic engines or to be used directly within the combustion development process.

In this chapter we focus on techniques based on Mie scattering for imaging sprays and artificial tracer particles added to the flow field to enable 2-dimensional flow-velocity mapping through particle image velocimetry (PIV). The measurement of single vector fields in engines by PIV has been achieved for some time initially in a plane parallel to the head gasket [3] and subsequently imaging a bore centerline plane via a corrective optic to remove the distortion generated in imaging through a glass barrel [4]. However, the characterization of cyclic variability and turbulence in in-cylinder flows is not routine. Cyclic variability was quantified initially from single-point laser Doppler anemometry data [1] utilizing the time series of information produced and temporal filtering to isolate the high-frequency turbulence from the bulk-flow components. However, single-point techniques measuring one or two components of the velocity vector are very time consuming to apply for planar or volumetric mapping in an engine, particularly when a range of operating conditions need to be compared. Full-field approaches such as PIV and particle tracking velocimetry (PTV) offer the opportunity to break through the data-acquisition bottleneck, significantly reducing rig test time and transferring the issue to time-efficient data processing. However, single snapshots of the velocity vector field as obtained from a typical double-frame PIV camera are not sufficient to separate the cyclic variability in a particular engine cycle from the turbulence in that cycle as a time series of information is traditionally needed for such an analysis. Initial attempts at achieving a full-field analysis of cyclic variability in engines was made in water-analogy rigs operating at a speed that was compatible with the cameras available at the time [5]. More recently, the use of higher-speed PIV systems, operating at image rates of a few 100 Hz have been used in conjunction with proper orthogonal decomposition to enhance the temporal resolution [6].

The work presented in this chapter concerns the quantification and data reduction of cyclic variability information in spray structure and planar velocity fields obtained from high-speed imaging in a single-cylinder optically accessed engine. The work is an expansion of a previous paper concerning enhanced data-processing techniques to give high validation rates in the PIV data obtained such that an automatic analysis of cyclic variability was possible [7]. Here, we include cyclic variability processing of high-speed Mie scattering images of direct injection gasoline sprays and examine the potential impact of recent developments in laser and camera hardware on the data obtainable from high-speed PIV in engines.

2 Experimental Systems

2.1 Optical Setup

The high-speed imaging equipment used in the experiments consisted of a Kodak 4540 CCD camera and 20 W average power copper-vapour laser. These hardware items are straightforwardly synchronized using the camera to define a pulse train sent to the laser to produce a pulse at the start of each camera frame. The useful framing rates available are 4.5, 9 and 13.5 kHz with spatial resolutions of 256×256, 256×128 and 128×128 pixels, respectively. The higher framing rates have been used herein and give an image and therefore pulse separation of approximately 111 and 74 µs. The disadvantage of this arrangement is that the interframe time and pulse separation for crosscorrelation analysis are always equal. Consequently, for the typical fields of view used in these experiments the PIV system has an upper velocity limit of approximately $20\,\mathrm{m\,s^{-1}}$ corresponding to a maximum particle displacement of ~ 4 pixels with 16-pixel interrogation regions used for crosscorrelation analysis. The upper velocity limit means that the system is compatible with in-cylinder flows from inlet valve closure through to the end of the compression stroke at engine speeds up to 2000 rpm. A further consequence of the long pulse separation coupled with the highly 3-dimensional nature of the flow field means that relatively thick, 1 to 2 mm, laser lightsheets must be used to increase the probability of exposing the same particles in consecutive frames. Out-of-plane motion was the primary cause of poor quality crosscorrelation

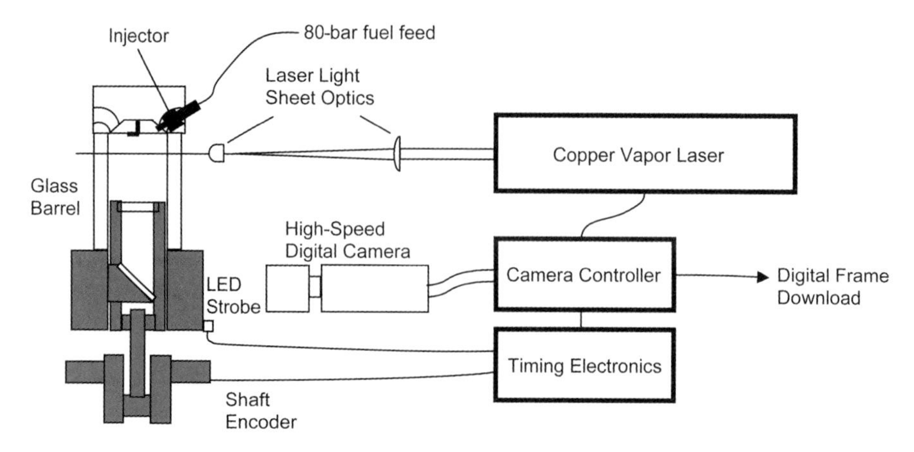

Fig. 1. Schematic representation of the optical system and engine

results from these experiments with the maximum out-of-plane displacement exceeding the design guidelines derived from previous Monte Carlo simulations [8]. It was found to be beneficial to use short focal length lenses to form the laser sheet in order to expand the beam quickly through narrow access points in order to illuminate the desired plane in the measurement volume. A typical schematic of the setup is given in Fig. 1.

2.2 Optical Engine

The optical engine used a production geometry 4-valve cylinder head and operated at a compression ratio of 11 : 1. Optical access was provided by a full-stroke fused-silica barrel and a window within the engine piston crown. The piston itself is elongated with one set of piston rings at the top running on the optical barrel and a further set of rings at the bottom running in a conventional metal barrel. A slot is machined into the piston body to enable a fixed mirror to be mounted directly beneath the piston crown optical window such that planes parallel to the head gasket can be imaged, see Fig. 1. The engine was operated at speeds up to 2000 rpm with run time limited to ~ 30 s due to temperature build up in the piston rings running on the optical barrel. Lubrication was not used between the rings and the barrel in order to prevent optical distortions in either the laser lightsheet or the images obtained. The engine incorporated a prototype hollow-cone fuel injector running from a fuel rail maintained at 80 bar via a camshaft-driven rotary fuel pump with the injector tip positioned in the combustion chamber between the inlet valves. The piston crown incorporated a bowl, slightly greater than half the cylinder bore diameter, designed to guide the directly injected spray towards the spark plug.

The engine timing was monitored using a crankshaft encoder and custom timing electronics provided crank-angle-resolved triggers for the fuel-injection system and high-speed camera. To provide a direct crank-angle reference in the captured image sequences an LED was mounted at the edge of the image field and triggered at a defined point within every engine cycle.

2.3 Seed-Particle Selection

The compromise in pulse separation and hence laser lightsheet thickness coupled with the available pulse energy of $\sim$ 2–3 mJ mean that conventional oil seeding cannot be satisfactorily imaged. Therefore, larger hollow microspheres were employed as a seeding material, Expancel 551 DE 40 from [9]. The thin-shelled particles had a mean diameter of 35 micrometers and a specific gravity of $\sim$ 0.04 giving a 95 % amplitude flow-following accuracy for a 1-kHz flow fluctuation rate in air at standard pressure and temperature, from [10]. This fluctuation rate corresponds to a swirl ratio of 30 in an engine operating at 2000 rpm, which is approximately a factor of 3 higher than that in typical GDI engines. In addition, the flow-following ability improves during the compression stroke owing to the increase in fluid density and viscosity. Care was needed in using the acrylonitrile microballoons to minimize the buildup of static charge and hence reduce the likelihood of the particles becoming trapped between the piston rings and the optical barrel, thereby damaging the barrel or leading to a catastrophic failure of the engine. The seed density was therefore controlled leading to approximately 3–4 particles per interrogation region of 16 $\times$ 16 pixels.

3 High-Speed Spray Imaging

To understand the temporal development of a GDI spray requires imaging along multiple planes to include data from the 3 orthogonal axes. With the laser sheet along the spray axis the penetration of the spray as a function of time and the spray cone angle may be measured, see Fig. 2. In this figure, images are shown at a temporal resolution of 5 crank angle degrees (CAD) and from two engine runs: the first using a lens F# of 5.6 in order to show the early spray propagation where the spray is dense and the second with an F# of 2.8 to bring out the sparse liquid spray detail towards the end of the compression stroke. The early images, up to 70 CAD before top dead centre (BTDC) at the end of the compression stroke, show both the main spray on the right-hand side coming from the injector at the upper right of each image and a reflection of the spray from the inside left surface of the optical barrel. In the early stage of injection the prespray may be observed (at 90 and 85 CAD BTDC). The main spray appears to slow from 85 to 75 CAD BTDC; high-speed imaging on a plane orthogonal to the main spray propagation direction reveals that the spray is in fact moving rapidly away from the original spray axis as it is

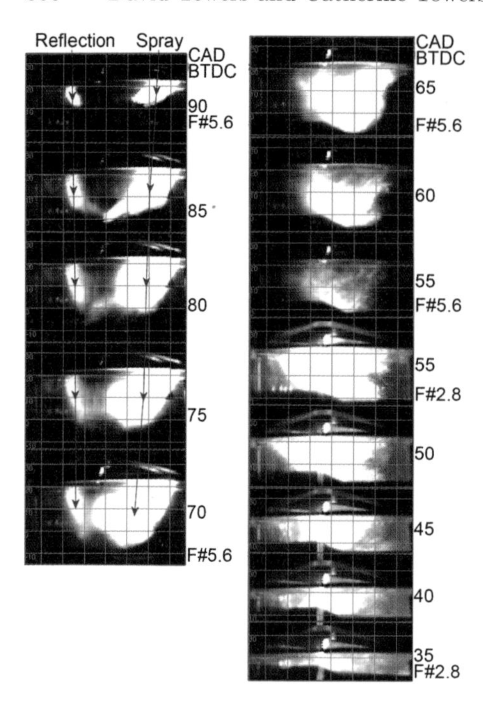

Fig. 2. Temporal development of a GDI spray imaged along the centerline of the injector, start of injection 95 CAD BTDC compression stroke. The images are at 5 CAD intervals, from 90 to 55 CAD BTDC at a lens F# of 5.6 and from 55 to 35 CAD BTDC at an F# of 2.8

influenced by the in-cylinder axial swirl generated at the intake ports. The change in lens F# allows the remaining liquid-phase fuel to be imaged up to the typical ignition timing at 30–35 CAD BTDC and the fuel cloud is seen to reside centrally beneath the spark plug in an ideal location for ignition. The spark plug can be identified as an increasingly bright region directly above the head-gasket plane.

Imaging in a vertical plane (see Fig. 1) orthogonal to the main spray propagation direction reveals the affect of axial swirl on spray location, see Fig. 3. The left-hand side of Fig. 3 shows the temporal development of the spray with maximum axial swirl ratio (2.8) and at an engine speed of 2000 rpm. It can clearly be seen that the spray arrives left of center with part of the spray falling to the left of the piston bowl. In contrast, at the maximum axial swirl ratio minus 40 % the spray propagation is initially more central and is contained within the piston bowl, thereby maintaining the location of the fuel cloud with respect to the spark plug. These results are consistent with the operation of a thermodynamic engine with the same geometry; at maximum axial swirl significantly higher hydrocarbon emissions were produced compared to those with 40 % less axial swirl ratio. It should be noted that Figs. 2 and 3 refer to different cylinder-head and piston-crown geometries.

Spray imaging by Mie scattering is also useful in identifying problem areas when a GDI spray is introduced during the induction stroke for engine operation with a homogeneous fuel–air mixture. Figure 4 shows the view

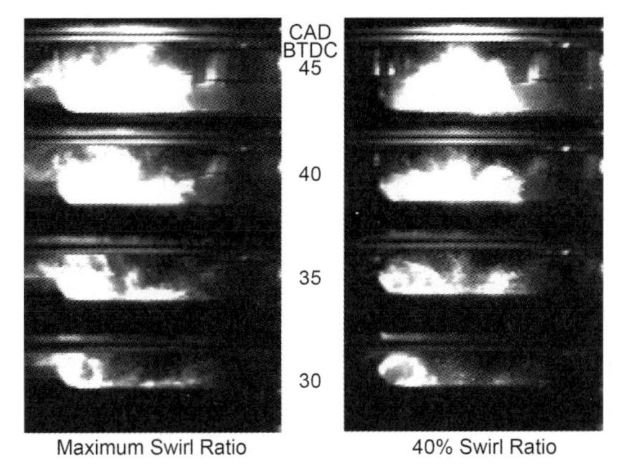

Fig. 3. GDI spray arrival in the piston bowl for maximum and 40 % axial swirl levels, the imaged plane is orthogonal to the direction of spray propagation. The images are at 5 CAD intervals

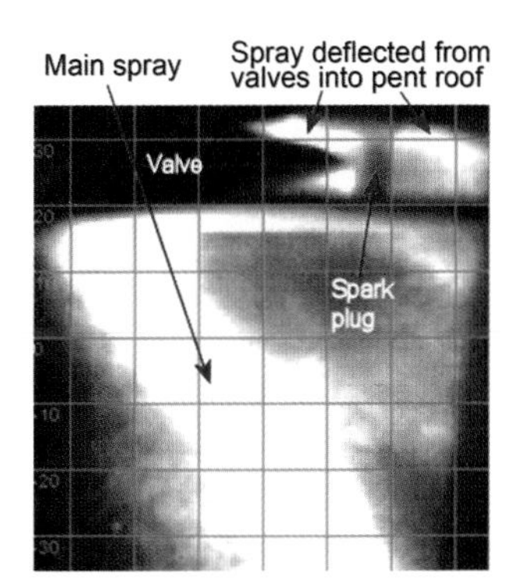

Fig. 4. GDI spray following injection on the induction stroke for homogeneous engine operation with the start of injection at 60 CAD ATDC at the beginning of the induction stroke and the image is shown at 75 CAD ATDC

along the bore and spray centerline where the inlet valves are open and the main spray can be seen to propagate below the valves. In this case, the start of injection was set at 60 CAD after top dead center (ATDC) at the beginning of the induction stroke, and the image is shown at 75 CAD ATDC. However, parts of the spray impinge on the back of the valves, and are deflected up into the head of the engine wetting the spark plug and the exhaust valves, again leading to poor hydrocarbon emission performance and potential ignition problems. Note that Fig. 4 was recorded from the opposite side of the engine compared to Fig. 2, hence the spray is seen to derive from the top left of the image.

3.1 Statistical Analysis of Cyclic Repeatability of Spray Propagation

A more detailed analysis of the spray under the influence of various experimental parameters, e.g., start of injection (SOI) timing, injection duration and in-cylinder flow swirl, can be made using a statistical analysis. A preprocessing stage may be performed by capturing an image sequence without the spray and this sequence subtracted from each cycle of spray images. Then, a threshold operation is performed to identify pixels where fuel is present. An adaptive approach to threshold selection is to calculate the grayscale corresponding to the image background, I_{b}, e.g., from the peak of an image histogram, and then using the measured maximum intensity, I_{Max}, the threshold is found as: $I_{\mathrm{t}} = I_{\mathrm{b}} + p(I_{\mathrm{Max}} - I_{\mathrm{b}})$ where p is a percentage. The most stable results have been found using $p = 10\,\%$. This operation is performed on each image in a sequence from a single injection event and repeated for many engine cycles. A new binary image sequence is generated where each pixel is white when:

$$\frac{\sum_{i=1}^{n}[IF(I(x,y,t,i) > I_{\mathrm{t}}) = 1,\ \mathrm{ELSE}\ 0]}{n} > \mathrm{PR}\,,$$

where (x, y) refer to a particular pixel, t is the time within an engine cycle normally expressed in CAD, i is the cycle number, n is the number of cycles of data obtained and PR is a defined probability. Most important to the development of the combustion system is to know where fuel is always present, i.e., to obtain the image sequence with $\mathrm{PR} = 1$. A representative set of results are summarized in Fig. 5 obtained from a plane parallel and just beneath the head gasket. Each row of the image corresponds to images at a particular crank angle. The columns are for a particular probability, from the left, $\mathrm{PR} = 0.5$, 0.8, 1. The variability in the spray field can be seen between the columns where towards the end of the compression stroke the size of the fuel cloud with $\mathrm{PR} = 0.5$ is significantly larger than that with $\mathrm{PR} = 1$. The data may be reduced further to extract the distance between the center of the fuel cloud and the spark plug for a particular probability as a means of comparing different engine configurations. Four engine setups are compared in this way in Fig. 6 with $\mathrm{PR} = 1$. Clear differences can be observed between the different configurations. The data show that the optimum configurations are with maximum swirl and an early start of injection (97 CAD BTDC) and with 40 % less swirl and later injection (73 CAD BTDC). Further data, similar to Fig. 3, showed that with early injection some of the fuel was lost outside the piston bowl. Hence, from a combination of the statistical analysis and the high-speed image sequences the optimum configurations can be determined.

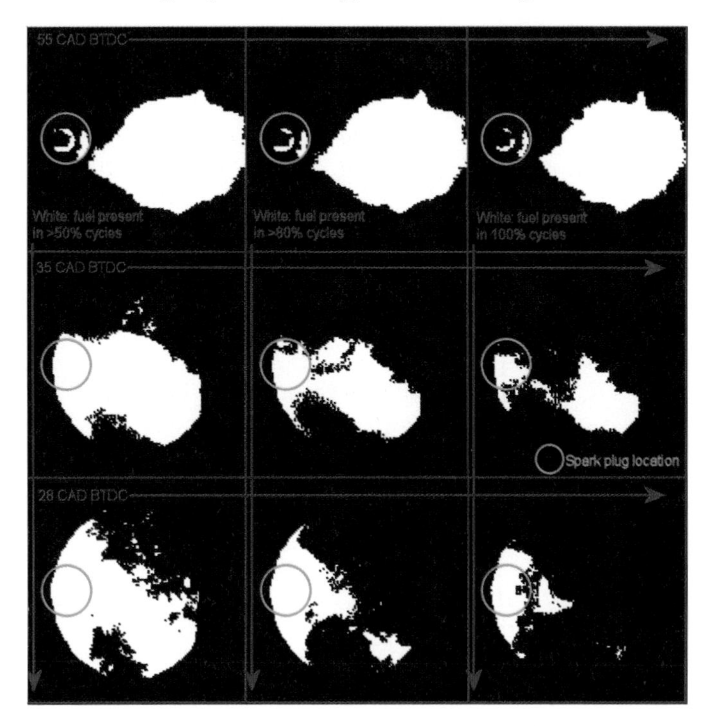

Fig. 5. Binary images showing the probability of fuel presence and resolved through the injection cycle

4 High-Speed PIV

The high-speed, temporally resolved particle images are processed to obtain the corresponding velocity vector fields using a normalized cross covariance correlation algorithm [11]. The analysis is problematic owing to the low seeding density that must be used for compatibility with the optical engine and that the analysis must be fully automated owing to the large quantity of vector maps that must be obtained in order to determine cyclic variability – vector fields resolved at 1 CAD resolution in many engine cycles. To improve the reliability of the cycle-resolved vector fields the product of crosscorrelation functions from consecutive image pairs is determined for each interrogation region [7]. For the particular results shown here, temporal sequences of 4 crosscorrelation functions were found to be optimal, representing a compromise between improved signal-to-noise ratio and the average timescale for flow similarity. The vector fields were postprocessed using a median filter to remove outliers, with 2nd or 3rd peak vectors substituted if appropriate and based on the fit with the surrounding vectors. It is estimated that some outlier vectors remained at the edge of the measurement field after the postprocessing operations and could be observed in $\sim 10\,\%$ of the vector maps in any engine cycle.

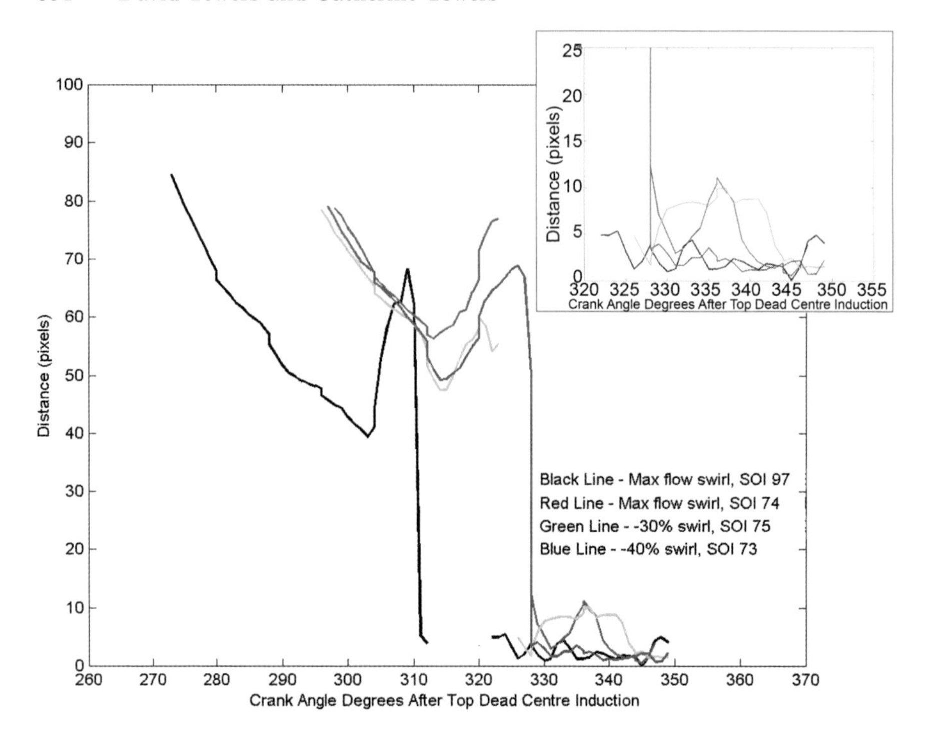

Fig. 6. Distance of fuel cloud from the spark plug for 4 engine configurations, *inset:* detail at the end of the compression stroke

4.1 Cyclic Variability Analysis of Temporally Resolved PIV Data

The meaning of cyclic variability is illustrated in Fig. 7. At each interrogation region a temporally resolved velocity is obtained and is shown schematically as the black line in Fig. 7. This velocity can be thought of as being made up of a number of components: a velocity that represents the average over all engine cycles, a term for the cyclic variability and a term for flow turbulence. The flow turbulence can be considered to be a high-frequency fluctuation and hence may be removed using appropriate temporal or Fourier-domain filtering resulting in a single cycle mean velocity at that point, shown as the red line in Fig. 7 [1]. By taking the average of the single cycle mean velocities over a representative number of engine cycles gives the ensemble mean velocity at the point, shown as the blue line in Fig. 7. The difference between the ensemble and single cycle means gives the cyclic variability for that particular engine cycle; graphically this is the difference between the red and blue lines in Fig. 7. To obtain a statistically representative value the individual cyclic variability values are averaged over a number of cycles to generate a scalar result at each interrogation region and resolved through time.

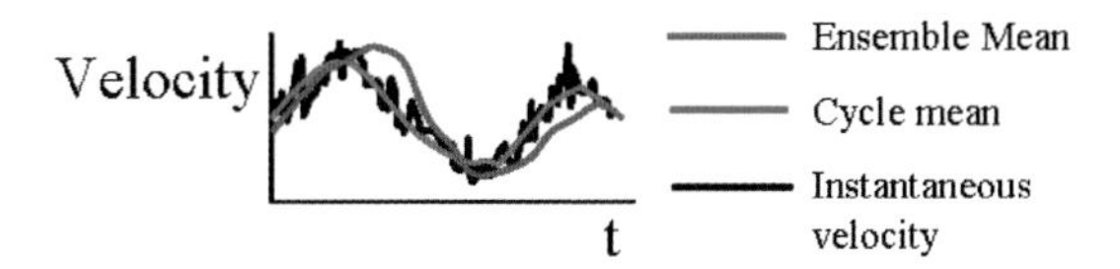

Fig. 7. Illustration of the ensemble-average velocity, the single cycle mean velocity and the instantaneous velocity from a single cycle

In this work the measured temporally resolved velocity at each interrogation region has been lowpass filtered in the temporal domain with a cutoff frequency at $\sim 300\,\text{Hz}$. This particular cutoff frequency was selected such that the smallest resolvable eddies would be removed [7]. The removal of the turbulence component was found to also remove most of the remaining outlier vectors as the outliers were typically present in single vector maps. Therefore, the outliers do not contribute to the calculated cyclic variability measure.

4.2 Results

The data presented were obtained at an engine speed of 2000 rpm, using a prototype cylinder head and piston crown that were practically identical in geometry to those being used in thermodynamic engine tests. By means of a butterfly valve in one of the two inlet port tracts, the in-cylinder axial swirl level could be varied. Two settings were evaluated, referred to as A and B, corresponding to maximum axial swirl and a medium level. Representative data from a single engine cycle during the compression stroke is shown in Fig. 8. The images show the temporal flow development in 5 CAD increments from 260 to 315 CAD BTDC. It can be seen that there is excellent spatial consistency across the vector fields. Regions of high-velocity tend to occur because of locally high-velocity jets of air that are moving through the measurement plane.

To summarize the results of the statistical analysis, Fig. 9 shows the ensemble-average vector fields at crank angles of 275 and 300° BTDC and the scalar average cyclic variability distributions at the same crank angles for engine configurations A and B. It can be seen that the cyclic variability maps also show good spatial consistency, indicating that the dominant variability effects occur in specific areas that are larger than the interrogation regions. Fifteen engine cycles of data were used for the analysis for each engine configuration. Whilst this number of cycles is relatively low, it has been identified that the mean flow velocity after 15 cycles at any particular location remains within $1\,\text{m}\,\text{s}^{-1}$ of that evaluated from the first 10 engine cycles. Therefore, the ensemble-averaged velocity field has been determined to an uncertainty of $1\,\text{m}\,\text{s}^{-1}$. In contrast, the scalar cyclic variability maps show peak values in excess of $4\,\text{m}\,\text{s}^{-1}$ and hence the cyclic variability data is statistically relevant even though the number of cycles analyzed is comparatively low.

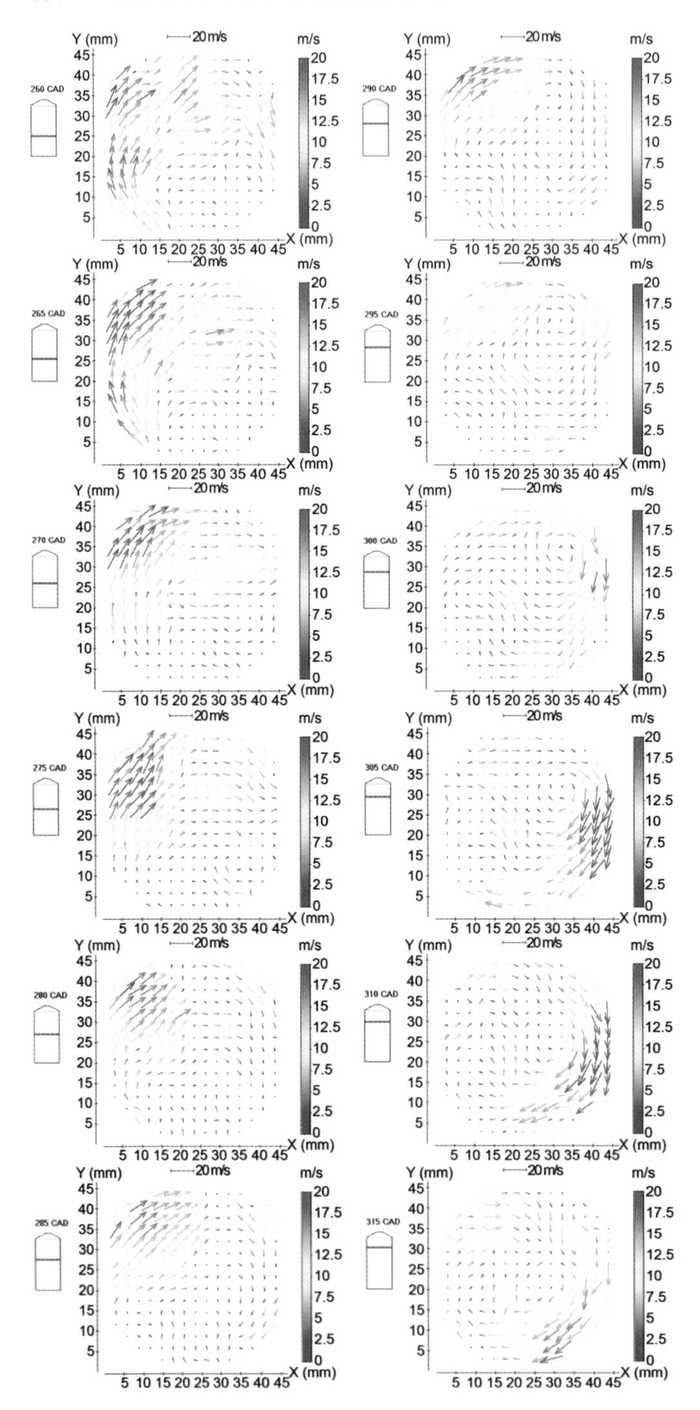

Fig. 8. Vector fields at 5 CAD increments from a single engine cycle for engine configuration B

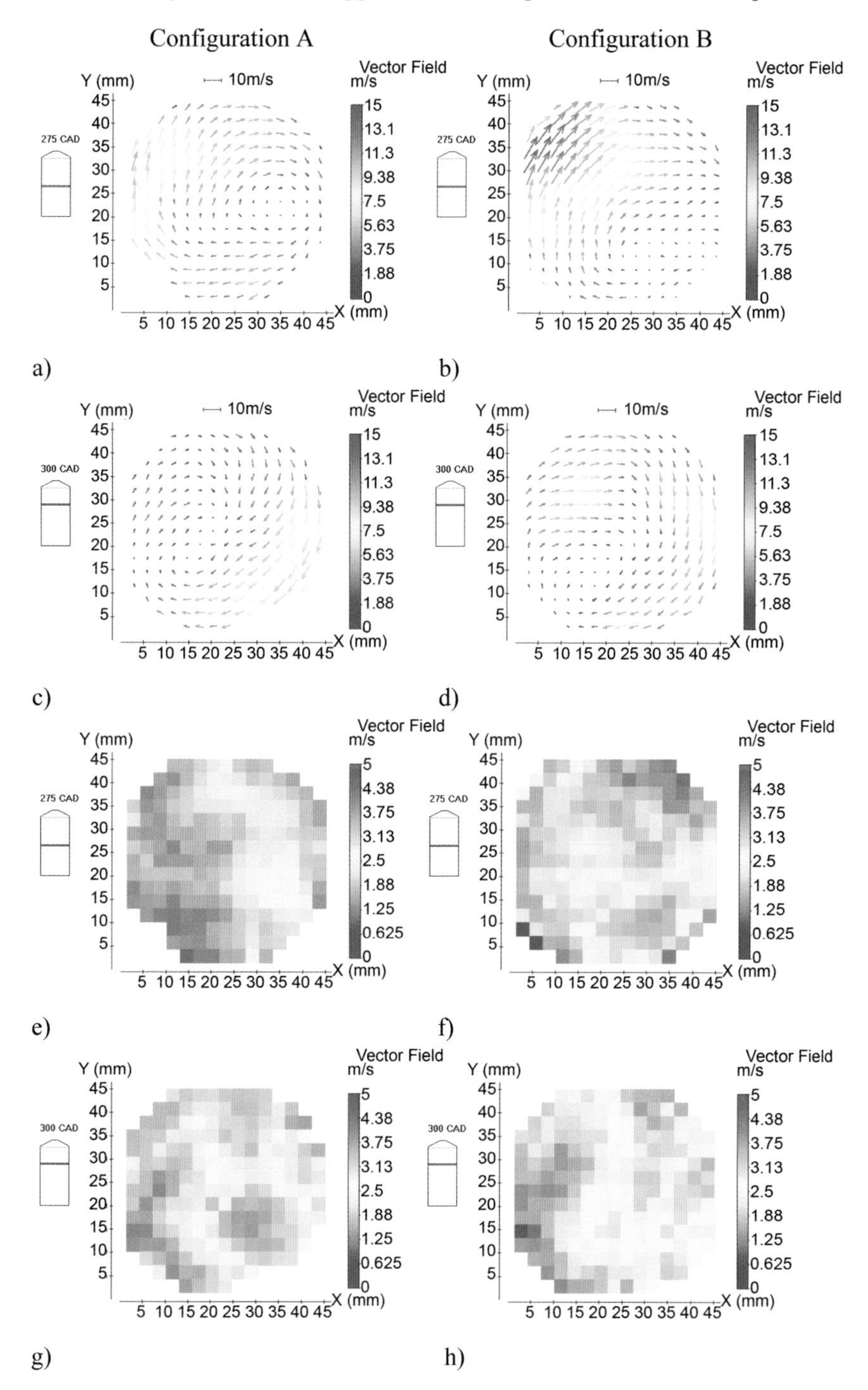

Fig. 9. Vector fields, (**a**–**d**), and cyclic variability distributions, (**e**–**h**), for 2 crank angles, 275 and 300 BTDC, and for engine configurations A and B

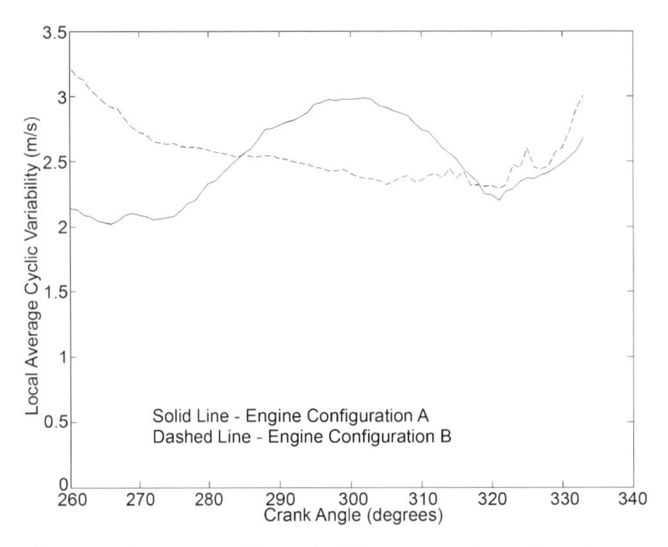

Fig. 10. Average cyclic variability across the region that most affects spray transport for engine configurations A and B

The data show interesting trends between the two engine configurations. There are increased levels of cyclic variability in the lower right of the distributions in Figs. 9f,g corresponding to 275 CAD BTDC with configuration B and 300 CAD BTDC with configuration A. Therefore, the cyclic variability is seen to peak at different points in the engine cycle for the two configurations. Given that a typical start of injection timing for the spray occurs around 290 CAD BTDC and that the spray propagates directly into the region of higher cyclic variability, it would be expected that configuration B would lead to less variability in spray transport than configuration A. Furthermore, when looking at the ensemble mean vector field for configuration B at 300 CAD BTDC it can be seen that the flow speed is considerably lower than for configuration A and hence the spray is less likely to be shifted away from the center of the combustion chamber for configuration B. These two effects combine to give a combustion system that is expected to give greater reliability in configuration B. Indeed, this conclusion has been borne out in the results of thermodynamic engine testing where the covariance of the indicated mean effective pressure (IMEP) – a value representing how smooth the engine is operating – shows lower levels of variability for configuration B than for configuration A. The cyclic variability data may be reduced to a single line graph for comparison of the two configurations by taking a mean value of variability over the region that most affects spray transport. The result for configurations A and B is shown in Fig. 10 and exhibits the same trends as have been described above.

5 Future Prospects in High-Speed PIV

Since the time when these experiments were performed there have been significant developments in both laser and camera hardware for time-resolved measurements. For example, dual-cavity solid-state lasers offering $> 10\,\mathrm{kHz}$ repetition rate are now commercially available. Such systems offer considerable advantages over the use of a single copper-vapour laser in that the pulse separation for crosscorrelation analysis is then independent of the pulse repetition rate. Hence, higher flow velocities could be measured and hence a more complete analysis of the full induction and compression stroke could be performed. Furthermore, the beam quality is considerably higher from a solid-state laser allowing the generation of thinner lightsheets with greater energy density. Therefore, the use of more conventional oil-drop seeding (rather than large hollow microspheres) is expected to be successful leading to the potential of extracting the per-cycle flow turbulence in addition to the variability. Similar developments have taken place in high framing rate digital cameras. The Kodak4540 system used in this work has a pixel clock of 200 MHz, whereas modern cameras offer over 10 times this value, to 3 GHz, giving a 512×512 pixel image at 10 kHz framing rate. The increased spatial resolution can be used to give a greater number of interrogation regions and hence vectors within the measurements produced. At the same time, more sophisticated optical systems are being developed for 3D particle metrology from a single view [12] or using multiple views [13] that ultimately offer the potential for genuinely 4D quantitative measurements in such systems. The increased capability of new high-speed PIV hardware will also enable measurements of flow variability in the presence of the directly injected spray, thereby enabling the influence of the spray on the mean and variability of the flow field to be determined.

6 Conclusions

In this chapter, the use of high-speed, or temporally resolved imaging for the analysis of in-cylinder sprays and flows in internal combustion engines has been presented. The benefit of having the temporally resolved image sequence has been exploited in order to obtain a statistical analysis of both the spray development and vector-field variability. Furthermore, it has been shown that by suitable data analysis full-field spray and velocity data can be reduced to the temporal variation of a single parameter in order to quantify the essential characteristics of the combustion system and hence enable engineering decisions to be made.

The hardware available at the time of these experiments led to the requirement for specialized data-processing algorithms in order to reliably and automatically quantify the vector fields. To a large extent these difficulties may now be overcome using state-of-the-art lasers and cameras. Therefore, it is expected that temporally resolved PIV and the analysis of nonrepeatable cyclic phenomena or genuinely transient behavior in spray and flow systems should become more widespread.

References

[1] J. B. Heywood: Fluid motion within the cylinder of internal combustion engines – the 1986 Freeman scholar lecture, Trans. ASME, J. Fluids Eng. p. 109 (1987)

[2] C. Arcoumanis, J. H. Whitelaw: Fluid mechanics of internal combustion engines - a review, in *Proc. Instn. Mech. Engrs.* (1987) pp. 201, C1

[3] D. L. Reuss, R. J. Adrian, C. C. Landreth, D. T. French, T. D. Fansler: Instantaneous planar measurements of velocity and large-scale vorticity and strain rate in an engine using particle image velocimetrypaper, in *SAE Cong. and Exposition* (1989) paper # 890616

[4] M. Reeves, C. P. Garner, J. C. Dent, N. A. Halliwell: Study of barrel swirl in a four valve optical I.C. engine using particle image velocimetry, in *Proc. 3rd Int. Symp. on Diagnostics and Modelling of Combustion in Internal Combustion Engines* (1994) pp. 426–436

[5] W.-C. Choi, Y. G. Guezennec: Measurement of cycle-to-cycle variations and cycle-resolved turbulence in an IC engine using a 3-D particle tracking velocimetry, JSME Int. J. B **41**, 991–1003 (1998)

[6] P. Druault, P. Guibert, F. Alizon: Use of proper orthogonal decomposition for time interpolation from PIV data, application to the cycle-to-cycle variation analysis of in-cylinder engine flows, Exp. Fluids **39**, 1009–1023 (2005)

[7] D. P. Towers, C. E. Towers: Cyclic variability measurements of in-cylinder engine flows using high speed particle image velocimetry, Meas. Sci. Technol. **15**, 1917–1925 (2004)

[8] R. G. Keane, R. J. Adrian: Theory of cross-correlation analysis of PIV images., Appl. Sci. Res. **49**, 191–215 (1992)

[9] Boud Minerals & Polymers: (2004) URL: `http://www.boud.com/expancel.htm`

[10] A. Melling: Tracer particles and seeding for particle image velocimetry, Meas. Sci. Technol. **8**, 1406–1416 (1997)

[11] H. Huang, D. Dabiri, M. Gharib: On errors of digital particle image velocimetry, Meas. Sci. Technol. **8**, 1427–1440 (1997)

[12] C. E. Towers, D. P. Towers, H. I. Campbell, S. J. Zhang, A. H. Greenaway: Three-dimensional particle imaging by wavefront sensing, Opt. Lett. **31**, 1220–1222 (2006)

[13] G. Elsinga, F. Scarano, B. Wieneke, B. W. van Oudheusden: Tomographic particle image velocimetry, Exp. Fluids **41**, 933–947 (2006)

Index

PIV in the Car Industry: State-of-the-Art and Future Perspectives

Davide Cardano, Giuseppe Carlino, and Antonello Cogotti

Aerodynamics and Aeroacoustics Research Center, Pininfarina, Italy.
g.carlino@pininfarina.it

Abstract. This contribution provides an overview on the applicability of particle image velocimetry (PIV) in an industrial environment as a full-scale automotive wind tunnel and its potentiality in this specific field. Experiences developed in the last nine years in the Pininfarina Wind Tunnel are summarized here. Emphasis is placed on the experimental aspects and the specific problems and unique solutions adopted in this specific field.

1 Introduction

Aerodynamics and aeroacoustics are two highly relevant aspects of car development, since they are directly linked to passengers' safety and comfort and to car fuel consumption. A detailed knowledge of both time-averaged and time-resolved flow structures is required to answer questions posed by aerodynamic and aeroacoustic issues. Furthermore, Computational Fluid Dynamics (CFD) in the automotive field is extending its importance even in unsteady simulations, but it requires validation with experimental measurements.

All these aspects and the growing need to shorten the car development time suggest the use of PIV as a useful experimental technique in automotive wind tunnels. See [1] for further details and references about PIV.

2 PIV in the Car Industry: Requirements

In a full-scale automotive wind tunnel, multihole pressure probes and hot wires are traditionally used to investigate the properties of the flow field around the car body. Even if these probes are extremely accurate, their main drawback is that they are often not able to measure all the three velocity components simultaneously and, furthermore, they are basically point-by-point measurement devices. They can be used to measure extended areas of the flow field with the time-consuming flying-probe technique, by mounting them on a traversing gear, see [2]. In principle, stereoscopic particle image velocimetry is able to overcome these issues, nonetheless, it has to deliver overall better performance and more information than other conventional

A. Schroeder, C. E. Willert (Eds.): Particle Image Velocimetry,
Topics Appl. Physics **112**, 363–376 (2008)

measurement techniques. In order to achieve this result, PIV has to satisfy some operational and performance requirements that usually are in conflict with each other. A PIV system in a full-scale automotive wind tunnel has to measure the three velocity components over large areas of the flow field (typically the order of magnitude is about $1\,\mathrm{m}^2$). The flow field around a vehicle involves wide areas and three-dimensional structures of different scale magnitudes, as wakes and vortices, which are usually responsible for a large amount of energy loss. Therefore, the spatial resolution also has to be quite high, of the order of few millimeters. Moreover, the flow-field structures are usually present in areas with problematic optical access around the vehicle body. Therefore, the PIV system has to be compact and light to be mounted on a traversing system in order to ease the investigation of these areas. While waiting for some technical improvements, a compromise among these issues is currently necessary.

The use of the PIV technique in full-scale automotive wind tunnels is strongly conditioned by the industrial environment itself. In such a facility, the economical and safety aspects are prevalent. For these reasons, the PIV system has to be easy and fast to setup. It is better to have the possibility to do an "offline" calibration outside the wind tunnel test section and, once in the wind tunnel, to measure many areas around the car body, changing as little as possible the system's configuration.

High operational costs of wind tunnels and more and more restricted car time development need fast measurements and fast postprocessing techniques. The improving of computer hardware, with larger memories and faster processors, is moving to the right direction, but the overall performance is still slow for the car industry that requires results "in real time".

All the devices necessary for the measurements have to be motorized and remotely controlled by an operator staying outside the test section, mainly for safety and operational reasons. During the measurement process, the test section has to be completely inaccessible and its windows covered by laser screens.

Laser reflections can give problems not only from a safety point of view but also for image-acquisition reasons. The treatment of this problem will not be addressed in this chapter. Nonetheless, it can be noticed that any successful technique adopted to solve this issue should take into account the size of full-scale tested cars and the peculiarities of industrial environment where people need to work in the wind tunnel test section with no risk to their safety.

The distribution of the PIV tracer particles can be a challenging task too. Sometimes, it is difficult to have a good and homogeneous particle distribution inside a large industrial environment, such as a full-scale automotive wind tunnel, but it is a necessary condition in order to obtain an optimal evaluation of the flow field. Obviously the adopted tracer particles have to be nontoxic, small to better follow the flow but big enough to be detected from a long distance between the observed large area and the recording cameras.

In the Pininfarina Wind Tunnel, olive-oil droplets are injected at the inlet of the contraction area.

3 PIV in the Pininfarina Wind Tunnel

In 1997 Pininfarina became a partner of the PivNet1 Framework as the task manager for the car industry, with the aim of developing the PIV technique in a full-scale automotive wind tunnel. In 1999 Pininfarina, in cooperation with CIRA, hosted a workshop and a demonstration to check the possibility of performing PIV measurements in a full-scale automotive wind tunnel. The demonstration consisted of measuring a longitudinal wake behind a rear view mirror. It was successful, even though the measurements were only two-dimensional and the measurement area was limited to $0.25 \times 0.25\,\mathrm{m}^2$. The instrumentation used was quite bulky, nevertheless, it was confirmed that the PIV measurements were feasible in a complex industrial environment such as a full-scale automotive wind tunnel. See [3] for further details.

In 2000, given the PivNet1 experience and the foreseeable progresses in technology, the Pininfarina Aerodynamics and Aeroacoustics Research Center conceived a 3D-PIV system as a single and compact probe where laser and cameras are enclosed in a unique frame. The Pininfarina 3D-PIV probe was built in the 2001–2005 time frame and was presented at the PivNet2 Workshop held in Pininfarina in November 2005, [4–8]. This workshop marked the line between feasibility and operation of PIV in the car industry on full-scale automotive testing. It has been shown that PIV is now an advanced measuring technique that can be successfully used in the car-development process.

The main topic of interest of the PivNet2 Pininfarina Workshop was to establish a link between the aerodynamic flow field and the aeroacoustic sources of vehicles. These are subjects where theoretical investigation is still relevant and PIV measurements can provide useful data and information about the flow field where noise sources are present. In order to check the possibility to achieve results about the described tasks, a demonstration has been shown. A longitudinal wake behind a rear-view mirror has been measured by the Pininfarina 3D-PIV probe and by 2D time-resolved PIV (see [9]), while the external aeroacoustic noise sources have been measured by the Pininfarina 66-microphone array. Although some qualitative and quantitative results have been shown, at present, fundamental research is very much needed to establish a more solid link between PIV measurements and acoustic-sources identification, [10, 11].

3.1 The PF 3D PIV Probe

Pininfarina Aerodynamics and Aeroacoustics Research Center developed a compact stereoscopic PIV probe composed of two CCD cameras and a pulsed

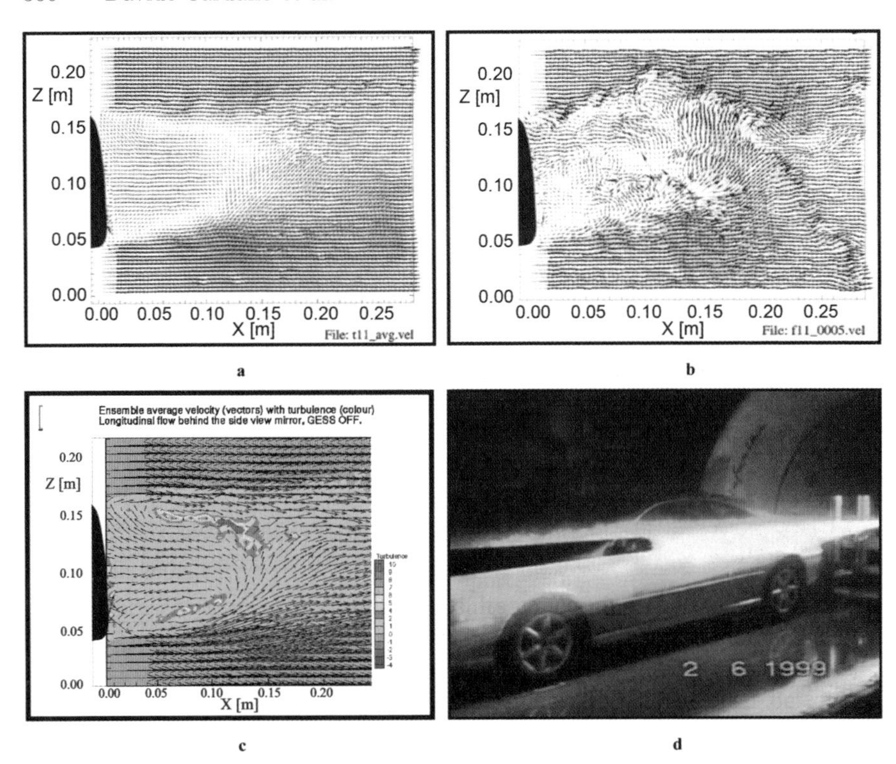

Fig. 1. Longitudinal wake behind a car rear-view mirror measured during the PivNet1 Pininfarina Workshop; (**a**) mean velocity field; (**b**) instantaneous velocity field; (**c**) turbulence flow field; (**d**) measurement setup

Fig. 2. Demonstration during the PiVNet2 Workshop held in Pininfarina in November 2005

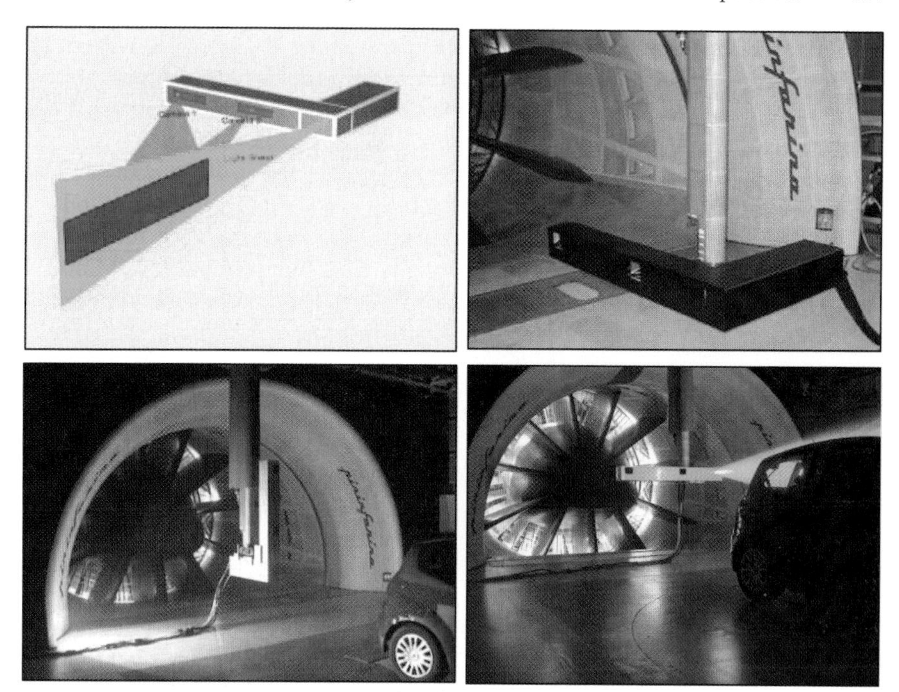

Fig. 3. The Pininfarina 3D-PIV Probe

Nd:YAG laser installed inside a unique aluminum L-shaped frame. All the hardware instrumentation installed in the probe, like Scheimpflug camera angle, focus, aperture and mirrors, is motorized and remotely controlled, while the camera signal is linked by fiber-optic cables to the computer frame grabbers.

Such a system is able to measure the three velocity components of the flow and can be easily and quickly mounted on the main traversing system of the wind tunnel, already present in the test section, in order to investigate different measurement planes around the body, even in areas with problematic access. One of the main advantages of such an arrangement is that the camera angles do not change while moving or rotating the probe with the traversing system. Therefore, only one offline calibration, performed outside the wind tunnel, is required, with a considerable saving of time.

In Fig. 3, the L-shaped Pininfarina 3D-PIV Probe is shown, while its technical data are summarized in Table 1.

3.2 PIV Application: Rear Wake

A typical application of 3D-PIV in a full-scale automotive wind tunnel is the measurement of the flow field of the rear wake of a vehicle. In the experi-

Table 1. PF-3D PIV probe technical data

Probe dimensions (m)	0.8 × 1.45 (L-shaped)
Probe weight (kg)	50
Camera angles to the normal of the light sheet plane (deg)	50 and 70
Max laser energy (mJ)	2 × 200
Laser umbilical length (m)	10
Cameras resolution (pixels)	2 k × 2 k
Camera frame rate (fps)	16
Seeding type	Olive-oil droplets

Fig. 4. Experimental setup to measure the three velocity components of the flow in the near-wake region of a test car in the Pininfarina full-scale automotive wind tunnel. The Pininfarina probe is mounted on the main traversing system. The image of the laser lightsheet is shown for the sake of clarity

ment hereafter presented, the car was mounted in the Pininfarina full-scale wind tunnel on the struts and the tests performed with the Ground Effect Simulation System, see Fig. 4.

The PIV system was mounted on the main traversing system, aligned in order to be at 0.1 m downstream of the vehicle (near-wake), and with the center of the acquisition plane at $Y = -0.5$ m and $Z = 0.5$ m (wind tunnel standard coordinates). The seeding distribution system was located at the wind-tunnel contraction inlet on a traversing system in order to properly optimize the seeding density with respect to the measurement plane.

The calibration was performed offline, before mounting the probe to the traversing system. This is possible since the angles between the cameras and the laser lightsheet remain fixed. The calibration surface was subsequently optimized with the acquired images of the seeding particles.

All the measurements were performed at $V_W = 100$ km/h, with full simulation of ground effect, i.e., spinning wheels, running belts and tangential blowing. In order to avoid laser-light reflections both the wind tunnel and the tested car were properly treated. 640 images were grabbed at a sustained sampling frequency of 7.5 Hz (sampling time about 85 s) and $\Delta t = 30\,\mu$s.

Standard recursive analysis of interrogation windows was performed, with a final interrogation window size of 32 × 32 pixels. The corresponding spatial resolution is 7.5 mm. The 3D final vectors have been validated with a median and standard-deviation filter.

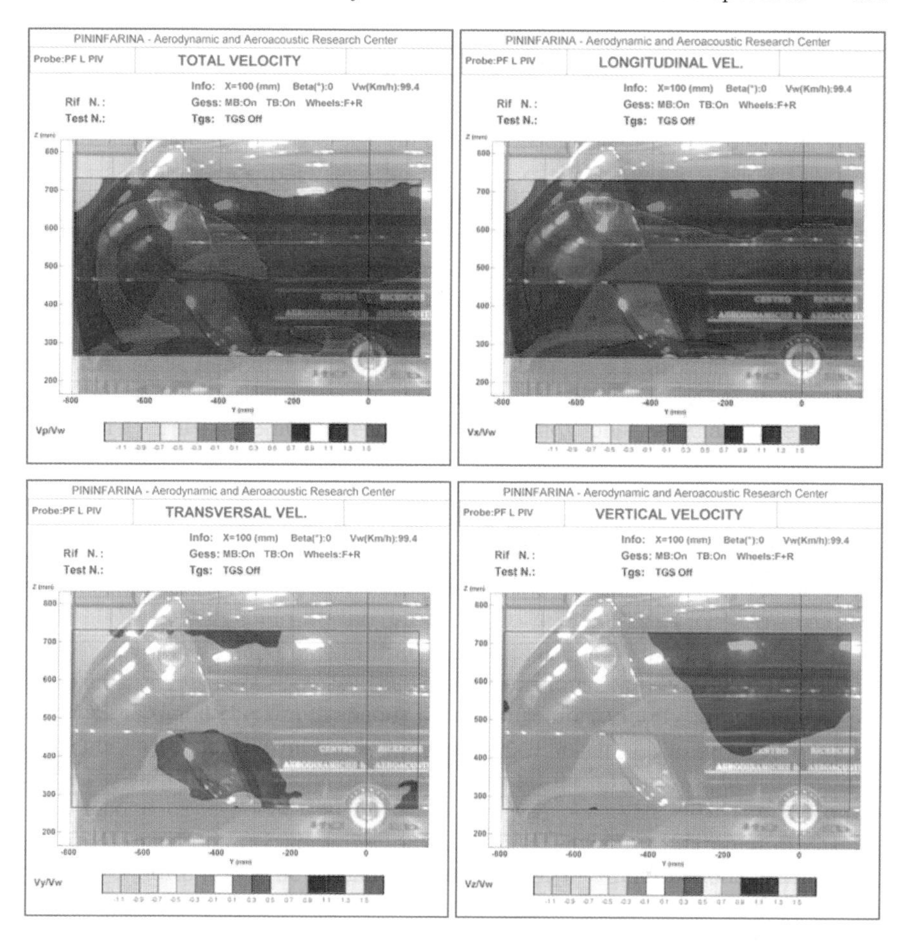

Fig. 5. Velocity maps of the near-wake of test car in the Pininfarina full-scale automotive wind tunnel. The velocities have been normalized to the free-stream velocity, $V_{\mathrm{W}} = 100\,\mathrm{km/h}$. Measurement area is about $0.9 \times 0.45\,\mathrm{m}$

In Fig. 5 the maps for mean total velocity and the corresponding three components are shown.

The measurement area is about $0.9 \times 0.45\,\mathrm{m}$. With a single measurement run, it is indeed possible to measure a meaningful portion of the rear wake. In order to fully cover the rear wake it would have been necessary to repeat the measurement process with the probe properly centered along different positions. This kind of patchwork could be easily performed with the traversing system without even stopping the wind tunnel.

As expected, in the maps in Fig. 5 it is possible to observe a deficit of velocity in the explored portion of the wake of the vehicle. In the longitudinal velocity map some areas of reverse flow are even noticeable. In the vertical

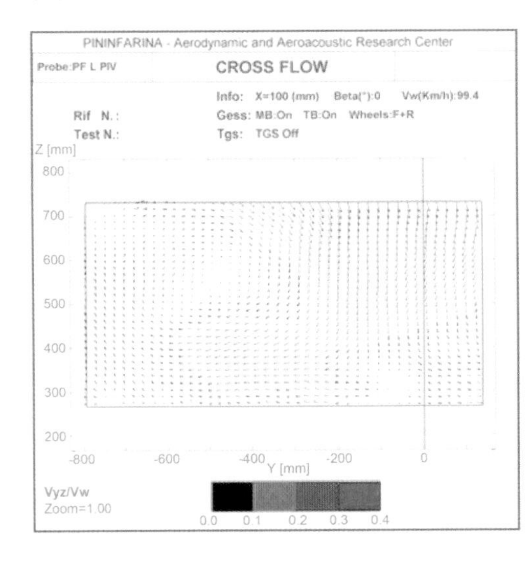

Fig. 6. Crossflow velocity map of the near wake of a test car in the Pininfarina full-scale automotive wind tunnel. A downwash vortex is located at about $Y = -0.5\,\mathrm{m}$ and $Z = 0.55\,\mathrm{m}$

velocity map, an extended region of negative velocity is manifest, showing the typical behavior of the near-wake region.

In Fig. 6 the crossflow map shows the presence of a downwash vortex characteristically shed by passenger cars.

In Fig. 7 the normalized transversal velocity V_Y, is plotted at constant height over ground along three lines immediately below, at the center and above the vortex core. It can be noticed that at the vortex core height the transversal velocity changes its direction, as expected.

The measurement of the flow field of the rear wake is only one of the possible applications of stereoscopic PIV in a full-scale automotive wind tunnel. In Figs. 8–11 some other examples are shown.

Another important application consists in adopting PIV measurements to investigate the flow characteristics of the wind-tunnel test section. The Turbulence Generation System (TGS) is a device able to simulate road conditions where the flow upstream of the tested car is conditioned by ambient wind, the wake generated by another car or continuous yawing of the flowCL [12, 13]. The Pininfarina 3D-PIV probe has been successfully used to characterize the turbulent flow produced by the different TGS configurations, see Fig. 12. For further details see [14].

4 Conclusions and Future Perspectives

PIV could be a useful, nonintrusive technique for aerodynamic and aeroacoustic car development. It provides the chance to get more information about the whole flow field around the vehicle, measuring in large areas both time-averaged and time-resolved phenomena.

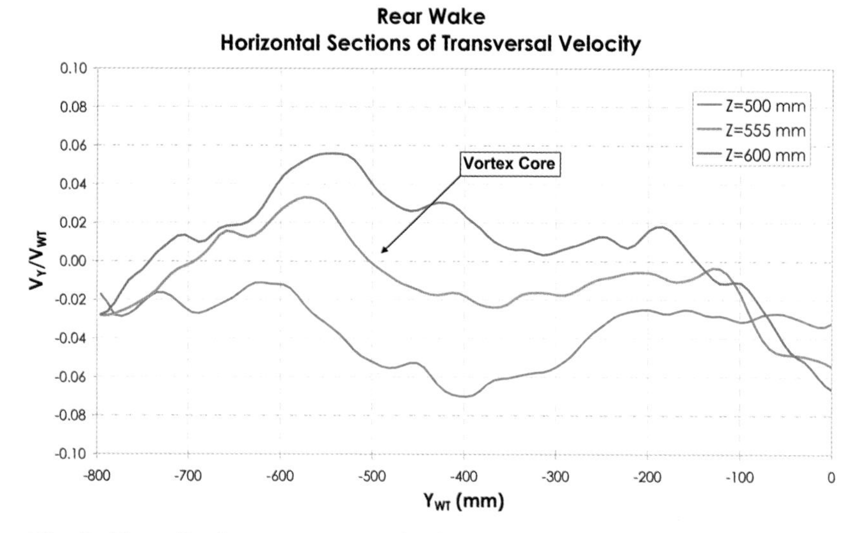

Fig. 7. Normalized mean transversal velocity at three different heights over ground. The vortex core is located at $Y = -0.5\,\mathrm{m}$ and $Z = 0.55\,\mathrm{m}$

However, at present, the adoption of PIV in the automotive car industry is still very limited to a few early adopters, mainly working on scale models. A full-scale automotive wind tunnel is still a challenge for PIV measurements. Typical issues involve the large-area field of view, the spatial resolution and optical accessibility of some measurement areas. Besides, given the cost of such industrial facilities, measurements and results have to be reached in a very short time, despite the mentioned difficulties.

The Pininfarina Aerodynamics and Aeroacoustics Center conceived its 3D-PIV probe in order to solve some of these issues and currently such a PIV system is one of the most advanced in the car industry field. Measurements performed by Pininfarina have been shown in this chapter. PIV is still a complex technique that needs skills and expertise not easily found in the industry. From the technological standpoint the PIV industrial application requires a "plug and play" PIV system and, in order to achieve this target, some improvements are still necessary in all the PIV aspects: hardware, calibration, acquisition and postprocessing techniques.

Hardware progresses are necessary mainly for the application of Time-Resolved PIV (TR PIV). High-frequency measurements are very useful in automotive wind tunnels in order to study fluctuant quantities of the flow field around the car body, but to achieve this task, TR PIV has to be able to measure the three velocity components over large areas.

A very useful tool for aerodynamic and aeroacoustic studies on car development can be given by the interaction of numerical simulation and PIV experimental measurements, especially about the investigation of turbulent and instantaneous flow field. Currently, a fair level of agreement has been

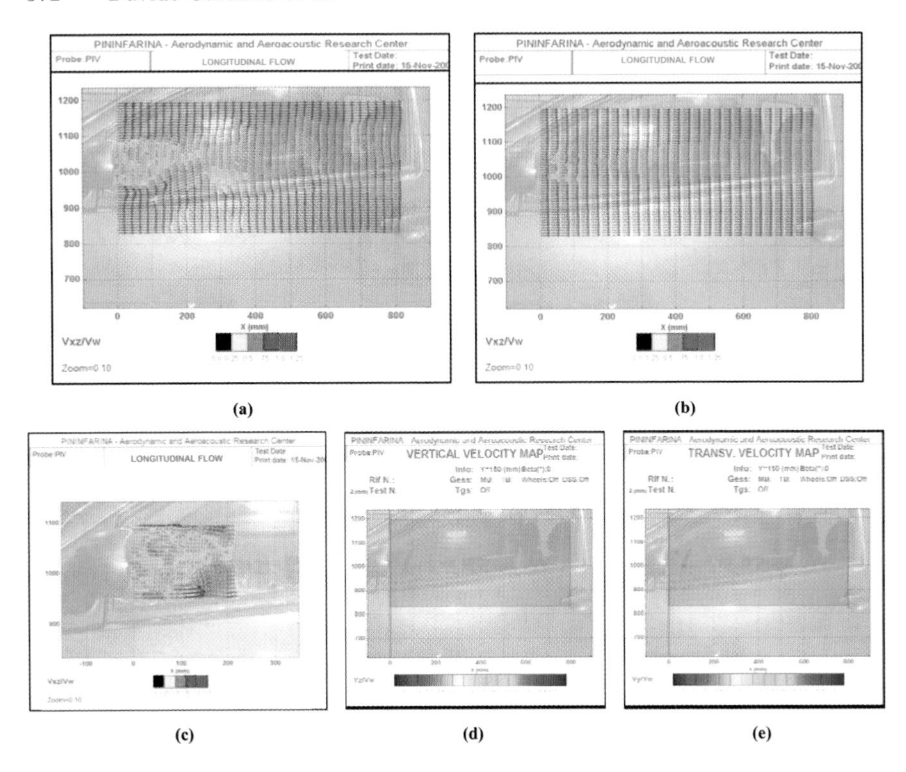

Fig. 8. Longitudinal wake behind a car rear-view mirror. Plane XZ (in wind-tunnel coordinate system) and $Y = -0.15\,\mathrm{m}$ from the left-side window: (**a**) Instantaneous velocity field; (**b**) mean velocity field; (**c**) detail of instantaneous velocity field; (**d**) vertical velocity field; (**e**) transversal velocity field. All velocities are normalized to the free-stream velocity

found between these two types of methods and PIV is often used to validate CFD results. In the future, a stronger link is needed with even more complex domain and geometry to gain enhanced knowledge about turbulent and unsteady phenomena, as is more and more required in the automotive field.

Acknowledgements

The authors wish to acknowledge the precious support of the Pininfarina Wind Tunnel personnel for their collaboration in the conduction of test session and the organization of the PivNet Workshops held in Pininfarina. In particular, we recognize the contribution of people from the PF workshop for the development of many devices.

We also would like to acknowledge all the members of the PivNet2 Network for the important and direct information exchange.

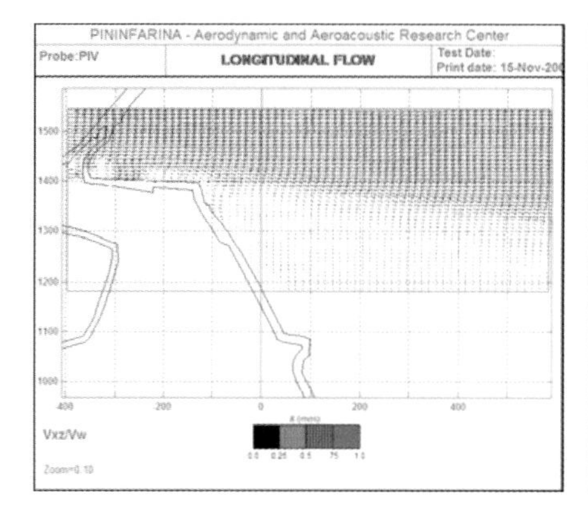

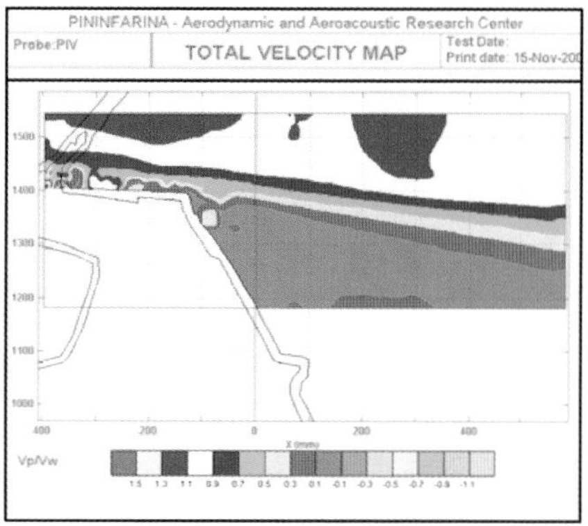

Fig. 9. Mean velocity field of a longitudinal wake behind the car in the plane XZ and $Y = -0.1\,\mathrm{m}$ (wind-tunnel coordinate system). Longitudinal flow (*left*) and total velocity map (*right*). All velocities are normalized to the free-stream velocity

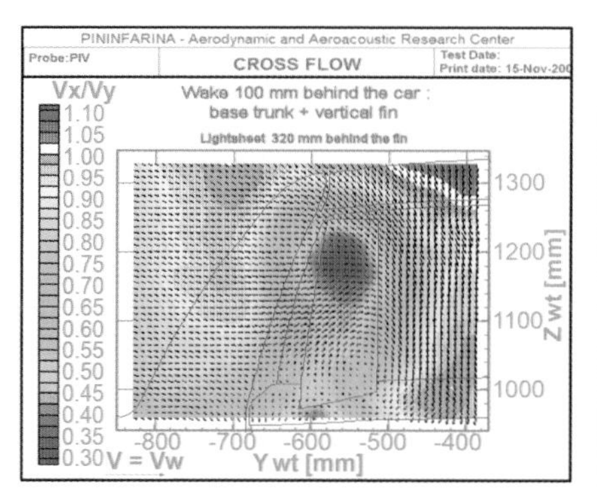

Fig. 10. Crossflow of a wake 0.1 m behind car with vertical fin. Mean velocity field (*left*). Car configuration (*right*). All velocities are normalized to the free-stream velocity

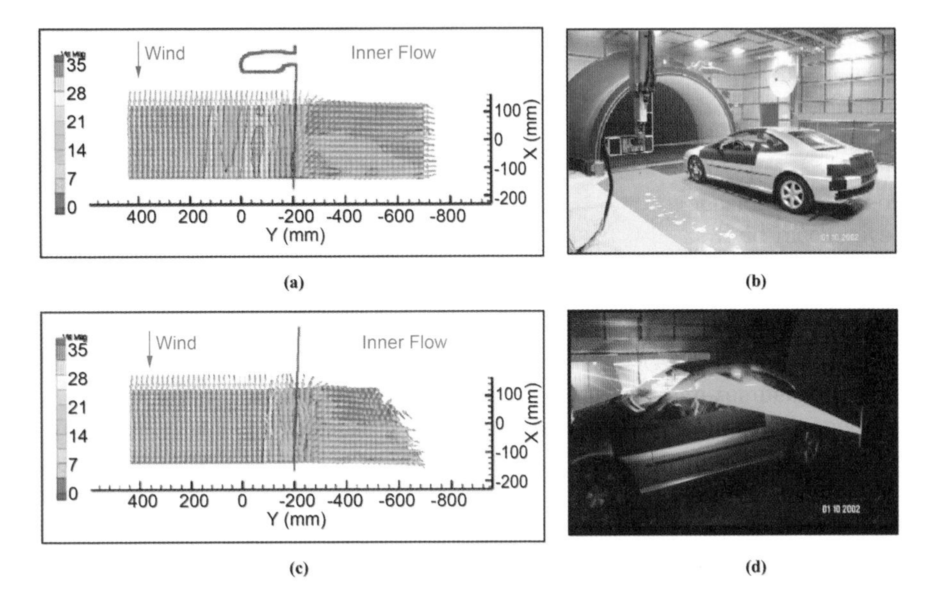

Fig. 11. Mean velocity field in a longitudinal area of the front left-side window, with glass lowered. Plane XY, $Z = 1.1\,\mathrm{m}$: (**a**) downstream of the RVM; (**c**) $Z = 1.2\,\mathrm{m}$: higher than RVM. (**b**,**d**) Measurement setup

Fig. 12. The Turbulence Generation System (TGS) and PIV measurements setup adopted to investigate the turbulent flows generated by the TGS

References

[1] M. Raffel, C. E. Willert, J. Kompenhans: *Particle Image Velocimetry: A Practical Guide* (Springer, Berlin, Heidelberg 2000)

[2] A. Cogotti: Flow-field surveys behind three squareback car models using a new "fourteen-hole" probe, in *Int. SAE Cong.*, SAE paper 870243 (1987)

[3] A. Cogotti, F. De Gregorio: Presentation of flow field investigation by PIV on a full-scale car in the Pininfarina wind tunnel, in *Int. SAE Cong.*, SAE paper 2000-01-0870 (2000)

[4] A. Cogotti, D. Cardano, G. Carlino, F. Cogotti: Aerodynamics and aeroacoustics of passenger cars in a controlled high turbulence flow: Some new results, in *Int. SAE Cong.*, SAE paper 2005-01-1455 (2005)

[5] A. Cogotti: PIV and aeroacoustics at Pininfarina, in *Workshop on Particle Image Velocimetry in Car Industry – Pininfarina* (2005)

[6] D. Cardano: The Pininfarina PIV-3D probe: Toward a user friendly system, in *Workshop on Particle Image Velocimetry in Car Industry – Pininfarina* (2005)

[7] F. Cogotti: Aeroacoustics by microphone arrays: Some examples regarding a rear view mirror, in *Workshop on Particle Image Velocimetry in Car Industry – Pininfarina* (2005)

[8] G. Carlino: PIV at Pininfarina: Some investigation behind a rear view mirror, in *Workshop on Particle Image Velocimetry in Car Industry – Pininfarina* (2005)

[9] C. Westergaard: Evaluation of new high speed, high performance camersa, in *Workshop on Particle Image Velocimetry in Car Industry – Pininfarina* (2005)

[10] A. Heider: Investigation of a turbulent free jet and a proposal of its noise reconstruction using MSPIV, in *Workshop on Particle Image Velocimetry in Car Industry – Pininfarina* (2005)

[11] A. Schroeder: Trailing edge noise source reconstruction by means of TR-PIV, in *Workshop on Particle Image Velocimetry in Car Industry – Pininfarina* (2005)

[12] A. Cogotti: Generation of a controlled level of turbulence in the Pininfarina wind tunnel for the measurement of unsteady aerodynamics and aeroacoustics, in *Int. SAE Cong.*, SAE paper 2003-01-0430 (2003)

[13] A. Cogotti: Update on the Pininfarina "turbulence generation system" and its effects on the car aerodynamics and aeroacoustics, in *Int. SAE Cong.*, SAE paper 2004-01-0807 (2004)

[14] G. Carlino, D. Cardano, A. Cogotti: A new technique to measure the aerodynamics response of passenger cars by a continuous flow yawing, in *Int. SAE Cong.*, SAE paper 2007-01-0902 (2007)

Index

Evaluation of Large-Scale Wing Vortex Wakes from Multi-Camera PIV Measurements in Free-Flight Laboratory

Carl F. v. Carmer[1], André Heider[1], Andreas Schröder[1], Robert Konrath[1], Janos Agocs[1], Anne Gilliot[2], and Jean-Claude Monnier[2]

[1] Inst. of Aerodynamics and Flow Technology, German Aerospace Center (DLR), Bunsenstr. 10, 37073 Göttingen, Germany
andreas.schroeder@dlr.de

[2] Applied Aerodynamics Department, ONERA-DAAP, 5 bd. Paul Painlevé, F-59045 Lille CEDEX, France
anne.gilliot@onera.fr

Abstract. Multiple-vortex systems of aircraft wakes have been investigated experimentally in a unique large-scale laboratory facility, the free-flight B20 catapult bench, ONERA Lille. 2D/2C PIV measurements have been performed in a translating reference frame, which provided time-resolved crossvelocity observations of the vortex systems in a Lagrangian frame normal to the wake axis. A PIV setup using a moving multiple-camera array and a variable double-frame time delay has been employed successfully. The large-scale quasi-2D structures of the wake-vortex system have been identified using the Q_{W} criterion based on the 2D velocity gradient tensor $\boldsymbol{\nabla}_H \boldsymbol{u}$, thus illustrating the temporal development of unequal-strength corotating vortex pairs in aircraft wakes for nondimensional times $t\,U_0/b \lesssim 45$.

1 Introduction

Vortex sheets shed off the wings of large aircraft roll up and organize into a multiple-vortex system in the aircraft wake. Such large-scale coherent vortical structures are usually very stable and energetic, and may pose a potential hazard to following aircraft. As a consequence, regulatory separation distances have to be met during takeoff and landing that limit the airport capacities. The alleviation of wake vortices remains an important issue in commercial aviation. In order to develop effective strategies for an early decay of the large-scale structures in wake-vortex systems, the development and possible forcing of their cooperative instabilities have been studied analytically and numerically (e.g., [1]), whereas quantitative experimental evidence is scarcely found.

In the wake near field, sheets of corotating wing vortices undergo mutual interaction and roll up into the strongest vortex – usually the wing-tip vortex. Due to such merging processes one predominant large-scale vortical structure remains on either side of the wing. These wing vortices constitute a stable pair of equal-strength counterrotating vortices in the wake far field finally prone to medium- and long-wavelength instabilities, e.g., of Crow mode. However,

A. Schroeder, C. E. Willert (Eds.): Particle Image Velocimetry,
Topics Appl. Physics **112**, 377–394 (2008)

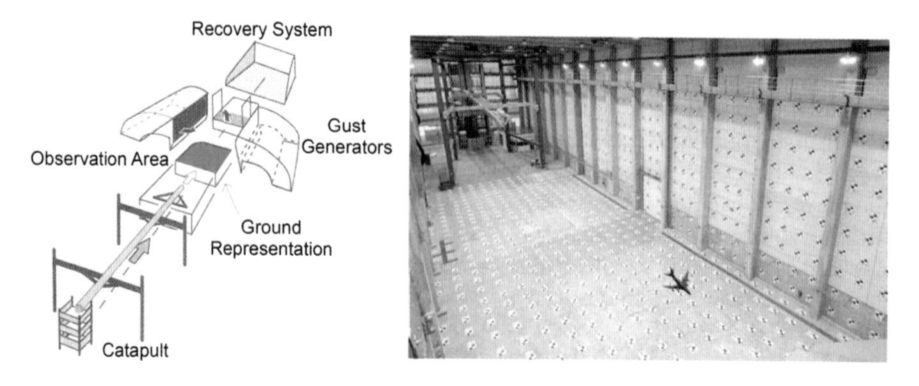

Fig. 1. The B20 catapult bench in the free-flight laboratory of ONERA, Lille [www.onera.fr]

a vortex induced, for instance, by an inboard corner of a flap may interact with the mature wing vortex to form an unequal-strength counterrotating vortex pair. This may lead to a more rapid decay and early disintegration of the wing vortices – a favorable condition regarding air-traffic safety.

Laboratory investigations employ the particle image velocimetry (PIV) technique with a specifically adjusted measurement system to quantitatively observe the flow in model-aircraft wakes. Section 2 describes the moving multiple-camera PIV setup employed in the free-flight catapult facility of ONERA, Lille. A brief overview of the applied vortex-identification scheme is presented in Sect. 3. Its application to the investigation of large-scale vortical structures is illustrated in Sect. 4 for the evolution of unequal-strength corotating vortex pairs induced by an aircraft model.

2 Moving Multiple-Camera PIV in Aerial Wake of Flying Aircraft Model

PIV measurements in an aircraft wake have been performed in the free-flight laboratory of ONERA in Lille, France, [2, 3] collaboratively by expert teams of ONERA and DLR in the framework of the EU-FP5 project AWIATOR [4, 5] in February 2005. With respect to a preceding AWIATOR measurement campaign in November 2003 [6], the measurement system (especially the camera array) and the experimental configuration have been optimized so as to also access the immediate wake near field $x/b \approx 1$ and follow the vortex development with the best available resolution up to $x/b \gtrsim 40$.

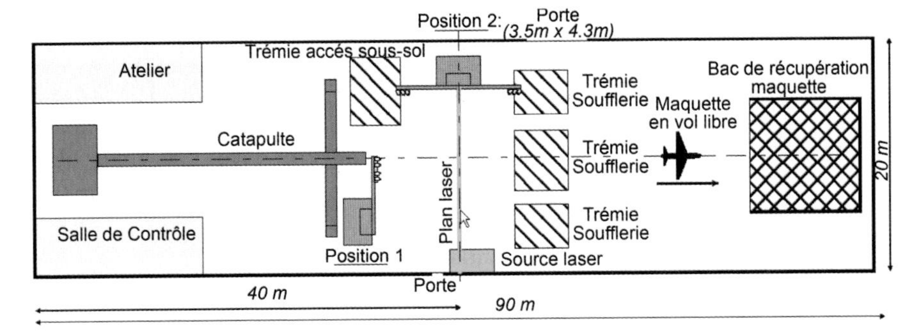

Fig. 2. Top view of B20 catapult bench and of main measurement devices [4]

2.1 Free-Flight Laboratory

The B20 catapult bench [2] – a unique free-flight laboratory of ONERA in Lille, France – is installed in a 90 m long hall with a cross section of 20 m by 20 m (cf. Fig. 1). Using the B20 catapult bench, an aircraft model with a wing span of more than 2 m was catapulted into free flight at exactly predefined flight conditions with a launching speed of $U_0 \approx 25$ m. After some 50 m of free flight the model was safely caught in a recovery box filled with soft foam plastic. Additional installations allow simulation of environmental influences (crosswind, ground proximity, thermal convection) on a large laboratory scale.

The present experimental investigation aimed at characterizing the development of wake-vortex systems for various high-lift configurations with different inboard and outboard flap settings. Ranging from the near-field well into the far field, the experiments were thought to provide a database to validate CFD calculations. Hence, the experiments took place in still, unbounded air; ambient influences were excluded. Figure 2 displays the main features of the experimental setup: the catapult to accelerate the model into flight through the seeded air (blue), the optical system to acquire images of the wake flow (red, "position 1"), the laser lightsheet to illuminate the flow (green).

2.2 Measurement Setup

A multiple-camera MonoPIV system [2, 3, 6] has been especially adapted to study descending wake-vortex systems induced by a freely flying scale model in the free-flight laboratory. 12 high-resolution cameras with $1280 \times 1024\,\mathrm{px}^2$ 12-bit CCD chips were mounted to a common camera rack (cf. Fig. 3). Ten cameras were equipped with lenses with focal length $f = 180$ mm and two with $f = 135$ mm lenses, the f number was 2.8 for all lenses. The camera rack was attached to a single vertical rail (cf. Fig. 4) that allowed us to smoothly

Fig. 3. The array of 12 cameras [2] and their fields of view

Fig. 4. The traversing system for the multiple-camera array [2]

traverse the cameras downwards for 9 m with a speed of up to 1 m/s in order to track the descending vortices.

Two double-oscillator Nd:YAG lasers (Quantel Brilliant-B) providing $2 \times 320\,\text{mJ}$ at $\lambda = 532\,\text{nm}$ with 5 ns pulses supplied the lightsheet. The lightsheet optics were fixed to a second traversing system that allowed us to move the lightsheet downwards simultaneously with the cameras (cf. Fig. 5). Two seeding generators containing five 12-hole Laskin nozzles each were employed to produced DEHS droplets with an average diameter of $1\,\mu\text{m}$ that were released into the air with a blower connected to the seeding generators. Thus, a homogeneous distribution of seeding particles was realized along the whole trajectory of the aircraft model in front of the lightsheet.

The cameras were oriented in the flight direction along the wake axis and normal to the lightsheet. The area of observation of almost $2\,\text{m}^2$ – resolved at 5120×3072 pixel^2 – covered the right side of the aircraft wake. Hence, the optical arrangement allowed us to obtain the crossflow velocity components of the right-hand wake vortices – mainly consisting of an unequal-strength

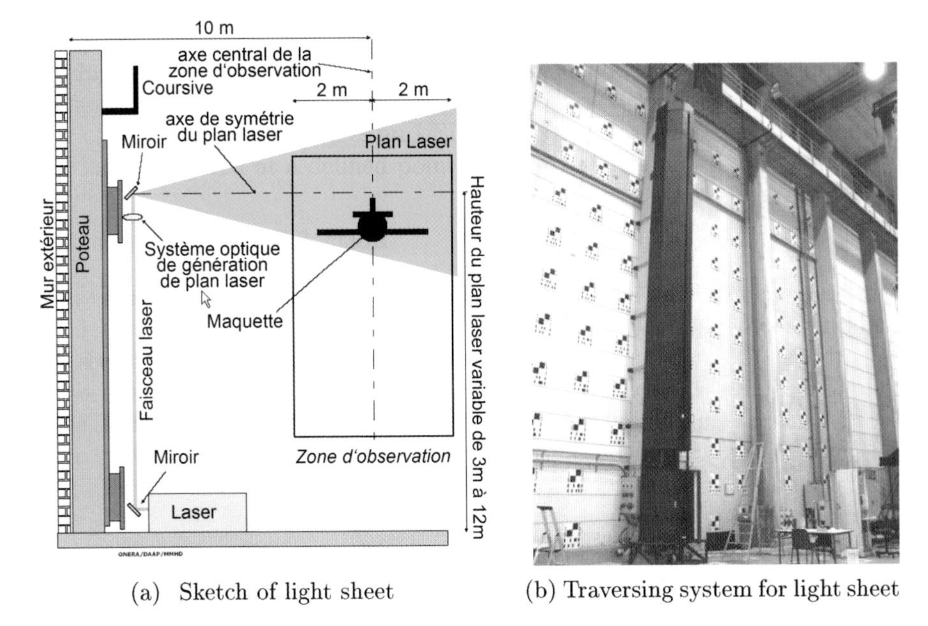

(a) Sketch of light sheet

(b) Traversing system for light sheet

Fig. 5. The lightsheet oriented normal to flight direction and its traversing system [2]

corotating vortex pair – with high spatial resolution. Synchronized PIV double images of the camera rack have been recorded with a frequency of 3 Hz over a duration of 13.3 s.

Crucially for this measurement setup – a programmable sequencer controlled the full synchronization of the two lasers, of the traversing system of the lightsheet optics, of the camera traverse, and of the cameras themselves. Additionally, for the PIV double images the pulse delay time has also been increased during a measurement run from initially 0.4 ms to 2.0 ms in order to account for the decreasing azimuthal velocity as the vortices decay.

For the well-controlled experimental conditions provided by the B20 catapult bench, the traversing multiple-camera PIV setup led to a very satisfying repeatability not only for runs of both measurement campaigns, November 2003 [6] and February 2005 [3, 7], but also between measurement runs of both campaigns [3].

Table 1 summarizes the main devices of the moving multiple-camera PIV system employed in the free-flight test facility.

3 Vortex Identification

In order to extract coherent vortical structures from a turbulent flow, advanced identification schemes are based on the analysis of the velocity gra-

Table 1. Overview of moving multiple-camera PIV system

PIV system	Moving multiple-camera MonoPIV system with 90° offaxis optical arrangement; obtain inplane velocity components in cross section of aircraft model wake; measurement frequency 3 Hz
Multiple-camera array	12 high-resolution cameras with total size of CCD sensors of 15 MPx and field of view of 2 m^2; mounted on common carrier rack; attached to vertical traversing system
Laser lightsheet	Horizontally illuminate cross section normal to flight direction; vertically traversable optics; produced by 2 double-oscillator Nd:YAG (Quantel Brilliant-B) pulsed independently at 10 Hz with 2 × 320 mJ at 532 nm with 5-ns pulses
Pro-grammable sequencer	Synchronization of laser pulses and camera recording with traversing systems of lightsheet and of camera array; adjustment of laser double-pulse delay
Seeding	DEHS droplets with average diameter of 1 μm from 2 generators with Laskin nozzles

dient tensor $\nabla \boldsymbol{u}$ evaluated locally [8, 9]. Use has been made of a positive second invariant Q of $\nabla \boldsymbol{u}$ [10], of a conjugate pair of complex eigenvalues of $\nabla \boldsymbol{u}$ (discriminant Δ criterion [11, 12], swirling strength [9, 13]), and of the eigenvalue λ_2 of $\boldsymbol{S}^2 + \boldsymbol{\Omega}^2$ [8].

For vortex identification from sectional PIV data we will utilize Q_{W} values, which has been proven an efficient eduction scheme for predominately two-dimensional vortical flows (e.g. [14, 15]). As has been demonstrated [16], the quantities of a second invariant Q, the Weiss Q_{W} values, the squared swirling strength λ_{ci}^2, and the discriminant D_2 are all directly proportional for a two-dimensional velocity gradient tensor.

The velocity gradient tensor of a three-dimensional velocity field can be decomposed into its symmetric and antisymmetric components,

$$\nabla \boldsymbol{u} = \boldsymbol{S} + \boldsymbol{\Omega} = \tfrac{1}{2}(\nabla \boldsymbol{u} + \nabla \boldsymbol{u}^t) + \tfrac{1}{2}(\nabla \boldsymbol{u} - \nabla \boldsymbol{u}^t)\,, \tag{1}$$

i.e., into the tensors of rate-of-strain and of vorticity, respectively. The Q criterion – a positive second invariant Q of $\nabla \boldsymbol{u}$ given as $Q = 1/2[(\nabla \cdot \boldsymbol{u})^2 - \mathrm{tr}(\nabla \boldsymbol{u}^2)]$ – for incompressible flow is defined as [10]

$$Q = \frac{1}{2}(|\boldsymbol{\Omega}|^2 - |\boldsymbol{S}|^2) > 0\,, \tag{2}$$

where $|\boldsymbol{\Omega}|^2 = \mathrm{tr}(\boldsymbol{\Omega}\boldsymbol{\Omega}^t)$ and $|\boldsymbol{S}|^2 = \mathrm{tr}(\boldsymbol{S}\boldsymbol{S}^t)$ are the Euclidian norms of the vorticity tensor and of the rate-of-strain tensor. Hence, Q constitutes a local indicator for the dominance of the rotation rate compared to the strain rate. Though $Q > 0$ does not guarantee the necessary condition of a local pressure minimum within the region of positive Q, in most cases the pressure minimum is implicitly satisfied [8, 14].

In the present study, temporally resolved velocity fields across the wake axis are used to compute the two-dimensional velocity gradient tensor $\nabla_H \boldsymbol{u}$, and to deduce further flow quantities like the axial vorticity component ω_x. Vortical structures in a wake-vortex system are identified based on the Weiss formulation of the Q value [17], given as

$$\begin{aligned} Q_{\mathrm{W}} &\equiv \left(\frac{\partial v}{\partial y} - \frac{\partial w}{\partial z}\right)^2 + \left(\frac{\partial w}{\partial y} + \frac{\partial v}{\partial z}\right)^2 - \left(\frac{\partial w}{\partial y} - \frac{\partial v}{\partial z}\right)^2 \\ &= \theta_1^2 + \theta_2^2 - \omega_x^2 , \end{aligned} \tag{3}$$

where θ_1^2 is a squared combination of normal strain rates, θ_2^2 the squared doubled shear strain rate, and ω_x^2 the squared axial vorticity.[1] Negative values of Q_{W} indicate flow regions where the magnitude of vorticity exceeds the magnitude of the strain terms. Regions with negative Q_{W} correspond to flow regions with a positive second invariant of $\nabla_H \boldsymbol{u}$ [10], if the normal strain term θ_1^2 is neglected. Then, $Q_{\mathrm{W}} < 0$ is associated with vorticity-dominated regions of the flow, and $Q_{\mathrm{W}} > 0$ with shear-dominated regions. The former regions in general identify vortical flow structures (e.g., [18]), in a wake-vortex system they will identify vortex cores.

4 Data Visualization and Analysis of Unequal-Strength Corotating Vortex Pairs

4.1 PIV Evaluation of Multiple-Camera Images

From the calibration images the exact relative positions of the 12 camera fields were obtained. Based on their relative positions, for each instant of recording time the "A" and "B frames" of the 12 PIV double images were amalgamated into two single large images prior to any PIV analysis of particle displacements [7]. Figure 6 illustrates the amalgamation of PIV images for the A frames of 12 simultaneous double images taken from a given measurement flight. As can be concluded from two gaps between neighboring images, a slightly overlapping alignment of the camera view fields is not a trivial task. Also, erroneous light intrusion – presumably from modified lens mounts – is visible in some of the camera images.

Regarding the seeding of the flow, the seeding particles in general are distributed homogeneously. However, a closer look reveals a reduced (but still sufficiently high) particle density in the cores of the large-scale vortices in the immediate near wake. Moreover, larger soft foam particles released

[1] The Weiss formulation [17] of the two-dimensional Q criterion shows an opposite sign with respect to the general definition of a positive second invariant [10]. In this chapter we use the two-dimensional Q_{W} criterion of Weiss and the associated sign convention.

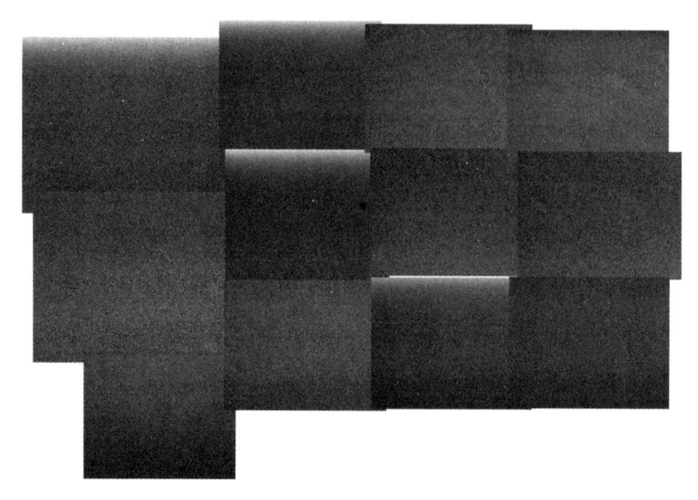

Fig. 6. The amalgamation of 12 simultaneous camera images into a single large image (dewarped A frames of double images at $t^* = 4.5$)

from the recovery box are observable that may lead to spurious vectors in the PIV evaluation.

The recorded images of the twelve cameras – though the image planes and the lightsheet were coplanar in general – could suffer from slight misalignments and lens distortion. They have been dewarped to reconstruct the image space onto a Cartesian grid using a ratio of first-order polynomial mapping functions [19, 20] for each individual camera. During the image mapping procedure the native resolutions of 2.5 px/mm and 3.3 px/mm for the 135 mm and 180 mm lenses, respectively, were enlarged by roughly 30 % to a common resolution.

Using state-of-the-art PIV displacement algorithms, flow fields of the transverse and spanwise velocity components were evaluated from the recorded PIV images in a plane perpendicular to the wake axis. Figure 7 depicts an exemplary planar distribution of the crossflow velocity components evaluated from a 15-MPixel multicamera PIV double image in a translating reference frame. The PIV evaluation software PIVview 2.4 by PivTec employed a multipass multigrid interrogation method with window deformation [19–21]. The correlation algorithm utilized a multiple (Hart) correlation method. The image-interpolation scheme worked with subpixel image shifting (image deformation) involving cubic B-splines. Peak detection was based on Whittaker reconstruction. The multigrid interrogation was performed by desampling of the images due to binning of neighboring pixels [19]. The size of the final correlation window was $32 \times 32\,\text{px}^2$ and 50 % overlap, and resulted in time-resolved flow fields with more than 60 000 vectors. For outlier detection normalized median filtering [22] and replacement by lower-order peaks was

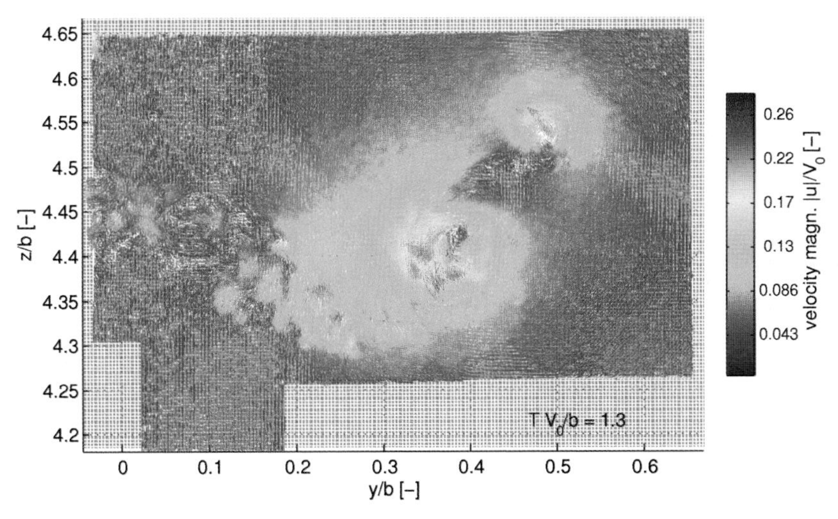

Fig. 7. Vector field of the crossvelocity components in the near-wake-vortex system evaluated from 15-MPixel multicamera PIV in a moving reference frame

employed, additionally, maximum changes in velocity magnitude and flow direction were set in the postprocessing.

In Fig. 7 color-coded vectors indicate the magnitude of the crossvelocity components. The ordinate z/b shows the height of the model above the ground normalized by the wing span; the abscissa y/b shows the spanwise distance from the fuselage. The wing tip is located at $y/b = 0.5$ on the right-hand side, the fuselage centers on the left at $y/b = 0$.

In the immediate near wake of the model at a nondimensional time $t^* = tU_0/b = 1.3$ the corotating pair of a strong outboard-flap vortex and a weaker wing-tip vortex is found on the right-hand side of Fig. 7. Closer to the fuselage (i.e., $y/b \leq 0.3$) numerous minor vortices – also with opposite sense of rotation – can be identified that were issued from the wing flaps and other control devices.

4.2 Velocity-Data Analysis of Wake-Vortex System

Q_{W} values computed from the velocity fields applying the Weiss formulation (3) are given as blue isolines in Fig. 8. For the crossvelocity field, exemplified at $tU_0/b = 4.5$, only 4 % of the vectors are displayed for clarity. Figure 8b shows a closeup of the corotating vortex pair found in Fig. 8a. Dashed lines represent values $Q_{\mathrm{W}} > 0$, and full lines represent values $Q_{\mathrm{W}} < 0$, indicating strain-dominated and vorticity-dominated flow regions, respectively. The latter regions will identify vortex cores in the given wake-vortex systems. Threshold values Q_{LCS} of negative Q_{W} close to zero (bold red lines) delineate the vortex cores from the shear flow. An absolute definition of Q_{LCS} is employed here, given as a multiple of a measure for the background Q_{W}

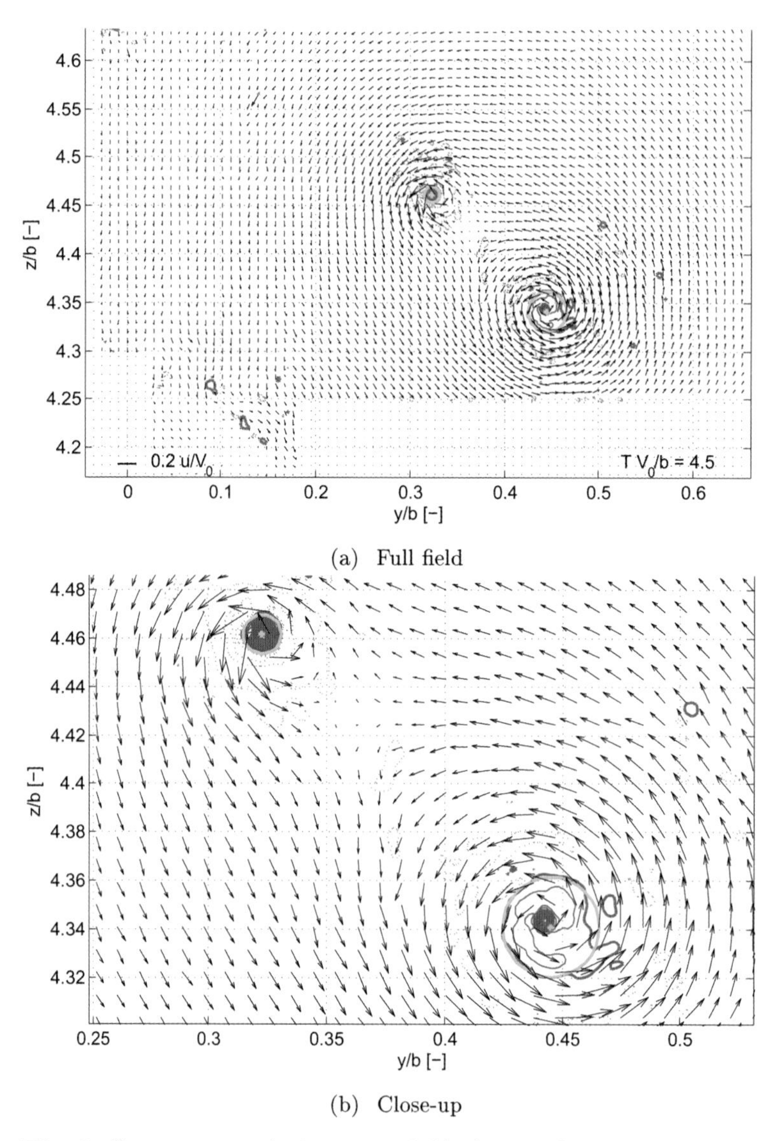

(a) Full field

(b) Close-up

Fig. 8. Cross-stream velocity vector field of unequal-strength corotating wing vortex pair (4 % of vectors displayed), and isocontours of Q_{W} values (*bold red contours* of threshold Q_{LCS} delineating possible large vortices)

fluctuation resulting from measurement and numeric noise sources. For each valid wake-vortex core, having a sufficient size and strength, red + symbols denote the location of the relative minimum values of Q_{W} as an indication of the vortex-center positions.

The centroid of vorticity, indicated by green + symbols, was evaluated for each valid vortex core within its Q_W-based boundary, i.e., for regions with $Q_W < Q_{LCS}$. The sum of the axial vorticity within a vortex was redistributed to a circular top-hat distribution centered around the vorticity centroid (bold green line) as to assume a solid-body rotation of the vortex core.

The temporal development of wake-vortex systems and its spatial coherence can be visualized from isosurfaces of Q_W values or of ω_x values in three-dimensional flow fields. Since, in the present investigation, the wake vortices develop rapidly, the temporal resolution of the measurements (i.e., 3 Hz) only allows to display isocontours in two-dimensional distributions. Figure 9 depicts the Q_W-value distribution of the wake flow field at specific time instances t after the passage of the aircraft model – more precisely, of its rear wing tips – through the lightsheet. Blue isolines denote negative Q_W values found in vorticity-dominated regions of the wake flow. The smallest negative value employed for an isocontour is the threshold value Q_{LCS} delineating vortical structure candidates. Standardized by the aircraft speed U_0 and the wing span b, cross sectional data is displayed at nondimensional times $t^* = tU_0/b = 1.3$, 4.5, 7.8, 11.0, 14.3, 17.5, 20.7 in the extended wake near field and transition to the far field.[2] If one assumes Taylor's hypothesis of frozen turbulence to hold for the evolution of large vortical structures, the nondimensional time t^* would correspond to a nondimensional longitudinal distance $x^* = x/b$.

Immediately after the passage of the model at $t^* = 1.3$ the corotating pair of a strong outboard flap vortex and a weaker wing-tip vortex on the right-hand side completed the generation process, i.e., the roll-up from the sheet vortices that are still visible closer to the fuselage at $y/b \leq 0.3$ (cf. also Fig. 7).

The sequence of cross sectional data at times $t^* = 1.3$, 4.5, 7.8, 11.0 represents a full counterclockwise orbit of the wing-tip vortex around the outboard flap vortex up to $t^* = 11.0$. A dashed red line connects the vorticity-centroid positions of the strong flap vortex at the different instances of time. During the orbital cycle the weaker vortex approaches the stronger vortex. Consecutively, from the merging of both vortices a strong single vortex remains until $t^* = 20.7$, which shortly afterwards enters a stage of accelerated decay.

The axial vorticity is shown color coded in Fig. 10 for four instants in time $t^* = 1.3$, 4.5, 20.7 and 36.9 with the underlying cross sectional velocity field showing only about 6 % of the vectors. The nondimensional vorticity $\omega_x^* = \omega_x b/U_0$ is displayed at a fixed color scale in a clipped range of $-4 \leq \omega_x^* \leq 4$. The further notation has been given with Fig. 8, except that the bold isolines of Q_{LCS} and the + symbols of local Q_{min} are depicted in white here.

At $t^* = 1.3$ the roll-up of the vortex sheets is still in progress. The primary flap vortex displays a kidney shape, some small vortices are strained between the vortex pair, rolling up and merging into the stronger vortices

[2] PIV measurements extended up to $t^* > 120$ well into the far field.

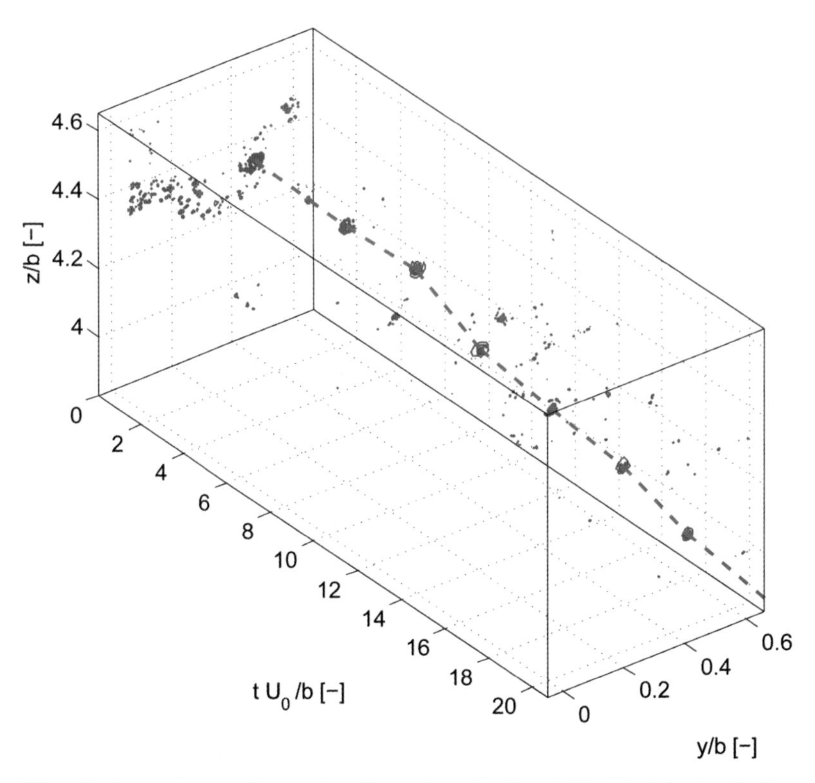

Fig. 9. Isocontours of negative Q_W values in the right-sided (extended) near field of an aircraft model wake are shown in 7 cross sectional slices at $t^* = tU_0/b = 1.3, \ldots, 20.7$

soon. A multitude of minor vortices both with positive and negative circulation that originated from the wing–fuselage junction and from control devices are being advected by the downwashing flow behind the aircraft model. The fully developed vortices at $t^* = 4.5$ and 20.7 before and after the merger of the main vortex pair display almost circular vortex cores of concentrated vorticity that are clearly identified by the Q_W criterion. At $t^* = 20.7$ the remaining single vortex begins to decay significantly. Its sharply peaked vorticity pattern develops into a patchy structure containing less gross vorticity distributed over a growing core area (Fig. 10d).

Beyond $t^* \approx 50$ the core has already disintegrated leaving a loose ring of weak vortices located around the former vortex core. Compared to the occurrence of three-dimensional vortex-pair interactions of "Crow" instabilities at $t^* \gtrsim 100$, the observed disintegration of the vortex core also occurs much earlier. Axial core instabilities induced by rapid acceleration or deceleration of the model during launching or landing, a so-called "end effect" [23], usually give rise to core disintegration by vortex bursting. Such axial instabilities have been observed to enter the PIV measurement plane at $t^* \approx 45$ for the

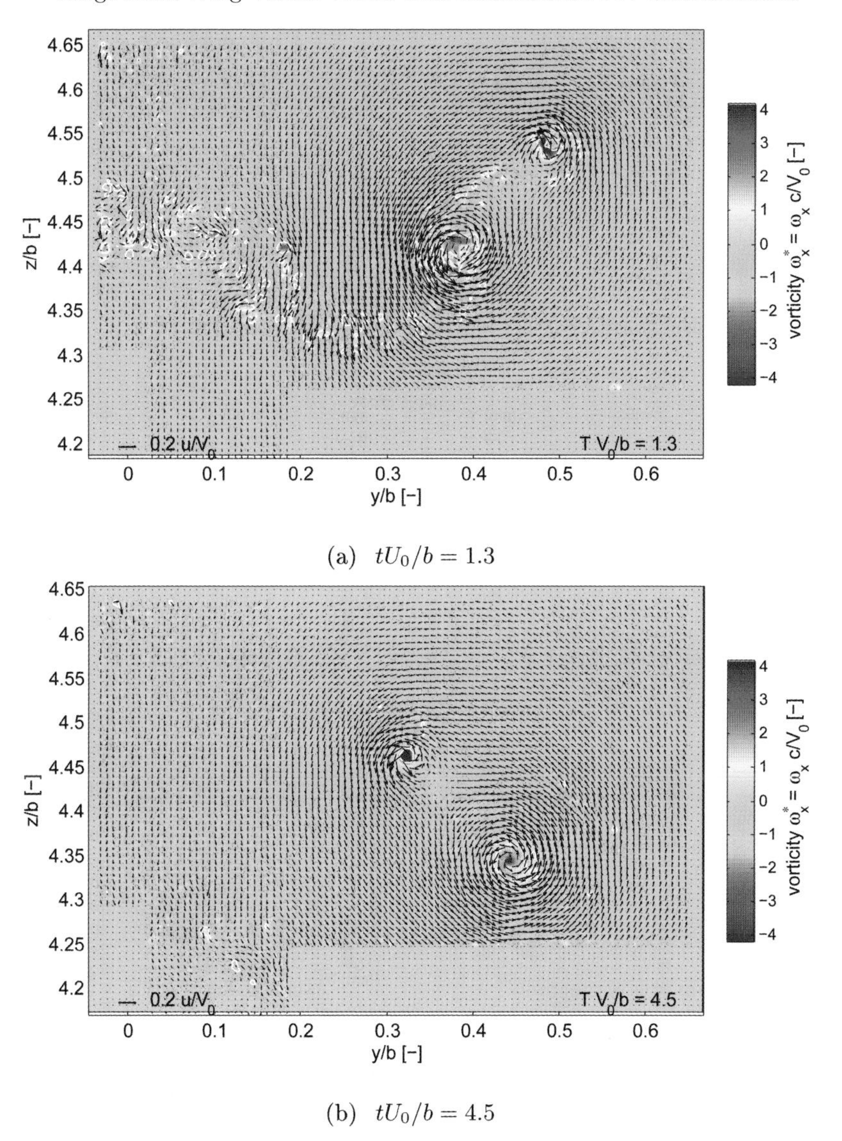

(a) $tU_0/b = 1.3$

(b) $tU_0/b = 4.5$

Fig. 10. Axial vorticity distribution of unequal-strength corotating vortex pair with underlying cross-stream velocity field (about 6 % of vectors displayed)

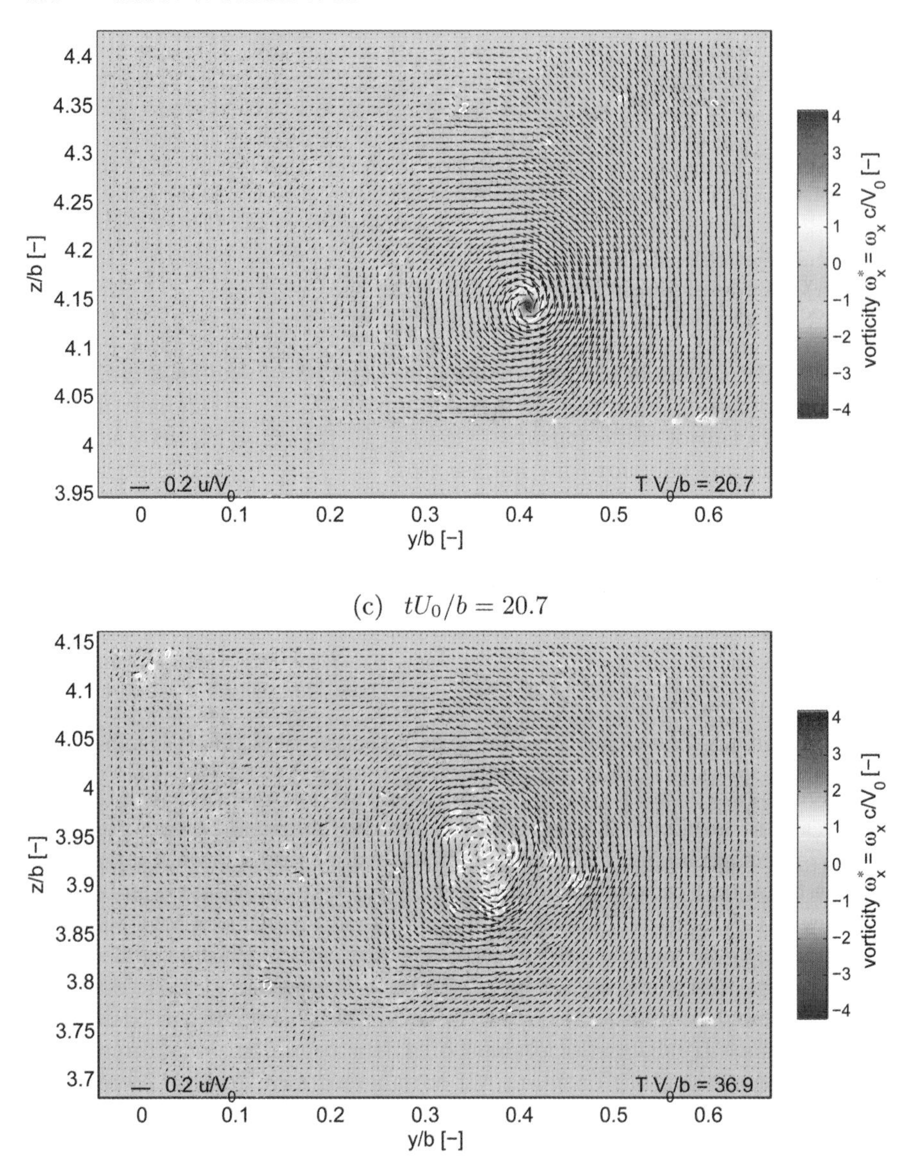

(c) $tU_0/b = 20.7$

(d) $tU_0/b = 36.9$

Fig. 11. Axial vorticity distribution (continued)

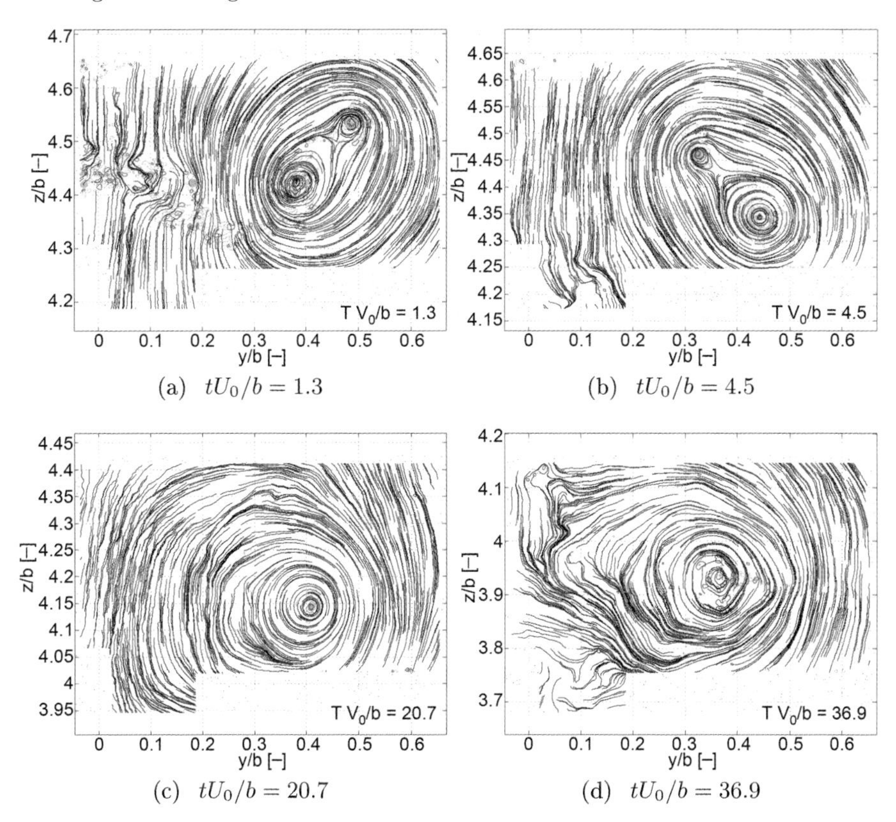

(a) $tU_0/b = 1.3$

(b) $tU_0/b = 4.5$

(c) $tU_0/b = 20.7$

(d) $tU_0/b = 36.9$

Fig. 12. Streamlines of unequal-strength corotating vortex pair

present configuration [4]. Hence, their occurrence constitutes the upper validity limit for the observation of an unbounded evolution of the wake-vortex system.

In Figs. 12a–d streamlines of the crossvelocity fields are displayed for times $t^* = 1.3$, 4.5, 20.7 and 36.9, respectively. In the moving reference frame, the streamlines of the corotating vortex pair indicate a circular shape of the viscous vortex cores and an elliptical strained shape of the outer vortex regions (Figs. 12a,b). Later the single vortex wake is associated with a more circular field (Fig. 12c). The circular motion is maintained also in the patchy core of the decaying vortex (Fig. 12d).

5 Summary and Conclusions

Multiple-vortex systems of aircraft wakes have been investigated experimentally in a unique large-scale laboratory facility, i.e., in the free-flight B20 catapult bench of ONERA, Lille. 2D/2C PIV measurements have been performed in a translating reference frame, which provided time-resolved crossvelocity observations of the vortex systems in a Lagrangian frame normal to the wake axis. A multiple-camera array has been assembled by 12 high-resolution cameras. A programmable sequencer synchronized the cameras, the lightsheet, and its associated traversing systems; it also allowed us to realize a variable pulse delay time of the lasers. Based on $\nabla_H \boldsymbol{u}$ the large-scale quasi-2D vortical structures have been identified using the Q criterion employing the Weiss formulation (3), thus illustrating the temporal development of unequal-strength corotating vortex pairs in aircraft wakes up to vortex ages $t^* \lesssim 120$. However, an early disintegration of the vortex core that can be ascribed to axial instability due to "end effects" limited the period of undisturbed development of the wake vortices to $t^* \lesssim 50$.

In order also to obtain the longitudinal velocity component u_x, stereo-PIV measurements will be performed in a translating reference frame. This allows us to access the axial flow in the vortex cores and the out-of-plane motion of the vortices. The duration of the measurements will be extended to significantly higher nondimensional times t^* for wake-vortex systems developing free of "end effects" in order to observe long-wavelength instabilities and viscous decay of the vortex systems. Also, medium-wavelength instabilities and mutual interaction of counterrotating vortex pairs have to be assessed, for which the wake flow is no longer two-dimensional on larger flow scales. For this purpose, volumetric or scanning PIV techniques have to be employed. Vortex identification schemes have to reflect the three-dimensional variability of the vortical structures.

Acknowledgements

CFvC and RK are indebted to the joint ONERA/DLR expert group comprising J. Agocs, E. Brunel, K. Christou, P. Coton, J.-P. Evard, C. Fatien, H. Frahnert, C. Geiler, A. Gilliot, A. Heider, H. Mattner, J.-C. Monnier, A. Schröder, P. Simon, and D. Thibaut for preparing and conducting the catapult measurements. E. Coustols and A. de Bruin provided helpful information and remarks during the preparation of the chapter. The comments made by the reviewers were greatly appreciated. The support of the European Commission cofunding the AWIATOR project in its 5th framework program is acknowledged.

References

[1] L. Jacquin, D. Fabre, D. Sipp, V. Theofilis, H. Vollmers: Instability and unsteadiness of aircraft wake vortices, Aerosp. Sci. Technol. **7**, 577–593 (2003)

[2] A. Gilliot, C. Geiler, J.-C. Monnier, C. Fatien, A. Schröder, H. Vollmers: Adaption des moyens de mesures PIV au B20, in *9e Congrès Francophone de Vélocimétrie Laser* (2004)

[3] A. Gilliot, A. Schröder: *Report of ONERA B20 catapult tests (PIV measurements – February 2005 campaign)*, Technical Report D 112 - 25, AWIATOR (2005)
[4] E. Coustols: Testing in wind tunnels, towing tank and B20 catapult, in *AWIATOR Technical Workshop, 5/6 Oct 2005, Bremen, Germany* (2005)
[5] G. Schrauf, F. Laporte: Awiator wake vortex characterization methology, in *KATnet 2005, Bremen, Germany* (2005)
[6] H. Vollmers: *Analysis report of velocity fields in the wake of an A340 model in the catapult*, Technical Report DLR-IB 224-2004 C 16, German Aerospace Center (DLR) (2004)
[7] C. F. v. Carmer, A. Heider, R. Konrath: *Wake vortex analysis of multiple–camera PIV measurements at B20 catapult bench*, Technical Report DLR-IB 224-2007 C 05, German Aerospace Center (DLR) (2007)
[8] J. Jeong, F. Hussain: On the identification of a vortex, J. Fluid Mech. **285**, 69–94 (1995)
[9] P. Chakraborty, S. Balachandar, R. J. Adrian: On the relationships between local vortex identification schemes, J. Fluid Mech. **535**, 189–214 (2005)
[10] J. C. R. Hunt, A. A. Wray, P. Moin: *Eddies, stream, and convergence zones in turbulent flows*, Report CTR-S88, Center for Turbulence Research (1988)
[11] M. S. Chong, A. E. Perry, B. J. Cantwell: A general classification of three-dimensional flow fields, Phys. Fluids A **2**, 765–777 (1990)
[12] U. Dallmann, A. Hilgenstock, S. Riedelbauch, B. Schulte-Werning, H. Vollmers: On the footprints of three-dimensional separated vortex flows around blunt bodies, in *67th Meeting of the FDP Symposium on Vortex Flow Aerodynamics*, vol. 494, AGARD Conf. Proc. (1991) pp. 9.1–9.13
[13] J. Zhou, R. J. Adrian, S. Balachandar, T. M. Kendall: Mechanisms for generating coherent packets of hairpin vortices in channel flow, J. Fluid Mech. **387**, 353–396 (1999)
[14] Y. Dubief, F. Delcayre: On coherent-vortex identification in turbulence, J. Turbul. **1**, 1–22 (2000)
[15] C. F. v. Carmer: *Shallow Turbulent Wake Flows: Momentum and Mass Transfer due to Large-Scale Coherent Vortical Structures*, vol. 2005/2, Schriftenreihe am Institut für Hydromechanik der Universität Karlsruhe (TH) (Universitätsverlag Karlsruhe, Karlsruhe, Germany 2005)
[16] C. F. v. Carmer, R. Konrath, A. Schröder, J.-C. Monnier: Identification of vortex pairs in aircraft wakes from sectional velocity data, Exp. Fluids pp. 1–13 (2007) (under review)
[17] J. Weiss: The dynamics of enstrophy transfer in two-dimensional hydrodynamics, Physica D **48**, 273–294 (1991)
[18] A. K. M. F. Hussain: Coherent structures and turbulence, J. Fluid Mech. **173**, 303–356 (1986)
[19] C. Willert: Stereoscopic digital particle image velocimetry for application in wind tunnel flows, Meas. Sci. Technol. **8**, 1465–1479 (1997)
[20] PivTec, Göttingen: *PIVview2C/3C version 2.4, user manual* (2006)
[21] M. Raffel, C. E. Willert, J. Kompenhans: *Particle Image Velocimetry. A Practical Guide* (Springer, Berlin Heidelberg New York 1998)
[22] J. Westerweel, F. Scarano: Universal outlier detection for PIV data, Exp. Fluids **V39**, 1096–1100 (2005)

[23] F. Bao, H. Vollmers, H. Mattner: Experimental study on controlling wake vortex in water towing tank, in *20th International Congress on Instrumentation in Aerospace Simulation Facilities, 2003: ICIASF '03* (2003) pp. 214–223

Index

Aerodynamic Performance Degradation Induced by Ice Accretion. PIV Technique Assessment in Icing Wind Tunnel

Fabrizio De Gregorio

LMSA/IWTU laboratory,
Centro Italiano Ricerche Aerospaziali (CIRA),
Capua, Italy
f.degregorio@cira.it

Abstract. The aim of the present chapter is to consider the use of PIV technique in an industrial icing wind tunnel (IWT) and the potentiality/advantages of applying the PIV technique to this specific field. The purpose of icing wind tunnels is to simulate the aircraft flight condition through cloud formations. In this operational condition ice accretions appear on the aircraft exposed surfaces due to the impact of the water droplets present in the clouds and the subsequent solidification. The investigation of aircraft aerodynamic performances and flight safety in icing condition is a fundamental aspect in the phase of design, development and certification of new aircrafts. The description of this unusual ground testing facility is reported.

The assessment of PIV in CIRA-IWT has been investigated. Several technological problems have been afforded and solved by developing the components of the measurement system, such as the laser system and the recording apparatus, both fully remotely controlled, equipped with several traversing mechanism and protected by the adverse environment conditions (temperature and pressure). The adopted solutions are described.

Furthermore, a complete test campaign on a full-scale aircraft wing tip, equipped with moving slat and deicing system has been carried out by PIV. Two regions have been investigated. The wing leading-edge (LE) area has been studied with and without ice accretion and for different cloud characteristics.

The second activitiy was aimed at the investigation of the wing-wake behavior. The measurements were aimed to characterize the wake for the model in cruise condition without ice formation and during the ice formation.

1 Introduction

Aggregation of ice on the exposed aerodynamic component of the vehicles is a major hazard for flying. The flow-field perturbation due to ice accretion, the degradation of the aerodynamic performance or the choking of the engine inlets are critical aspects for the performance and safety of airplanes and helicopters. The investigation of the icing-accretion problem requires a big effort in the experimental area as well as in the numerical simulation. In the past and nowadays extended flight campaign have been performed in order to better understand the composition of the clouds. The clouds are mainly

A. Schroeder, C. E. Willert (Eds.): Particle Image Velocimetry,
Topics Appl. Physics **112**, 395–417 (2008)

characterized by two parameters, the liquid water contents (LWC), indicating the concentration of water and by the mean volume diameter (MVD), reporting the mean size of the droplets. These two parameters together with the air static temperature, the static pressure and the flight condition and the geometry of the affected surfaces govern the type of ice formation and the relative hazard degree that can be encountered. Depending on the thermal balance on the model, two main types of ice shapes are found: a smooth surface one, but usually showing protuberances, corresponding to glaze ice or an opaque one with rough surface corresponding to rime ice.

The effect of the influence of the ice accretion on the aerodynamic behavior has been investigated by performing several flight tests. One of the earliest successful attempts to measure the effect of ice accretion was that of *Preston* and *Backman* [1]. In one flight an increment of 81 % of parasite drag was seen and almost the complete loss of control.

A large data-set has been obtained by NASA Glenn Research Center on the Twin Otter aircraft as summarized by *Bragg* et al. [2]. Tests in natural icing as well as tests with simulated icing aimed to investigate the behavior of the stability derivatives have been performed. From the results of the NASA research, it is clear that ice accretion affects the longitudinal and lateral static stability and control of the aircraft. This effect occurs even at low angle of attack and high power setting, conditions typical of cruise. A typical reduction in stability or control was 10 %. There is also evidence that the effects of ice are more significant at large angle of attack near stall where significant early flow separation occurs due to the ice.

Flight tests are extremely expensive and potentially dangerous so many investigations are conducted in ground facilities. For studying the effect of the ice formation on the aerodynamic coefficients and on the stability derivatives, tests are conducted in aerodynamic wind tunnels mounting dummy ice shapes on the models. These activities are anticipated by tests in icing facilities in order to evaluate the ice-formation shape for the selected flying condition.

Bragg and *Gregorek* [3] predicted the aerodynamic effects of ice on a typical subsonic airfoil by means of analytical and experimental investigations. The analytic method predicted the size and shape of rime-ice accretion, and the simulated ice shape was tested in the wind tunnel with a smooth surface and with a roughness equivalent to the natural rime-ice accretion. In this way the effects of ice shape and roughness on the aerodynamic coefficients were isolated. *Bragg* and *Coirier* [4] conducted an experimental program to measure the aerodynamic characteristics of a NACA 0012 airfoil with simulated glaze-ice shape. The model was also instrumented with a distribution of surface taps to provide detailed information around the ice shapes, where flow separation occurs. As expected, airfoil lift and drag were severely affected by the ice shape. *Cuerno* et al. [5] investigated the flow field over a clean NACA 0012 and simulated rime- and glaze-ice accretions at low Reynolds number (Re = 140 000), using laser Doppler velocimetry: velocity profiles on the upper surface at several angles of attack were provided. The existence of separation bubbles at angles of attack much lower than in the clean airfoil, for both rime- and glaze-ice shapes, was reported.

De Gregorio et al. [6] studied a NACA 0012 airfoil affected by rime-, mixed and glaze-ice shapes at 200 000 Reynolds number. Together with load measurements, PIV investigation was applied in order to evaluate by the flow characterization the lift and drag coefficient. Comparison between 3 component balance results and aerodynamic coefficient obtained by PIV results presented remarkable agreement.

Starting from this experience the basic idea was to apply the PIV technique directly during an icing test in the CIRA IWT facility in order to evaluate directly the aerodynamic behavior, skipping subsequent aerodynamic tests. By measuring the flow field around and downstream of the model it is possible to obtain information on the aerodynamic characteristics, on the effectiveness of the de/anti-icing systems, on the cloud-droplets trajectories together with the flow streamlines and their perturbation during the ice-accretion phase. Starting from the droplet trajectories it is possible to evaluate the collection-efficiency parameter that governs the amount of ice accretion on the exposed surface. Typically during icing, running the ice-accretion shape can be evaluated only at the end of the run, information about the evolution of the accretion in time is lost. This information is fundamental for CFD code validation. If performing PIV measurement in an industrial wind tunnel is still a challenge, the utilization in an icing wind tunnel presents some additional difficulties. Together with the other classical problems the researchers have to deal with: low air temperature, 100 % of air relative humidity (moisture and corrosion problems), static pressure conditions. In icing conditions the PIV measurements are carried out using as seeding particles the droplets of the generated clouds. This produces some limitations on the quality of the seeding and on the applicability for some cloud conditions. A preliminary test campaign has been conducted in order to demonstrate the assessment of PIV in the CIRA-IWT testing the extreme operating condition of the tunnel in terms of cloud, temperature, speed and static pressure.

Subsequently, an experimental test campaign has been carried out on a full-scale aircraft wing tip. The region in proximity of the wing leading edge has been investigated. Measurement in cruise condition (retracted slat) and in the high-lift condition (extended slat) have been performed in order to evaluate the disturbance and eventually the flow separation due to the ice formation. In particular the collection efficiency has been investigated.

Measurements have been devoted to characterize the flow field downstream of the wing for wake characterisation. Starting from the initial condition of a clean model, the flow-field measurements have been conducted, during the icing-formation process in order to evaluate the drag increment due to the ice formation. This chapter presents the experimental setup, the technological problems that have been afforded and solved, the performed measurements and the obtained results.

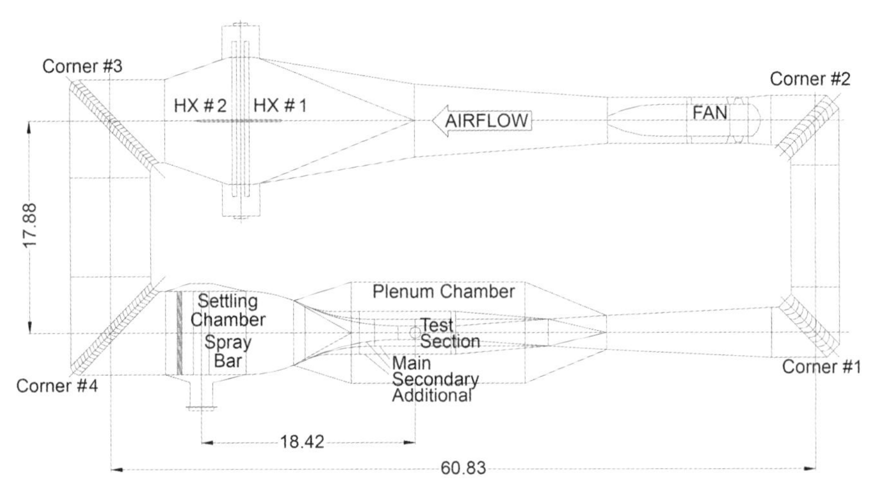

Fig. 1. CIRA-IWT layout

2 Experimental Apparatus

The experimental apparatus, including the test facility, test models and instrumentation are described in the following.

2.1 Icing Wind Tunnel (IWT)

The CIRA IWT has been in operation since 2002. The IWT facility is a closed-loop, refrigerated wind tunnel with three interchangeable test sections (TS) and one open jet. A sketch of the IWT layout is shown in Fig. 1. The IWT is fan driven by a 4-MW motor. The fan drive is located downstream of corner 2, in the back leg. Downstream of the fan diffuser, a twin-row heat exchanger (HX) is located to provide low-temperature operation capability and to compensate the fan heating. The facility settling chamber is fitted with a honeycomb for reducing large-scale eddies and ensures flow straightening.

Downstream of the honeycomb, an interchangeable section provides the possibility to install either: 1. a spray-bar module for generating the artificial cloud for icing tests or 2. a screen module when lower turbulence airflow is necessary for high-quality aerodynamic testing.

Downstream of the fixed contraction, the test section leg is made up of two interchangeable components (movable contraction + test section) and one adjustable component (collector diffuser).

The maximum speed achievable in the IWT depends upon the selected configuration. Mach number up 0.4 can be achieved using the main test section ($2.25 \times 2.35\,\mathrm{m}^2$), Mach 0.7 in the secondary test section ($1.15 \times 2.35\,\mathrm{m}^2$) and Mach 0.25 in the additional test section ($3.60 \times 2.35\,\mathrm{m}^2$). The test chamber walls are slotted in order to allows larger models.

Fig. 2. Icing blade in the secondary TS

The minimum temperature achievable is −32 °C in the main test section and in the additional test section. The secondary test section can get down to −40 °C.The HX is also capable of controlling the air relative humidity (RH) in a range from 70 % to 100 %, through a cooling–heating cycle and steam injection. In addition, an evacuation/pressurization air system allows the pressure to be regulated from 39 000 Pa (abs), corresponding to an altitude of 7000 m, up to 145 000 Pa (abs).

2.2 Test Model

During the first phase, PIV assessment, the tests have been performed in the empty secondary test section and in the wake of the icing-blade probe. This probe, composed by a steel blade of prefixed sizes, is aimed to measure the LWC value of the cloud by measuring the thickness of the ice accreted on the blade. The blade was mounted on the side wall at the centerline height as shown in Fig. 2. The blade is shielded by a cover while the requested cloud conditions are obtained.

The measurement campaign has been carried out on a full-scale tip wing. The wing has been mounted in the additional TS (Fig. 3). The model represents the extremity of a swept wing with 3.18 m span wise and 32° degree

Fig. 3. Swept wing in the additional TS

of sweep angle. The model is equipped with a movable slat with a hot-air de-icing system (piccolo tube). For the present work the de-icing system was not activated. The model was mounted on a motorized turntable allowing an incidence angle in the range from $-15°$ to $+15°$.

2.3 PIV System

The icing wind tunnel, being pressurized, is equipped with a large plenum chamber containing the test section. All the PIV recording and illuminating instrumentations were installed inside the plenum. The plenum chamber presents the same characteristics of total temperature, total pressure and humidity of the test section. For this reason, the PIV instrumentation has been protected using casings that maintain the temperature and pressure condition in the range of the operating parameters.

The illuminating system is composed of two Nd:YAG head resonators each providing 200 mJ pulse energy at 532 nm wavelength with a 10-Hz repetition rate. Each laser head has a dedicated power supplier. The power suppliers, together with the control unit of the temperature resistance and motor control, are located inside a protective casing. This one is connected to the laser head by means of a 5-m long umbilical cable that is thermally isolated. In

Fig. 4. Laser power-supply unit

this way, the laser head can be easily located close to the selected optical access, while the power-supply casing, which is much heavier, is located on the plenum walk way. The recombined laser beams are directed inside a mechanical optical arm of 2.5 m length by means of an alignment mirror. At the end of the arm the optics generating the lightsheet are mounted inside a dedicate housing. The optical head is thermally controlled in order to avoid moisture on the exit windows. The alignment mirror and the spherical lens to adjust the lightsheet focusing distance are equipped with servomotors remotely controlled by a PC. Figure 4 shows the power supply unit on the wind-tunnel passage way, Fig. 5 shows the laser head with the optical arm mounted on the ceiling of the additional TS and Fig. 6 illustrates a detail of the optical head mounted outside one of the windows present on the ceiling of the TS.

The recording system is composed of two 1280×1024 pixel resolution CCD cameras. The cameras are located inside special housings that are controlled in temperature. The cameras are equipped with remote focusing systems. Nitrogen is supplied to the windows in order to avoid moisture problems.

Figure 7 shows the two camera housings mounted outside the side wall of the additional TS.

Fig. 5. Laser head

The instrumentation condition, the synchronization of the camera with laser pulses, the laser energy intensity, the alignment of the laser beam, the camera focusing and the focusing distance of the lightsheet are all parameters that are fully remotely controlled. Figure 8 illustrates the control and recording station of the PIV system, installed in the IWT control room.

For PIV measurement the seeding quality is a crucial aspect. Good signal-to-noise ratio is obtained for the correct density of (10 particles per interrogation windows), uniform concentration and small-size particles. For the icing test instead of the classical DEHS or olive-oil seeding particles, the water droplets of the cloud have been used. In this case, the cloud characteristics are defined by the values of LWC and MVD. It was found that the LWC parameter strongly influences the quality of the measurements. For high values of the cloud concentration the measurement area is blinded. Furthermore, high cloud concentration or the presence of strong wake induce moisture condensation on the optical windows. Heating or purge systems are necessary.

Fig. 6. Lightsheet head

3 Experimental Test

Hereinafter, two different experimental campaigns are described.

3.1 PIV Assessment in CIRA-IWT

The first campaign was aimed to demonstrate the applicability of PIV in CIRA-IWT and to determine the limits of the operating conditions in terms of cloud characteristics, static temperature, static pressure and wind speed. For this first entry, the secondary TS ($2.35 \times 1.15 \times 5\,\mathrm{m}^3$) has been selected due to the fact that it presents the extreme condition in terms of wind speed, and static temperature. The measurement zone has been set in line with the center longitudinal plane. The reference system has been set with the origin on the bottom centerline of the TS, the x-axis was directed versus the wind velocity, the z-axis toward the wall ceiling. The laser head was mounted on the test-chamber ceiling, and the lightsheet was vertically directed through an available window. An area of $420 \times 340\,\mathrm{mm}^2$ has been

Fig. 7. Camera housing

Fig. 8. PIV control station

acquired with a single camera with a 60-mm lens located outside the side wall.

Tests have been conducted at 100, 120 and 148 m/s wind speeds in free-stream conditions. In order to validate the protection system, tests have been performed at various static temperatures (−4 °C, −25 °C and 30 °C) and at two total pressures (110 000 and 53 846 Pa).

Particular care has been taken in investigating PIV system behavior for different cloud conditions. The droplet diameter (MVD) has been varied in the range from 15 to 250 μm, while the concentration (LWC) has been set in the range from 0.15 to 2.8 g/cm^3. Measurements in the wake of the icing blade probe have been performed as well in order to evaluate possible problems arising from the wake condition. For each test condition 150 instantaneous velocity fields have been acquired. The ensemble-average velocity field and relative turbulence level field has been evaluated by the total sample available for each test condition.

3.2 Performance Degradation Investigation

The second test campaign was aimed to investigate the real potentiality of PIV in the framework of an icing test. The test foresaw the installation of the largest test section (additional $2.35 \times 3.6 \times 8.3\,\mathrm{m}^3$) together with a full-scale wing tip with movable slat.

Two different model configurations have been simulated: high lift and cruise condition. For both configurations, tests with and without ice accretion have been conducted at, respectively, static temperatures of $-1\,°\mathrm{C}$ and $-5\,°\mathrm{C}$. A fixed LWC value of 1 g/cm^3 and two MVD values (10 and 27 μm) have been selected in order to investigate the influence on the behavior of the water-droplet trajectories. All the tests have been conducted at 60 m/s wind speed. PIV measurement has been carried out during ice accretion for 15 min with a 0.5-Hz frame rate.

The high-lift model configuration foresaw an angle of attack (AoA) of 3.5° and a 32° of slat-deflection angle. Two cameras were aimed at the slat-wing gap region. The measurement plane has been located orthogonal to the wing leading edge. One camera was equipped with a 60-mm focal lens covering all the region upstream of the slat, whereas the second camera mounting a 100-mm lens has been focused on the intersection region. The recording areas were, respectively, $340 \times 270\,\mathrm{mm}^2$ and $200 \times 160\,\mathrm{mm}^2$.

The reference system has been positioned with the origin in the intersection between the lightsheet and the slat trailing edge (TE), the x-axis directed orthogonal to the TE toward the main speed direction, the z-axis was positioned vertical toward the ceiling wall and the y-axis parallel to the TE. The two fields of view presented an overlap region. Figure 9 shows the location of the recording areas. The laser, as in the previous case, was located above the test chamber (Figs. 5 and 6).

The cruise configuration used a 0° of incidence angle and the slat was retracted. Also in this case, two cameras have been used. The cameras have been located at about two chords downstream of the trailing edge of the wing. In order to cover the entire wake region the cameras have been mounted vertically with an overlap in the recording region. The cameras mounted 60-mm focal length lenses providing a recording area of $276 \times 220\,\mathrm{mm}^2$ each. The region of interest has been located in the longitudinal centerline plane.

Fig. 9. PIV recording setup on the LE

Fig. 10. Recording camera and experimental setup in the wake region

Figure 10 shows the experimental setup adopted for the characterization of the wake.

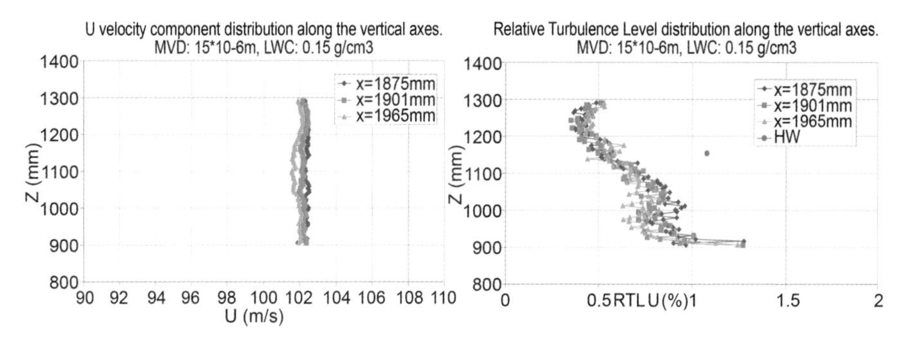

Fig. 11. U velocity component and RTL distribution vs. the z-axis in free-stream conditions

4 Results and Discussions

In the following, the main results are presented and discussed. As is possible to deduce from the test description from the previous section, a considerable amount of data was acquired. So far, in the following only an overview is reported.

4.1 PIV Assessment

PIV measurements have been performed for several cloud configurations in free-stream conditions. All the conditions of speed, static temperature and static pressure have been fulfilled. The only limitation encountered was on the composition of the cloud. A LWC value of $2.0\,\mathrm{g/m^3}$ has been found as the upper limit for good-quality PIV images. The tests performed at higher values of LWC generated clouds with concentrations that prevented detection of the particles in the measurements area. In Fig. 11 the ensemble-average U-component of the velocity field and the relative turbulence level versus the z-axis are reported. The test has been carried out at a 100-m/s wind tunnel speed, $15\,\mu\mathrm{m}$ of MVD and $0.15\,\mathrm{g/cm^3}$ of LWC. Comparison at different locations along the x-direction is reported. The U velocity component shows a good uniformity along the z-direction and for different stations along the x-axis. The relative turbulence level (RTL) behavior obtained with PIV is compared with a measurement point obtained by performing hot-wire anemometry measurements in the air. The slight underestimation is explained by the velocity lag induced by the mass of the particle characterized by a mean diameter of $15\,\mu\mathrm{m}$. This predicts a reduction of the velocity fluctuation.

The second configuration applied the PIV technique to a typical condition arising during the aerodynamic test, i.e., the wake characterization. In the case of icing tests, low temperature and high humidity, the wake is affected by condensation phenomena, as is visible in Fig. 12. The image clearly shows

Fig. 12. Flow-field picture of the wake behind the icing blade

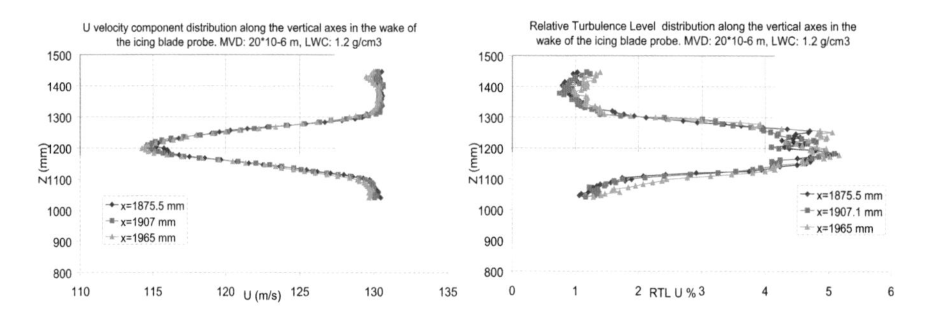

Fig. 13. U velocity component and relative TRL distribution vs. the z-axis in the wake region

a white zone induced by the wake, this brighter background can affect the quality of the signal-to-noise ratio.

Figure 13 presents the U-component of the ensemble-average velocity field and the RTL versus the z-direction measured in the wake of the icing blade. The results correspond to a test condition of 125 m/s, $-30\,^{\circ}$C static temperature, cloud characteristic of 20 μm of MVD and 2.0 g/cm^3 of LWC. The diagrams show the typical loss of momentum induced by the wake. The RTL starting from the undisturbed value of 1 % increases in the wake up to a value of 5 %.

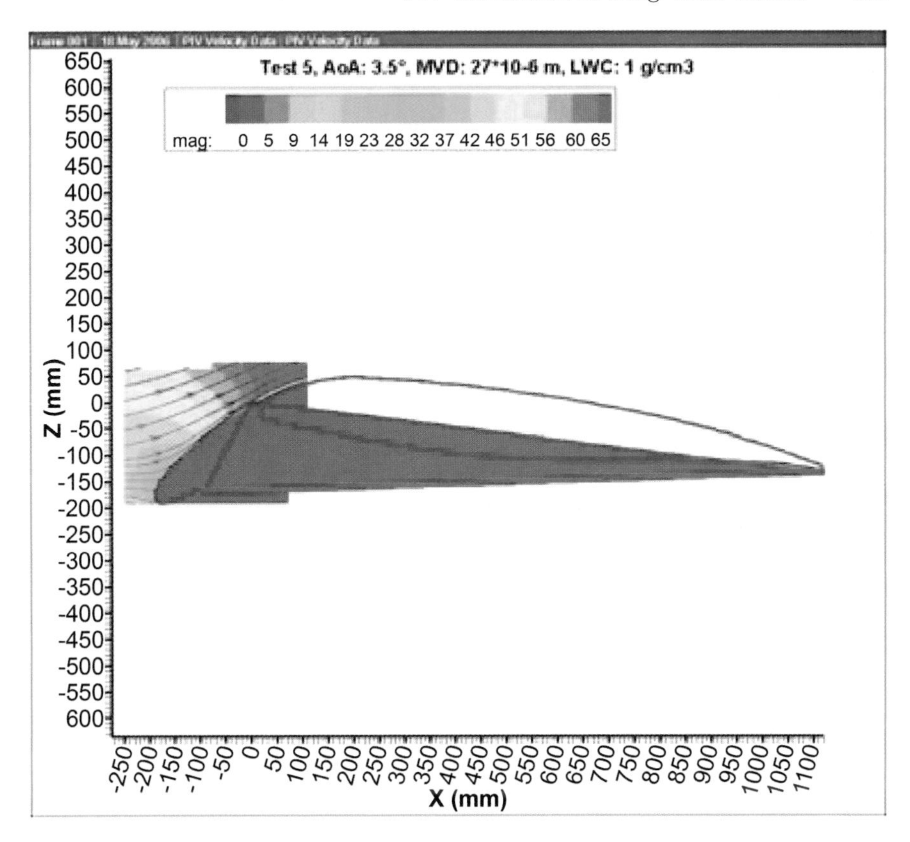

Fig. 14. PIV measurement areas

4.2 Performance Degradation Investigation

The first measurements have been carried out at a $-1\,^{\circ}$C temperature value in order to avoid ice formation on the model and 60 m/s of wind velocity. The droplet trajectories, on the leading-edge zone of the model in the high-lift configuration, have been measured for two different values of droplet diameter (15 and 27 µm MVD). In Fig. 14 the model zone investigated by the PIV measurement are visible. On the geometry of the model with 3.5° of AoA and −32° of slat deflection the ensemble-average velocity field obtained on 150 instantaneous samples has been superimposed. A detail of the PIV results is shown in Fig. 15. The figure presents the droplet trajectory evaluated on the mean velocity field.

Starting from the trajectories it is possible to evaluate the global collection efficiency (E) defined as the ratio between the mass water droplets impinging and the mass water seen in the body-projected area. This value is obtained by integrating on the leading surface of the model the local collection efficiency

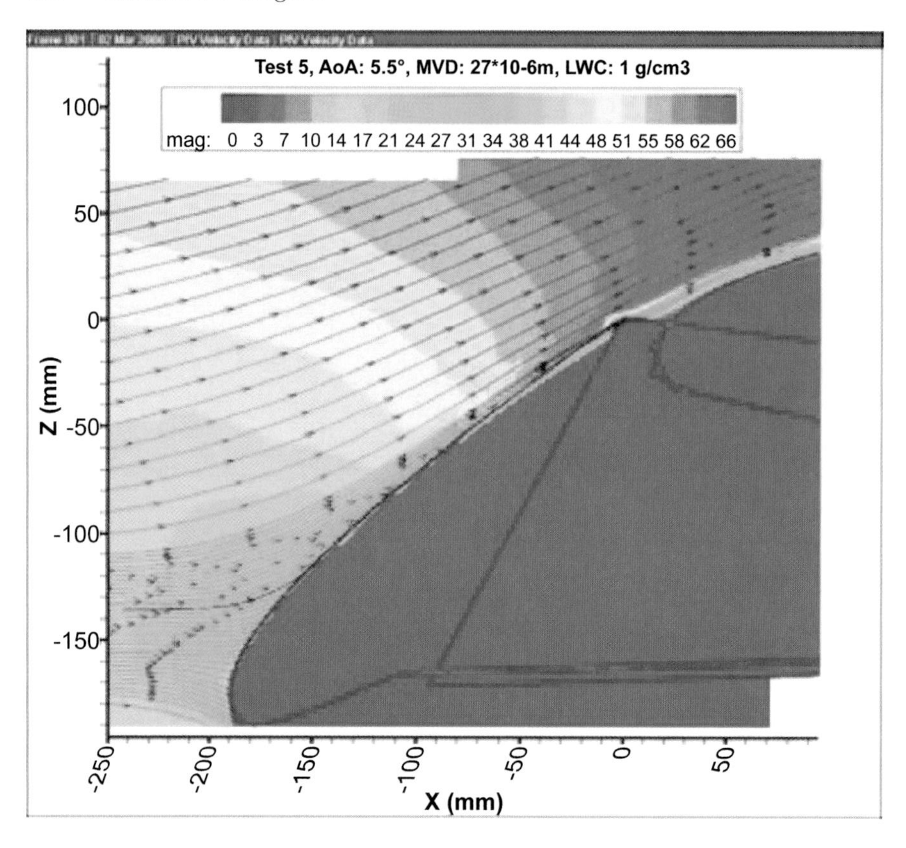

Fig. 15. Droplet trajectories and velocity-field color map

(β) defined as the ratio between the space (d_{z}) between two side streamlines upstream of the body in the undisturbed zone and the distance on the body surface where the trajectories impact (d_{s}), i.e., $\beta = d_{\mathrm{z}}/d_{\mathrm{s}}$. The local value of the collection efficiency has been evaluated for 15 μm and 27 μm MVD. The results are shown in Fig. 16.

Beta has been evaluated on the slat surface, imposing the origin in the slat TE and moving toward the LE. The diagram presents, as expected, an increment of the value of beta moving toward the leading edge. The curve relative to the larger droplet diameter shows a higher value of beta. This is due to the inertial force of the droplet, the trajectories are less influenced by the presence of the model and consequently a larger ice accretion occurs. The local collection efficiency obtained by PIV has been compared with the CFD results obtained with multi-ice CIRA code [7]. The agreement between the CFD and experimental result is remarkable (Fig. 17).

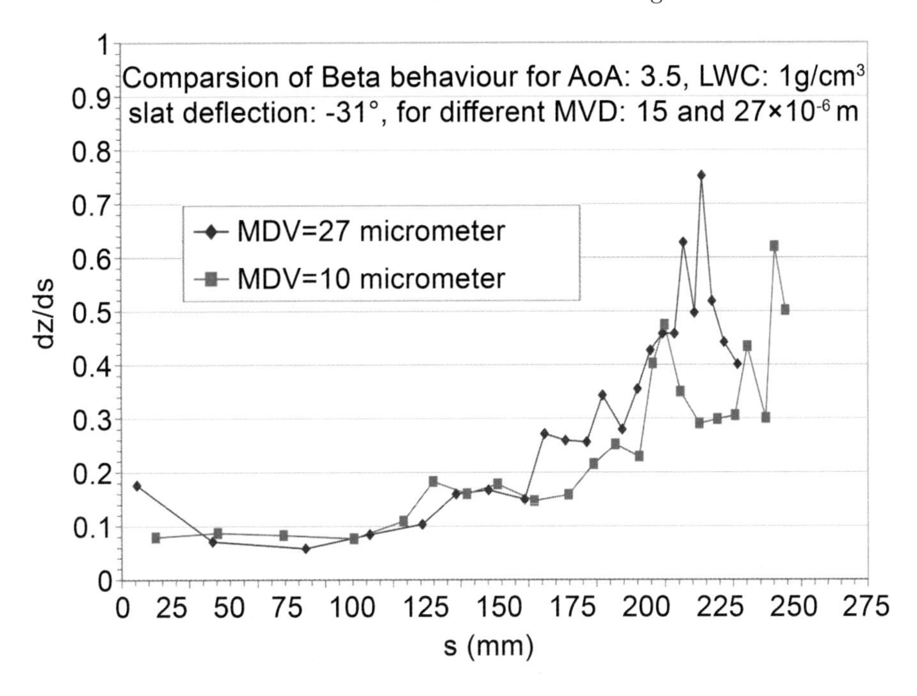

Fig. 16. Beta vs. surface coordinate

In order to allow the ice formation on the model the static temperature has been dropped to $-5\,^{\circ}$C. The conditions selected: 27 µm of MVD and 1.0 g/cm^3 of LWC, correspond to a glaze-ice shape.

This type of ice accretion is the worst that can occur in terms of aerodynamic-performance degradation, altering the profile shape and in terms of PIV recording because the ice is bright and highly reflective, as shown in Fig. 18. The detection of the ice-shape profile is quite challenging due to the nonuniformity of the ice formation spanwise, so it occurs that ice formation covers the measurement area and strong reflections inside the ice affect the quality of the PIV images close to the ice formation. Figure 19 shows the formation of ice after 15 min of flight through the cloud. PIV data have been recorded with a frequency of 0.5 Hz. In Fig. 20 the results obtained by the analysis of two PIV images recorded, respectively, after 5 and 15 min from the beginning of the ice accretion on the wing. The disturbance on the streamlines due to the ice presence is clearly visible (for comparison see the results shown in Fig. 15).

The zone colored blue in front of the slat profile represents the ice-accretion shape obtained by tracing the PIV image. From the two color maps the increasing of the ice thickness with time is also visible. The tracing of the ice shape is still affected by a large error due to the strong laser-light scattering, but it is a big step forward because information on the ice shape

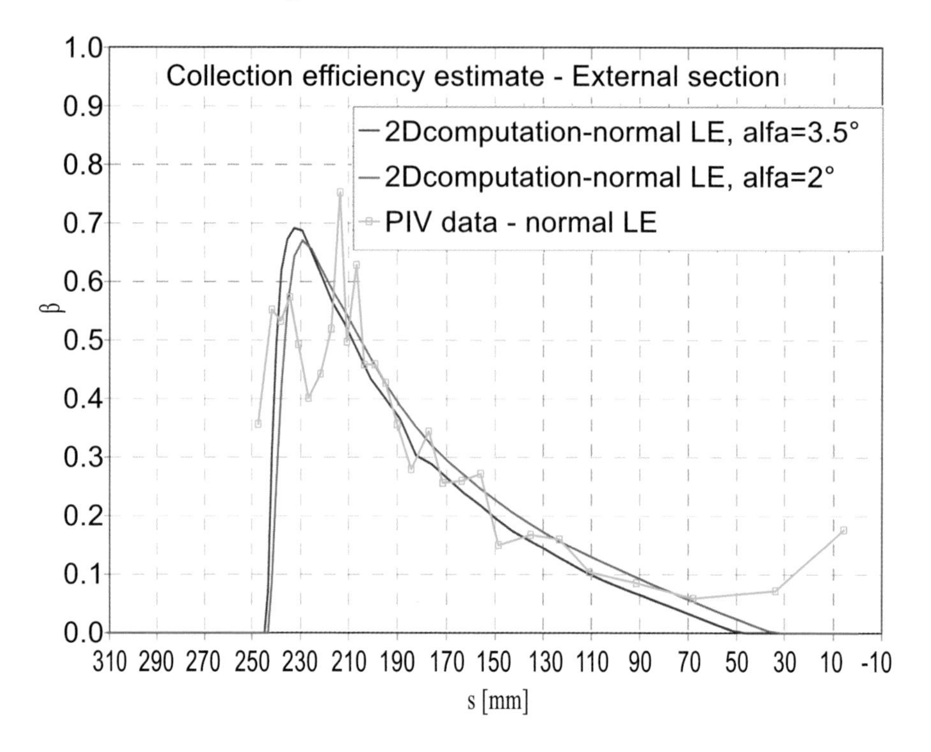

Fig. 17. Beta: CFD and PIV data comparison

is possible to obtain only at the end of the run and by a manual tracing. Activities are in progress for increasing the quality of the recorded image.

The wake measurements have been performed with the model in cruise configuration. The test conditions were the same as the high-lift configuration. Measurements during the ice-accretion phase have been carried out for 15 min with a 0.5-Hz recording frequency.

Figure 21 shows the wake behavior at different instants during the ice accretion. In each diagram the value of the wake obtained for the case of clean profile averaging on 150 samples is compared with the single wake obtained by the instantaneous velocity fields at different instants. The diagrams clearly show that in accordance with the rising run time the loss of wake velocity increases due to the accretion of ice formation on the model. The typical separated flow behavior occurs starting after 500 s of the beginning of ice formation, it is indicated by the double peak typical of the vortex shedding induced by the flow separation.

Using the wake-flow-field characterization a first attempt to measure the drag coefficient has been carried out by means of the loss of momentum. A simplified method for 2D profiles has been used due to the fact that the third component of velocity was not available.

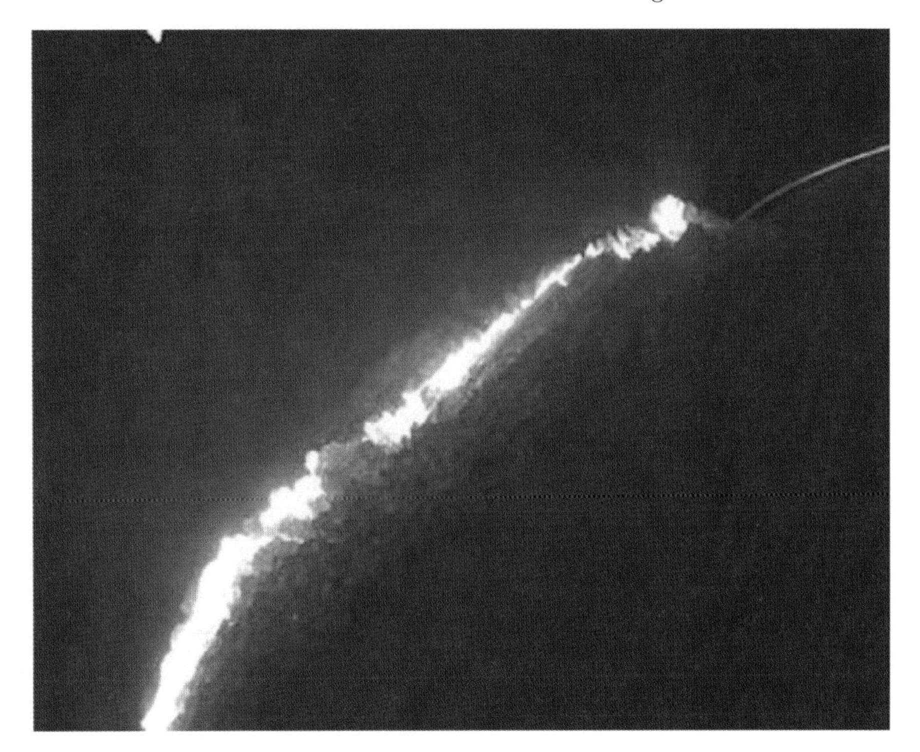

Fig. 18. PIV image during ice accretion

Figure 22 presents the drag coefficient (CD) behavior at different instants since the start of the ice accretion. The scattering of the data is due to the fact that the CD has been obtained on a single instantaneous velocity field, which is highly unsteady. The diagram presents a uniform trend, as the time passes, and consequently the ice accretes, the drag coefficient increases almost linearly.

5 Conclusions

The assessment of the PIV technique in an icing wind tunnel has been demonstrated for the first time. The measurements were effective in the following range of application: Wind speed up to 150 m/s, MVD range: 15 to 250 μm, LWC range: 0.15 to 2.0 g/cm^3, Altitude: 0 to 6000 m on s.l., static temperature: 0 °C to −30 °C.

Several measurement campaigns have been performed in order to investigate the range of applicability of PIV and the potentiality of the technique. The PIV technique allowed us to evaluate, using the water cloud as seeding,

Fig. 19. Wing after the accretion of the ice

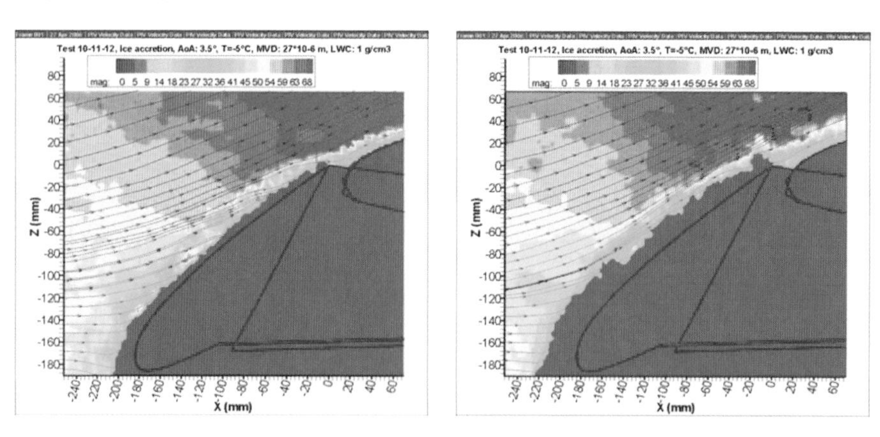

Fig. 20. PIV results obtained, respectively, after 5 and 15 min from the start of the ice accretion

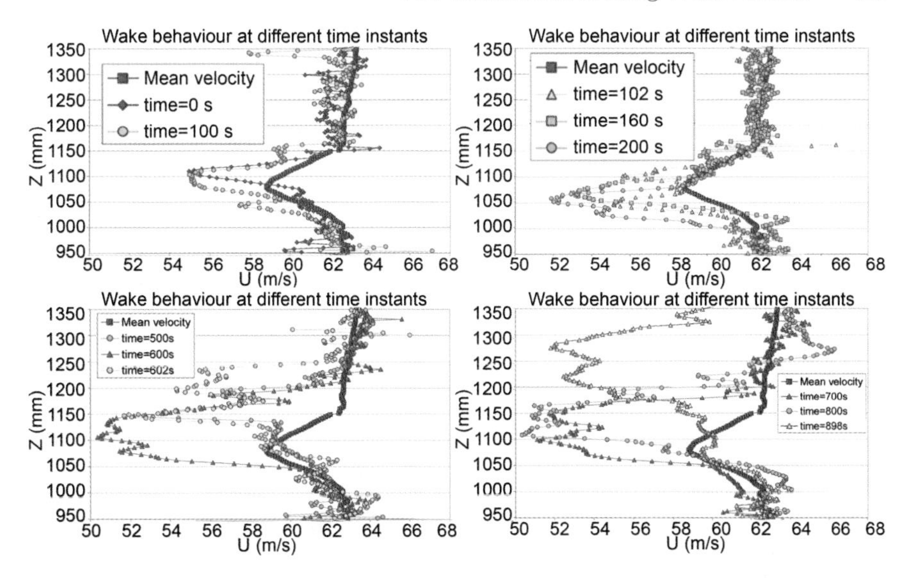

Fig. 21. Wake behavior at different time instants during ice accretion

the droplet trajectories and the collection efficiency on the model. Using standard PIV it is possible to evaluate the flow streamlines and in this way it is possible to measure the different behavior with the water-droplet trajectories. These measurements provide unique data for CFD validation.

The wake measurement allows us to measure, by means the loss of momentum, the increasing of the drag coefficient due to the ice accretion.

Some problems are still present, the strong light scattering due to the ice formation on the model affects the quality of the result in the proximity of the model. The background noise due to the water droplets running back on the model provides a spurious vector in the final results. Methods to eliminate or to reduce the laser-light scattering are under study. If successful, together with the improvement of the quality of the PIV results, it may also be possible to measure quantitatively the ice accretion on the model continuously, whereas nowadays it is only possible at the end of the test run to perform a manual tracing of the ice formation.

Acknowledgements

This work has been performed in the framework of PIVNET 2 thematic network and partially founded by the project "ACADEMIA" sponsored by the Italian University and Scientific Research Ministry. The author wishes to acknowledge the valuable contribution enabling the success of this investigation by: the IWT crew and in particularly by M. Bellucci, G. Esposito and M. Lupo.

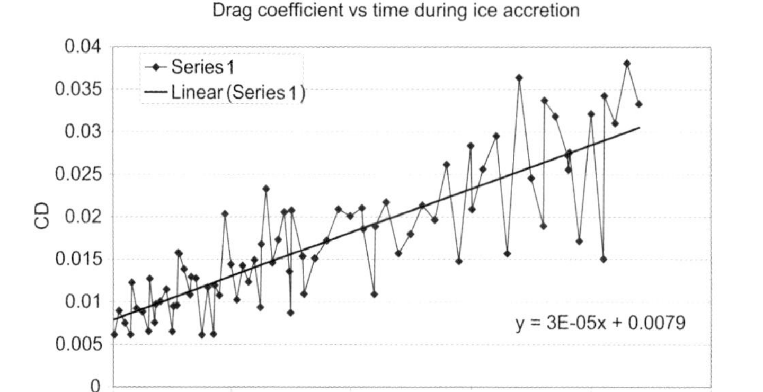

Fig. 22. CD behavior versus time accretion

References

[1] G. M. Preston, C. C. Blackmann: Effects of ice formation on airplane performance in level cruise condition, NACA **TN 1598** (1948)
[2] M. B. Bragg, T. Hutchison, J. Merret, R. Oltman, D. Pokhariyal: Effect on ice accretion on aircraft flight dynamics, AIAA **2000-0360** (2000)
[3] M. B. Bragg, G. M. Gregorek: Aerodynamic characteristics of airfoils with ice accretions, in *AIAA 20th Aerospace Sciences Meeting* (1992)
[4] M. B. Bragg, W. J. Coirier: Aerodynamic measurements on an airfoil with simulated glaze ice, in *AIAA 24th Aerospace Sciences Meeting* (1986)
[5] C. Cuerno, R. Martinez-Val, A. Perez: Experimental diagnosis of the flow around a NACA 0012 airfoil with simulated ice, in *Proc. 7th Conf. of Laser Anemometry Advanced and Application* (1997) pp. 747–754
[6] F. De Gregorio, A. Ragni, M. Airoldi, G. P. Romano: PIV investigation on airfoil with ice accretion and resulting performance degradation, in *Proc. 19th ICIASF* (2001) pp. 94–105
[7] G. Mingione, V. Brandi: Ice accretion predictions on multielement airfoils, J. Aircraft **35**, 240–246 (1998)

Index

Analysis of the Vortex Street Generated at the Core-Bypass Lip of a Jet-Engine Nozzle

José Nogueira, Mathieu Legrand, Sara Nauri, Pedro A. Rodríguez, and Antonio Lecuona

Dept. of Thermal Engineering and Fluid Mechanics,
Universidad Carlos III de Madrid, Spain
goriba@ing.uc3m.es

Abstract. The reduction of the noise generated by jet-engine aircrafts is of growing concern in the present society. A better understanding of the aircraft noise production and the development of predictive tools is of great interest. Within this framework, the CoJeN (Coaxial Jet Noise) European Project includes the measurement of the flow field and the noise generated by typical turbofan jet-engine nozzles. One of the many aspects of interest is the occasional presence of acoustic tones of a defined frequency, symptomatic of the presence of quasiperiodic coherent structures within the flow. This chapter analyzes the characteristics of a vortex street in the core-bypass lip of one of the nozzles under study. The measurements were made by means of advanced PIV techniques within the above-mentioned project.

1 Introduction

The experimental part of this work has been performed at the QinetiQ NTF (Noise Test Facility) in Farnborough, UK, one of the largest jet-noise wind tunnels in the world. Even for this large anechoic chamber (Fig. 1), the nozzle model sizes have to be downscaled by a factor of 10 with respect to the large commercial aircraft ones. This is in order to properly assess the whole plume characteristics within a reasonable scale, among other issues. Despite this general reduction in size, for a particular model of the nozzles under study, the thickness of the lip has not been downscaled. This allows study of the effects behind a full-scale lip, as the effects close to the lip are not related to the diameter of the nozzle. Besides that, the velocity and temperature conditions on the experiments are those of the real jet.

Due to the operating costs of the tests the measurement of the flow features at the nozzle lip had to be combined with the measurement of the whole nozzle velocity profile. This constrains the possibilities of spatial resolution at the lip because the camera view had to include the whole bypass section. This is a challenging situation for the PIV analysis algorithms. The whole camera field corresponds to the square depicted in Fig. 2. A rectangular section of 1135 by 297 pixels (49 by 13 mm) has been zoomed in to indicate the position of the vortex street.

The stability of the regular arrangement of vortices sometimes found in the wake of bluff bodies was first studied by *von Kármán* [1], finding only

A. Schroeder, C. E. Willert (Eds.): Particle Image Velocimetry,
Topics Appl. Physics **112**, 419–428 (2008)

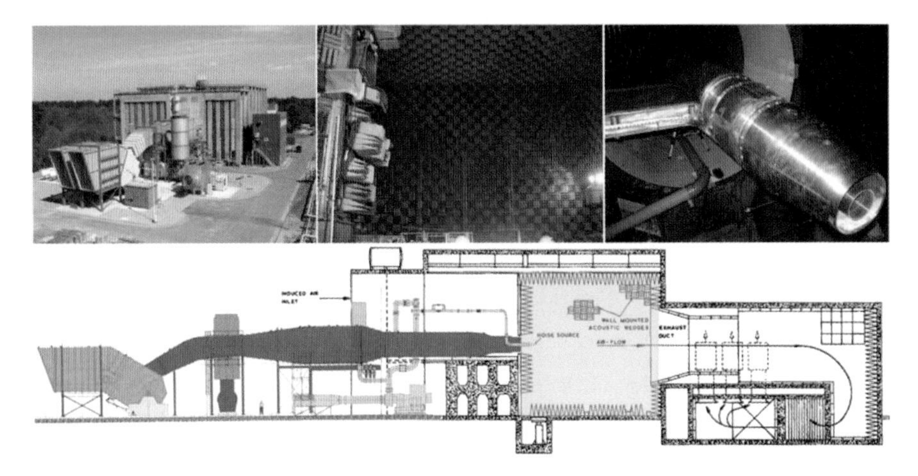

Fig. 1. QinetiQ Noise Test Facility used in the measurements campaign of the Coaxial Jet Noise European Project. The images above the blueprint represent from *left* to *right:* (**a**) The whole facility, (**b**) The almost cubic anechoic chamber shadowed below. The jet nozzle is not installed here. The blue nozzle up to the left corresponds to the flight stream blower, 8 m above the floor of the chamber. (**c**) Jet nozzle installed in the flight stream blower

one linearly stable configuration for isothermal flow. Later, it was shown that even for that configuration the arrangement is unstable if second-order terms are taken into account [2]. Since then many researchers have studied possible stabilizing effects that explain its persistence [3–5, among others]. This chapter describes an apparently unstable non-isothermal vortex street, located at a shear layer, producing significant effects such as acoustic noise. Despite its instability and the large Reynolds number ($\sim$ 7000 based on the lip thickness), the vortex-shedding mechanism is persistent enough to continuously generate a defined vortex street. Approximately nine counterrotating vortex couples can be observed before breakdown. For this nozzle, a defined acoustic tone was measured, being compatible with the measured flow structure.

The chapter gives and discuses the flow magnitudes that characterize the vortex street for a certain flow condition. This is offered as valuable information for the validation of numerical predictive tools aiming to reproduce this effect. In addition to that the performance of different PIV analysis algorithms are tested and compared when attaining information on the fine structure of the vortices.

2 Experimental Setup

As noted in the introduction the measurements have been performed at the QinetiQ NTF (Noise Test Facility) in Farnborough, UK. This facility has

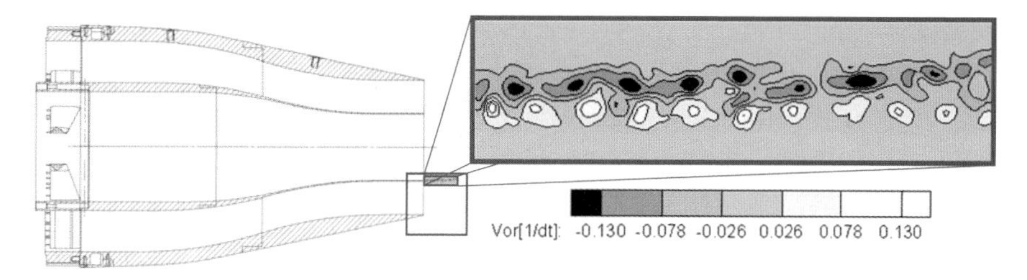

Fig. 2. Example of a measurement on a zoomed image of the vortex street. Its size and location are also depicted in respect to the nozzle. $dt = 1.3 \times 10^{-6}$ s

the capability of producing large free jets thermally and kinetically stabilized within 0.1 % margins. The case analyzed in this chapter corresponds to the coaxial nozzle depicted in Fig. 2, being the diameter of the core section 100 mm and the diameter of the bypass section 200 mm. Although the core and bypass outlets are initially coplanar, the thermal expansion during operation makes the core outlet lip protrude a couple of millimeters with respect to the bypass one.

The core flow corresponds to an outlet velocity of 400 m/s and a static temperature of 560 °C, the bypass had a velocity of 303 m/s and a static temperature of 53 °C, all of them constant over time. The lip between both flows is the one producing the vortex street under study. It has a thickness of 1 mm and is made of solid stainless steel with a smooth surface and axisymmetric construction. Its thickness is almost constant up to 50 mm upstream and then thickens and widens to achieve the required rigidity and smooth flow entrance. The tip ending corresponds to a simple half-circle finishing. The Reynolds number of the flow using the thickness of the lip is $\sim 4.6 \times 10^3$ for the hot flow and $\sim 2 \times 10^4$ for the cold one.

The flow was seeded with commercial 0.3-micrometer nominal size titanium dioxide particles using a fluidized bed followed by a cyclone separator. The acquisition system for the PIV measurements consists of a double-cavity 380 mJ per pulse Nd:YAG laser form QUANTEL and a LAVISION FlowSense camera of 2 k by 2 k pixels. The repetition rate was around 1 frame per second.

The magnification of the images is 0.0443 mm/pixel. The time between laser pulses is $dt = 1.3 \times 10^{-6}$ s, assuring high-quality PIV images.

The temperature difference between the flows corresponding to the core and bypass of the jet-generated optical deflections. Thanks to the proximity between the measured structures and the temperature gradient zone, these effects where small. Theoretical estimations indicate positional errors smaller than 0.05 mm. In addition, the short time between laser pulses freezes the flow structures that may modify the optical deflection. A very similar deflection is expected for each particle image in a particle pair, reducing the error

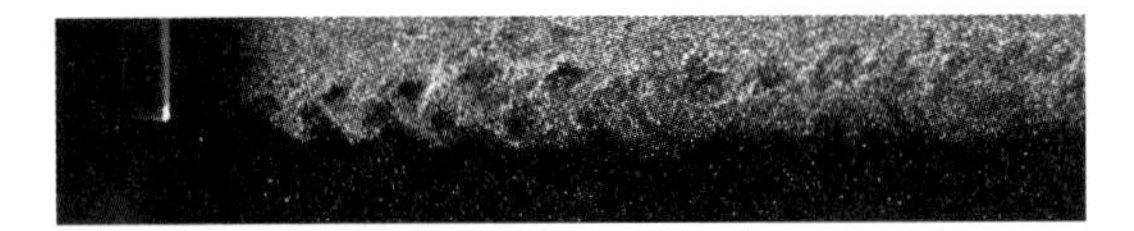

Fig. 3. Visualization of the vortex street due to differences in seeding density (15×76 mm)

for velocity calculations to less than 0.01 mm (7.7 m/s). This value is similar to PIV errors coming from other sources and allows reasonable measurements.

3 Vortex-Street Characteristics and Possible Dynamics

Some images were obtained with different seeding densities between the core and bypass. This allows visualization of the vortex street. Figure 3 shows one of these images. It should be pointed out that the vortices visualized in this way depict the history of rolling that the tracer particles have experienced, and not the actual motion of the flow. This makes visualization easy because the perturbations on the relative motion between vortices or their breakup causes an effect on the image that is delayed until their time integral amounts as much distortion as the rolling they have experienced from the beginning. In this sense, it is more correct to conceptualize the visualization of the vortex street as its history rather than as an indication of the particular instantaneous vorticity.

It should be mentioned that this is not a regular Kármán vortex street. The flows at both sides of the lip have a significant velocity difference. Consequently, the vortices at each side have a different intensity and, owing to this, induce different velocities to each vortex sheet. It can be said that the configuration is therefore unstable. Nevertheless, the measurements reported in this work show that the persistence of the vortex-shedding mechanism is strong enough to produce a defined vortex-street.

In order to analyze the overall characteristics of the vortex street a low-resolution PIV study has been performed over 300 images. The selected interrogation window size was 64×64 pixels (2.8 mm) and the grid step size was 16 pixels (75 % overlapping). The advantage of such an arrangement is that the fine structure of the vorticity is averaged and straightforward algorithms can be used to identify the main vortices. The detection of vortices was based on the presence of a local maximum in vorticity. To check that the vorticity detected this way was a swirling and not a shear one, several images were analyzed at higher resolution. In all the cases the stream lines around these vorticity peaks were closed circles. In consequence, more complex criteria, like λ_2 [6], were not used. Vorticity peaks closer than 2.1 mm were merged. On doing this, images like the ones depicted in Fig. 4 were obtained. Visual inspection of the images showed erroneous identifications (missing vortices) of the order of 2 %.

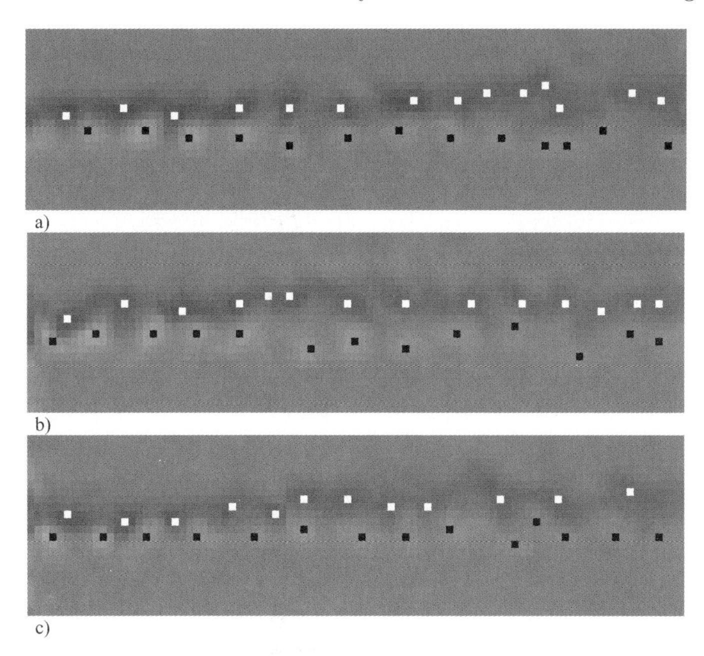

Fig. 4. Some vorticity plots showing identification of the vortices within the vortex street. The lower part of the image is for lower main stream velocity. Vorticity contour plots with appropriated scale are presented in Fig. 8

Statistical analysis of the identified vortices showed that the mean distance between vortices, in the axial direction, corresponds to 4.5 mm with a standard deviation of 1.7 mm for the upper vortex row and 4.7 mm with a standard deviation of 1.6 mm for the lower one. The particular distances measured are depicted in Fig. 5 together with a second-order polynomial fit curve. Each vortex was larger than two grid steps. This allowed for a parabolic fitting that reduces the location uncertainty below the grid step size (0.7 mm). Nevertheless, intervortex distances smaller than 2.8 mm could be underevaluated due to the large size of the interrogation window. The higher-resolution analysis in Sect. 4 verifies that this interrogation window size is appropriate for the estimation of Fig. 5.

The absolute velocity of these vortices (approximated by the velocity at the vorticity maximum) corresponds to the plots depicted in Fig. 6.

With these mean distances and velocities, the frequency of the vortices varies from $\sim$ 53 kHz for the lower vortex row at 5 mm from the lip to $\sim$ 72 kHz for the upper row at 65 mm from the lip. This is coherent with the noise spectrum obtained by QinetiQ for this nozzle, which showed a peak at $\sim$ 60 kHz, having a width of 20 kHz. This peak rises 4 dB over a smooth curve for the rest of the nearby frequencies.

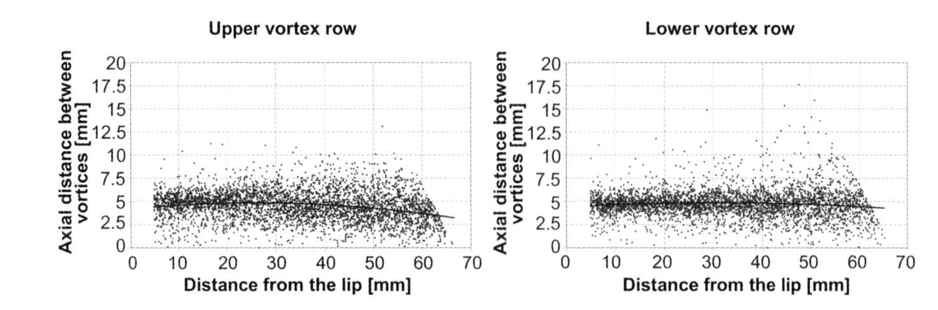

Fig. 5. Distances measured between consecutive vortex pairs

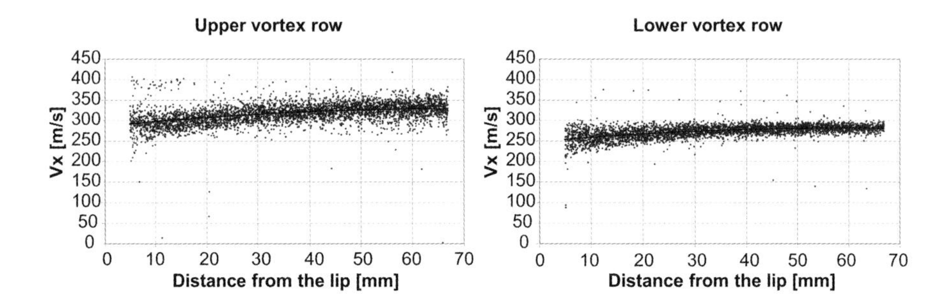

Fig. 6. Axial velocities measured for each vortex

At the nozzle lip, the shedding frequency is the same for the upper and lower rows of vortices. Besides that, it has been measured that each of these two rows has a different mean vortex velocity (18 % of difference). This raises the question of how can both rows have such a similar mean distance between vortices. One possible answer seems to be related to an apparent vortex division. Paying attention to Fig. 4a it can be observed that, according to the difference in vortex velocity plotted in Fig. 6, when the lower row travels 4 intervortex distances, the upper one travels 4.5. This means that around the 4th vortex couple, the vortices are almost aligned instead of staggered. Further downstream it seems like there is a rearrangement, where a vortex of the upper row is divided into two. This happens twice in Fig. 4b, but Fig. 4c shows that this is not a fixed rule and consequently further study is needed. Sudden vortex division does not seem to be a correct explanation, a higher-resolution processing of the PIV images allows for a more detailed analysis of this issue in Sect. 4.

With the aim to focus back on the geometrical characteristics, the width of the street is depicted in Fig. 7. It has been measured as the transversal distance between the positive vortices and the midpoint between the two adjacent negative ones.

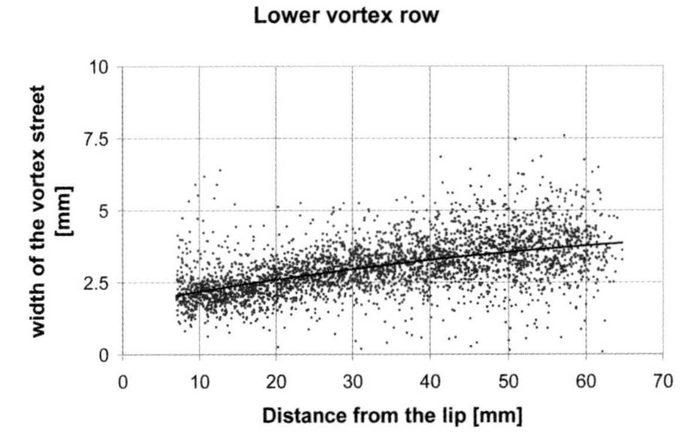

Fig. 7. Vortex-street width displayed as a function of the distance from the lip

The ratio between the width and the distance between vortices varies between ~ 0.45 and ~ 0.84, clearly different from the 0.28 corresponding to a Kármán vortex street.

4 Vortex-Street Fine Detail

The apparent division of the vortices noted in the previous section could correspond to the evolution of the vorticity along the vortex street or to the generation of two vortices of the same sign at the nozzle lip for some cycles of the shedding mechanism. In both cases, the occurrence in the low-resolution analysis seems very sudden and distinct to the neighboring vortex arrangement. Some further analysis has been done searching for the fine structure of the vorticity. Figure 8 shows the analysis performed in the case of Fig. 4a. The initial spatial resolution used in the analysis corresponds to interrogation windows of 2.8 mm of side length, where the measurement is averaged. The vorticity plot is depicted in Fig. 8a. It is just a surface plot of Fig. 4a where the correspondence of vortices can be easily checked. It was obtained with a conventional PIV enhanced with image-distortion correction. A review of these methods can be found in [7]. The second and third cases correspond to the advanced method LFC-PIV [8]. In this case, top-hat smoothing of 3 by 3 vectors [9] has been applied between iterations to reduce high spatial frequencies and establish a clear resolution-limiting value. In this way, for Figs. 8b,c, the spatial resolution has been improved down to 1.2 and 0.6 mm, respectively. It can be seen that the vortex division in Fig. 4a is not sudden and three-dimensional interactions between vortices could happen from the very early stages.

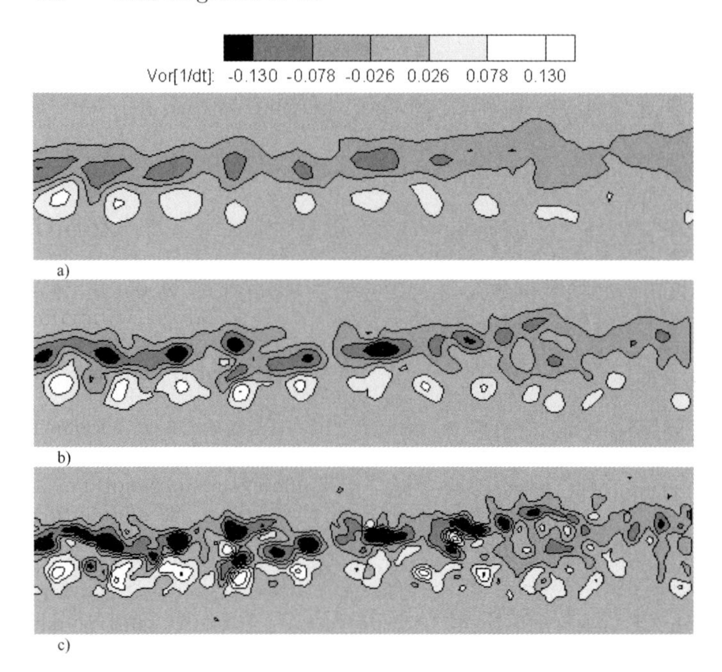

Fig. 8. Vorticity contour plots of Fig. 4a with increasing resolution. ($dt = 1.3 \times 10^{-6}$ s)

5 Conclusions

PIV has been shown to offer a valuable measuring technique in assessing nonsteady flows even when the sampling is random. The statistical analysis of the results gives the main parameters of the nonregular vortex street under study.

A particular vortex street emerging from the lip of a round nozzle has been characterized. The velocities at both sides are different as well as the temperature. In spite of that, a remarkable coherence length is evident both in the visualization images and in the vorticity fields.

The data indicates that the shedding mechanism of the vortex street is persistent. This leads to the generation of acoustic noise 4 dB above the rest of the jet noise at its frequency. This shedding persistence is enough to generate the coherent structure responsible for the noise even in a situation where the vortex street is not stable.

The frequency associated with this structure is not likely to scale with the nozzle when referring to a real-size engine. Its generation seems to be associated with the lip geometry whose thickness has been kept the same as the size of the real engine.

A higher spatial resolution does not always offer all the advantages. Vortex identification is best performed with a smaller than highest spatial resolution.

As spatial resolution is increased, by decreasing the interrogation window distance [10] and applying an advanced PIV algorithm, more detail is obtained. This allows identification of vortex breakdown and separation. The image analysis performed is at the limit of the PIV technique, thus offering information on its capabilities.

Acknowledgements

This work has been partially funded by the COJEN, Specific Targeted RESEARCH Project EU (Contract No. AST3-CT-2003-502 790); the EUROPIV 2 European project (CONTRACT No.: GRD1-1999-10 835); the Spanish Research Agency grant DPI2002-02 453 "Técnicas avanzadas de Velocimetría por Imagen de Partículas (PIV) Aplicadas a Flujos de Interés Industrial" and under the PivNet 2 European project (CONTRACT No: GTC1-2001-43 032).

We would like to thank the well-known researchers F. de Gregorio, C. Kähler and C. Willert for their kind advice on the wind-tunnel experimental setup.

References

[1] Von Kármán: Flüssigkeits- u. Luftwiderstand, Phys. Zeitschr. **XIII**, 49 (1911)
[2] N. Kochin: On the instability of van Kármán's vortex streets, Doklady Academii Nauk SSSR **24**, 19–23 (1939)
[3] H. Aref, E. D. Siggia: Evolution and breakdown of a vortex street in 2 dimensions, J. Fluid Mech. **109**, 435–463 (1981)
[4] P. G. Saffman, J. C. Schatzman: Stability of a vortex street of finite vortices, J. Fluid Mech. **117**, 171–185 (1982)
[5] J. Jimenez: On the linear-stability of the inviscid Karman vortex street, J. Fluid Mech. **178**, 177–194 (1987)
[6] J. Jeong, F. Hussain: On the identification of a vortex, J. Fluid Mech. **285** (1995)
[7] F. Scarano: Iterative image deformation methods in PIV, Meas. Sci. Technol. **13**, R1–R19 (2002)
[8] J. Nogueira, A. Lecuona, P. A. Rodríguez: Local field correction PIV, implemented by means of simple algorithms, and multigrid versions, Meas. Sci. Technol. **12**, 1911–1921 (2001)
[9] T. Astarita: Analysis of weighting windows for image deformation methods in PIV, Exp. Fluids **43**, 859–872 (2007)
[10] J. Nogueira, A. Lecuona, P. A. Rodríguez: Limits on the resolution of correlation PIV iterative methods. Fundamentals, Exp. Fluids **39**, 305–313 (2005)

Index

PIV Measurements in Shock Tunnels and Shock Tubes

M. Havermann, J.Haertig, C. Rey, and A. George

French-German Research Institute of Saint-Louis (ISL),
5 Rue du Général Cassagnou, F-68 301 Saint-Louis, France

Abstract. Shock tubes and shock tunnels generate compressible flows that are characterized by short time durations and large gradients. Therefore, these flows are a challenging application for all kinds of measurement systems. Additionally, a high information density is desirable for each experiment due to short measurement times. During recent years, particle image velocimetry (PIV) was therefore extensively tested and successfully applied to different flow configurations in the shock-tube department at ISL. It turned out that one difficulty common to all shock-tube or tunnel applications is a good timing and triggering of the PIV system. An appropriate seeding is another crucial factor. The latter is particularly difficult to manage because, in contrast to continuous facilities, no assessment of the seeding quality is possible before and during the experiment. Several measurement results are shown and explained in the chapter.

1 Introduction

Because of continuous progress in numerical simulations, more accurate and comprehensive experiments are required for comparison and validation purposes. For the study of compressible flows, shock tubes are experimental facilities that are used for unsteady-flow research and, in the case of shock tunnels, for realization of high Mach number flows. An ideal measurement system for these short-duration facilities is of a nonintrusive nature and yields a high information density. Both requirements are fulfilled by the particle image velocimetry (PIV) technique, which has been assessed since 2000 at ISL for several flow configurations in both shock tunnels and shock tubes. The first experiments dealt with nozzle calibration measurements in a shock tunnel at Mach numbers of 3.5 and 4.5 [1]. The studies were continued by PIV measurements of attached and detached shock waves around sharp and blunt bodies at Mach 4.5 [2]. On increasing the flow velocity in the shock tunnel to Mach 6, new experiments with an improved PIV hardware were performed at the higher Mach number [3]. An improved PIV analysis method based partly on ISL shock-tunnel data was suggested by [4]. A summary of high-speed supersonic PIV measurements at ISL was given during the PivNet 2 workshop in 2005 by [5].

It should be mentioned that PIV was already applied to hypersonic flows by other researchers (Mach 6 wind-tunnel flow [6] and, more recently, to a

A. Schroeder, C. E. Willert (Eds.): Particle Image Velocimetry,
Topics Appl. Physics **112**, 429–443 (2008)

Mach 7 Ludwieg tube flow [7]). Because of the lower stagnation temperatures of these facilities compared to a shock tunnel, the absolute flow velocities were also lower than the velocities obtained at ISL.

More recently, PIV was applied at ISL to study unsteady vortex flows generated by an open-end shock tube [8]. In the following sections, some selected and partly unpublished data of shock-tube-related PIV measurements performed at ISL are presented.

2 PIV System

A double-frame digital PIV system was used during all experiments. The light source consisted of a frequency-doubled Nd:YAG double-pulse laser (Quantel Twins) with a nominal pulse energy of 150 mJ. A laser lightsheet typically 200 mm wide and 0.2 mm thick was created perpendicular to the flow axis by means of an optical arrangement of cylindrical and spherical lenses, illuminating the flow from above. The imaging CCD camera was mounted on the horizontal axis to view the illuminated flow-field. The IDT sharpVISION 1300 DE camera (1280×1024 pixels) is capable of acquiring two images with a minimum time separation of 0.4 µs. A Nikon zoom objective was used to modify the field of view. In the shock-tunnel facility, the laser and camera synchronizer were triggered by a heat-flux sensor located on the tube wall. The synchronizer separation time was checked by a fast-response photodiode and timing errors were found to be less than 1 %. The PIV images were analyzed after the experiment by a crosscorrelation algorithm using either the IDT software with adaptive interrogation window sizes down to 24×24 pixels or using an inhouse standard crosscorrelation code.

3 Shock-Tunnel Experiments

3.1 ISL Shock-Tunnel Facility STA

A shock tunnel is a short time duration wind tunnel consisting of a shock tube connected to a supersonic nozzle and a test chamber (Fig. 1). The ISL shock tube has an inner diameter of 100 mm and is divided into a 2.7-m long, high-pressure driver tube and an 18.4-m long, low-pressure-driven tube. The driver gas is a mixture of hydrogen and nitrogen at a pressure of up to 50 MPa, whereas the driven or test gas consists of pure nitrogen at a pressure of up to 0.5 MPa. The two sections are separated by a steel diaphragm with a thickness of up to 4 mm. The pressure in the high-pressure section is increased until the diaphragm bursts, which leads to the formation of a shock wave in the low-pressure tube. At the end of the driven tube, the shock wave is reflected at the beginning of the convergent entrance section of the nozzle. The shock travels back through the accelerated test gas, which is decelerated and further heated

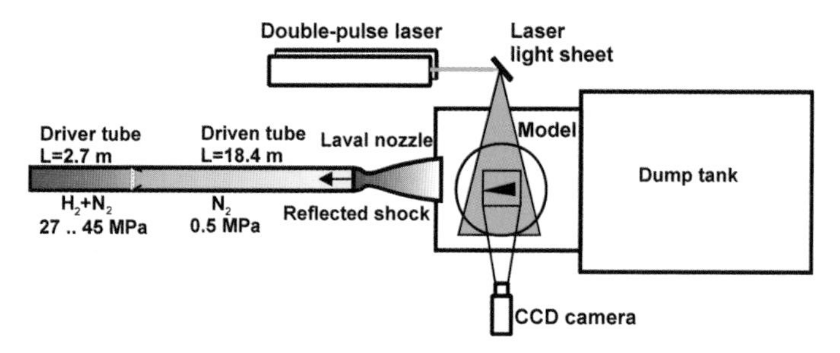

Fig. 1. ISL shock-tunnel facility STA

and pressurized. As a result, a highly heated and compressed stagnation gas volume is created at the nozzle entrance for some milliseconds. This gas is then expanded in the convergent-divergent nozzle to supersonic or hypersonic velocity. Conical and contoured Laval nozzles with nominal Mach numbers between 3 and 6 are available. The flow is stationary during the test time of approximately 2 ms. After each shot, the freestream flow conditions are recalculated using a one-dimensional shock-tube code, which requires the measured shock-wave speed in the driven tube as an input.

The test chamber contains the models to be studied and catches the shock-tube gases after the experiment. Additionally, the test chamber has optical access from three sides to apply flow visualization and laser-based methods. The lightsheet is mirrored into the test section from the top and the PIV pictures are taken from the front, as can be seen in Figs. 1 and 2.

To withstand the high stagnation temperatures of up to 1700 K after the shock reflection at the nozzle entrance, solid particles were chosen as seeding material. Either Al_2O_3 or TiO_2 particles with nominal diameters of 0.3 μm were used. Both tracers worked well, but the Al_2O_3 particles gave better pictures of individual particles. The particles were dried before use and dispersed by a fluidized-bed seeder into the low-pressure tube together with the nitrogen test gas just before the experiment started.

3.2 Experimental Results

3.2.1 Cylinder Flow at Mach 6

Several examples of PIV measurements in the ISL shock-tunnel facilities were already presented in earlier publications, as for example nozzle flows at Mach 3.5 and 4.5 in [1], and Mach-6 flows around wedges and spheres in [3]. Several critical issues were addressed, such as the correct timing and pulse separation time, which was lowered to 0.4 μs for the high flow velocities of nearly 2 km/s encountered in Mach-6 flows. One critical issue in short-duration supersonic flows, however, remains the seed-particle performance (homogeneous seeding and flow-following ability). In particular, strong ve-

Fig. 2. Picture of the PIV system mounted at the STA test chamber

Table 1. Flow conditions for Mach-6 cylinder flow

Mach (-)	p_0 (Pa)	T_0 (K)	p_∞ (Pa)	T_∞ (K)	u_∞ (m/s)
5.93	2.08×10^7	1622	1.21×10^4	219	1792

locity gradients like shock waves should be well resolved by a good measurement.

In the following, the flow around a circular cylinder with a diameter of 100 mm and a span of 200 mm exposed to a Mach-6 flow in the shock tunnel is discussed.

The experimental flow conditions with the stagnation (index 0) and free-stream (index ∞) values are given in Table 1.

A differential interferogram taken of a circular cylinder exposed to Mach-6 conditions in the ISL shock tunnel indicates the density gradient and shows very clearly the formation of a detached shock wave with a bow shape (Fig. 3). A description of the differential interferometry method can be found in [9].

The rectangular area marked in the picture corresponded to the area measured with the PIV system in a second experiment (111×89 mm, pulse separation: 0.4 µs; optical calibration factor: 86.9 µm/pixel). The corresponding PIV picture is depicted in Fig. 4. The particle seeding (Al_2O_3) is mostly homogeneous and the bow shock is clearly indicated by an increased particle density.

The crosscorrelation performed with the IDT software for a correlation window size of 24×24 pixels gives the horizontal velocity depicted in Fig. 5. It can be seen that the free-stream velocity is well represented as well as the velocity drop caused by the bow shock wave.

On the stagnation streamline the bow shock has normal shock properties. Therefore, it is interesting to compare the velocity evolution along the streamline with the velocity drop caused by a normal shock. This velocity drop can be easily calculated for a perfect gas using the Rankine–Hugoniot

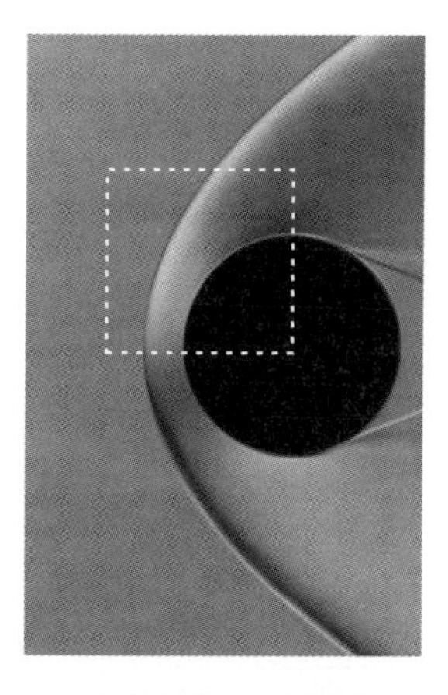

Fig. 3. Differential inteferogram of cylinder in a Mach-6 flow

Fig. 4. PIV image of cylinder in Mach-6 flow

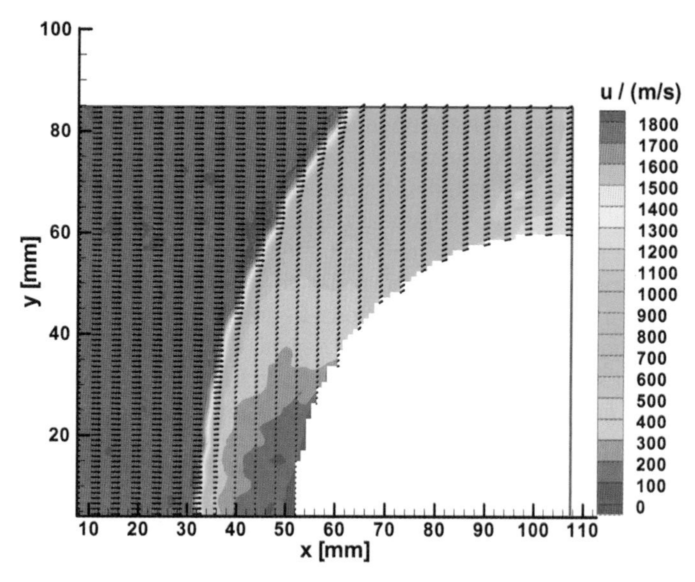

Fig. 5. Horizontal velocity field and velocity vectors obtained by PIV

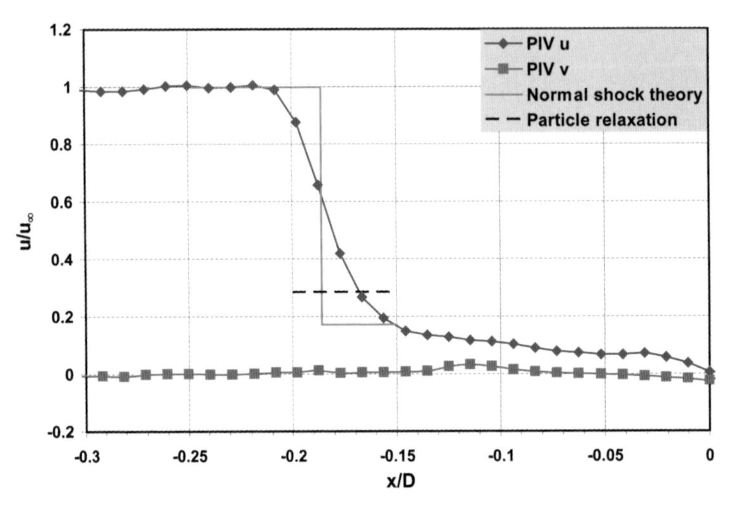

Fig. 6. Stagnation-point streamline of Mach-6 cylinder flow ($x/D = 0$ at the cylinder surface)

relations. Figure 6 shows that the vertical velocity component v is nearly 0, whereas the horizontal velocity component undergoes the normal shock deceleration. For comparison, the theoretical velocity drop according to the normal shock theory is included in Fig. 6, at the location where the shock was measured by PIV.

The PIV measurement exhibits, however, a shock wave that is much broader than theoretically predicted. One reason is the particle velocity relaxation due to the velocity slip, which is estimated in the following. The particle relaxation length can be conveniently defined by a $1/e^2$ criterion, if the Stokes drag law is used and a particle relaxation time τ is defined as

$$\tau = \frac{\rho_p d_p^2}{18\mu}, \tag{1}$$

where ρ_p denotes the particle density, d_p its diameter, and μ the fluid viscosity.

The normalized theoretical particle velocity at this length is included in Fig. 6 as a dotted line. The particle relaxation length is estimated from this criterion by measuring the distance from the theoretical normal shock location to the intersection of the particle relaxation velocity and the measured velocity. The estimation is of the order of $\Delta x/D = 0.02$, which corresponds to 2 mm. A theoretical estimation of the relaxation length for Al_2O_3 particles with a diameter of 0.2 µm yields a relaxation length of 1.8 mm, confirming the measured value. However, the measurement also shows a velocity drop upstream of the shock-wave location, which is physically not possible. The reason for this phenomenon is that PIV measures the displacement Δx of particles during a finite time Δt. Let u_1 and u_2 be the velocities upstream and

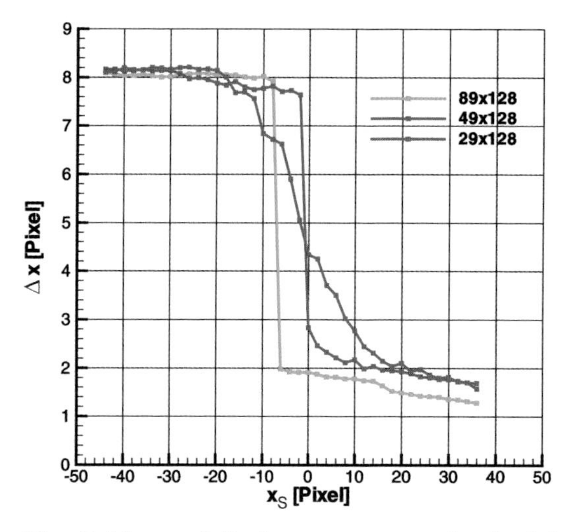

Fig. 7. Measured displacements across the shock for different interrogation window sizes

downstream of the shock. Neglecting the particle's inertial lag, the displacement of a particle located at x_S upstream of the shock at $-u_1 \cdot \Delta t < x_S < 0$ consists of a displacement $\Delta x_1 = -x_S$ during a time interval $\Delta t_1 = x_S/u_1$ and a displacement Δx_2 during a time interval $\Delta t_2 = \Delta t - \Delta t_1$ at speed u_2. The displacement Δx during Δt is therefore: $\Delta x = u_2 \cdot \Delta t - (1 - u_2/u_1) \cdot x_S$. This means that the measured velocity is a decreasing ramp of width $u_1 \cdot \Delta t$, (i.e., the upstream displacement) upstream of the shock location. This is a kind of negative (predicted) lag.

For a finite-width interrogation window, this effect is increased by the greater particle concentration in the low-speed side of the shock and is only partly cancelled by the inertial lag of the particle. Using the ISL image processing software, it is found that for a small window width the correlation peak corresponds to the mean displacement. For larger window widths there are two distinct populations of particles giving two correlation peaks and it is possible to get the correct shock position and velocity jump (49 pixels). Even with a very large interrogation window width of 89, the velocity jump across the shock is better represented than for a small width of 29. Figure 7 shows the displacement found for a time interval $\Delta t = 0.4\,\mu s$ and window widths of 29, 49 and 89 pixels when a secondary correlation peak of 0.5 is allowed.

3.2.2 Lateral Jet-Crossflow Interaction

Another measurement example of shock-tunnel flows being studied at ISL is the interaction of a supersonic jet emanating vertically from a cone-cylinder model (diameter: 50 mm, length: 500 mm) with a horizontal hypersonic crossflow. Phenomenologically, the supersonic jet is bent downstream by the cross-

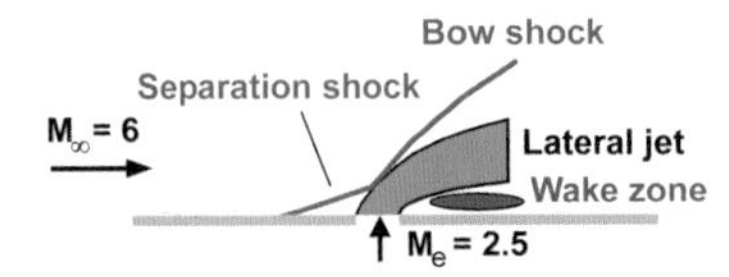

Fig. 8. Schematic flowfield of a supersonic jet in a hypersonic crossflow

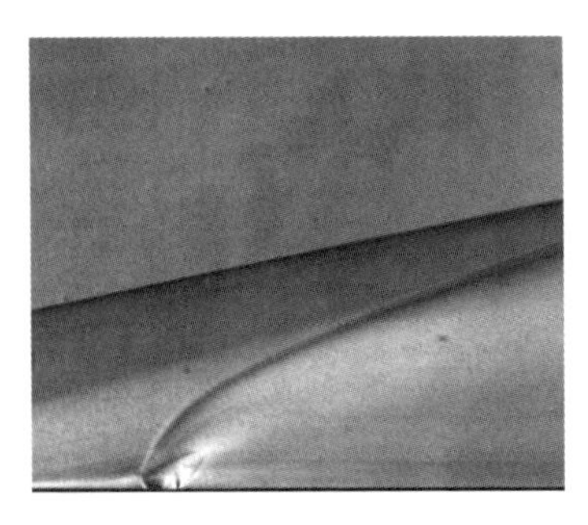

Fig. 9. Differential interferogram (exposure time: 75 µs) of a supersonic jet in a hypersonic crossflow

Fig. 10. PIV image of lateral jet in crossflow

flow and a bow shock is formed in front of it. The bow shock interacts with the boundary layer on the body leading to a separation shock and a wake region downstream of the jet exit (Fig. 8).

The flow field was again first visualized using a differential interferometer with an exposure time of 75 µs, so that only time-averaged structures are visible (Fig. 9).

For the PIV measurement, both the shock-tunnel flow and the lateral jet flow were seeded with Al_2O_3. The particles were put into the jet-stagnation chamber and the jet was triggered about 10 ms before the shock-tunnel flow was established to assure quasisteady conditions for both flows. The corresponding PIV picture (Fig. 10) shows a rather inhomogeneous particle seeding in the jet zone and a highly turbulent jet flow, which becomes visible by the instantaneous measurement nature of the PIV method. The highly turbulent structure of the flow field is confirmed by a differential interferogram with a short exposure time in the 1 µs range (Fig. 11).

Fig. 11. Short-time exposure (1 µs) differential interferogram

Table 2. Flow conditions for Mach-6 lateral jet flow

Mach (-)	p_0 (Pa)	T_0 (K)	p_∞ (Pa)	T_∞ (K)	u_∞ (m/s)
5.92	2.11×10^7	1652	1.23×10^4	225	1810

For this PIV case the pulse separation was set to 0.6 µs and the optical calibration factor corresponded to 65 µm/pixel. The PIV data was analyzed with the inhouse software for a rather large correlation window size of 64×64 pixels (50 % overlap). As mentioned before even with a large correlation window size a good shock resolution can be obtained in regions with inhomogeneous particle seeding. The resulting vertical velocity component is plotted in Fig. 12. Compared to the smooth velocity distribution from the time-averaged CFD simulation (Fig. 13), the PIV result exhibits more turbulent regions and some zones that are not correctly measured due to the inhomogeneous seeding.

4 Shock-Tube Experiments

4.1 Vortex-Ring Shock Tube

Vortex rings are interesting fluid-dynamic objects characterized by a rotating mass of fluid that conserves its form for larger propagation distances. Even if the vortex-ring shape might be regarded as simple, vortex rings are quite difficult to study due to free fluid surfaces without fixed borders and due to their unsteady formation process. A small shock tube was built at the ISL for the study of compressible vortex rings. The shock tube shown in Fig. 14 has a length of 2.2 m and an inner diameter of 47 mm. The high-pressure section is separated from the low-pressure section by thin plastic film diaphragms, which are punctured at a defined pressure by a needle mechanism.

Vortex rings generated by a shock tube have their origin in unsteady shock-wave diffraction at the tube opening, which induces a high level of vor-

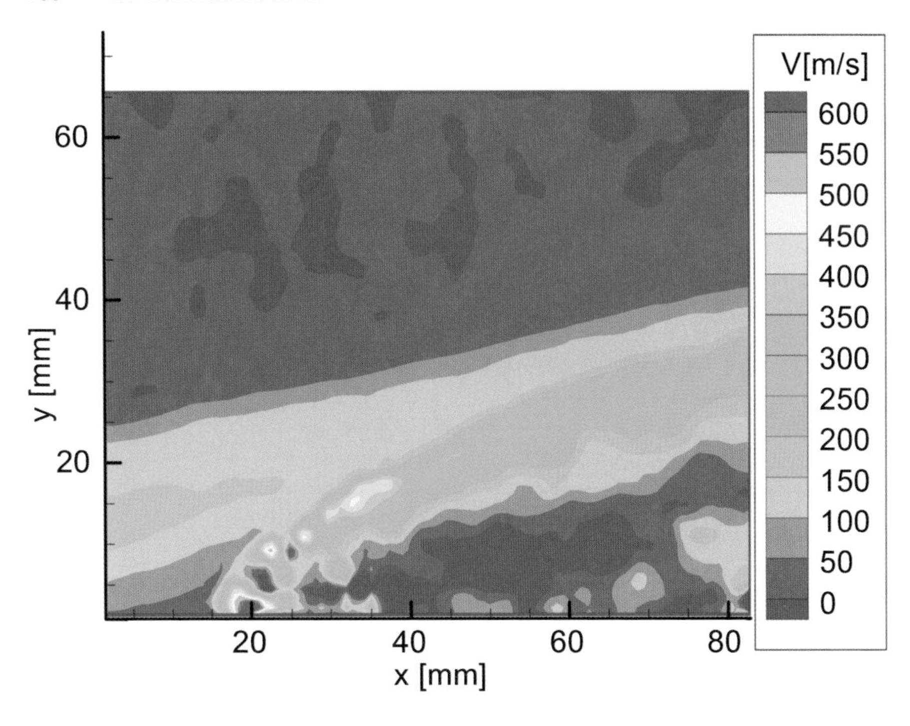

Fig. 12. Vertical velocity component obtained from PIV measurement

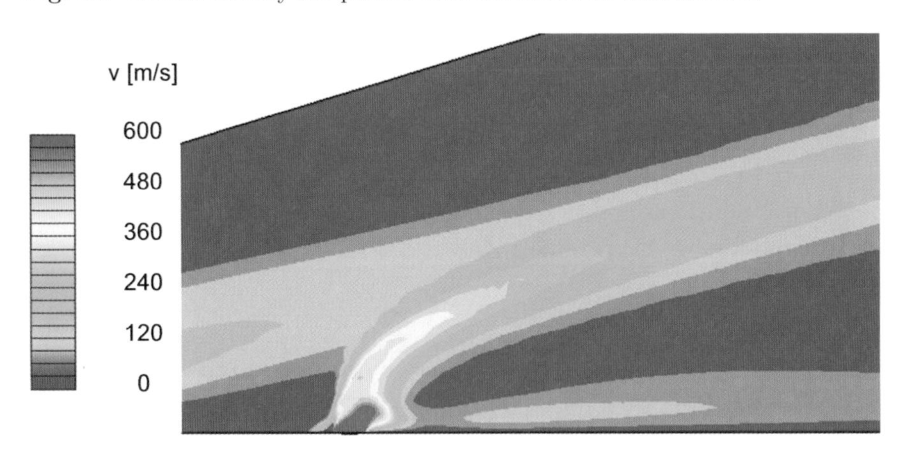

Fig. 13. Vertical velocity component obtained from CFD simulation

Fig. 14. Picture of the small shock tube for vortex-ring studies

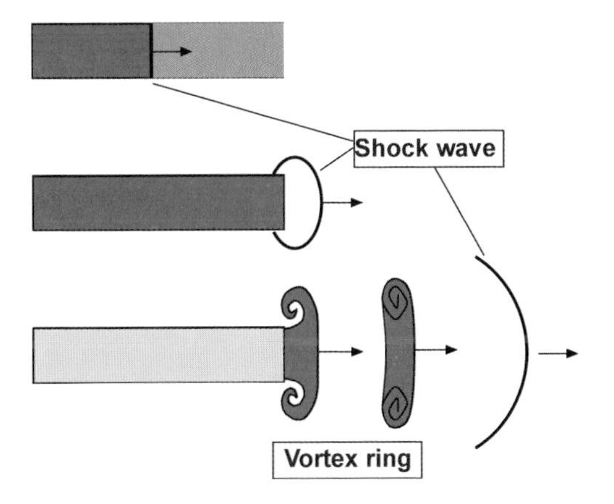

Fig. 15. Generation of a compressible vortex ring by an open-end shock tube

ticity into the emanating flow. As a result, a highly rotating flow is generated that develops to a separated vortex ring after flow detachment from the tube. The vortex-ring generation process by a shock tube is sketched in Fig. 15.

The experimental setup with the shock tube and the PIV system is depicted in Fig. 16. Particle seeding was accomplished by a smoke generator burning incense resin that gives a typical smoke particle diameter of approximately 1 μm. Both the open-end tube and the ambient air near the opening were seeded before each experiment. The PIV system was triggered by a pressure transducer located close to the shock-tube opening. A similar study was conducted by *Arakeri* et al.[10].

4.2 Experimental Results

A Schlieren picture of a vortex ring is shown in Fig. 17. The ring is well formed and already detached from the shock-tube opening, propagating from the left

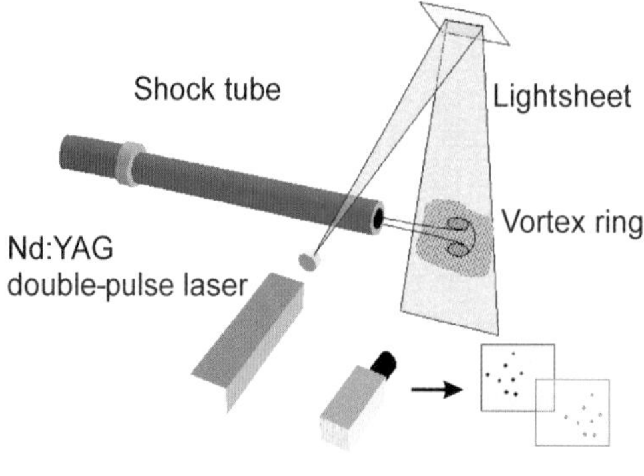

Fig. 16. PIV setup

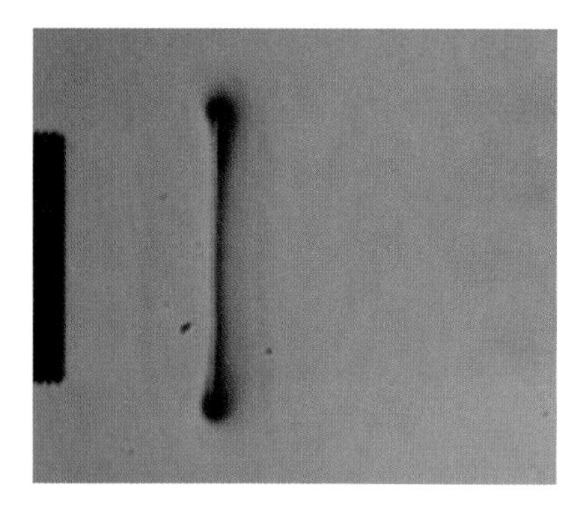

Fig. 17. Schlieren picture of vortex ring

to the right at a velocity of about 50 m/s. The vortex inner structure is not visible due to the integral nature of the Schlieren method.

The corresponding PIV image (Fig. 18) shows the vortex cores in the lightsheet plane as bright circles. Around the cores, the particle density is reduced compared with the ambient air, indicating a region with a higher velocity and a lower density. The pulse-separation time was set to 5 µs and the optical calibration factor was 24 µm/pixel.

The PIV data analysis was performed for a 36 × 36 pixel interrogation window size using the inhouse software. Despite a low particle concentration in the core region a velocity measurement was obtained. First, the horizontal velocity component in the laboratory coordinate system is plotted (Fig. 19). Masked regions are automatically blanked by selecting a certain noise level of the correlation function. The vectors indicate the strong rotation of the flow around the vortex cores. If the propagation velocity of the vortex ring is subtracted from the original data, the velocities in a vortex-fixed coordinate system are given, which allows the plot of streamlines to be made (Fig. 20). The vortex centers were determined by the center of closed-loop streamlines in the frame of reference moving with the local vortex propagation velocity.

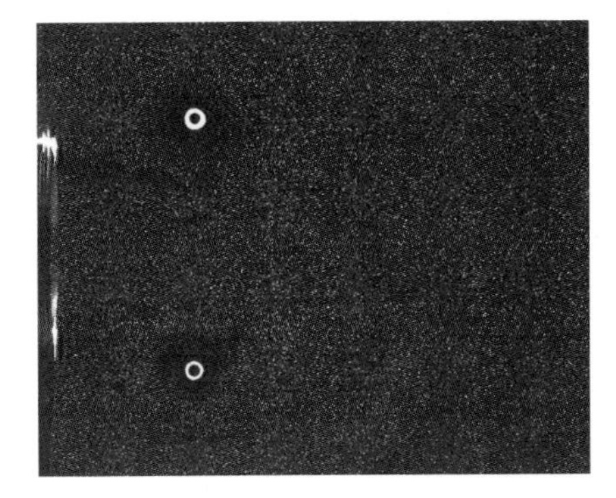

Fig. 18. PIV image of vortex ring

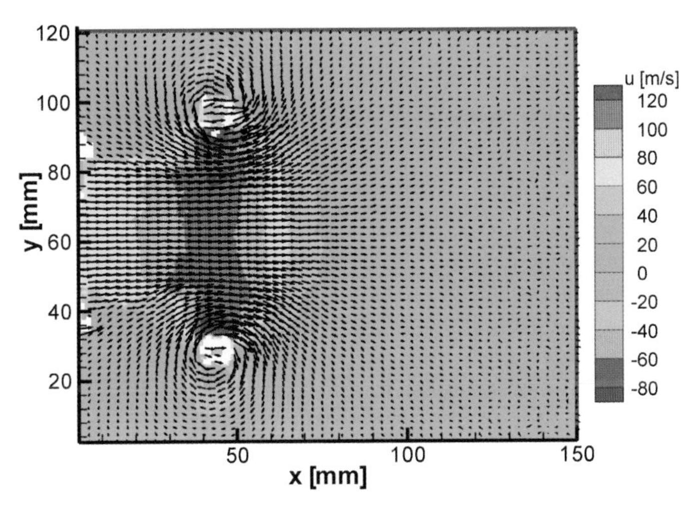

Fig. 19. Horizontal velocity component and vector field for the laboratory-fixed frame

Again, the rotation around the vortex core is clearly visible and quantitative data from the circulation becomes available.

5 Conclusions

PIV has been applied to a large variety of simple and complex flows in short-duration facilities (shock tunnel and shock tube) at ISL. For rather simple flow fields with a low level of turbulence the instantaneous PIV results represent the flow field very well even with rather large interrogation window sizes. Velocities up to nearly 2 km/s have been measured in a shock tunnel with an accuracy of a few per cent comparing the experimental results with analytical and theoretical values. This confirms the ability of PIV to be used for CFD validation even in a very high velocity range.

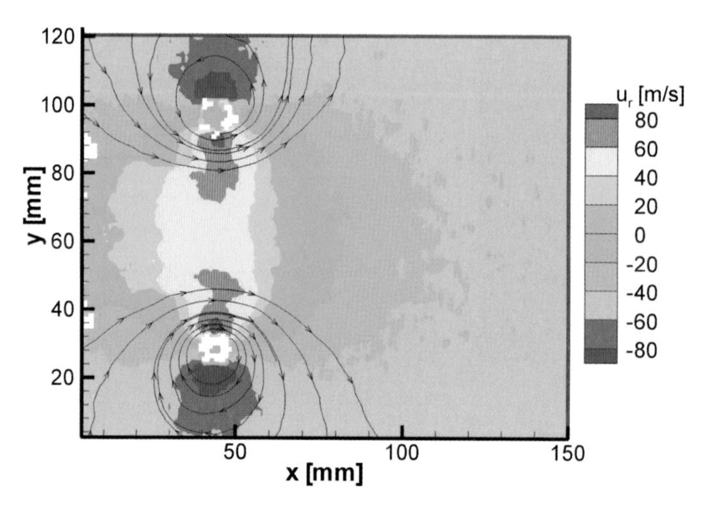

Fig. 20. Horizontal velocity component and streamlines for the vortex-fixed frame

Together with a correct timing, a crucial element to success is a homogeneous particle seeding, which cannot be checked before a shock-wave experiment, however. Even if the free stream particle distribution is homogeneous, the PIV data analysis can be disturbed by the high gradients of particle density across shocks, because an unphysical shock thickening upstream of the shock was observed.

It further turned out that more complex flow fields with a high level of turbulence are difficult to measure with just a single PIV experiment. If powerful lasers with higher repetition rates become available, even the complex flows could be better addressed with high-speed PIV, so that useful experimental data in highly turbulent high-speed flows will be available for CFD-code validation.

References

[1] J. Haertig, M. Havermann, C. Rey, A. George: PIV measurements in Mach 3.5 and 4.5 shock tunnel flow, in *Proc. 39th Aerospace Sciences Meeting and Exhibit* (2001) AIAA paper 2001-0699

[2] M. Havermann, J. Haertig, C. Rey, A. George: Particle image velocimetry (PIV) applied to high-speed shock tunnel flows, in *Proc. 23rd Int. Symp. on Shock Waves* (2001)

[3] M. Havermann, J. Haertig, C. Rey, A. George: Application of particle image velocimetry to high-speed supersonic flows in a shock tunnel, in *Proc. of the 11th Int. Symp. on Applications of Laser Techniques to Fluid Mechanics* (2002)

[4] F. Scarano, J. Haertig: Application of non-isotropic resolution PIV in supersonic and hypersonic flows, in *5th Int. Symp. on Particle Image Velocimetry* (2003)

[5] J. Haertig, M. Havermann, C. Rey, A. George: Application of particle image velocimetry to high-speed supersonic flows, in *Proc. of the PIVNET II International Workshop on the Application of PIV in Compressible Flows* (2005)
[6] W. M. Humphreys, R. A. Rallo, W. W. Hunter, S. M. Bartram: Application of particle image velocimetry to Mach 6 flows, in *Proc. of the 5th Int. Conf. of Laser Anemometry, Advances and Applications*, vol. 2052 (Int. Society for Optical Engineering (SPIE) 1993)
[7] F. F. J. Schrijer, F. Scarano, B. W. van Oudheusden: Application of PIV in a Mach 7 double-ramp flow, Exp. Fluid **41**, 353–363 (2006)
[8] J. Haertig, C. Rey, M. Havermann: PIV measurements of compressible vortex rings generated by a shock tube, in *Proc. 13th Int. Symp. on Applications of Laser Techniques to Fluid Mechanics* (2006)
[9] R. Memmel, J. Straub: Differential interferometry, in F. Mayinger (Ed.): *Optical Measurements. Techniques and Applications* (Springer, Berlin, Heidelberg 1994) pp. 75–90
[10] J. H. Arakeri, D. Das, A. Krothapalli, L. Lourenço: Vortex ring formation at the open end of a shock tube: A particle image velocimetry study, Phys. Fluids **16**, 1008–1019 (2004)

Index

Overview of PIV in Supersonic Flows

Fulvio Scarano

Aerospace Engineering Department, Delft University of Technology,
The Netherlands
f.scarano@tudelft.nl

Abstract. The extension of particle image velocimetry to supersonic and hypersonic wind-tunnel flows has been achieved in the last decade. This was mainly possible with the advent of short interframing-time CCD cameras with temporal resolution allowing to obtain correlated particle images at flow velocities exceeding 1000 m/s. The most challenging aspects of PIV experiments in supersonic flows are still recognized as the seeding-particle-selection and seeding-distribution techniques. Also, the optical access for illumination and imaging require a specific attention since pressurized facilities offer limited optical access. The presence of shock waves in supersonic flows introduces regions where particle tracers slip with respect to the surrounding flow. Moreover, the particle seeding density becomes strongly nonuniform and particle-image blur can occur as a result of the strong refractive index variations. The present chapter reviews the physical and technical problems of PIV experiments and discusses the potential of such techniques on the basis of recent experiments performed in high-speed wind tunnels: double compression ramp at Mach 7 and shock-wave turbulent boundary interaction at Mach 2.

1 Introduction

The technological relevance of compressible flows in both scientific and industrial research is well acknowledged, since it constitutes an established branch of science and a field of ongoing research and development for industrial, mostly aerospace, applications. Experimental high-speed investigations rely as a rule on wind-tunnel facilities producing a controlled uniform high-speed stream in which the model under investigation is immersed. Once the first step of realizing controlled test conditions is achieved, a second one of equal importance must be accomplished, consisting of the actual measurement of the relevant flow quantities. Among the techniques that can be employed for supersonic flow measurements, nonintrusive optically based techniques have received large attention due to their promising performances. Starting from qualitative techniques such as shadowgraphy and Schlieren methods [1], the experimental community has developed quantitative methods such as holographic interferometry [2, 3], laser Doppler velocimetry [4, 5], laser-induced fluorescence [6, 7] and particle image velocimetry [8]. The present chapter gives a detailed discussion about the current status of PIV capabilities in supersonic and hypersonic wind-tunnel flows.

A. Schroeder, C. E. Willert (Eds.): Particle Image Velocimetry,
Topics Appl. Physics **112**, 445–463 (2008)

Supersonic flows are characterized shock by waves that challenge measurement techniques, where a large flow deceleration occurs across a very thin region. Also, boundary layers and shear layers can be relatively thin due to the limited turbulent mixing typical of the compressible regime [9] and the high values of the Reynolds number for high-speed flows. Flow compressibility introduces two additional difficulties, namely the large spatial variation of the particle-seeding density and the inhomogeneous refractive index throughout the flow field. The latter results in a distortion of light propagation for optically based measurement techniques [10].

The application of PIV in compressible flows was pioneered by *Moraitis* and *Riethmuller* [11] and *Kompenhans* and *Höcker* [12] who used the photographic recording technique combined with image-shifting methods. The advent of frame-straddling CCD cameras and short-duration nanosecond-pulsed double-cavity Nd:YAG lasers turned the technique into a versatile tool finally operated not only by the developers' research groups. Investigations range from supersonic jets [13, 14] to transonic turbomachinery flows [15, 16]. Turbulence measurements in compressible mixing layers were more recently performed with PIV by [17]. Supersonic turbulent wakes as well as shock-wave turbulent boundary-layer interaction were investigated by the author and coworkers [18–21]. The technique was first extended to the hypersonic flow regime by *Haertig* et al. [22] at ISL and further by *Schrijer* et al. [23] at TU Delft.

2 Flow Seeding and Imaging

The main requirements for the flow seeding particles is that they provide an accurate flow tracing still at a sufficient particle-seeding density. The flow-tracing capability of particles of diameter d_p and a specific gravity ρ_p is usually quantified through the particle relaxation time τ_p. The theoretical behavior for small spherical particles may be reduced to the modified Stokes drag law [24]. Given the relatively low value of the Mach number and Reynolds number based on the particle diameter, the modified drag relation that takes into account rarefaction effects yields the following expression for the relaxation time

$$\tau_\mathrm{p} = d_\mathrm{p}^2 \frac{\rho_\mathrm{p}}{18\mu_\mathrm{f}} (1 + 2.7\mathrm{Kn}_\mathrm{d}) \,, \tag{1}$$

where Kn_d is the Knudsen number obtained as the ratio between the molecular mean free path and the diameter of the particle tracers d_p. A detailed study on the particle motion in supersonic flows is provided by [25].

In several flow conditions the Knudsen number is a function of the particle Reynolds number and relative Mach number between the particle and the surrounding fluid $M_{\Delta V}$, following the relation $\mathrm{Kn}_\mathrm{d} = 1.26\sqrt{\gamma} M_{\Delta V}/\mathrm{Re}_\mathrm{d}$, [26] where γ is the gas specific heat ratio ($\gamma = 1.4$ for air). The Reynolds number of the flow around the slipping particles is

$$\mathrm{Re_d} = \frac{\rho_\mathrm{f} \cdot \Delta V \cdot d_\mathrm{p}}{\mu_\mathrm{f}}, \tag{2}$$

where ρ_f and μ_f are the fluid density and dynamic viscosity, respectively. The flow-tracing accuracy within a given experiment is then related to the Stokes number S_k expressing the ratio between τ_p and the characteristic flow timescale τ_f. Recalling that the most critical conditions are always encountered across shocks where $S_\mathrm{k} \gg 1$, the most critical conditions in turbulent flows are met when particle tracers are immersed in thin shear layers, for instance downstream of sharp separation points [27]. *Samimy* and *Lele* [28] suggested the following expression for the flow timescale in a free shear layer of thickness δ

$$\tau_\mathrm{f} = 10 \cdot \frac{\delta}{\Delta V} \tag{3}$$

and performed a particle dynamics computation associating the RMS slip velocity with the Stokes number. In conclusion, a Stokes number of 0.1 corresponds to a RMS slip velocity of approximately 1 %, which may be considered as an acceptable error in view of the other sources of error associated with the measurement technique.

3 Experimental Assessment of Particle Response

The measurement of the particle response time with PIV is technically challenging for numerous reasons. In many cases, the particle properties cannot be directly evaluated: scanning electron microscopy of solid particles only yields the particle properties when lying on a physical support and particle agglomeration introduces an uncertainty on the effective particle size when introduced in the flow [29]. Moreover, the effective particle density can only be predicted with a rough approximation for porous seeding materials. Liquid droplets may undergo condensation-evaporation from their production to the measurement section, which affects the particle size. It is therefore best practice to infer the dynamic particle response through a well-defined input flow test case. A planar oblique shock wave constitutes the commonly adopted test, which corresponds to a sharp velocity falling edge, through which particle tracers decelerate gradually due to inertia. Such a procedure was adopted in previous investigations [4, 17, 18, 24, 30]. The particle relaxation time τ_p is then evaluated from the measurement of the relaxation length/time to the flow velocity behind the shock wave. It should be said, however, that the particle-velocity measurement across a shock wave may often be inaccurate due to the combined effect of numerous sources of error; the most important being:

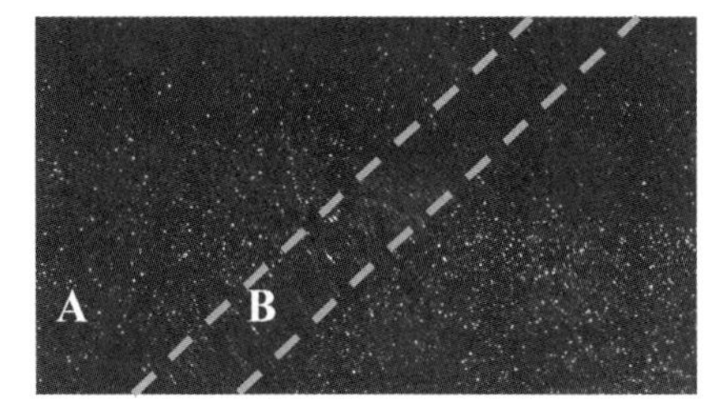

Fig. 1. PIV recording across a shock wave. Blur is visible at location B between the *dashed lines*

1. *Nonmonodispersed particle diameter distribution* introduces a different velocity lag for neighboring particles traveling even in the same region of space. This leads to a broadening of the velocity correlation peak or even to incorrect velocity measurements. Moreover, in actual experiments the relatively large size of PIV interrogation windows often covers the entire relaxation length [30] and introduces two peaks in the correlation map corresponding to the particle velocity upstream and downstream of the shock, respectively;
2. *Finite time separation between laser pulses.* In high-speed flows the temporal resolution is often limited by the minimum interframing time that is of the same order as τ_{p}. This will result in a smearing of the velocity distribution in proximity of the shock position. Turbulence in the flow free stream causes small unsteady fluctuations in the position of the shock. Ensemble-averaged measurements will be smeared by such an effect.
3. *Optical distortion due to nonuniform index of refraction* (spatial blur) may introduce an important bias error on the velocity and an RMS error due to particle image blur as investigated by [31]. Figure 1 shows the blurring effect encountered when measuring the flow across shock waves with the viewing direction aligned with the planar shock. The cross-correlation peak will not only broaden across the shock region, increasing the measurement uncertainty, but may also skew (Figs. 2 and 3), which causes a bias error as shown in Fig. 4. These effects can be reduced by arranging the viewing from the upstream side of the shock of a few degrees.

Despite the above-mentioned effects, several PIV experiments could be conducted, which were able to infer the particle-tracer response. For instance, in the situations depicted below the two-dimensional flow across shock waves measured by PIV returned a response time of approximately two microseconds with sub-micrometer sized titanium dioxide particles.

4 Online Seeding in Supersonic Wind Tunnel

The transonic-supersonic wind tunnel (TST) of the Aerodynamics Laboratories at Delft University of Technology is a blowdown-type facility generating flows in a Mach range from 0.5 to 4.2. The test section area is

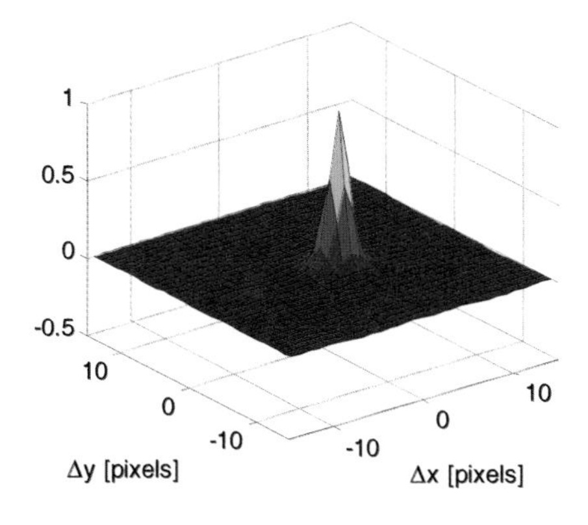

Fig. 2. Image crosscorrelation map of undistorted particle images at location A

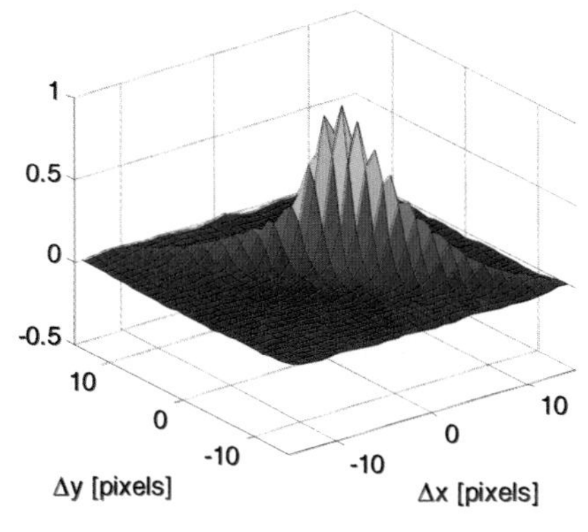

Fig. 3. Image crosscorrelation map of skewed particle images at location B

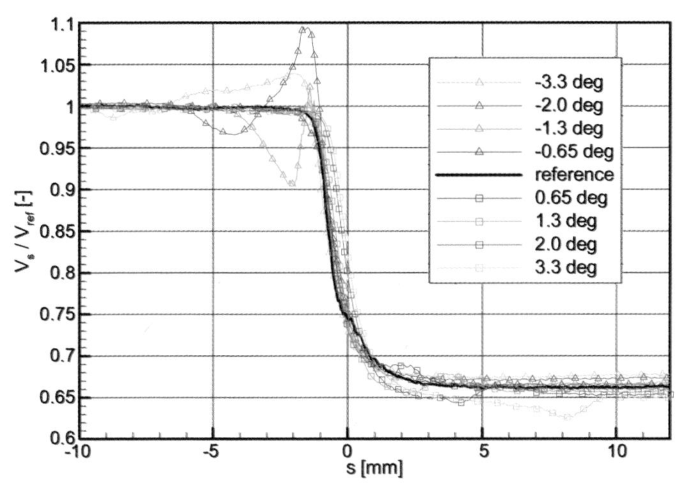

Fig. 4. Velocity measurements across the shock from different viewing angles (wrt shock plane, positive values indicate view from upstream of the shock)

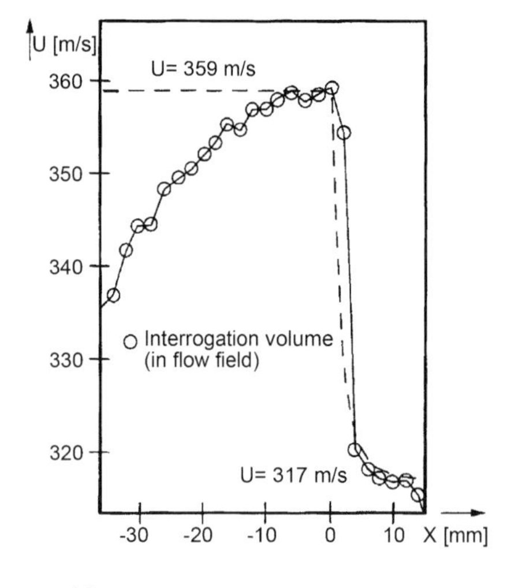

Fig. 5. PIV measurements, (d_p = 1.7 μm) and theory (*dashed line*). Flow across a shock produced by a bluff cylinder [30]

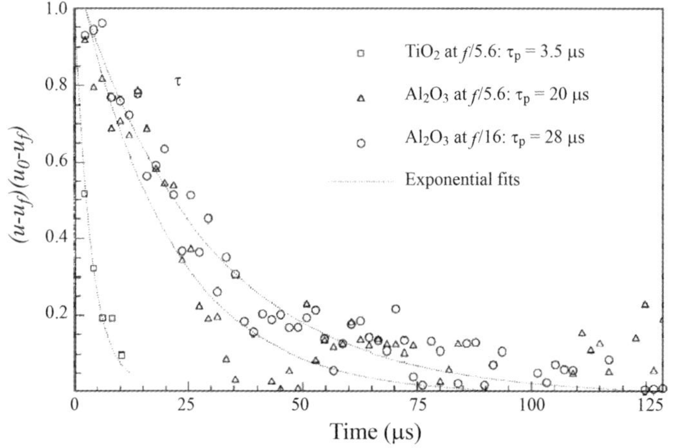

Fig. 6. Particle relaxation traces from different particle size and material. Oblique shock wave at $M = 1.8$, deflection $\theta = 10$ deg [17]

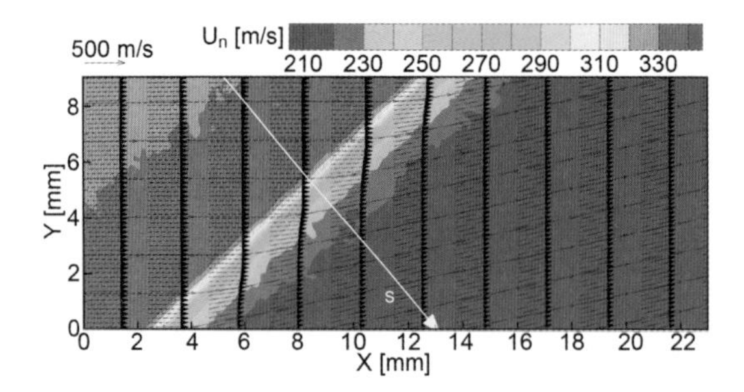

Fig. 7. Mean velocity field across oblique shock wave ($M = 2$, deflection $\theta = 11.3$ deg). Normal velocity component contours. In yellow is the shock-normal abscissa [18]

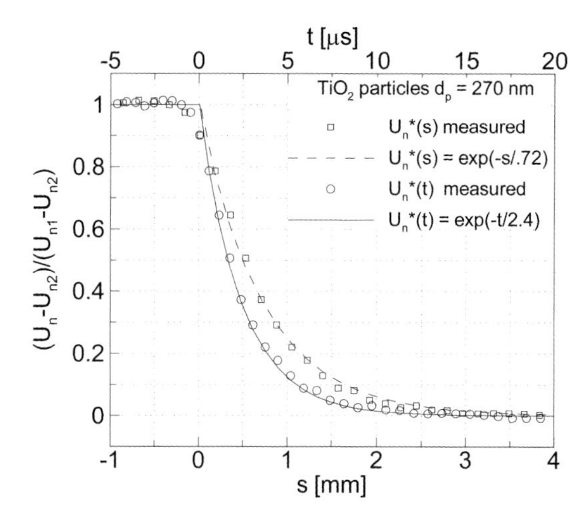

Fig. 8. Normalized velocity profile along abscissa s from Fig. 7

300(W) $\times$ 270(H) mm^2 and operates at values of the unit Reynolds number ranging from 38×10^6 to $130 \times 10^6\,\mathrm{m}^{-1}$ with a typical model length of 10 cm. In the transonic and supersonic ranges, the Mach number is set by means of a continuous variation of the throat section and flexible upper and lower nozzle walls. The wind-tunnel run time is about 300 s. A fraction of the flow is seeded online in the settling chamber by means of a seeding distributor array one third the size of the settling chamber. The seeded flow within a streamtube allows PIV measurement in the test section over a region of approximately 10 cm diameter (Fig. 9). Under operating conditions, the seeded flow exhibits a mean particle concentration of about 10 particles/mm^3 estimated from PIV recordings. The interference of the seeding device onto the free-stream turbulence assessed with hot-wire anemometry (HWA) returned a barely detectable turbulence intensity increase (0.2 %) with a background turbulence of 1 %.

The particle-laden flow is produced by entraining titanium dioxide (TiO_2) particles with a high-pressure cyclone separator operated at 10 bar. The particle median diameter estimated by electron microscopy is $d_\mathrm{p} = 0.4\,\mu\mathrm{m}$, with a specific gravity estimated at $\rho_\mathrm{p} = 1.0 \times 10^3\,\mathrm{kg \cdot m^{-3}}$. An example of the seeding conditions achieved in the TST-27 facility is shown in Fig. 14. Not all the flow is seeded and the interface of the seeded stream tube is clearly turbulent, which introduces undesired seeding intermittency for some cases.

5 Storage-Tube Seeding in Hypersonic Wind Tunnels

At higher Mach number in short-duration facilities (run time of milliseconds) the seeding technique becomes paradoxically more straightforward. This is

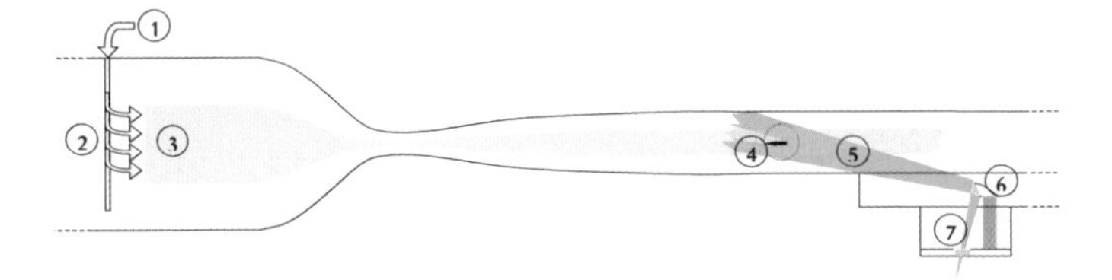

Fig. 9. Schematic of the seeding distribution device and lightsheet delivery in the TST wind tunnel. (**1**) Air + seeds inlet from seeding generator; (**2**) multiorifice distribution pipe in the settling chamber; (**3**) seeded air stream; (**4**) camera optical window and model; (**5**) laser sheet; (**6**) reflecting prism; (**7**) laser-light optical window

Fig. 10. Hypersonic Test Facility Delft (HTFD) with installed PIV system

due to the fact that the flow in the storage tube can be seeded offline, removing the need of a large mass flow rate through the seeder. PIV experiments in the Mach range 4 to 6 and 6 to 10 were performed in the Shock Tube of ISL (Saint Louis) and in the Hypersonic Test Facility Delft (HTFD), respectively (Figs. 10 and 11). In these conditions, the freestream can be seeded completely and rather uniformly, as shown in Figs. 12 and 13.

6 Seeding Concentration

The seeding concentration for compressible flows strongly reveals the flow features, which generally degrades the quality of the PIV measurement tech-

Fig. 11. ISL shock tunnel with PIV system (*Haertig* et al. 2002)

Fig. 12. PIV picture of TiO_2 particles in HTFD at Mach 7

Fig. 13. PIV picture of TiO_2 particles in ISL shock tunnel at Mach 4.5

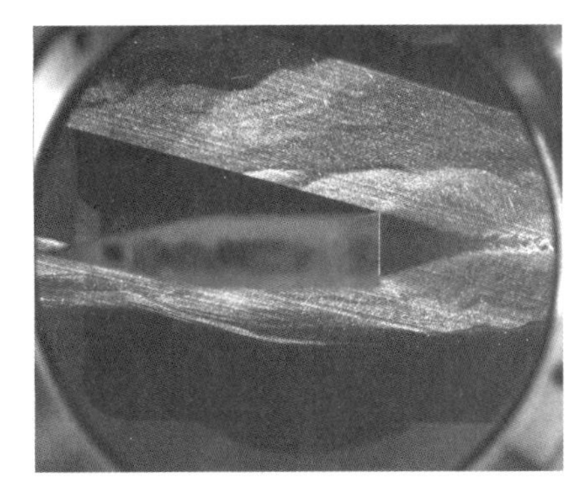

Fig. 14. Supersonic flow around a wedge-plate model in TST-27 at Mach 2

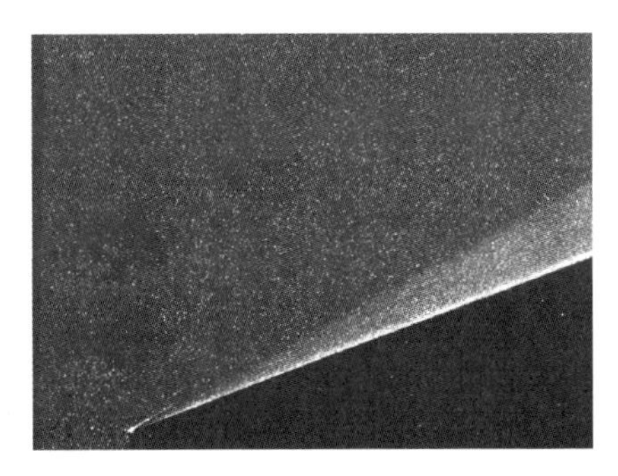

Fig. 15. Supersonic wedge flow in the ISL shock tube at Mach 6 [32]

nique. There are several causes leading to nonhomogeneous seeding concentration within high-speed flow experiments:

- inhomogeneous seeding dispersion upstream of the test section (mostly occurring for online seeding with poor mixing, Fig. 14);
- compressibility effects such as shocks, expansions and viscous layers (particle tracers concentration follows the thermodynamic density, Figs. 15 and 17); particle-tracer ejection from rotational flow regions (vortices, separated flow, boundary layers, Figs. 14 and 16).

The effects of compressibility cannot be avoided; therefore imaging devices used for supersonic flows should have a high dynamic range in order to detect a small light intensity and at the same time not saturate.

7 Hypersonic Compression Ramp Flow

The application of PIV in a Mach 7 flow over a two-dimensional double compression ramp is described here. The application is related to the study of atmospheric re-entry vehicles from orbital missions. The aerodynamic phenomenon of shock–shock and shock-wave–boundary-layer interaction plays an

Fig. 16. Double ramp flow at Mach 7

Fig. 17. Flow around a sphere at Mach 6

important role especially in the proximity of concave walls and at the insertion of control surfaces. Such interactions cause high local heat and pressure loads affecting the performance of control surfaces or the structural integrity of the vehicle. A generic representation of a control surface configuration is the 2D ramp flow, where a compression ramp induces an adverse pressure gradient triggering boundary-layer separation and a further interaction with the external flow through the separation shock. At reattachment on the ramp a system of compression waves emerges from the boundary layer, eventually coalescing into a shock [29].

The overall phenomenon is accompanied by high localized surface heat transfer, as reported in [33, 34].

Most studies were conducted for moderate ramp angles ($\alpha < 30°$) where an attached straight shock solution is possible and the associated interaction pattern is relatively simple and well understood. Conversely, the flow over a ramp with 15–45° deflection presents a conceptually different situation since no attached shock solution is able to realize the imposed flow deflection. Therefore, a curved shock characterizes the flow compression over the ramp and the resulting shock interaction exhibits a more complex pattern. The complexity of the phenomenon is further increased by the unsteady behavior of the separated flow region and of the shocks in the proximity of the

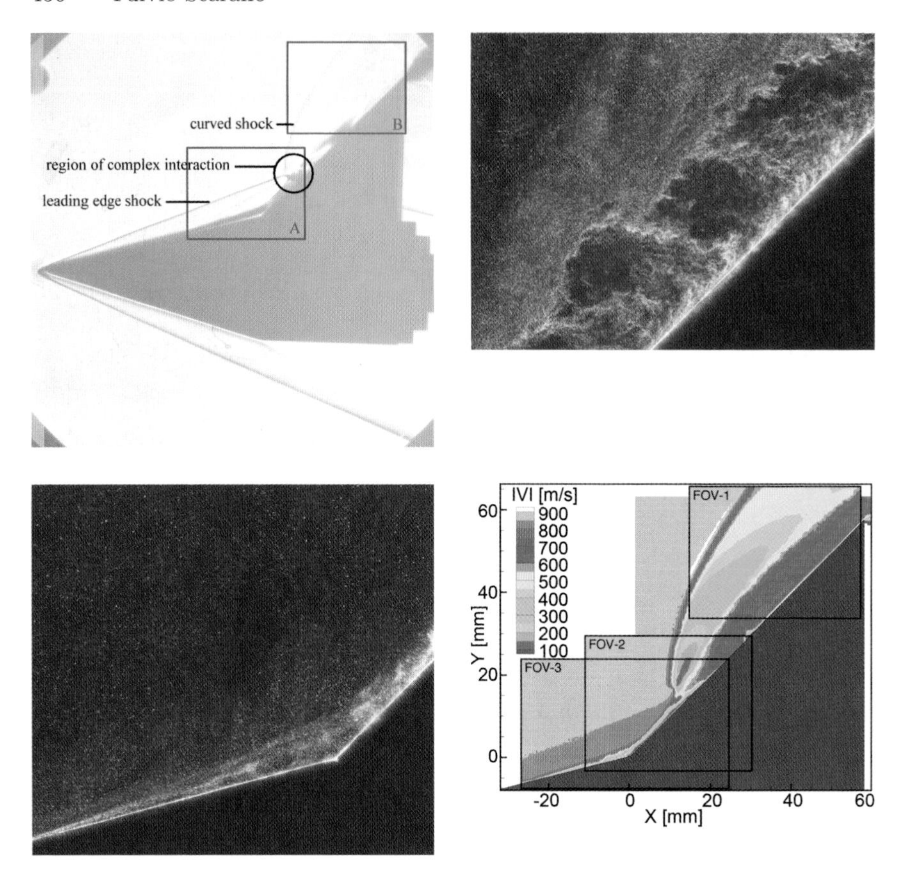

Fig. 18. Schlieren photograph of the double-ramp flow (*top left*); PIV recording in the separation region (*bottom left*); PIV recording detail downstream of the interaction (*top right*); combined mean velocity field (*bottom right*)

interaction. This difficulty also applies to numerical techniques attempting the simulation of such flow because of the large uncertainty introduced by the modeling of unsteady fluctuations by turbulence models. As a result, a considerable discrepancy has been reported between experiments and CFD results [35]. PIV experiments conducted at Mach 7 in the HTFD facility aim at the measurement of the velocity field associated to such flow phenomena. The reservoir conditions are $T_0 = 740\,\mathrm{K}$ and $P_0 = 100\,\mathrm{bar}$ for temperature and pressure respectively. A fast-acting valve puts the high-pressure side in communication with a vacuum vessel at $P_\mathrm{v} = 0.2\,\mathrm{mbar}$ through a convergent-divergent axial symmetric nozzle. The test-section diameter is $D = 35\,\mathrm{cm}$. The flow is seeded with 400-nm diameter TiO_2 particles illuminated by a 400 J/pulse Nd:YAG laser with pulse separation of 1 µs. The fluid velocity in the free stream is about 1040 m/s, and the particle-tracer relaxation time and length is 2 µs and 2 mm, respectively. The compressible flow features

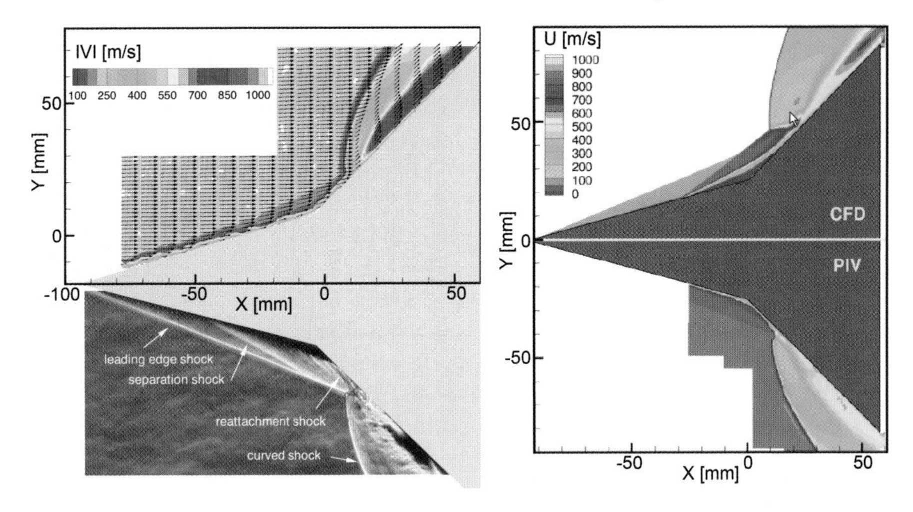

Fig. 19. Comparison of PIV and Schlieren (*left*, [29]. CFD results versus experiments (*right*)

are visualized with Schlieren and surface heat transfer is measured by means of quantitative infrared thermography (QIRT). The comparison between the Schlieren visualization and PIV shows that the overall shock pattern is captured with both techniques (Fig. 19, left), however, weak shocks cannot be clearly identified by the velocity measurement due to the limited dynamic range of single PIV snapshots. Conversely, the velocity field past the reattachment and downstream of the quasinormal shock can only be accessed by PIV with a quantitative description of the postshock flow velocity. The comparison with a 2D numerical simulation based on a Navier–Stokes solver and Menter shear stress turbulence model, returns a remarkable agreement concerning the shock pattern and the postshock velocity distribution. However, a significant discrepancy is observed concerning the extent of flow separation and its interaction with the external flow. In the CFD results, the sharp separation generates a compression shock interacting with the curved shock downstream away from the wall as opposed to the experimental results suggesting a much thinner separation and a weaker separation shock that basically merges with the leading-edge shock prior to encountering the ramp shock. This difference is also ascribed to the finite span of the wind-tunnel model opposed to the 2D flow simulation.

8 Shock-Wave–Boundary-Layer Interaction

The present section describes the experimental study by PIV of a classical problem of high-speed aerodynamics, namely the interaction between a fully developed turbulent boundary and a compression shock impinging on it (SWBLI). Several investigations have been reported on the subject for more than fifty years, as discussed in a review [36], yet some questions are left unanswered and predictive tools for the flow properties at the interaction are not yet completely reliable. The boundary layer developing along the upper wall of the divergent nozzle of the TST-27 wind tunnel is perturbed by a planar oblique shock wave generated by a two-dimensional wedge [21]. The wedge is placed in the free stream and imposes a flow deflection of 10 degrees. A schematic of the experimental setup is given in Fig. 20, top left. The Schlieren photograph in Fig. 20, top right, illustrates the phenomenon: a turbulent boundary layer is detected as the slightly darker corrugated region close to the wall; the impinging shock interferes with the boundary layer and is turned into a normal shock approaching the sonic line towards the wall. The effect of the adverse pressure gradient is transmitted upstream along the boundary layer and a system of compression waves is formed upstream of the impinging shock (see also Fig. 20, bottom). The unsteady nature of the flow originates from the interaction of the turbulent structures in the boundary layer and the foot of the reflected shock. Moreover, for strong interactions the wall flow between the reflected and impinging shock locally separates and turbulence fluctuations are amplified along the free shear layer producing large-scale turbulent structures. Although the Schlieren photograph allows observation of the overall pattern of the interaction, the details of the flow close to the wall require a planar quantitative measurement technique to be fully described.

The PIV recordings are analyzed with a resolution-adaptive method [37], refining the window size close to the solid wall. The windows are also stretched along the wall direction, thus maximizing the spatial resolution along the wall-normal direction. From the instantaneous velocity distribution (Fig. 21) the complex dynamics of the wall flow becomes more apparent: the turbulent boundary layer subject to the negative pressure gradient tends to separate; the shear layer lifted off the surface undergoes further instability and generates large coherent structures; the shear layer delimits a region of reverse flow at the wall. Such behavior occurs intermittently, alternating separation events with attached flow condition. The instantaneous separation seems to correlate with a locally increased shape factor for the incoming boundary layer (emptier velocity profile) associated with the local occurrence of a low-speed streak. Conversely, for a fuller velocity profile in the incoming flow the interaction does not exhibit flow separation. Given the probability of reverse flow for this specific configuration no region of separated flow is observed in the mean flow pattern. An inflection point is observed in the velocity profile close to the wall in the region of interaction, as also reported by [21].

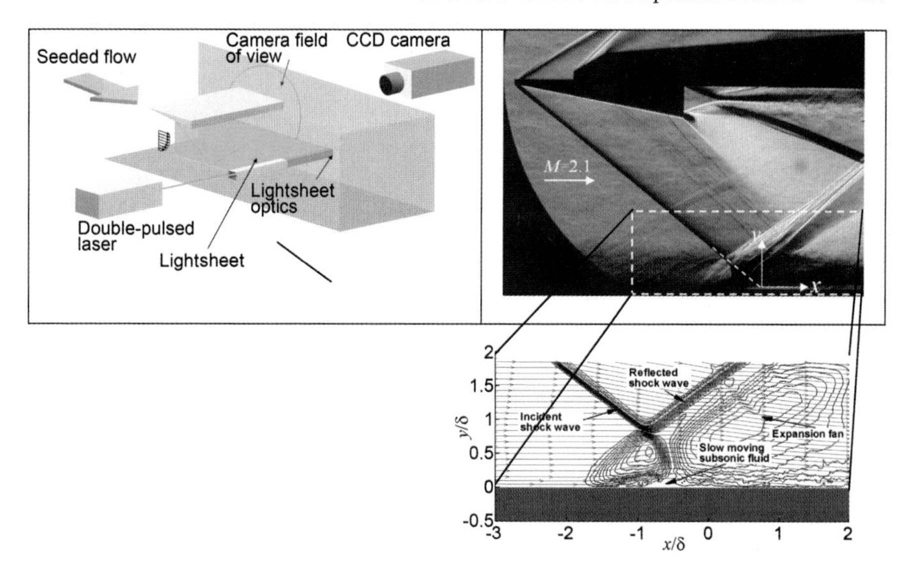

Fig. 20. *Top left:* SWBLI experiment with test section, illumination and imaging arrangement. *Top right:* Schlieren photograph (50 ns exposure time). *Bottom:* SWBLI flow schematic based on PIV data

9 Conclusions

The developments of PIV in the high-speed flow regimes result nowadays in the possibility to investigate supersonic and hypersonic flows even in the turbulent regime and for three-dimensional configurations. The physical properties of particle tracers need a careful assessment in order to limit the error due to nonideal flow tracking. The oblique shock-wave test represents an appropriate method yielding unambiguous evaluation of the particle response time, which is approximately 2 μs for submicrometer solid particle tracers. The seeding distribution in high-speed facilities presents more difficulties with respect to the low-speed and especially closed-loop wind tunnels. Seeding has to be produced under pressurized conditions and transported into the settling chamber. The operation of PIV for short-duration facilities like high-pressure discharge tubes is simplified by the fact that offline seeding of the working fluid can be applied. In the hypersonic flow regime, the limiting factors for the measurement accuracy are the finite particle response and the large variations in the seeding density resulting from the large compressibility effects. Temporal separation of the exposures of less than a microsecond is often required for flow velocities of the order of 1000 m/s. The application of PIV to supersonic turbulent boundary layer flows interacting with shock waves shows that accurate mean and fluctuating velocity statistics can be obtained, which compare with the qualitative level obtained in the low-speed regime. Finally, the study of compression ramps in the hypersonic regime yields quantitative

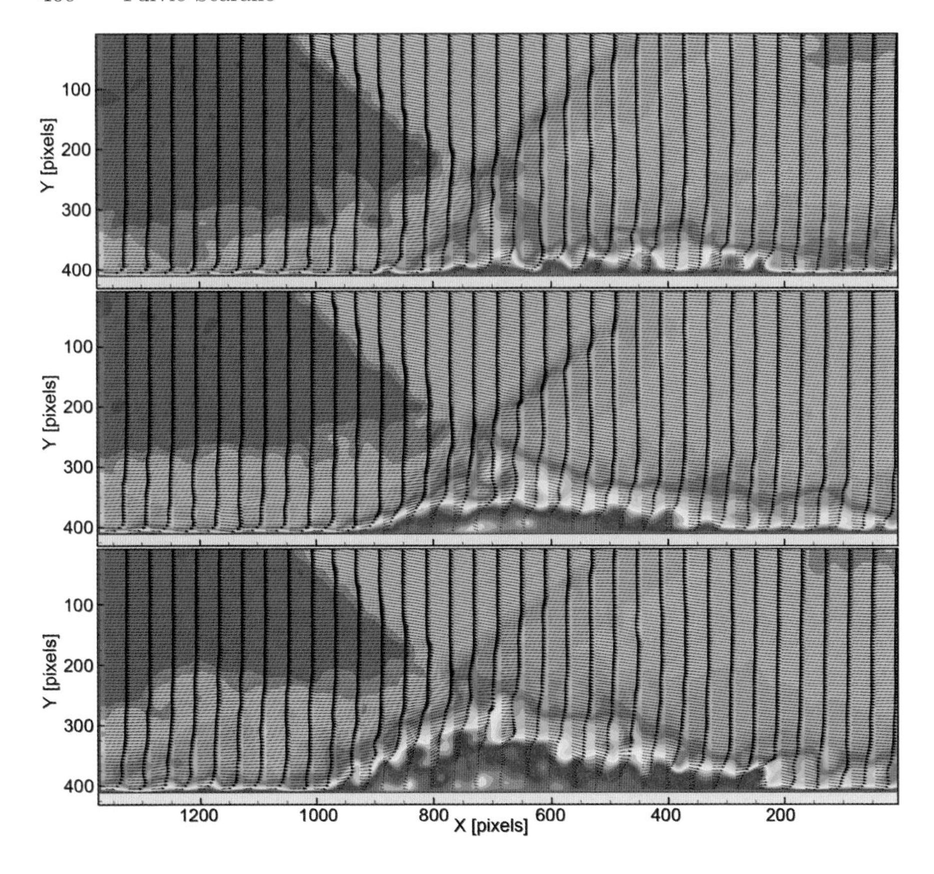

Fig. 21. Three instantaneous (uncorrelated) velocity fields. Velocity vectors and streamwise velocity contours

velocity information, which can be compared with the flow field computed by CFD with the perspective for a more extensive comparison between these means of investigation.

Acknowledgements

B. van Oudheusden, F. Schrijer, G. Elsinga, R. Humble and R. Theunissen have contributed to the research conducted at TU Delft.

This work is supported by the Dutch Technology Foundation STW under the *VIDI – Innovation impulse* scheme, Grant DLR.6198.

References

[1] G. Settles: *Schlieren & Shadowgraph Techniques* (Springer, Berlin, Heidelberg 2001)
[2] D. Kastell, G. Eitelberg: A combined holographic interferometer and laser-schlieren system applied to high temperature, high velocity flows, in *Instrumentation in Aerospace Simulation Facilities* (ICIASF '95 Record, International Congress 18–21 1985) pp. 12/1–12/7
[3] S. R. Sanderson, H. G. Hornung, B. Sturtevant: The influence of non-equilibrium dissociation on the flow produced by shock impingement on a blunt body, J. Fluid Mech. **516**, 1–37 (2004)
[4] V. A. Amatucci, J. C. Dutton, D. W. Kuntz, A. L. Addy: Two-stream, supersonic, wake flow field behind a thick base. General features, AIAA J. **30**, 2039 (1992)
[5] R. Benay, B. Chanetz, B. Mangin, L. Vandomme, J. Perraud: Shock wave/transitional boundary–layer interactions in hypersonic flow, AIAA J. **44**, 1243–1254 (2006)
[6] N. T. Clemens, M. G. Mungal: A planar Mie scattering technique for visualizing supersonic mixing flows, Exp. Fluids **11**, 175 (1991)
[7] N. T. Clemens, M. G. Mungal: Large scale structure and entrainment in the supersonic mixing layer, J. Fluid Mech. **284**, 171 (1995)
[8] R. J. Adrian: Twenty years of particle image velocimetry, in *12th Int. Symp. Appl. of Laser Tech. Fluid Mech.* (2004)
[9] A. J. Smits, J. P. Dussauge: *Turbulent Shear Layers in Supersonic Flow* (AIP Press 1996)
[10] G. E. Elsinga, B. W. van Oudheusden, F. Scarano: Evaluation of optical distortion in PIV, Exp. Fluids **39** (2005)
[11] C. S. Moraitis, M. L. Riethmuller: Particle image displacement velocimetry applied in high speed flows, in *Proc. 4th Int. Symp. Appl. of Laser Anemometry to Fluid Dyn.* (1988)
[12] J. Kompenhans, R. Höcker: Application of particle image velocimetry to high speed flows, in Riethmuller, M. L. (Eds.): *Particle image displacement velocimetry*, VKI Lecture Series (1988)
[13] A. Krothapalli, D. P. Wishart, L. M. Lourenço: Near field structure of a supersonic jet: 'on-line' PIV study, in *Proc. 7th Int. Symp. Appl. of Laser Tech. Fluid Mech.* (1994)
[14] Lourenço, L. M.: Particle image velocimetry, in M. L. Riethmuller (Ed.): *Particle Image Velocimetry*, VKI Lecture Series **1996-03** (1996)
[15] M. Raffel, H. Höfer, F. Kost, C. Willert, J. Kompenhans: Experimental aspects of PIV measurements of transonic flow fields at a trailing edge model of a turbine blade, in *Proc. 8th Int. Symp. Appl. of Laser Tech. Fluid Mech.* (1996) paper No. 28.1.1
[16] M. P. Wernet: Digital PIV measurements in the diffuser of a high speed centrifugal compressor, in (1998) AIAA Paper No. 98-2777
[17] W. D. Urban, M. G. P. Mungal: Velocity measurements in compressible mixing layers, J. Fluid Mech. **431**, 189 (2001)
[18] F. Scarano, B. W. Van Oudheusden: Planar velocity measurements of a two-dimensional compressible wake, Exp. Fluids **34**, 430–441 (2003)

[19] F. Scarano: PIV image analysis for compressible turbulent flows. Extension of PIV to the hypersonic flow regime, particle tracers assessment, in *Lecture Series on Advanced measuring techniques for Supersonic Flows* (von Kármán Institute for Fluid Dynamics, Rhode-Saint Genèse 2005)
[20] F. Scarano: Quantitative flow visualization in the high speed regime: Heritage, current trends and perspectives, in *Proc. 12th Int. Symp. On Flow Vis.* (2006) paper No. 2.1
[21] R. A. Humble, F. Scarano, B. W. van Oudheusden, M. Tuinstra: PIV measurements of a shock wave/turbulent boundary layer interaction, in *Proc. 13th Int. Symp. Appl. of Laser Tech. Fluid Mech.* (2006)
[22] J. Haertig, M. Havermann, C. Rey, A. George: Particle image velocimetry in Mach 3.5 and 4.5 shock-tunnel flows, AIAA J. **40**, 1056 (2002)
[23] F. F. J. Schrijer, F. Scarano, B. W. van Oudheusden: Application of PIV in a hypersonic double-ramp flow, in (2005) AIAA paper No. 2005-3331
[24] A. Melling: Tracer particles and seeding for particle image velocimetry, Meas. Sci. Technol. **8**, 1406 (1997)
[25] G. Tedeschi, H. Gouin, M. Elena: Motion of tracer particles in supersonic flows, Exp. Fluids **26**, 288 (1999)
[26] S. A. Schaaf, P. L. Chambre: *Fundamentals of Gas Dynamics* (Princeton Univ. Press 1958)
[27] J. L. Herrin, J. C. Dutton: Effect of a rapid expansion on the development of compressible free shear layers, Phys. Fluids **7**, 159 (1995)
[28] M. Samimy, S. K. Lele: Motion of particles with inertia in a compressible free shear layer, Phys. Fluids A **3**, 1915 (1991)
[29] F. F. J. Schrijer, F. Scarano, B. W. van Oudheusden: Application of PIV in a Mach 7 double-ramp flow, Exp. Fluids **41** (2006)
[30] M. Raffel, C. Willert, J. Kompenhans: *Particle Image Velocimetry, a Practical Guide* (Springer, New York 1988)
[31] G. E. Elsinga, B. W. van Oudheusden, F. Scarano: The effect of particle image blur on the correlation map and velocity measurement in PIV, in *Optical Engineering and Instrumentation, SPIE Annual Meeting* (2005)
[32] M. Havermann, J. Haertig, C. Rey, A. George: Application of particle image velocimetry to high-speed supersonic flows in a shock tunnel, in *11th Int. Symp. on Applications of Laser Techniques to Fluid Mechanics,* (2002)
[33] W. L. Hankey, Jr, M. S. Holden: Two-dimensional shock-wave boundary layer interactions in high speed flows, in *AGARDograph*, vol. 203 (1975)
[34] J. Délery, J. G. Marvin: Shock-wave boundary layer interactions, in *AG 280, AGARD* (1986)
[35] K. Sinha, M. J. Wright, G. V. Candler: The effect of turbulence on double-cone shock interactions, in (1999) AIAA paper No. 99-0146
[36] D. S. Dolling: Fifty years of shock wave/boundary layer interation research: What next?, AIAA J. **39** (2001)
[37] R. Theunissen, F. F. J. Schrijer, F. Scarano, M. L. Riethmuller: Application of adaptive PIV interrogation in a hypersonic flow, in *Proc. 13th Int. Symp. Appl. of Laser Tech. Fluid Mech.* (2006)

Index

PIV Investigation of Supersonic Base-Flow–Plume Interaction

Bas W. van Oudheusden and Fulvio Scarano

Dept. of Aerospace Engineering, Delft University of Technology, The Netherlands
{b.w.vanoudheusden,f.scarano}@tudelft.nl

Abstract. An experimental flow study of an axisymmetric body in a supersonic stream with an exhaust nozzle operating in the supersonic regime has been performed by means of particle image velocimetry. The "FESTIP" model was investigated at free-stream Mach numbers of 2 and 3. The exhaust jet is slightly underexpanded at the nozzle exit at Mach 4. The measurements cover the base region and wake of the model. The PIV technique employs solid submicrometer particle tracers and the particle-image recordings are interrogated with an adaptive multigrid window-deformation technique. Statistical flow analysis provides mean velocity and turbulence intensity for the cases with and without an exhaust jet. The latter case is used for a quantitative comparison of the experimental results with a 3D numerical flow simulation performed with the LORE code.

1 Introduction

The interaction between the base flow and an exhaust plume is relevant to the development of next-generation reusable launchers and re-entry vehicles. One of the critical areas is concerned with the proper modeling of the base-flow aerodynamics. Depending on the afterbody configuration and flow regime, the interaction between the external flow, the separated base flow and the exhaust plume may give rise to important mechanical and thermal loads on the nozzle external surface and the vehicle base. In particular, asymmetric effects and unsteady flow phenomena may cause severe side loads. The current investigation aims at the characterization of the complex turbulent interaction occurring in proximity of the base between the external supersonic stream and the supersonic exhaust jet. The presence of an exhaust jet will be studied for underexpanded cases ($p_{\text{exit}} > p_{\text{amb}}$), which is of special interest in relation to launcher performance at higher altitudes. The exhaust jet is characterized by the pressure ratio $N = p_{\text{t, jet}}/p_{\infty}$, which for the cases presented here is such that the jet is overexpanded with respect to the free-stream static pressure. It should be realized, however, that in the presence of a supersonic coflow the effective ambient pressure that governs the post-exit behavior of the jet plume is the base pressure, which is substantially lower than the free-stream pressure. This makes the studied nozzle flow effectively slightly underexpanded. The behavior and topology of the base flow in presence of jet-plume interaction is of particular importance in view of base

A. Schroeder, C. E. Willert (Eds.): Particle Image Velocimetry,
Topics Appl. Physics **112**, 465–474 (2008)

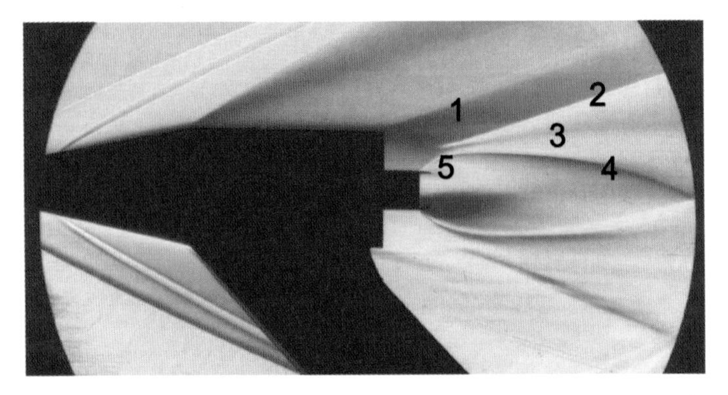

Fig. 1. Schlieren visualization of the FESTIP model with underexpanded exhaust jet ($M_\infty = 3$, $N = 400$). Flow features: 1 expansion fan, 2 plume shock, 3 shear layer, 4 barrel shock, 5 expansion fan of jet flow

pressure and possibly convective transport of hot combustion gases towards the base.

The model considered in the present study, illustrated in Fig. 1, was defined in the framework of the FESTIP (Future European Space Transportation Investigations Programme) research program [1, 2]. A series of experiments have been performed previously on this configuration [3, 4], using surface and field probe pressure measurements as well as Schlieren flow visualizations. These techniques do not allow a quantitative characterization of the separated flow immediately behind the model base itself, which is a region of particular interest. The objective of the present work is to provide detailed experimental evidence of the flow topology in this region, by means of particle image velocimetry (PIV). In addition, a validation will be made of previous CFD computations [5], carried out with the LORE code developed at the Aerospace Department of Delft University in conjunction with ESA [6]. One of the important features of the flow revealed by comparing experiments and CFD simulations is the strong three-dimensional effect introduced by the presence of the supporting sting.

2 Experimental Arrangement

The FESTIP model is an axisymmetric blunted cone-cylinder with a sharply truncated base [4]. The conical forebody and cylindrical afterbody are of similar length, the total model length is 187 mm. The base diameter is 50 mm. To ascertain a turbulent boundary layer a tripping wire (0.15 mm diameter) is applied at about 40 % of the cone length. From the center of the base a nozzle protrudes. Its outer surface is a circular cylinder, the inner surface is a conical nozzle with a divergence of 15°. The exit diameter is 17 mm, which is one third of the base diameter; the design exit Mach number is 4. The jet

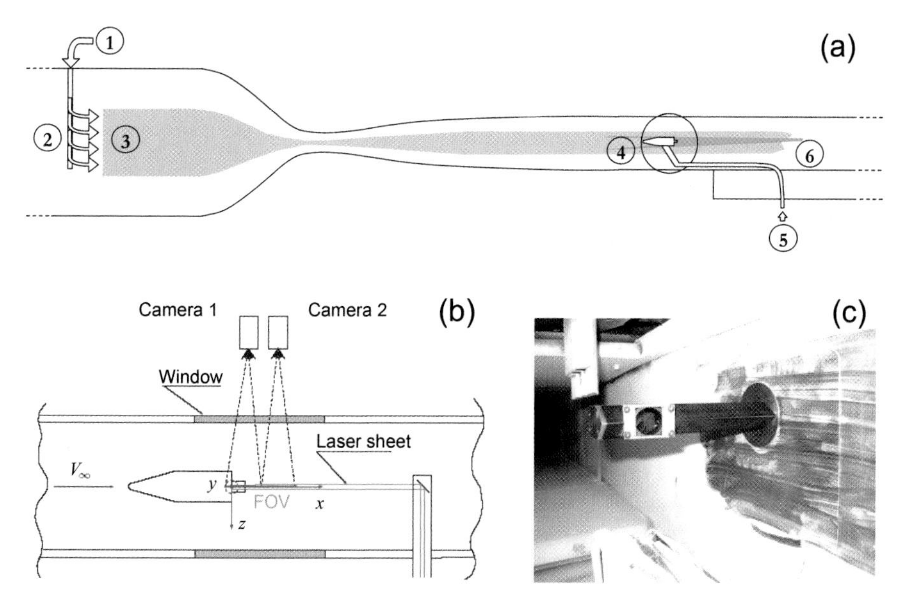

Fig. 2. Schematic of the wind tunnel and PIV setup. (**a**) Particle-tracer injection system: 1 seeding inlet, 2 distributor, 3 seeded flow in settling chamber, 4 test section, 5 pressurized air supply line for jet, 6 laser lightsheet; (**b**) two-camera extended view; (**c**) sealed laser-light transmission device

is supplied with compressed air at ambient temperature. A relatively large sting (hosting air-feed tubing inside) supports the model from the bottom.

The experiments were carried out in the transonic-supersonic wind tunnel (TST-27) of the High-Speed Aerodynamics Laboratories at Delft University of Technology. The facility is a blowdown-type and can generate flows in a Mach range from 0.5 to 4.2 in the test section (280 W $\times$ 270 H mm^2). In the transonic and supersonic range, the Mach number is set by means of a continuous variation of the throat section and flexible upper and lower nozzle walls. Dry air stored at 40 bar in a 300-m^3 storage vessel allows intermittent operation of the wind tunnel for a duration of 300 s. The tunnel operates at values of the unit Reynolds number ranging from 38×10^6 to $130 \times 10^6\,\mathrm{m}^{-1}$. The test matrix for the experiments considered in this investigation is given in Table 1, where p_t is the free-stream total pressure and $\mathrm{Re_L}$ the Reynolds number based on the model length.

Figure 1 shows a color Schlieren photograph of the model installed in the wind tunnel and operating with an exhaust jet. Figure 2 describes some details of the PIV apparatus installation in the wind tunnel. The integration of PIV in a supersonic wind tunnel requires the consideration of several technical aspects, such as seeding generation and dispersion, particle illumination and recording [7]. Fine-grade titanium dioxide (TiO_2) solid particles of 50 nm

Table 1. Test conditions

M_∞	p_t(bar)	Re_L	M_{jet}	$N = p_{t,jet}/p_\infty$
1.98	1.99	5×10^6	3.96	70
1.98	1.99	5×10^6	no jet	–
2.98	5.73	9×10^6	3.96	80
2.98	5.73	9×10^6	no jet	–

mean diameter are dispersed in the settling chamber of the wind tunnel, generating a seeded stream of approximately 10 cm diameter in the test section (Fig. 2a). The exhaust jet air flow is seeded with the same particle tracers using a high-pressure cyclone generator. The particle tracers' response time is approximately 2 µs, yielding a relaxation length of 1 mm.

The laser lightsheet, introduced into the test section from downstream, is provided by a Nd:YAG double-cavity laser emitting 400 mJ pulses (duration 6 ns) at a rate of 10 Hz. The pulse separation is 1 µs, which allows the particle tracers to travel 0.5 mm between pulses. Two CCD cameras (1280×1024 pixels, 12 bit) with interline frame transfer are used to record the particle image pairs. Figures 2b,c show the camera and laser sheet arrangement for the measurements. Two cameras are used to extend the field of view in the wake region (FOV $= 160 \times 70\,\text{mm}^2$).

Figure 3 shows an instantaneous PIV recording with superimposed instantaneous velocity vector field at Mach 2. An overall high seeding density is achieved, however, strong variations in the particle-tracer distribution is observed, with low levels in the separated base flow and lack of uniformity in the free stream due to limitations in the distribution system. Indeed, achieving appropriate seeding conditions in high-speed experiments is one of the current challenges in PIV. The particle-tracer pattern allows a first visualization of the flow features. Special care is required also in terms of image analysis in view of the particle-image-pattern deformation across shear layers and shock/expansion waves. The image interrogation is performed by an advanced window-deformation iterative multigrid crosscorrelation algorithm [8]. The resulting measurement grid is 320×120 points with a spatial resolution of about 2 grid points/mm. Each dataset is composed of 500 independent velocity fields that were recorded at 3.3 Hz.

3 Experimental Results

Figure 4 displays the time-averaged flow at a free-stream Mach number of 2 (top) and 3 (bottom). The mean velocity and the velocity fluctuations are represented by means of velocity vectors and turbulence intensity color contours, respectively. The essential flow features are identical for both Mach numbers. The supersonic flow expands around the base and is deflected

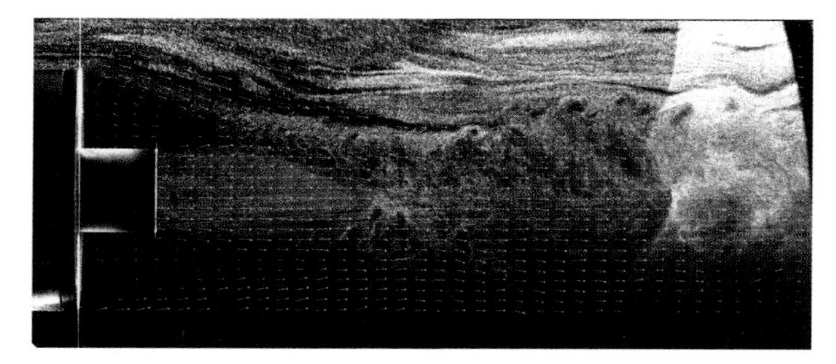

Fig. 3. PIV snapshot at Mach 2. The particle-seeding pattern allows visualization of the flow turbulence in the shear layers, exhaust plume and terminating normal shock downstream. The base region shows a poor particle seeding. The instantaneous velocity vector distribution is represented with one vector every four for clarity

through a Prandtl–Meyer fan (cf. the Schlieren visualization in Fig. 1). Flow separation occurs at the base edge and a region of reversed flow is revealed between the base and nozzle, which cannot be observed in the Schlieren visualization. The velocity at the jet exit is 630 m/s, while the flow continues expanding along the plume (underexpanded jet). The expansion is terminated at the barrel shock. An annular low-momentum region that separates the external flow and the jet is observed, which can be interpreted as the wake of the base region. It has been formed by the merging of the two shear layers that originate from the base edge and the nozzle edge, respectively. The interaction of the shear layers occurs about one nozzle diameter downstream of the exit. Approximately 80 mm downstream of the base an expansion region is generated from the interaction of shock waves with the shear layers. At Mach 2 a strong flow deceleration occurs at the end of the observed domain, which indicates the presence of a normal shock terminating the supersonic flow region. The shock is not an element of the base flow, but is due to the effect of backpressure in the wind tunnel at this particular low tunnel stagnation pressure (2 bar). The normal shock oscillates back and forth over a distance of about 1 cm and it is able to affect the separation and expansion of the flow on the lower side of the model, where an increased level of fluctuation is found. This behavior is not observed in the absence of the jet or at a higher value of the free-stream Mach number.

The free-stream turbulence level is measured to be about 1 % of the free-stream velocity, which agrees with hot-wire data. The turbulence within the external shear layer remains limited to 15 %, while a higher activity is observed across the jet shear layer close to the exit. The turbulence of the two shear layers can be distinguished in the interaction region as well as some distance downstream. On the lower side of the model a significantly higher turbulence level is observed, which is predominantly due to the wake of the model support and to the unsteady behavior of the separated shear layer referred to above.

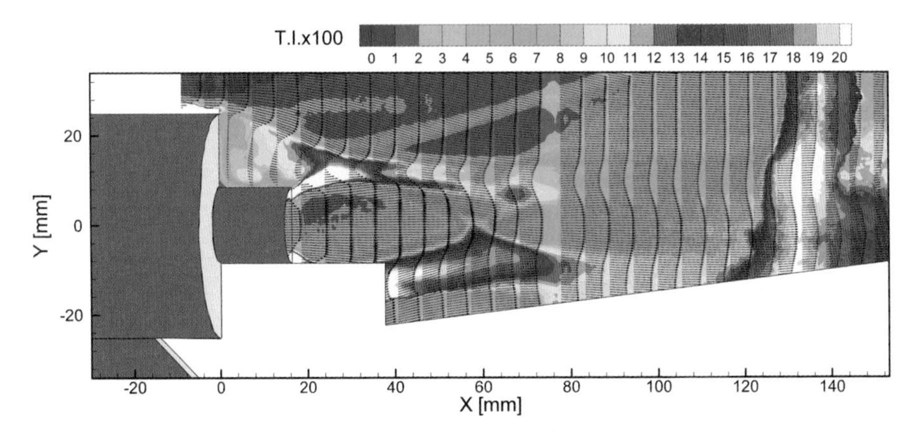

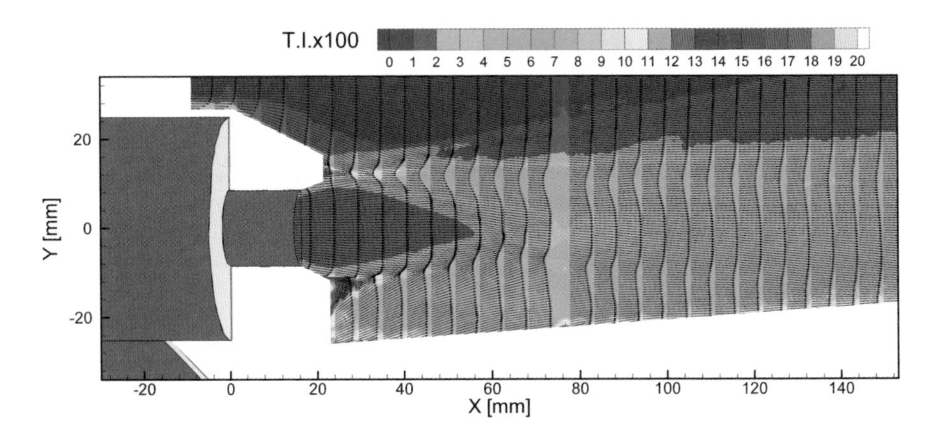

Fig. 4. Mean velocity (vectors) and inplane turbulence intensity $\sqrt{\frac{1}{2}(\overline{u'^2}+\overline{v'^2})}/U_\infty$ (*background color*). *Top:* Mach 2, jet-pressure ratio $N = 70$. *Bottom:* Mach 3, $N = 80$

A PIV image analysis at higher resolution (3 grid points/mm) allows evaluation of the development of the compressible shear layer along the separated flow region and provides additional detail of the flow structure inside this region immediately behind the base. The details in Figs. 5a,b return the mean flow topology with recirculation. The high momentum stream at the jet exit is observed to feed a counterrotating structure. A saddle point is visualized at the downstream end of the two recirculating regions ($x = 30$ mm, $y = 12$ mm). The details in Figs. 5c,d reveal turbulent activity in the attached boundary layer prior to separation. The boundary-layer turbulence intensity drops across the expansion, to slightly build up along the separated shear-layer. A maximum of about 22 % is returned close to the reattachment point. The jet shear-layer turbulence exhibits a higher peak, also due to the

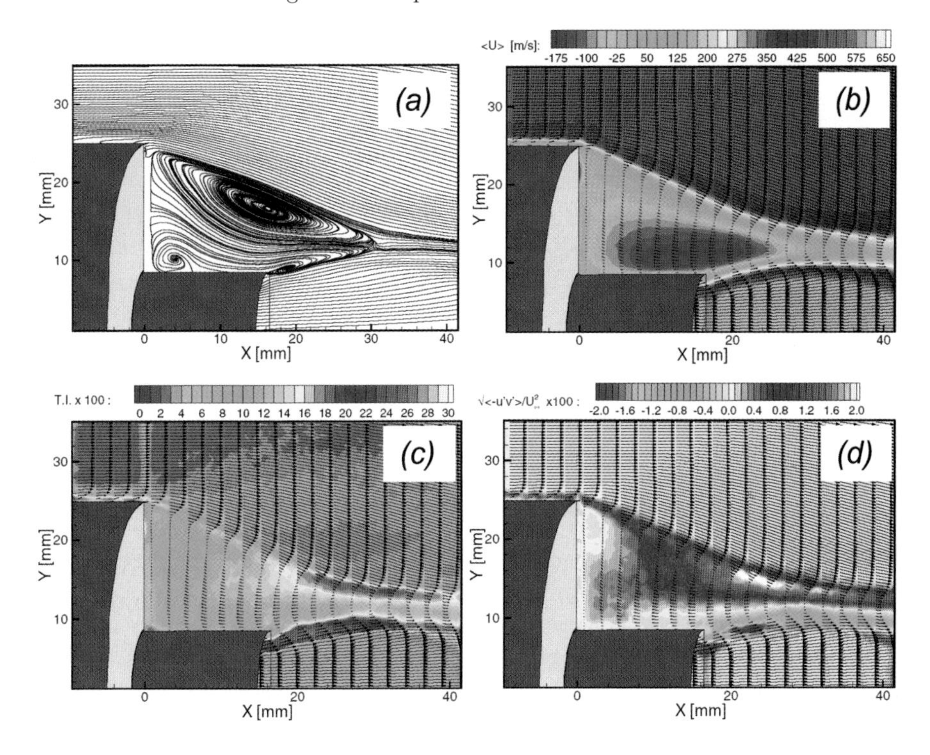

Fig. 5. Detail of the separated flow close to the base after a high-resolution PIV analysis: (**a**) streamlines, (**b**) velocity vectors and horizontal velocity component contours, (**c**) turbulence intensity, (**d**) Reynolds shear stress

higher exit velocity. The two turbulent regions are still distinct at two jet exit diameters downstream.

4 Comparison with CFD

The present PIV results have been compared with flow simulations performed with the CFD code LORE [5,6], which is a Navier–Stokes solver developed for high-temperature gas dynamics. The governing equations are discretized with a cell-centered finite-volume method provided with a second-order upwind flux-difference splitting scheme. Multiblocked coarse and fine 3D grids are generated of the model, support and flow regions. Two turbulence models of the two-equation type were applied: k–ω and SST (Menter's shear-stress transport model). The results of the latter (with compressibility correction) proved to be closest to the experimental pressure data [2]. Therefore, these computational results are used for a comparison with the PIV data. The flow case considered is that for Mach 3 without an exhaust jet.

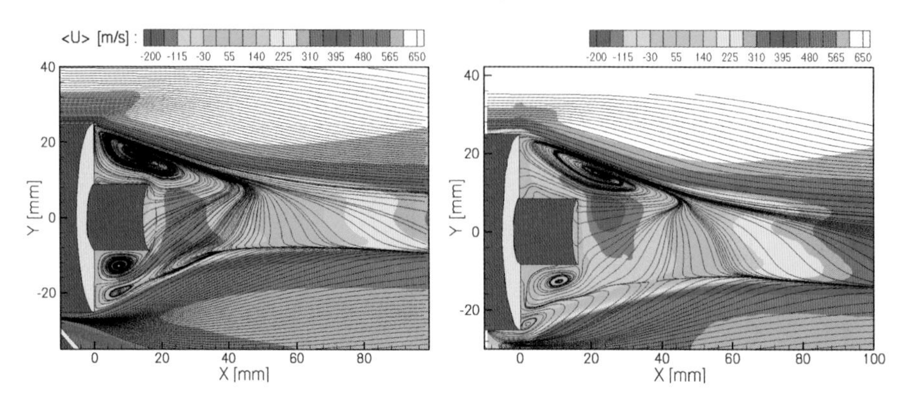

Fig. 6. Mean velocity and streamlines pattern at Mach 3 without an exhaust jet. *Left:* CFD. *Right:* experiments

Figure 6 shows the mean horizontal velocity component and streamlines as obtained by CFD (left) and PIV (right). A good agreement is found on the flow topology, particularly regarding the upper side of the base, where one main recirculation and a saddle point at the base are observed. A similar agreement holds for the flow structure further downstream. Some differences can be detected in the separated flow topology in the region below the nozzle, as well as in the location of maximum reversed flow. Overall, the agreement may be considered acceptable in view of the strong 3D and highly unsteady effect of the model support wake in which case the RANS modeling of turbulence is usually inaccurate. On the other hand, the lack of illumination and particle seeding may give some concern for the reliability of the PIV experiments in those regions as well.

For a further quantitative comparison Fig. 7 shows the distribution of the mean velocity component U and the turbulent kinetic energy (TKE), as a function of the vertical coordinate y at distinctive locations behind the model. The TKE is taken in the PIV data as $\frac{1}{2}(\overline{u'^2} + \overline{v'^2})$ and completed with $\overline{w'^2}$ in the CFD. The computed U-profiles show a somewhat smaller shear layer growth than those experimentally observed. Also, the contraction toward the model axis is more pronounced in the former. Near the axis the magnitude of the flow reversal is very comparable at $x < 30\,\mathrm{mm}$. Further downstream the velocity recovery predicted by CFD is considerably less than in the experiments.

The computed TKE displays in the upper shear layer a higher peak level than in the experimental results, while also the peak location is more inward (consistent with the stronger contraction). In the central region of the wake, behind the nozzle, the computational TKE is systematically higher than the experimental values. The experiments show a decay of TKE with downstream distance, whereas the CFD displays an increase, which may possibly be an in-

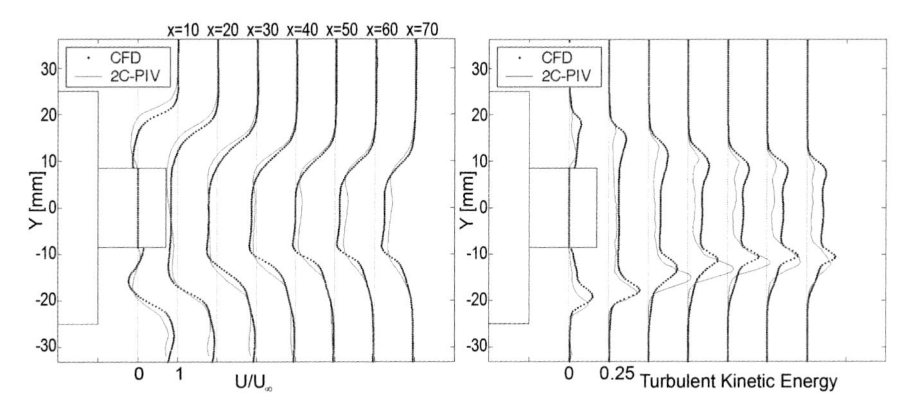

Fig. 7. Horizontal velocity component (*left*) and turbulent kinetic energy (*right*) behind the model (exhaust jet off)

dication that the turbulent dissipation is underestimated in the RANS model. For the lower shear layer the peak values of PIV are much higher and more pronounced than in the CFD. Taking into account the significant unsteadiness in the actual flow field at the bottom side of the model due to the presence of its support, this discrepancy may not be surprising, in view of the fact that such large-scale unsteady phenomena are not properly modeled by a RANS approach. Further downstream, theory and experiment show a closer agreement. Here again, a slower relaxation of the peak distribution obtained by CFD is observed in comparison with the experiments.

5 Conclusions

The experimental investigation shows that base flow–plume interaction phenomena can be accurately studied by means of PIV. The technique is suitable to describe high-speed turbulent flows (up to 700 m/s) and the choice of solid TiO_2 particle tracers would allow the utilization of the technique possibly also at higher temperatures (e.g., for heated jet simulations). The flow past the FESTIP model was found to be strongly affected by the presence of the large support, which breaks up the axial symmetry and introduces a highly turbulent wake. In spite of several deficiencies in the CFD models as well as in the PIV experiments, the agreement between the two approaches is good, certainly in terms of flow topology. However, presently a flow field with a high degree of turbulent unsteadiness may be too demanding for both. In both cases, flow determination close to the base is affected by a larger uncertainty: PIV lacks particles in the separated region with signal drop, while the numerical simulation shows a stronger dependence upon the chosen turbulence model. Further developments are expected to extend the technique to high-temperature experiments.

References

[1] E. Hirschel: Vehicle design and critical issues; FESTIP technology development work in aerothermo-dynamics for reusable launch vehicles, in *ESA/ESTEC FESTIP Workshop* (1996)

[2] P. G. Bakker, W. J. Bannink, P. Servel, Ph. Reijasse: *CFD Validation for Base Flows with and without Plume Interaction*, AIAA Paper 2002-0438 (2002)

[3] M. M. J. Schoones, W. J. Bannink, E. M. Houtman: *Base Flow and Exhaust Plume Interaction: Experimental and Computational Studies*, Delft University Press (Delft 1998)

[4] W. J. Bannink, E. M. Houtman, P. G. Bakker: *Base Flow/Underexpanded Exhaust Plume Interaction in a Supersonic External Flow*, AIAA Paper 98-1598 (1998)

[5] H. B. A. Ottens, M. I. Gerritsma, W. J. Bannink: *Computational Study of Support Influence on Base Flow of a Model in Supersonic Flow*, AIAA paper 2001-2638 (2001)

[6] L. M. G. F. M. Walpot: *Development and application of a hypersonic flow solver*, Ph.D. thesis, Delft Univ. Technology (2002)

[7] F. Scarano, B. W. van Oudheusden: Planar velocity measurements of a two-dimensional compressible wake, Exp. Fluids **34**, 430–441 (2003)

[8] F. Scarano, M. L. Riethmuller: Advances in iterative multigrid PIV image processing, Exp. Fluids **29**, S51–60 (2000)

Index

Developments and Applications of PIV in Naval Hydrodynamics

Fabio Di Felice and Francisco Pereira

INSEAN – Italian Ship Model Basin, Via di Vallerano, 139 - Rome, Italy
f.pereira@insean.it

Abstract. In this chapter, particle image velocimetry (PIV) is introduced as an essential tool for flow diagnostics in the field of naval hydrodynamics. The needs and requirements that the PIV technique is expected to meet in this particular sector are examined in Sect. 2. In Sect. 3, we review some of the most recent applications of experimental naval hydrodynamics and the technological solutions chosen to address the scientific challenges involved therein.

1 Introduction

Laser-based optical measurement techniques have, in recent years, become increasingly popular in naval research, and have been used in a number of instances ranging from ship wakes to propeller flows and underwater-vehicle hydrodynamics. This interest is being driven, in particular, by an increasing demand for faster and quieter boats, by new transportation regulations that enforce the respect of tighter environmental impact standards, as well as by new technological challenges, in particular in the area of marine propulsion (e.g., pod propulsors and waterjets). This multifaceted framework has given rise to a family of critical and specific problems connected with the hydrodynamic performance of hulls and appendices, with the hydroacoustical and structural interaction between the hull and the propulsion system and, in general, with the onboard comfort of passenger ships and with the optical- and noise-scattering characteristics of military units. The research and development operators of the field have put great effort into tackling the scientific and technological challenges posed by these issues. From the standpoint of marine hydrodynamics, the preferred and well-accepted strategy lies in a twofold approach where computational fluid dynamics (CFD), based on enhanced theoretical and numerical models, is concurrently supported by experimental fluid dynamics (EFD) for the purpose of model validation and development. Ultimately, the goal is to use the validated CFD models to provide complete documentation and diagnostics of the flow, and eventually provide guidelines for geometry optimization and validation.

Amongst the various techniques based on light scattering, laser Doppler velocimetry (LDV) is the best known and well-established method in naval applications [1–10]. It is nowadays routinely used in many naval research

A. Schroeder, C. E. Willert (Eds.): Particle Image Velocimetry,
Topics Appl. Physics **112**, 475–503 (2008)

organizations around the world. It provides valuable and accurate information on the mean and fluctuating velocity fields in a number of complex situations, such as the flow surveys of propellers and of ship wakes. For example, measurements by LDV have been a turning point in the analysis of the three-dimensional complex flow around rotors and propellers, providing quantitative insight on the roll-up process, the slipstream contraction and the tip-vortex evolution [4–7]. However, besides its undeniable advantages, LDV has some limitating issues related to its being, in nature, a point-measurement technique: LDV is unable to provide information on the spatial characteristics of large coherent structures, which are generally encountered in complex and separated flows, especially in ship flows (e.g., bilge keel roll-up, propeller wake, far wake); LDV can induce large errors in the intensity of unsteady vortical flow structures; the complexity of the system increases significantly if 3-component measurements are required; LDV requires extended periods of facility operation, especially in tow tanks where the limited length can heavily impair the productivity of the technique and induce a steep increase of the operational costs.

The relatively recent development of particle image velocimetry (PIV) has renewed the interest in the field of experimental naval hydrodynamics. In contrast to LDV, PIV has many advantageous features that make it popular: it makes possible the study of the spatial structure of the flow since it is a whole-field measurement technique that maps full planar domains; the recording time is very short and depends essentially on the technological characteristics of the acquisition and processing components; the analysis can be done a posteriori, which is a definitive advantage when high-cost facilities are being used; PIV is less demanding in terms of technical expertise; stereoscopic PIV (SPIV), the latest evolution of planar PIV, allows 3-component velocity measurements in a plane. In the particular field of naval hydrodynamics, PIV has seen its range of applications widely expanded in recent years: underwater vehicle flows, ship wake [11], ship roll motion [12], propeller flow [13], wave breaking, freak waves, free surface flows [14], bubbly flows, boundary layer characterization for drag reduction by microbubbles, sloshing flows, waterjets, etc. To address these many topics and face the variety of testing facility environments, specific instrumentation was quickly devised to incorporate the state-of-the-art PIV technology. INSEAN has pioneered this approach with the development of the underwater stereoscopic PIV probe [15], and paved the way with applications in towing tank and cavitation tunnel facilities that have demonstrated the validity of this new concept.

In the present work, we review the requirements that one should verify when using PIV for experimental naval hydrodynamics. A review of the existing systems and their use in typical industrial-type problems in the naval field is presented, illustrated with four cases of particular relevance: 1. unsteady flow around the bilge keel of a ship in damped roll motion, for CFD model development; 2. propeller flow survey in a cavitation tunnel, using phase-locked measurements, for the identification of noise sources; 3. wake survey of a large underwater vehicle model, for the purpose of design validation; 4. bubble transport in a propeller flow, for the purpose of noise reduction and propeller performance. These applications illustrate the many issues related to the use of PIV as a flow survey tool in a variety of facilities: experimental setup, calibration, large measurement areas, technique resolution, data processing and management.

2 Needs and Requirements

The PIV technique will typically be used in the framework of ship and propeller model testing, both in cavitation and towing-tank facilities. In most towing-tank institutions, the ship model length ranges from 3 to 8 m and the diameter of the propeller model is within the range 130 to 350 mm. These features provide the basic requirements for the size of the measurement area and for the spatial resolution of the PIV system. On one hand, the measurement area should cover the complete crossflow area of the propeller, with an additional 10 to 20 % margin to map the full slipstream. On the other hand, and together with the previous requirement, the system should be able to resolve the thin blade wake that is usually of the order of a few millimeters. For a long time, the resolution of the PIV sensors and/or the power of the laser sources would not allow these two conditions to be satisfied simultaneously. The only practical approach was to make a compromise between spatial resolution and measurement area or, alternatively, to patch multiple high-resolution areas into a larger one, at the cost of additional running time. These end-user needs have motivated the hardware manufacturers to improve the performance of the PIV components, especially in the area of the imaging sensors: the sensitivity is being continually enhanced and very high resolution (4 Mpixels and more) PIV cameras are now commercially available. On his side, the end-user has put the effort on the development of advanced PIV algorithms that would contrast the hardware limitations. The combination of techniques such as window offset, iterative grid refinement, window deformation and other algorithms now enables the reconstruction of velocity gradients from interrogation windows as small as 16×16 pixel^2, without compromising the accuracy.

The implementation of the hardware and software components has also evolved to meet the challenges of use in large experimental structures. The current trend, followed by most of the commercial and research institutes in the sector of marine hydrodynamics, demonstrates that underwater PIV is the only viable solution for applications both in large free-surface cavitation tunnels and in towing tanks. In the first case, the facility optical access ports consist usually of low-quality and thick windows, i.e., not designed to match the optical standards required by the PIV technique. Furthermore, the light requirements are also more demanding because of the relatively large distances between the measurement area and the section walls. More powerful laser sources are generally required (200 mJ per pulse or more). In addition, long focal length optics are needed to map an area with a sufficient spatial resolution. Hence, brighter, high-quality (i.e., low distortion and low aberration specifications) and more expensive lenses are necessary. In the tow tank case, the unavailability of windows makes the use of an underwater system a necessity. A PIV system onboard the model is not a practicable option as it would require optical ports on the ship hull itself.

In cavitation facilities, the presence of two-phase flows introduces new practical and technological challenges. The measurements in these instances have been so far performed in off-cavitation conditions to avoid camera blooming and consequent sensor damage. Cavitation manifests itself by the occurrence of gas/vapor structures of various types: submillimeter (10 to 100 μm) cavitation bubbles convected by the flow, large vapor cavities attached to the solid boundaries (e.g., sheet cavitation on a propeller), vapor filaments (e.g., cavitating vortex at the tip of a blade or cavitating turbulent structures in the blade shear wake). To map the two phases simultaneously, fluorescent particles have long been suggested and used in laboratory-scale setups. However, their use in large facilities has not yet been assessed because of their small size and low scattering efficiency, both factors hindering their detection from distant imaging systems, as well as their prohibitive cost and progressive efficiency decay over time. Safety issues related to their chemical characteristics, such as toxicity and chemical reactivity with other compounds present in the facility, are another major obstacle to their use in industrial- or commercial-type facilities.

3 State-of-the-Art

3.1 Surface-Ship Flows

Particle image velocimetry is particularly well suited, as opposed to single-point measurement techniques like LDV, for flows characterized by large spatial instabilities. In these situations, a whole-field measurement approach is the only way to efficiently address the problem. The application described hereafter illustrates such a situation with the case of the flow created by a ship hull in free roll decay.

Motion prediction of a ship in seaway is a complex task and is one of the main topics in current hydrodynamic research. Whereas, for most degrees of freedom, a small amount of information on the ship geometry and on the sea state are sufficient to predict the response of the ship, the roll motion is particularly difficult to estimate. Roll is, in fact, a lightly damped and lightly restored motion and the roll natural period of the conventional ship is very close to the richest region of the wave-energy spectrum. Very large amplitude of the roll motion can then occur, even in a moderate sea state, if the wave-frequency spectrum is narrow and tuned with the ship roll natural period. Ship capsizing is the extreme consequence of this phenomenon. The difficulties in predicting the roll damping of a ship depend not only on the nonlinear characteristics of the phenomenon but also on its strong dependence upon the presence or not of bilge keels and upon the forward speed of the ship. Therefore, the analysis of the induced vortical structures and their evolution with the roll angle, as well as the evaluation of forces and moments applied to the ship during the roll motion, are of primary relevance to the understanding of the dynamics of the ship. The difficulty of implementing this approach explains the strong development of the empirical methods [16, 17]. From the numerical point of view, simulations based on 2D viscous flow mathematical models [18, 19] have been extensively used in the past. Recently, simulations of increasingly complex ship 3D flows have been performed [20, 21]. For the development and the validation of CFD codes, much more detailed model-scale, surface-ship experimental data is required. Official issues from ITTC 1999 and ITTC 2003 require explicitly EFD and CFD to be more and more interlaced in order to provide reliable tools for a better understanding of the fluid physics. Experimental works for validating theoretical and numerical models have been performed [22]. Recently, LDV measurements on a frigate model in forced roll motion have been performed [23]. Here, we present a 2D PIV experimental study of the flow around the bilge keels of a ship model in free roll decay motion.

The experiments were carried out in the INSEAN towing tank no. 2. The tank is 250 m long, 9 m wide and 4.5 m deep, with a maximum carriage speed of 10 m/s. A 5720-mm length frigate model (INSEAN model C2340) comprising of bilge keels is tested. A sketch of the experimental setup is shown in Fig. 1. Two-component planar PIV measurements were performed using an underwater camera with a 1280×1024 pixel^2 resolution. The lightsheet was fired at a rate of 12.5 Hz, with an energy output of 200 mJ per pulse, by a two-head Nd:YAG laser coupled with underwater optics. Roll being the only motion investigated in this experiment, the ship model was locked to the dynamic trim at an initial roll angle of 10° by means of a magnet-based release mechanism. A TTL trigger sequence was generated by a synchronizing device to pilot the Nd-YAG lasers and the camera acquisition system. The first pulse of this sequence would deactivate the magnet, thus releasing the model and starting the roll cycle and the image recording sequence. The maximum recording frequency was 3 Hz, due to the limitation of the image acquisition bandwidth. This was not sufficient to resolve in time the flow-field evolution of the damped roll motion, which had a natural frequency of about 0.5 Hz. In order to obtain a mean evolution of the unsteady flow, each run has been repeated several times with a varying start delay. By using a delay generator between the synchronizer and the magnetic release system, the time interval to the start of the acquisition sequence was varied. Hence, the flow-field evolution was sampled at different roll angles by scanning the free decay time history four times and using a delay of 1/12 s between two consecutive acquisitions. This procedure allowed a final reconstruction of the ensemble

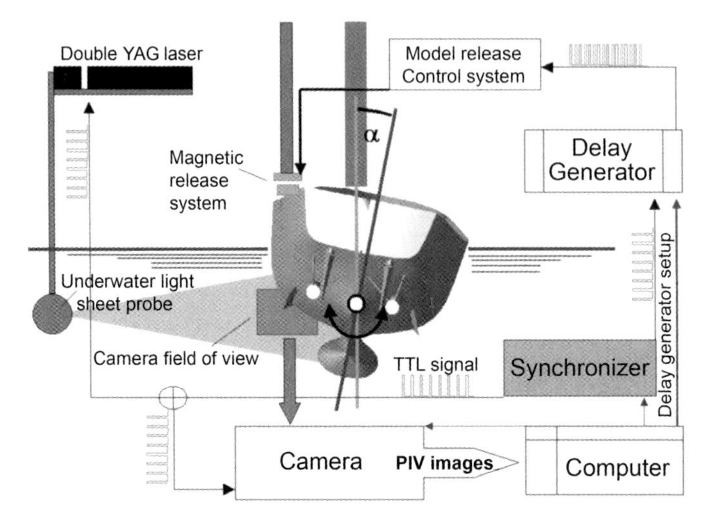

Fig. 1. Surface ship flow: experimental setup

averaged flow field at an effective sampling rate of 12 Hz. For each roll angle, 64 PIV measurements have been performed to evaluate the ensemble mean flow. The measurements presented here were performed at $x/L = 0.504$ (L is the model length) and for Froude numbers $\mathrm{Fr} = 0$ and $\mathrm{Fr} = 0.138$. The high repeatability of the roll-angle evolution allowed a quick convergence of the ensemble statistics.

The PIV analysis included the discrete offset technique and the iterative image-deformation method [24]. The displacement field obtained with the first analysis method was used to distort the images [25], upon which direct crosscorrelation was applied to calculate a corrected velocity field. This newly updated velocity predictor was subsequently used to redistort the original images and the process was repeated on an iterative basis until a convergence criterion could be verified. This type of iterative algorithm has been recently proved to be very efficient, accurate and stable [26]. Tests made on synthetic images showed that both the mean displacement error and the rms error were reduced by about one order of magnitude, whereas the velocity dynamic range was found to be 1.5 to 2.5 orders of magnitude larger than with the standard PIV analysis. This improvement is very important for the correct detection of the vortex cores and of the shear regions generated by the bilge keels. In the general case, it allows for the correct evaluation of velocity in crossflow measurements where the predominant out-of-plane velocity component requires very small time intervals between PIV frames, thus concentrating the displacement information in the subpixel range where errors tend to be important. The processing setup used in this case was composed by a 3-step discrete offset method with a final window size of 28×28 pixel^2 and a grid spacing between the vectors of 10 pixels (65 % overlap). The image deformation was then performed through 4 iterations with a local Gaussian weighting filter applied to the predictor field and a Gaussian subpixel interpolation fit.

The first series of experiments were performed at Fr = 0, for which the roll-damping contribution is small and is mainly induced by the vorticity shedding from the bilge keels. This behavior is well highlighted in Fig. 2, where the evolution of the vorticity during the first two cycles of the free roll decay motion is shown. The vortex shedding occurs in vortex packets with a spatial frequency proportional to the angular acceleration of the bilge keels. In particular, the vorticity field reproduces the well-known Kelvin–Helmholtz instability, also reported in numerical simulations [27] for a uniformly accelerated flat plate. The vortex packet is rolled up in a large vortex whose intensity decreases with the amplitude of the roll motion at every new cycle. When the roll motion changes direction, two strong counterrotating vortices are simultaneously present. Expansion and energy diffusion of the vortex structures are visible as the free roll progresses in time. This physical effect is further amplified by the ensemble-averaging process. During the successive cycles, in fact, the released vortices interact randomly with the previously generated flow structures causing a slightly different displacement at each new cycle.

The second series of measurements were performed at a speed of 1.024 m/s (Fr = 0.138). The vorticity shedding time history is shown in Fig. 3. The forward speed of the ship produces strong modifications of the flow behavior: the advection of the vorticity along the direction perpendicular to the measurement plane takes place and is accompanied by a boundary layer in the longitudinal direction that interacts with the vorticity field and causes its diffusion. Vorticity reaches 50 % higher values than the case at Fr = 0, inducing a stronger damping. The instability previously observed at Fr = 0 is not visible on the ensemble mean history. However, it has been observed in the instantaneous flow field, though with a random spacing that is therefore lost in the ensemble-averaging process where it appears in the form of a shear layer. This randomness is caused by the hull turbulent boundary layer. Here, the tip of the bilge keel is outside the boundary layer and the diffusion of the vortex structures starts essentially in the second cycle, where their intensity is lower. Extended results are reported in [12].

3.2 Propulsor Hydrodynamics

The first successful applications of particle image velocimetry to the propeller wake flow analysis appeared in the late 1990s [13, 28], in medium-scale facilities and for off-cavitation operation. The technique proved to be very well suited for this particular problem. Furthermore, the characteristic features of this flow, such as the strong velocity gradients and the presence of coherent vortex structures, have triggered the development of the iterative multigrid algorithms [24], which are now used on a standard basis in general PIV analysis.

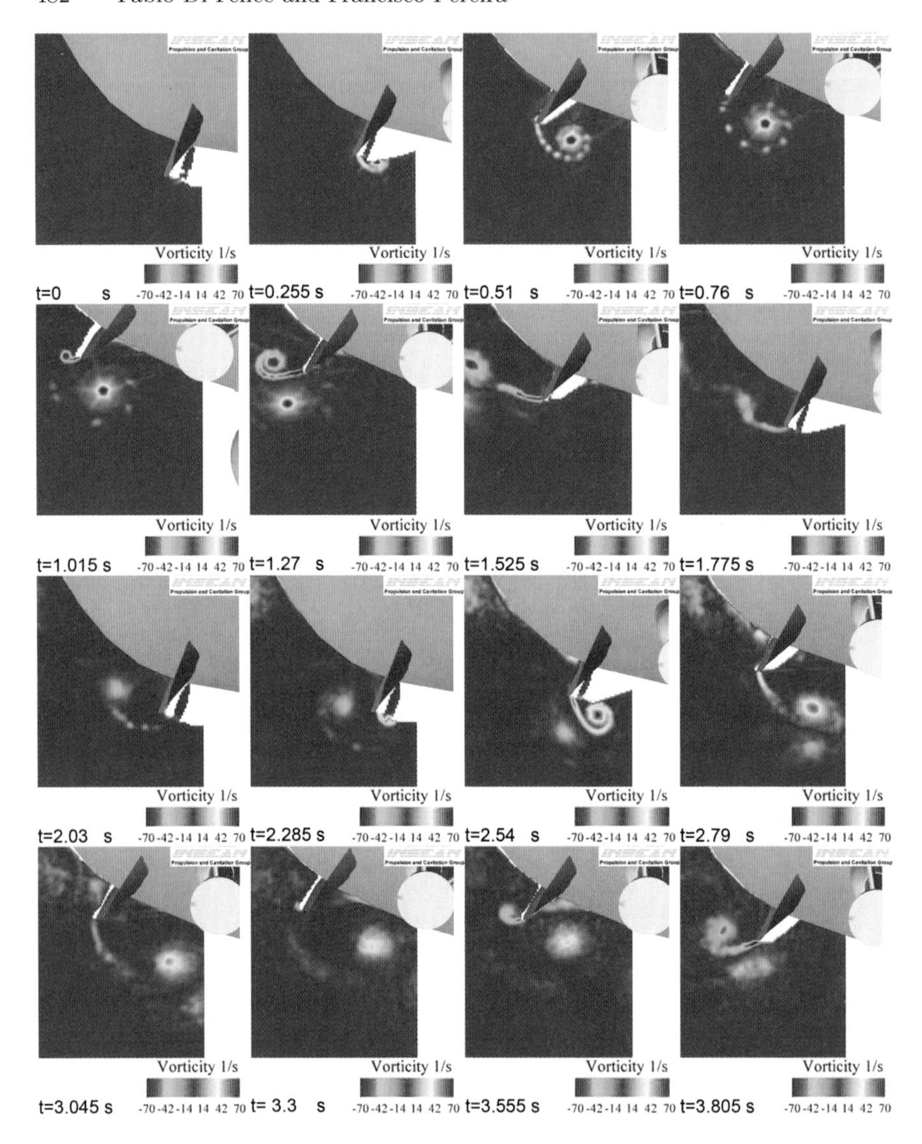

Fig. 2. Time evolution of vorticity during the free roll decay; $x/L = 0.504$, Fr $= 0$

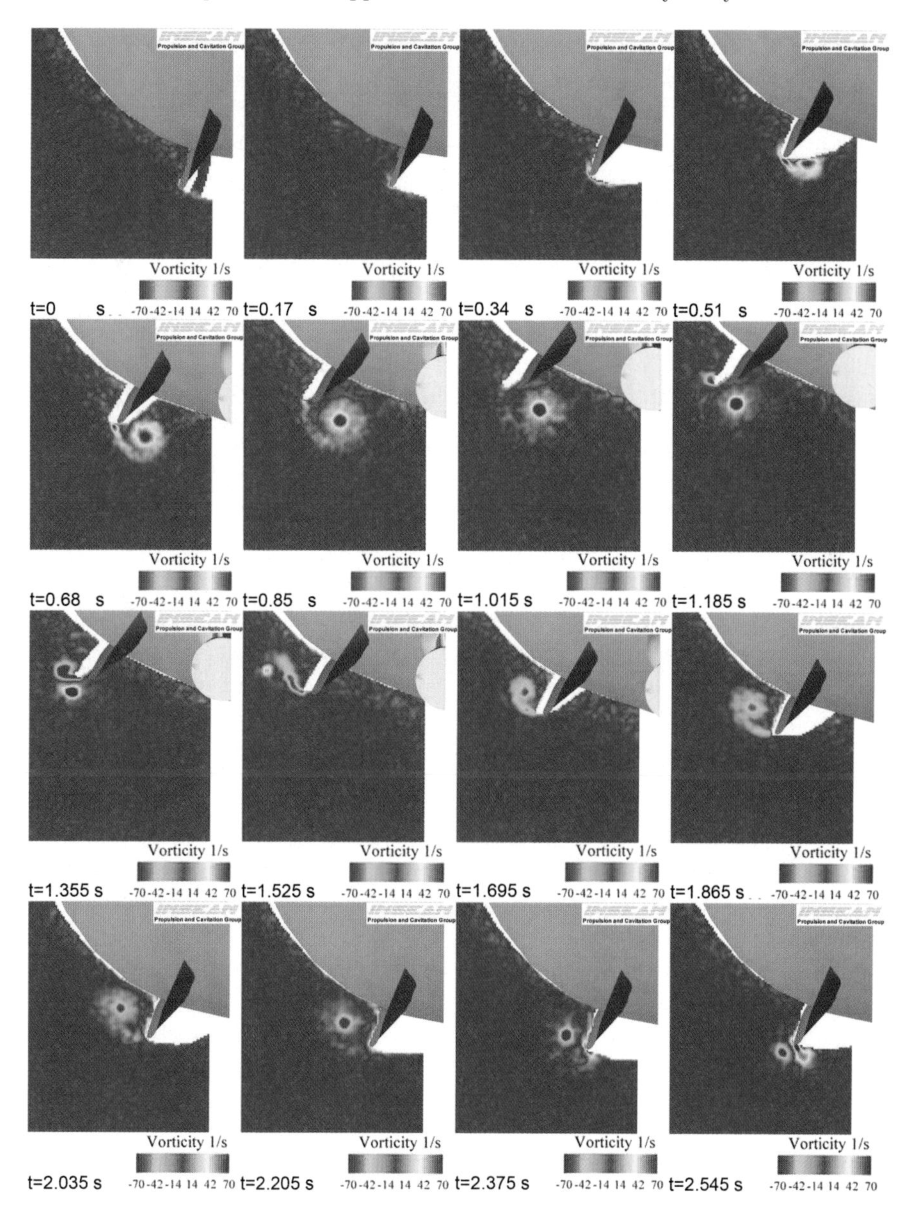

Fig. 3. Time evolution of vorticity during the free roll decay; $x/L = 0.504$, Fr = 0.138

In this section, we report the first application of the stereoscopic PIV technique to the wake flow survey of a propelled ship model in a large cavitation facility. This application presented a wide spectrum of practical problems that generally arise when working in conditions that depart from the laboratory-type experiment. It also gave us the opportunity to explore and develop a number of technological solutions, both on the setup and algorithmic sides, that have been specifically introduced during the operation of this experiment. Many of these solutions still represent today the state-of-the-art in the field.

The measurements have been carried out in the INSEAN circulating water channel, a free-surface cavitation tunnel with a test section 10 m long, 3.6 m wide and 2.25 m deep, capable of a 5.2 m/s maximum flow speed. The ship-model was a series 60 with a 6.096 m length, propelled by a five-blade MAU propeller (diameter: 221.9 mm).

Tests were performed at the propeller angular velocity of 6.7 rps and an upstream velocity of 1.22 m/s, corresponding to a Froude number $\mathrm{Fr} = 0.16$. Measurements were performed in three crossplanes orthogonal to the shaft and located downstream of the propeller disk at $x/L = 0.9997$, 1.0000 and 1.0187 (L is the model length). The experimental setup is shown in Fig. 4. Underwater optics were used to produce the lightsheet from a Nd:YAG laser pulsing at 10 Hz with a 200 mJ energy per pulse. The propeller angular position is given by a rotary 3600-pulse/revolution encoder to the synchronizer, which drives the two flashlamps and Q-switches of the two-head laser and the PIV cameras. The mean fluctuating velocity fields have been established on the basis of 129 PIV recordings phase-locked with the propeller angle, which was varied from 0 to 69° with a step of 3°.

The stereo-PIV method uses two cameras that image the flow field from two different directions. Each camera measures an apparent displacement, perpendicular to its optical axis. Combining the information from both views, it is possible to reconstruct the actual three-dimensional displacement vector in the plane. The angular displacement method is chosen, as opposed to the translation method [29]. The stereo-PIV system consisted of two 1280×1024 pixel^2 PCO cameras with a 12-bit depth resolution. One camera (*wet*) is placed underwater in a sealed container downstream of the ship model. The second camera (*dry*) is placed outside the test section of the tunnel and points at an angle to the measurement plane through a water-filled prism designed to minimize the optical aberrations caused by the nonperpendicularity of the optical axis to the air/glass/water interfaces. The angle between the wet and dry cameras ranges from 36° to 40°, depending on the measurement plane. The choice of an asymmetrical configuration was driven by a number of operational constraints linked to the test section size that impaired the practicality of having a symmetrical system with a second dry camera on the opposite side. Despite this latter arrangement being optimal in terms of accuracy [29], the current setup provides (through the wet camera view) a direct measure of the crossflow components as in a standard 2-component PIV

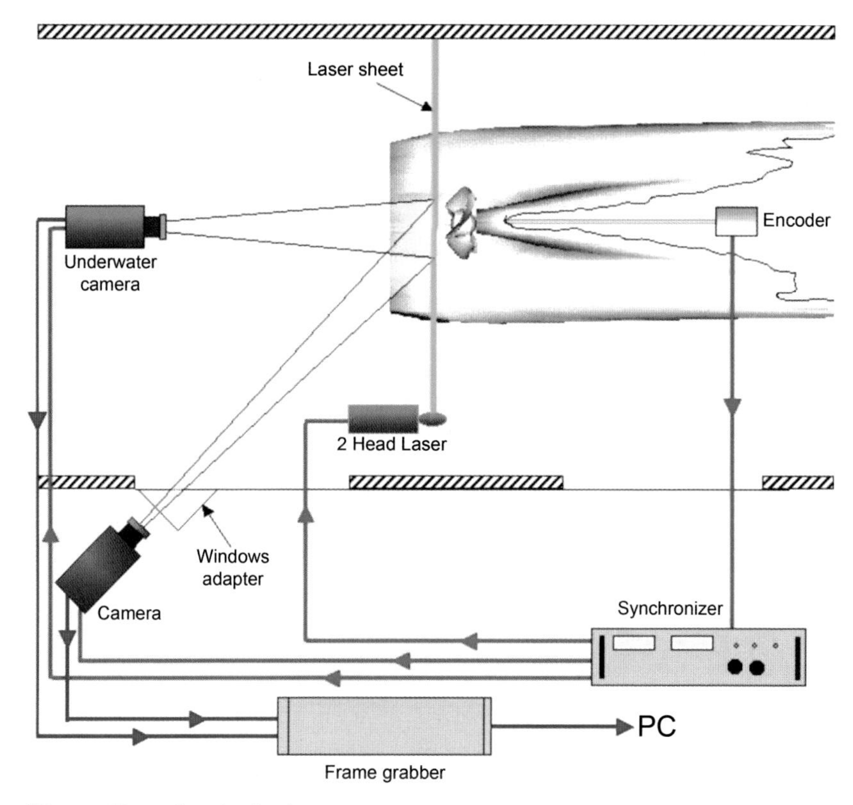

Fig. 4. Propulsor hydrodynamics: experimental setup

setup, thus ensuring optimal accuracy on the crossflow field. The errors in the three-dimensional reconstruction are in this case essentially confined to the estimate of the longitudinal component.

Each set of PIV images was analyzed using the following algorithmic procedure: window offset correlation method [30], recursive refinement of the interrogation grid and of the interrogation window size, adaptive Gaussian weight function [24], window overlap. This procedure increases both the dynamic range and the spatial resolution. A final validation procedure is applied to the vector field to remove any spurious vectors. A 32×32 pixel^2 window with a 75 % overlap factor was found to offer the best compromise in terms of spurious vector reduction and spatial resolution. This window size was equivalent to an area of $7 \times 7\,\mathrm{mm}^2$ in real space.

A mapping function, established through calibration, is applied to each view prior to the reconstruction of the three-dimensional vector field. This function performs a nonlinear perspective correction that compensates for the optical distortions introduced by the presence of multiple interfaces [31]. The calibration is done using a 20×20 dots grid target positioned in the measure-

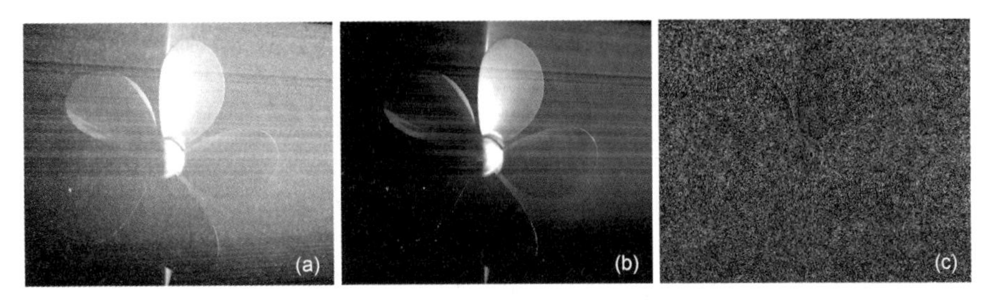

Fig. 5. Background removal

ment plane. The actual measurement area, corresponding to the overlapped region between the corrected views, is $250 \times 200\,\text{mm}^2$.

The PIV analysis is based on the scattering of flow markers, hence it is particularly sensitive to noise caused by direct and indirect illumination of boundaries. Although the effects can be minimized if the surfaces are made less reflective by means of light-absorbing coatings or painting, those are in many cases not sufficient. In our case, the measurement in the plane $x/L = 0.9997$ would present such a difficulty because of the close proximity of the lightsheet to the propeller, despite this latter being painted black. The light scattered by the flow markers tends to merge with that of the scattering noise, causing the PIV analysis to lock on the propeller features such as the hub and blade edges rather than on the seeding particles. The approach proposed here to address this particular problem consists in a preprocessing step where a mean image, relative to a specific angle, is subtracted from the instantaneous image. This procedure, illustrated in Fig. 5, allows the almost complete removal of any background information, such as the propeller, and greatly improves the PIV result.

The measurement uncertainty on the single camera is essentially associated to the evaluation of the particle displacement. A value lower than 0.1 pixel is generally reported for the sort of analysis performed here [32]. This is equivalent to approximately 4 cm/s in terms of velocity. For stereo-PIV, the three-dimensional vector reconstruction yields additional uncertainty, as outlined before. The configuration used here has been assessed on a test bench and compared with the more standard symmetrical arrangement. The θ angle between the wet camera and the dry camera has been varied from 10 to 70°. The errors, expressed in terms of standard deviation, were found to be smaller than 2.5 % of the upstream mean velocity for the in-plane components, with values fairly constant across the whole range of angles. The errors on the normal component were found to be minimum, and lower than 3 %, between 30 and 60°. A systematically lower error is found with the symmetrical configuration (i.e., $\theta = 104°$) on all components, as expected [29].

The propeller wake, observed at $x/L = 0.9997$, presents a number of features well outlined in Figs. 6 and 7, which represent, respectively, the

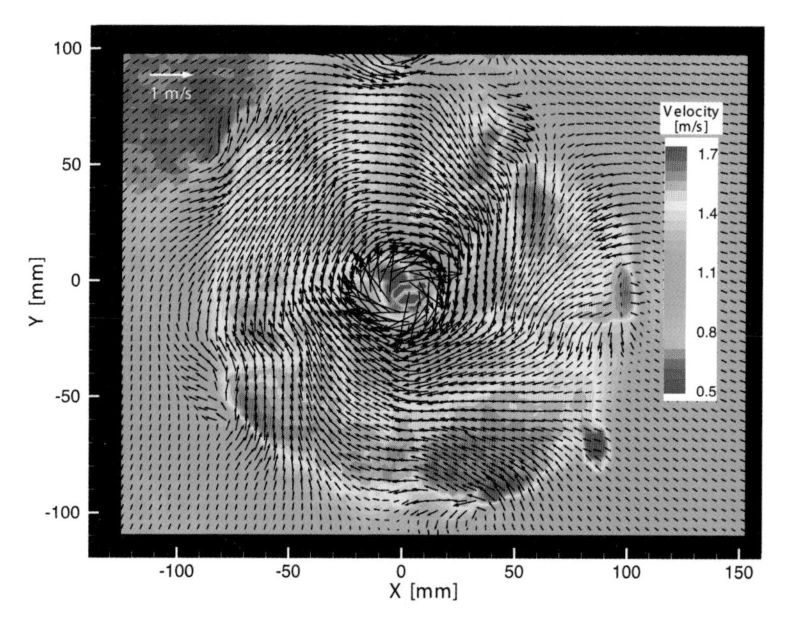

Fig. 6. Near-propeller wake: mean three-dimensional flow field at $x/L = 0.9997$; vectors represent the crossflow field (one every two vectors represented), contour represents the axial velocity magnitude

mean three-dimensional flow field (averaged over 129 instantaneous 3D vector fields) and the vorticity for θ varying from 0 to 63° with a step of 9°: wake asymmetry with maximum acceleration between $r/R = 0.6$ and 0.7; strong three-dimensional effects in the narrow wake of the hull at 12 o'clock during the blade sweep; trailing vorticity made of two layers of opposite sign overlapping at about $r/R = 0.7$ where hydrodynamic loading is maximum; wake strongly distorted and fragmented at 6 and 12 o'clock positions. For a complete discussion, we refer the reader to [33]. Note that Fig. 6 is the first published stereoscopic PIV measurement in naval hydrodynamics.

3.3 Underwater Ship Flows

The previous application served as a test ground to validate the concept of using stereoscopic PIV as a diagnostic tool in large-scale facilities. The requirements have been refined based on this initial experience and can be summarized as follows: flexibility, integrability, modularity, upgradeability. These qualities would make the instrument easily transportable from one facility to the other, easily configurable to adapt to various structural environments and various measurement requisites, and finally easily expandable and upgradeable to integrate the latest hardware developments (such as new sensors).

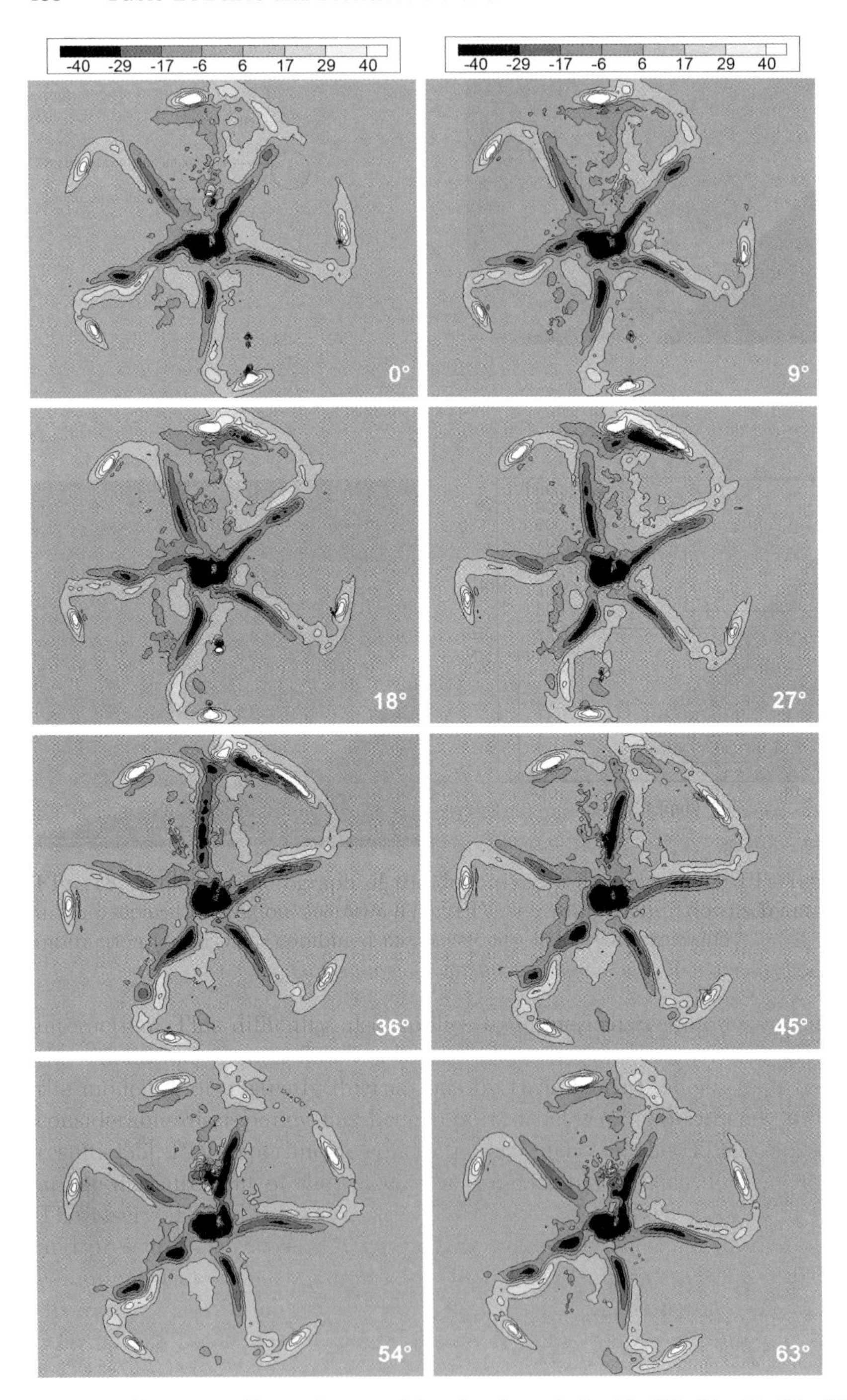

Fig. 7. Near-propeller wake: vorticity for θ = 0, 9, 18, 27, 36, 45, 54, 63° at $x/L = 0.9997$

The application presented here demonstrates the validity of the concept designed upon these prerequisites, and represents the first successful instance of stereo-PIV measurements in a towing tank for an industrial-type application. Furthermore, this experiment illustrates how to obtain high spatial resolution PIV data on very large areas.

A unique, highly modular and flexible underwater system for stereoscopic particle image velocimetry measurements has been devised at INSEAN for planar three-dimensional velocity measurements in large-scale facilities such as water tow tanks and tunnels [15]. The underwater stereo-PIV probe is designed in the form of modules that can be assembled into different configurations. Figure 8 shows the probe in its principal configuration, referred to as the design configuration. The probe, when fully assembled, forms a streamlined torpedo-like tube with an external diameter of 150 mm. The tube can be rotated about its axis in steps of 15°, and is rigidly linked to a bench through two hydrodynamically optimized struts. The whole system can be attached to the tow-tank carriage or to a traversing system. The components are: 1. two struts; 2. two waterproof-camera sections; 3. two camera mirror sections, opened to water; 4. a waterproof section for the laser lightsheet optics; 5. a waterproof section for the lightsheet mirror; 6. the nose and tail sections, which have a semiellipsoidal shape. The sections are made of stainless steel and are connected together through union couplings and waterproofing is guaranteed by pairs of O-rings on both sides of the couplings. The length of the probe in its longest configuration is 2070 mm. The stereoscopic system consists of two 2048 × 2048 pixels CCD cameras, with a 12-bit resolution. The aperture and the focus of the camera lenses are remotely controlled. Each camera head is mounted on a rotation platform that is also remotely controlled for the adjustment of the angle between the axis of the camera optics and the sensor normal axis. This angle, also known as the Scheimpflug angle, allows the camera to focus on a measurement plane that is not perpendicular to the camera optical axis. The camera housings are ventilated with nitrogen gas to avoid the formation of condensation on the optical parts and on the sensors. Humidity sensors are also fitted in the camera housings, primarily to monitor the presence of water. The laser light is delivered to the underwater sheet optics through the struts, inside an articulated arm. Mirrors within the arm elbows allow the beam to be correctly driven to the output optics. The laser optics consist of a set of changeable cylindrical and spherical lenses, which, respectively, expand the beam into a sheet and focus it onto the measurement plane. Each strut also hosts the respective camera cables, as well as those of the corresponding remote control motors and humidity sensor, and the ventilation hose. The laser subsystem features two Nd:YAG laser heads firing side by side at synchronized pulse rates up to 10 Hz and a maximum of 200 mJ energy per pulse.

The camera mirror sections are opened to water to avoid multimedia refractions, thus minimizing the optical aberrations thanks to the single orthogonal water/air interface. This interface consists of a high optical quality

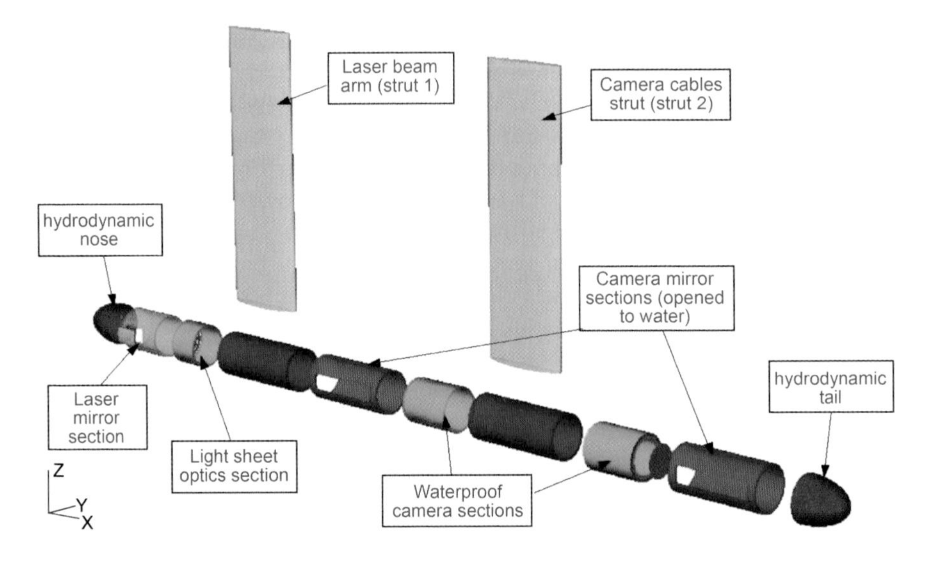

Fig. 8. INSEAN's underwater stereoscopic PIV probe

glass window, which also seals the camera section at one end. The mirrors can be manually translated along the probe axis, within their host section, and rotated to adjust the location of the measurement plane and to set the overlap between the two cameras for the stereoscopic configurations.

Figure 9 shows four possible configurations of the stereoscopic PIV system. In the main configuration, see Fig. 9A, the cameras are on the same side of the lightsheet. This setup is adequate for crossflow measurements along the hull of a ship model or in the near wake. The arrangement shown in Fig. 9B is suitable for use in the far wake and in the midsections of a ship model. The setup depicted in Fig. 9C allows the measurement in streamwise sections of the flow, whereas the arrangement of Fig. 9D is geometrically identical to that described in Sect. 3.2. Setups C and D have the advantage of providing one plane normal to the optical axis of at least one camera, therefore suitable for standard PIV analysis (inplane 2-component analysis).

The system is applied to the flow survey around a deeply submerged underwater vehicle model, referred to as UVM in the following sections. The experiments were carried out in the INSEAN towing tank no 1. The tank is 470 m long, 13.5 m wide and 6.5 m deep, with a maximum carriage speed of 15 m/s. The length of the model, made of glass-reinforced plastic, is 7.2 m. The UVM is equipped with a 7-blade propeller with a diameter of 530 mm. The body was painted black and the propeller black-anodized to minimize the laser-beam reflections, see Sect. 3.2.

The UVM is attached to a strut and is rotated by 90°, as shown in Figs. 10 and 11. This arrangement has the main advantage of minimizing the interference with the support and with the free surface. The model was set up

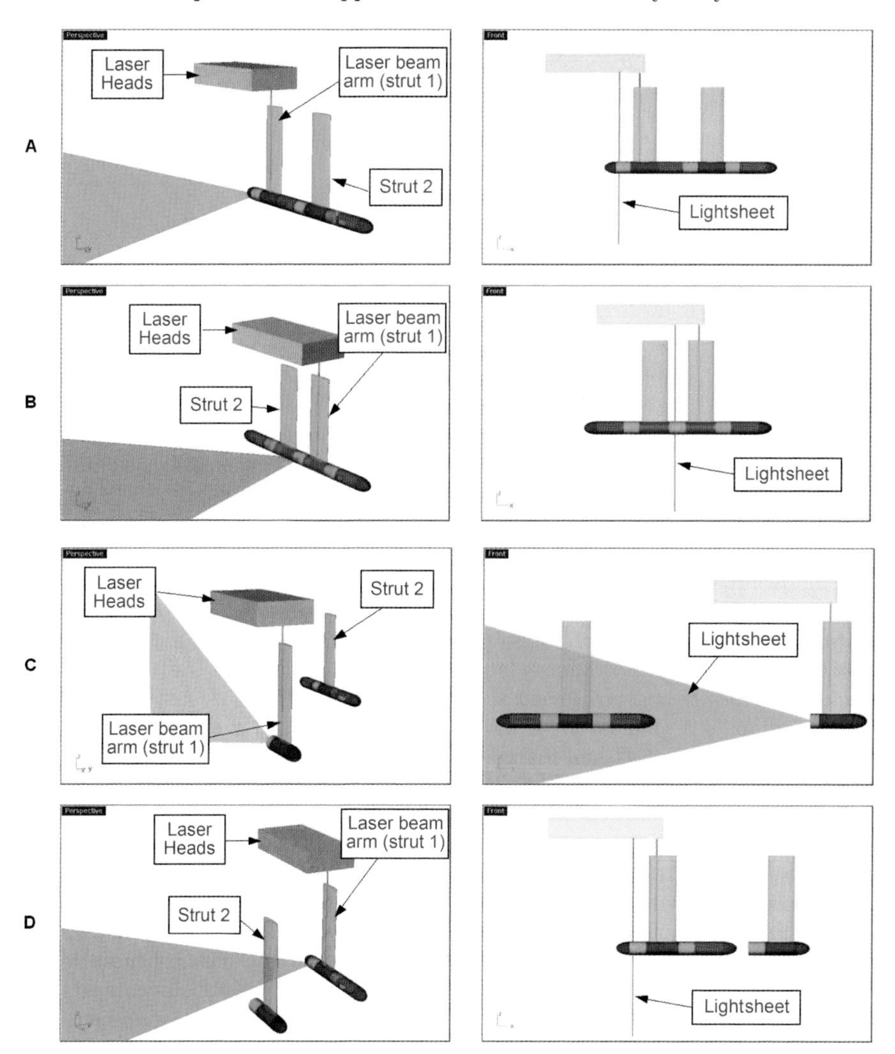

Fig. 9. Underwater PIV probe (from top to bottom; *left column* is perspective view, *right column* is side view): (**A**) design configuration; (**B**) symmetric configuration; (**C**) offset laser sheet configuration; (**D**) offset camera configuration

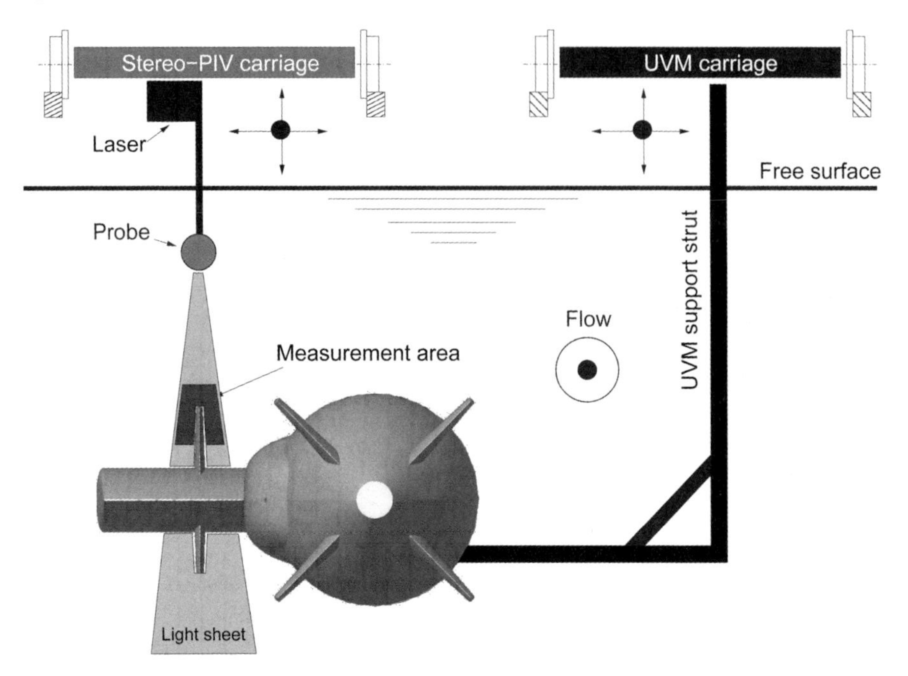

Fig. 10. Underwater vehicle setup

at a depth of 1600 mm (distance between the free-surface and the propeller axis). At this depth, the free-surface effects are not significant. Moreover, the distance from the bottom is large enough to also neglect the bottom-floor effects. A coarse 3-axis system allows the UVM to be positioned with respect to the stereo-PIV measurement system, which is accurately positioned by means of a remotely controlled 3-axis traverse system with a displacement range of 440 mm in the X-axis (horizontal crossflow), 760 mm in Y (vertical crossflow) and 760 mm in Z (longitudinal) and an accuracy of 0.01 mm on each axis. The traverse system is installed on secondary carriage rails, which are parallel to the main rails supporting the UVM. The laser heads are supported by the probe structure itself.

A $500 \times 450\,\text{mm}^2$ calibration target, made of a regular grid of 23×21 round dots spaced 20 mm apart from one another, was used for the stereo-PIV system calibration. The calibration was performed in water, with the underwater system mounted on the towing carriage and the target rigidly fixed to a wall structure of the tank. The target was accurately placed in the plane of the laser lightsheet, with its center located at a distance of 1100 mm below the probe axis. In contrast with the usual procedure where the target is translated with respect to the PIV system, the probe was moved by means of its own traversing system, in a direction normal to the target plane with steps of 1 mm from -5 to $+5$ mm. Having a target permanently in place during

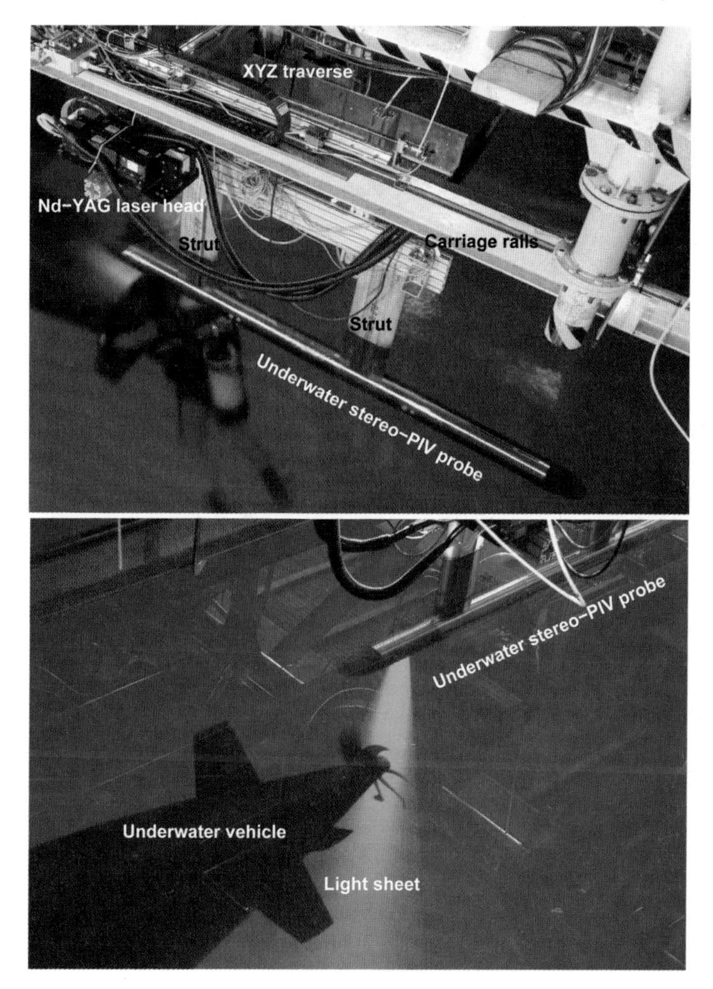

Fig. 11. Probe and traverse (*top*), UVM with propeller and lightsheet (*bottom*)

the measurements allowed regular checks of the system state to be made by simply moving the whole carriage to the calibration station, in order to repeat the calibration process. The measurement area of a stereo-PIV plane was found to be 325 mm wide and 250 mm high.

The flow around the UVM has been mapped along the hull and in the propeller near and far wake at the following cross sections: $x/L = 0.5$, $x/L = 0.625$, $x/L = 0.75$, $x/L = 1.05$ and $x/L = 1.5$, with L being the model length. To cover a representative area of the flow at each cross section, we devised a method where adjacent $325 \times 250\,\mathrm{mm}^2$ areas (or patches) were measured and assembled to provide the full plane dataset. Two different plane configurations were used for the hull and for the wake sections, see Fig. 12, respectively, 6×2 and 6×3 patches with an overlap varying between 25 and 30 %. The resulting

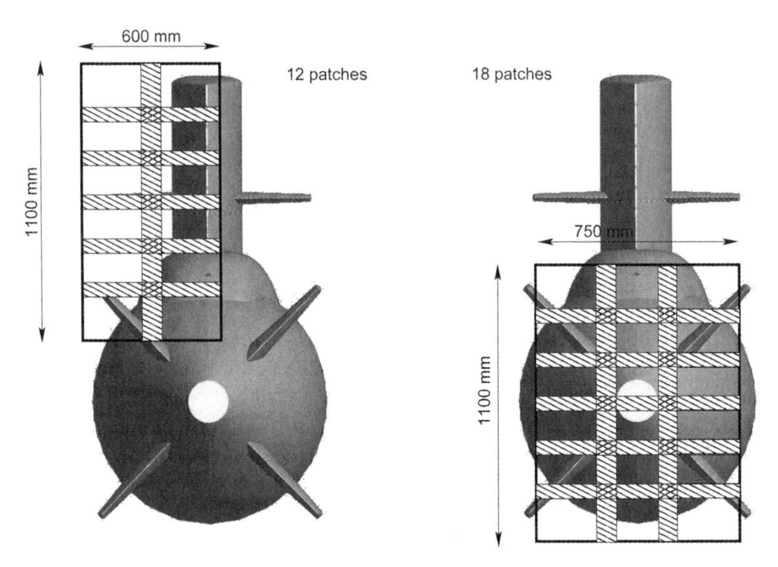

Fig. 12. Patch configuration for the hull (*left*) and wake (*right*) cross sections. The dashed areas represent the overlap regions

measurement areas were $550 \times 1000\,\mathrm{mm}^2$ and $740 \times 1000\,\mathrm{mm}^2$, respectively. The total number of vectors per plane is approximately 60 000 and 100 000, respectively, for the hull and for the wake sections, with a spacing of 2.5 mm in real space.

Each patch consists of a minimum of 200 PIV acquisitions, resulting in a total of 2400 and 3600 PIV recordings, respectively, for the hull planes and for the wake planes. The system was able to acquire 200 PIV recordings (one patch) per tow, at a frequency of 8 Hz. To reach this frequency, the data was first stored into volatile memory before being transferred onto the mass storage devices. Each recording represents 32 Mbytes in storage space, thus the total dataset (5 planes) represents about 0.5 Tbytes. The full dataset took four days to complete.

In practice, this procedure requires a precise positioning of the different patches relative to the model. This is done, for each measurement plane, by resetting the traverse/probe system to a known reference point on the UVM body. In our case, this point is materialized by calibrated pins placed at precise locations along the hull. The probe is positioned so that this reference point appears within the boundaries of the measurement plane. Its image location is used to define the reference origin relative to which the individual patches are repositioned for the final assembly.

Figure 13 represents an overview of the UVM where we report the velocity magnitude in four planes (x/L = 0.5, 0.625, 0.75, 1.05), normalized by the free-stream velocity. The insets provide a detailed view of some important features of the flow: twin counterrotating vortex structures generated at the

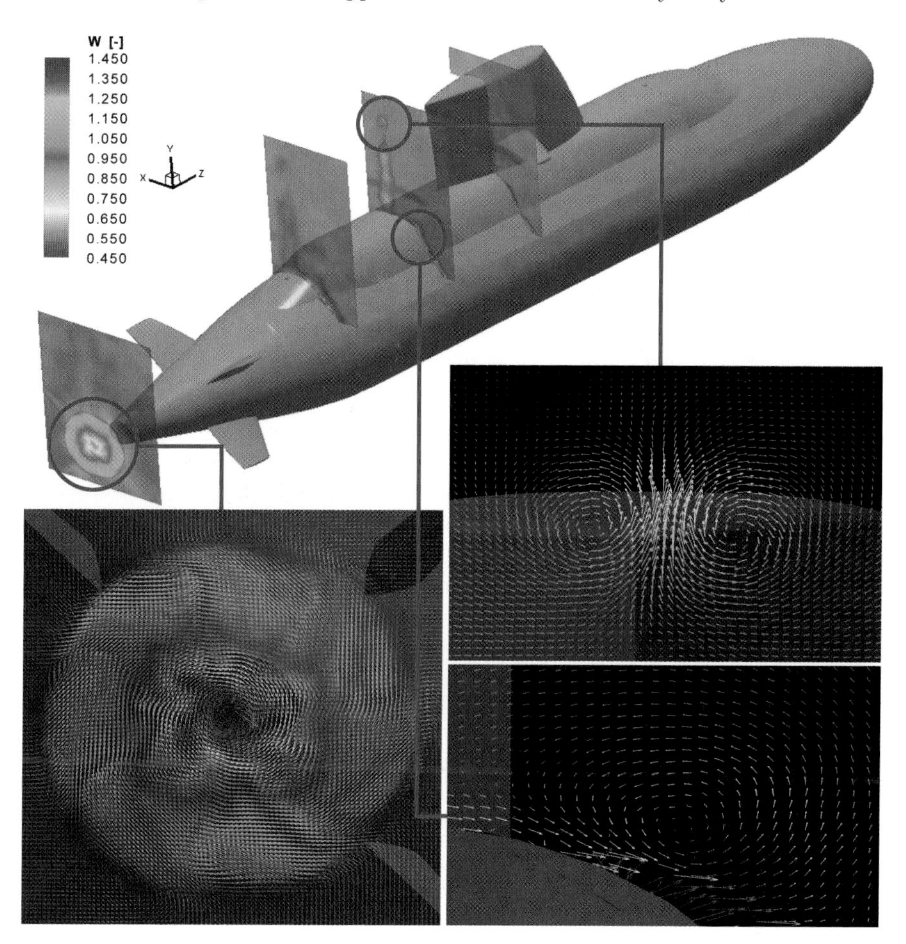

Fig. 13. Overview and detail of the velocity field; four crossflow velocity fields are represented: $x/L = 0.5$, 0.625, 0.75, 1.05.

top of the sail, horseshoe vortex produced at the base of the sail, propeller phase-averaged field showing the accelerated flow. The test was carried out at a carriage velocity of 2 m/s. Note that the measurements for the plane $x/L = 1.05$ have been performed in phase with the propeller rotation in order to capture the flow features linked to the blades. A rotation pulse was generated for each revolution to trigger the stereo-PIV image acquisition, hence generating a dataset phase-locked with the propeller angular position.

3.4 Two-Phase Bubble Flows

Two-phase flows are another class of flows of great interest that require extended measurement capabilities. In fact, such a flow generally requires that both phases be measured in order to have a comprehensive insight into the physical mechanisms. Furthermore, flow three-dimensionality is a common feature in complex (real) flows, and a 3-component, 2D information as provided by a standard stereoscopic PIV approach can only provide a partial view of the global problem. The current efforts in the PIV community are related to the extension of the PIV concept to the full 3D space. A number of intermediate solutions have been developed, such as the multilayer PIV [34], scanning PIV [35] or multiplane PIV [36]. Holography [37] and defocusing digital particle image velocimetry (DDPIV) [38, 39] provide true volumetric information.

The application presented here makes use of DDPIV, and illustrates its potential in a laboratory-scale experiment designed as a test ground to verify the validity of the technique for industrial-type applications. The approach is therefore similar to that followed with the development of the underwater stereoscopic-PIV probe. The experiment described thereafter attempts to address the issue of bubbly two-phase flows and paves the way towards a system capable of mapping simultaneously the flow field of both phases, as well as provide additional information on bubble-population characteristics such as size and density [40].

The DDPIV technique is a new approach to the three-dimensional mapping of flow fields, that naturally extends the planar PIV techniques to the third spatial dimension. The interrogation domain is a volume where 3D coordinates of fluid markers are determined prior to flow analysis. However, unlike PTV or stereo-based methods, DDPIV has one unique optical axis and is based on pattern matching rather than on stereoscopic matching of particle images. The other fundamental difference resides in the statistical evaluation of the particle displacement, which is here recovered by performing a 3D spatial correlation of particle locations. In planar digital particle image velocimetry analysis, either in its 2- or 3-component version, the pixel image crosscorrelation is the *de facto* standard evaluation tool for estimating the particle displacement [41]. For 3D situations, an implementation of the cross-correlation principle has been devised for DDPIV [38]. In this latter approach, the displacement is estimated by performing a three-dimensional spatial crosscorrelation between the particles' 3D locations. Hence, DDPIV differs radically from traditional PIV in that it requires first to identify all the particles in physical space, while PIV operates on the particle pixel images.

In a standard 2D imaging system, light scattered by a point source is collected through a converging lens and a single aperture, which is usually located on the lens axis. The DDPIV technique uses a mask with a multiplicity of off-axis pinhole apertures, arranged in a predefined geometrical pattern, to obtain a multiplicity of images from each scattering source. These images form the same identical geometrical pattern on the image plane, but scaled according to the depth location of the scattering source. For on-focus particles, the pattern is reduced to a point. Hence, the particle three-dimensional location can be determined, through simple ray-optics relationships, by mea-

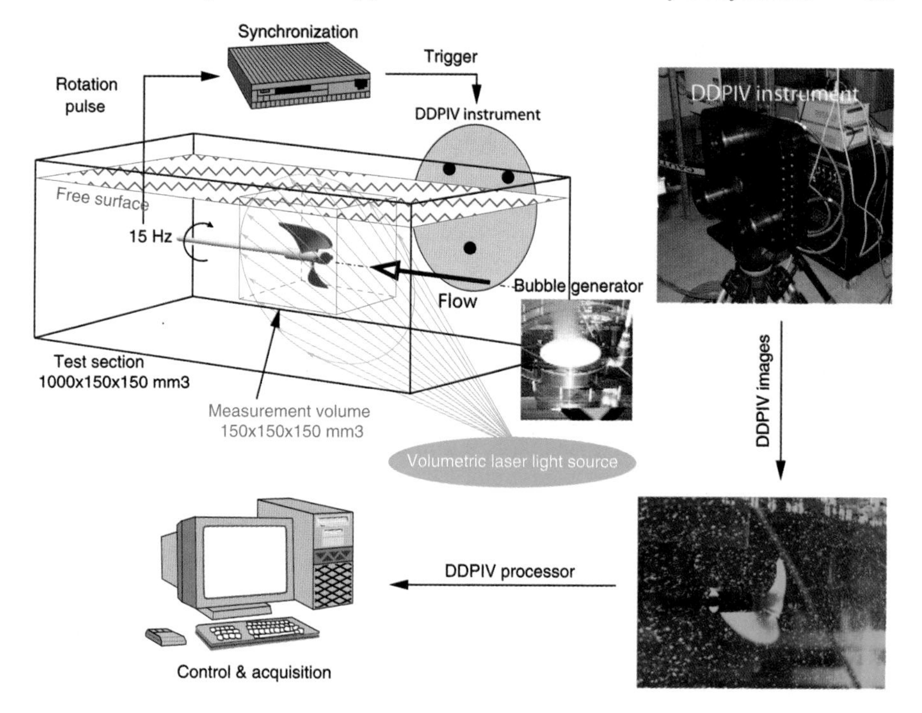

Fig. 14. Bubbly flows: experimental setup

suring the centroid and size of the pattern on the image plane. The reader will find a detailed description of the DDPIV principle and assessment in [38].

The defocusing DPIV is here applied to investigate the collective bubble interaction with the flow induced by a propeller and, more specifically, their interaction with the vortex system generated by the blades. The experiments are performed in a laboratory water channel, as shown in Fig. 14. The test section is $1000 \times 150 \times 150\,\mathrm{mm}^3$. The upstream flow velocity is 45 cm/s. A bubble generator, placed upstream in the convergent section, generates a dense cloud of submillimeter air bubbles, which are used as flow markers. The two-blade model boat propeller has a diameter of 67 mm. The rotation speed is set to 15 rps (3.16 m/s at the blade tip). The measurements are locked on the propeller rotation for phase averaging. Flow mean quantities (velocity and vorticity) and second-order statistics (Reynolds' tensor) are calculated on sequences of 200 velocity vector fields.

The DDPIV camera is a self-contained instrument that allows the capture of the three-dimensional information of the flow. It is composed of three apertures arranged according to a specific geometry. A 1600×1200 pixels CCD sensor, with an 8-bit resolution, is mounted behind each aperture to collect the light scattered by the flow markers (bubbles in our experiment). The camera used in this experiment has been designed to map a volume of

$120 \times 120 \times 120\,\text{mm}^3$ placed at a distance of about half a meter from the instrument. The camera records directly to disk sequences of full-resolution images at a frame rate of 15 Hz per DDPIV recording, corresponding to a sustained transfer rate of approximately 173 Mbytes/s. The measurement domain is illuminated by a 200 mJ Nd:YAG laser light source (NewWave Research®), the beam of which has been expanded into a diverging cone by means of a series of spherical and cylindrical lenses, see Fig. 14.

Sample results are shown in Fig. 15, showing the volumetric three-dimensional velocity vector field combined with the measurement of the bubble local density. The interrogation domain consists of $135 \times 81 \times 73$ voxels in X, Y and Z, respectively. The maximum bubble density, represented as a red isosurface on the plots, is found to correspond to the location of the tip vortex structures and secondary hub vortices. Extended results can be found in [42], which comprise void-fraction measurements.

4 Conclusion

The principal intent of this work was to give the reader the opportunity to have a general overview on the state-of-the-art of PIV applications in the field of experimental naval hydrodynamics. Our review suggests that the new technological challenges, posed by applications of ever-increasing complexity, are constantly motivating further progress on the hardware and software sides. It also suggests that the trend observed through the past years is decisively towards the full three-dimensional flow-field characterization. This trend is being followed by the principal actors in the field and has motivated the development of leading-edge technologies that are yet to be fully validated on real cases.

Acknowledgements

We acknowledge the past and current members of the INSEAN Cavitation and Propulsion group for their contribution to the PIV work: Mr. G. Aloisio, Mr. E. Binotti, Mr. A. Ciarravano, Dr. G. Calcagno, Mr. M. Colaprete, Mr. T. Costa, Mr. A. Dolcini, Dr. M. Falchi, Mr. S. Farina, Dr. M. Felli, Mr. G. Giordano, Mr. S. Grizzi, Prof. G. P. Romano ("La Sapienza" University), Mr. A. Testerini. We also thank Prof. M. Gharib and Dr. E. C. Graff from the California Institute of Technology (Pasadena, CA) for their contribution to the DDPIV work.

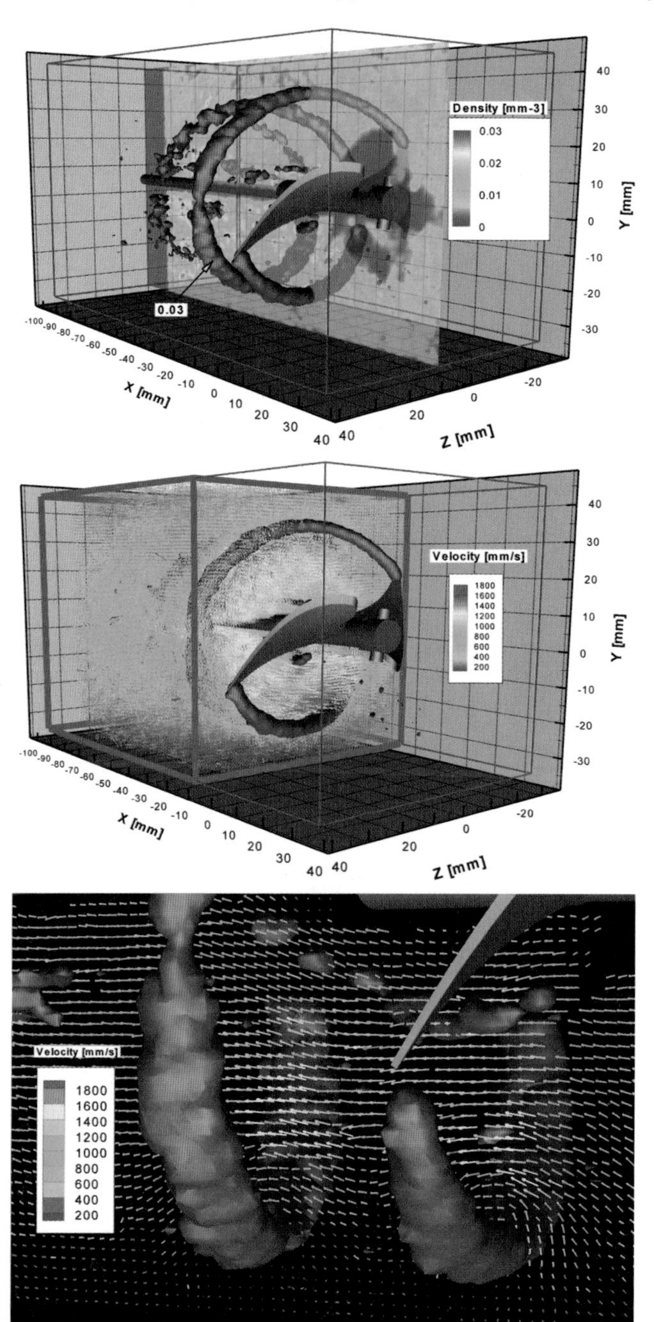

Fig. 15. Sample results of the bubble flow: bubble density in a meridian plane and isosurface at $0.03\,\mathrm{mm}^{-3}$ (*top*); volumetric velocity field (partial) and isosurface as above (*center*); slice in a meridian plane and isosurface as above (closeup on the tip vortex region, *bottom*)

References

[1] K. S. Min: *Numerical and Experimental Methods for Prediction of Field Point Velocities Around Propeller Blades*, Technical Report 78-12, MIT, Dept. of Ocean Engineering (1978)

[2] S. Kobayashi: *Experimental Methods for the Prediction of the Effects of Viscosity on Propeller Performance*, Technical Report 81-7, MIT, Dept. of Ocean Engineering (1981)

[3] A. Cenedese, L. Accardo, R. Milone: Phase sampling techniques in the analysis of a propeller wake, in *Proc. of the International Conference on Laser Anemometry Advances and Application* (Manchester, UK 1985)

[4] S. D. Jessup: *An Experimental Investigation of Viscous Aspects of Propeller Blade Flow*, Ph.D. thesis, The Catholic University of America, Washington DC (1989)

[5] C. Chesnack, S. D. Jessup: Experimental characterisation of propeller tip flow, in *Proc. 22nd Symposium on Naval Hydrodynamics* (ONR, Washington DC 1998)

[6] A. Stella, G. Guj, F. Di Felice, M. Elefante: Experimental investigation of propeller wake evolution by means of LDV and flow visualizations, J. Ship Res. **44**, 155–169 (2000)

[7] M. Felli, F. Di Felice, G. P. Romano: Installed propeller wake analysis by LDV: Phase sampling technique, in *Proc. 9th Int. Symp. on Flow Visualization* (Edinburgh, UK 2000)

[8] P. Esposito, F. Salvatore, F. Di Felice, G. Ingenito, G. Caprino: Experimental and numerical investigation of the unsteady flow around a propeller, in *Proc. 23rd Symposium on Naval Hydrodynamics* (ONR, Val de Reuil, France 2000)

[9] F. Di Felice, M. Felli, G. Ingenito: Propeller wake analysis in non-uniform inflow by LDV, in *Proc. of the Propeller and Shafting Symp.* (SNAME, Virginia Beach, US 2000)

[10] A. Stella, G. Guj, F. Di Felice: Propeller flow field analysis by means of LDV phase sampling techniques, Exp. Fluids **28**, 1–10 (2000)

[11] L. Gui, J. Longo, F. Stern: Towing tank PIV measurement system, data and uncertainty assessment for DTMB model 5512, Exp. Fluids **31**, 336–346 (2001)

[12] F. Di Felice, A. Dolcini, F. Pereira, C. Lugni: Flow survey around the bilge keel of a ship model in free roll decay, in *Proc. 6th Int. Symposium on Particle Image Velocimetry* (California Institute of Technology, Pasadena, CA, USA 2005)

[13] A. Cotroni, F. Di Felice, G. P. Romano, M. Elefante: Investigation of the near wake of a propeller using particle image velocimetry, Exp. Fluids **29**, S227–S236 (2000) suppl.

[14] D. Dabiri: On the interaction of a vertical shear layer with a free surface, J. Fluid. Mech. **480**, 217–232 (2003)

[15] F. Pereira, T. Costa, M. Felli, G. Calcagno, F. Di Felice: A versatile fully submersible stereo-PIV probe for tow tank applications, in *Proc. 4th ASME-JSME Joint Fluids Engineering Conference (FEDSM'03)* (ASME, Honolulu, HI (USA) 2003)

[16] Y. Ikeda, Y. Himeno, N. Tanaka: *A Prediction Method for Ship Roll Damping*, Technical Report 00405, Dep. of Naval Arch., Univ. of Osaka (1978)

[17] Y. Himeno: *Prediction of Ship Roll Damping - State of Art*, Technical Report 239, Dept. of Naval Arch. and Marine Eng, Univ. of Michigan (1981)
[18] R. Yeung, S. Liao, D. Roddier: On roll hydrodynamics of rectangular cylinders, in *Proc. Int. Off. and Polar Eng. Conf.* (Montreal, Canada 1998)
[19] D. Roddier, S. Liao, R. Yeung: On freely-floating cylinders fitted with bilge keels, in *Proc. Int. Off. and Polar Eng. Conf.* (Seattle, USA 2000)
[20] R. Broglia, A. Di Mascio: Unsteady ranse calculations of the flow around a moving ship hull, in *Proc. 8th NSH* (Busan, Korea 2003)
[21] A. Di Mascio, R. Broglia, R. Muscari, R. Dattola: Unsteady RANSE simulations of a manoeuvering ship hull, in *Proc. 25th Symposium on Naval Hydrodynamics* (ONR, St John's, Newfoundland, Canada 2004)
[22] R. Yeung, C. Cermelli, S. Liao: Vorticity fields due to rolling bodies in a free surface – experiment and theory, in *Proc. 21st Symposium on Naval Hydrodynamics* (ONR, Trondheim, Norway 1996)
[23] M. Felli, F. Di Felice, C. Lugni: Experimental study of the flow field around a rolling ship model, in *Proc. 25th Symposium on Naval Hydrodynamics* (ONR, S. John's, Newfoundland, Canada 2004)
[24] D. Di Florio, F. Di Felice, G. P. Romano: Windowing, re-shaping and re-orientation interrogation windows in particle image velocimetry for the investigation of shear flows, Meas. Sci. Technol. **13**, 953–962 (2002)
[25] H. T. Huang, H. F. Fiedler, J. J. Wang: Limitation and improvement of PIV; part II: Particle image distortion, a novel technique, Exp. Fluids **15**, 263–273 (1993)
[26] F. Scarano: Iterative image deformation methods in PIV, Meas. Sci. Technol. **13**, 1–19 (2002)
[27] P. Koumoutsakos, D. Shiels: Simulations of the viscous flow normal to an impulsively started and uniformly accelerated flat plate, J. Fluid Mech. **328**, 177–227 (1996)
[28] A. Cotroni, F. Di Felice, G. P. Romano, M. Elefante: Propeller tip vortex analysis by means of PIV, in *Proc. 3rd International Workshop on PIV* (UCSB, Santa Barbara, US 1999)
[29] A. K. Prasad, R. J. Adrian: Stereoscopic particle image velocimetry applied to liquid flows, Exp. Fluids **15**, 49–60 (1993)
[30] J. Westerweel: Fundamentals of digital particle image velocimetry, Meas. Sci. Technol. **8**, 1379–1392 (1997)
[31] S. M. Soloff, R. J. Adrian, Z. C. Liu: Distortion compensation for generalized stereoscopic particle image velocimetry, Meas. Sci. Technol. **8**, 1441–1454 (1997)
[32] M. Raffel, C. Willert, J. Kompenhans: *Particle Image Velocimetry* (Springer, bh 1998)
[33] G. Calcagno, F. Di Felice, M. Felli, F. Pereira: A stereo-PIV investigation of a propeller's wake behind a ship model in a large free-surface tunnel, Mar. Technol. Soc. J. **39**, 97–105 (2005)
[34] M. Abe, N. Yoshida, K. Hishida, M. Maeda: Multilayer PIV technique with high power pulse laser diodes, in *Proc. 9th Int. Symp. Appl. Laser Tech. Fluid Mech.* (Instituto Superior Técnico, Lisbon, Portugal 1998)
[35] C. Brücker: 3D scanning PIV applied to an air flow in a motored engine using digital high-speed video, Meas. Sci. Technol. **8**, 1480–1492 (1997)

[36] C. J. Kähler, J. Kompenhans: Fundamentals of multiple plane stereo particle image velocimetry, Exp. Fluids **29**, S070–S077 (2000)
[37] H. Meng, G. Pan, Y. Pu, S. C. Woodward: Holographic particle image velocimetry: From film to digital recording, Meas. Sci. Technol. **15**, 673–685 (2004)
[38] F. Pereira, M. Gharib: Defocusing digital particle image velocimetry and the three-dimensional characterization of two-phase flows, Meas. Sci. Technol. **13**, 683–694 (2002)
[39] F. Pereira, H. Stüer, E. Castaño-Graff, M. Gharib: Two-frame 3D particle tracking, Meas. Sci. Technol. **17**, 1680–1692 (2006)
[40] F. Pereira, M. Gharib: A method for three-dimensional particle sizing in two-phase flows, Meas. Sci. Technol. **15**, 2029–2038 (2004)
[41] R. J. Adrian: Particle-imaging techniques for experimental fluid mechanics, Annu. Rev. Fluid Mech. **23**, 261–304 (1991)
[42] F. Pereira, E. Castaño-Graff, M. Gharib: Bubble interaction with a propeller flow, in *Proc. 26th Symp. on Naval Hydrodynamics*, vol. 1 (Office of Naval Research, Rome, Italy 2006) pp. 299–308

Index

Index

Topics in Applied Physics

97 **Terahertz Optoelectronics**
By K. Sakai (Ed.) 2005, 270 Figs. XIII, 387 pages

98 **Ferroelectric Thin Films**
Basic Properies and Device Physics for Memory Applications
By M. Okuyama, Y. Ishibashi (Eds.) 2005, 172 Figs. XIII, 244 pages

99 **Cryogenic Particle Detection**
By Ch. Enss (Ed.) 2005, 238 Figs. XVI, 509 pages

100 **Carbon**
The Future Material for Advanced Technology Applications
By G. Messina, S. Santangelo (Eds.) 2006, 245 Figs. XXII, 529 pages

101 **Spin Dynamics in Confined Magnetic Structures III**
By B. Hillebrands, A. Thiaville (Eds.) 2006, 164 Figs. XIV, 345 pages

102 **Quantum Computation and Information**
From Theory to Experiment
By H. Imai, M. Hayashi (Eds.) 2006, 49 Figs. XV, 281 pages

103 **Surface-Enhanced Raman Scattering**
Physics and Applications
By K. Kneipp, M. Moskovits, H. Kneipp (Eds.) 2006, 221 Figs. XVIII, 464 pages

104 **Theory of Defects in Semiconductors**
By D. A. Drabold, S. K. Estreicher (Eds.) 2007, 60 Figs. XIII, 297 pages

105 **Physics of Ferroelectrics**
A Modern Perspective
By K. Rabe, Ch. H. Ahn, J.-M. Triscone (Eds.) 2007, 129 Figs. XII, 388 pages

106 **Rare Earth Oxide Thin Films**
Growth, Characterization, and Applications
By M. Fanciulli, G. Scarel (Eds.) 2007, 210 Figs. XVI, 426 pages

107 **Microscale and Nanoscale Heat Transfer**
By S. Volz (Ed.) 2007, 144 Figs. XIV, 370 pages

108 **Light Scattering in Solids IX**
Novel Materials and Techniques
By M. Cardona, R. Merlin (Eds.) 2007, 215 Figs. XIV, 432 pages

109 **Molecular Building Blocks for Nanotechnology**
From Diamondoids to Nanoscale Materials and Applications
By G.A. Mansoori, Th.F. George, L. Assoufid, G. Zang (Eds.) 2007, 229 Figs. XII, 426 pages

110 **Sputtering by Particle Bombardment**
Experiments and Computer Calculations from Treshold to MeV Energies
By R. Behrisch, W. Eckstein (Eds.) 2007, 201 Figs. XVIII, 508 pages

111 **Carbon Nanotubes**
Advanced Topics in the Synthesis, Structure, Properties and Applications
A. Jorio, G. Dresselhaus, M. Desselhaus (Eds.) 2008, 250 Figs. XXIV, 722 pages

112 **Particle Image Velocimetry**
New Developments and Recent Applications
A. Schroeder, C. E. Willert (Eds.) 2008, 335 Figs. XVIII, 512 pages

Printing: Krips bv, Meppel, The Netherlands
Binding: Stürtz, Würzburg, Germany